Propagation of Sound in Porous Media

Propagation of Sound in Porous Media:

Modelling Sound Absorbing Materials, Second Edition

Jean F. Allard

Université le Mans, France

Noureddine Atalla

Université de Sherbrooke, Qc, Canada

WILEY

A John Wiley and Sons, Ltd., Publication

This edition first published 2009

Registered office
John Wiley & Sons Ltd, The Atrium, Southern Gate, Chichester, West Sussex, PO19 8SQ, United Kingdom

For details of our global editorial offices, for customer services and for information about how to apply for permission to reuse the copyright material in this book please see our website at www.wiley.com.

Library of Congress Cataloguing-in-Publication Data
Allard, J.-F.
Propagation of sound in porous media : modelling sound absorbing materials / Jean F. Allard – 2nd ed. / Noureddine Atalla.
p. cm.
Includes bibliographical references and index.
ISBN 978-0-470-74661-5 (cloth)
1. Porous materials–Acoustic properties–Mathematical models. 2. Absorption of sound. 3. Sound–Transmission. I. Atalla, Noureddine. II. Title.
TA418.9.P6A42 2009
620.1′1694015118–dc22

2009028758

A catalogue record for this book is available from the British Library.

ISBN: 978-0-470-746615-0

Typeset in 10/12pt Times by Laserwords Private Limited, Chennai, India.

Contents

Preface to the second edition xiii

1 Plane waves in isotropic fluids and solids 1
- 1.1 Introduction 1
- 1.2 Notation – vector operators 1
- 1.3 Strain in a deformable medium 2
- 1.4 Stress in a deformable medium 4
- 1.5 Stress–strain relations for an isotropic elastic medium 5
- 1.6 Equations of motion 8
- 1.7 Wave equation in a fluid 10
- 1.8 Wave equations in an elastic solid 11
- References 13

2 Acoustic impedance at normal incidence of fluids. Substitution of a fluid layer for a porous layer 15
- 2.1 Introduction 15
- 2.2 Plane waves in unbounded fluids 15
  - *2.2.1 Travelling waves* 15
  - *2.2.2 Example* 16
  - *2.2.3 Attenuation* 16
  - *2.2.4 Superposition of two waves propagating in opposite directions* 17
- 2.3 Main properties of impedance at normal incidence 17
  - *2.3.1 Impedance variation along a direction of propagation* 17
  - *2.3.2 Impedance at normal incidence of a layer of fluid backed by an impervious rigid wall* 18
  - *2.3.3 Impedance at normal incidence of a multilayered fluid* 19

2.4 Reflection coefficient and absorption coefficient at normal incidence 19
2.4.1 Reflection coefficient 19
2.4.2 Absorption coefficient 20
2.5 Fluids equivalent to porous materials: the laws of Delany and Bazley 20
2.5.1 Porosity and flow resistivity in porous materials 20
2.5.2 Microscopic and macroscopic description of sound propagation in porous media 22
2.5.3 The Laws of Delany and Bazley and flow resistivity 22
2.6 Examples 23
2.7 The complex exponential representation 26
References 26

3 Acoustic impedance at oblique incidence in fluids. Substitution of a fluid layer for a porous layer 29
3.1 Introduction 29
3.2 Inhomogeneous plane waves in isotropic fluids 29
3.3 Reflection and refraction at oblique incidence 31
3.4 Impedance at oblique incidence in isotropic fluids 33
3.4.1 Impedance variation along a direction perpendicular to an impedance plane 33
3.4.2 Impedance at oblique incidence for a layer of finite thickness backed by an impervious rigid wall 34
3.4.3 Impedance at oblique incidence in a multilayered fluid 35
3.5 Reflection coefficient and absorption coefficient at oblique incidence 36
3.6 Examples 37
3.7 Plane waves in fluids equivalent to transversely isotropic porous media 39
3.8 Impedance at oblique incidence at the surface of a fluid equivalent to an anisotropic porous material 41
3.9 Example 43
References 43

4 Sound propagation in cylindrical tubes and porous materials having cylindrical pores 45
4.1 Introduction 45
4.2 Viscosity effects 45
4.3 Thermal effects 50
4.4 Effective density and bulk modulus for cylindrical tubes having triangular, rectangular and hexagonal cross-sections 54
4.5 High- and low-frequency approximation 55
4.6 Evaluation of the effective density and the bulk modulus of the air in layers of porous materials with identical pores perpendicular to the surface 57
4.6.1 Effective density and bulk modulus in cylindrical pores having a circular cross-section 57
4.6.2 Effective density and bulk modulus in slits 59

4.6.3 *High- and low-frequency limits of the effective density and the bulk modulus for pores of arbitrary cross-sectional shape* 60
4.7 The Biot model for rigid framed materials 61
4.7.1 *Similarity between G_c and G_s* 61
4.7.2 *Bulk modulus of the air in slits* 62
4.7.3 *Effective density and bulk modulus of air in cylindrical pores of arbitrary cross-sectional shape* 64
4.8 Impedance of a layer with identical pores perpendicular to the surface 65
4.8.1 *Normal incidence* 65
4.8.2 *Oblique incidence – locally reacting materials* 67
4.9 Tortuosity and flow resistivity in a simple anisotropic material 67
4.10 Impedance at normal incidence and sound propagation in oblique pores 69
4.10.1 *Effective density* 69
4.10.2 *Impedance* 71
Appendix 4.A Important expressions 71
Description on the microscopic scale 71
Effective density and bulk modulus 71
References 72

5 Sound propagation in porous materials having a rigid frame **73**
5.1 Introduction 73
5.2 Viscous and thermal dynamic and static permeability 74
5.2.1 *Definitions* 74
5.2.2 *Direct measurement of the static permeabilities* 76
5.3 Classical tortuosity, characteristic dimensions, quasi-static tortuosity 78
5.3.1 *Classical tortuosity* 78
5.3.2 *Viscous characteristic length* 79
5.3.3 *Thermal characteristic length* 80
5.3.4 *Characteristic lengths for fibrous materials* 80
5.3.5 *Direct measurement of the high-frequency parameters, classical tortuosity and characteristic lengths* 81
5.3.6 *Static tortuosity* 82
5.4 Models for the effective density and the bulk modulus of the saturating fluid 83
5.4.1 *Pride et al. model for the effective density* 83
5.4.2 *Simplified Lafarge model for the bulk modulus* 83
5.5 Simpler models 84
5.5.1 *The Johnson et al. model* 84
5.5.2 *The Champoux–Allard model* 84
5.5.3 *The Wilson model* 85
5.5.4 *Prediction of the effective density with the Pride et al. model and the model by Johnson et al.* 85
5.5.5 *Prediction of the bulk modulus with the simplified Lafarge model and the Champoux-Allard model* 85

5.5.6 Prediction of the surface impedance 87
5.6 Prediction of the effective density and the bulk modulus of open cell foams and fibrous materials with the different models 88
5.6.1 Comparison of the performance of different models 88
5.6.2 Practical considerations 88
5.7 Fluid layer equivalent to a porous layer 89
5.8 Summary of the semi-phenomenological models 90
5.9 Homogenization 91
5.10 Double porosity media 95
5.10.1 Definitions 95
5.10.2 Orders of magnitude for realistic double porosity media 96
5.10.3 Asymptotic development method for double porosity media 97
5.10.4 Low permeability contrast 98
5.10.5 High permeability contrast 99
5.10.6 Practical considerations 102
Appendix 5.A: Simplified calculation of the tortuosity for a porous material having pores made up of an alternating sequence of cylinders 103
Appendix 5.B: Calculation of the characteristic length Λ' 104
Appendix 5.C: Calculation of the characteristic length Λ for a cylinder perpendicular to the direction of propagation 106
References 107

6 Biot theory of sound propagation in porous materials having an elastic frame **111**
6.1 Introduction 111
6.2 Stress and strain in porous materials 111
6.2.1 Stress 111
6.2.2 Stress–strain relations in the Biot theory: The potential coupling term 112
6.2.3 A simple example 115
6.2.4 Determination of P, Q and R 116
6.2.5 Comparison with previous models of sound propagation in porous sound-absorbing materials 117
6.3 Inertial forces in the Biot theory 117
6.4 Wave equations 119
6.5 The two compressional waves and the shear wave 120
6.5.1 The two compressional waves 120
6.5.2 The shear wave 122
6.5.3 The three Biot waves in ordinary air-saturated porous materials 123
6.5.4 Example 123
6.6 Prediction of surface impedance at normal incidence for a layer of porous material backed by an impervious rigid wall 126
6.6.1 Introduction 126
6.6.2 Prediction of the surface impedance at normal incidence 126
6.6.3 Example: Fibrous material 129

Appendix 6.A: Other representations of the Biot theory 131
References 134

7 Point source above rigid framed porous layers 137
7.1 Introduction 137
7.2 Sommerfeld representation of the monopole field over a plane reflecting surface 137
7.3 The complex $\sin\theta$ plane 139
7.4 The method of steepest descent (passage path method) 140
7.5 Poles of the reflection coefficient 145
7.5.1 Definitions 145
7.5.2 Planes waves associated with the poles 146
7.5.3 Contribution of a pole to the reflected monopole pressure field 150
7.6 The pole subtraction method 151
7.7 Pole localization 153
7.7.1 Localization from the r dependence of the reflected field 153
7.7.2 Localization from the vertical dependence of the total pressure 155
7.8 The modified version of the Chien and Soroka model 156
Appendix 7.A Evaluation of N 160
Appendix 7.B Evaluation of p_r by the pole subtraction method 161
Appendix 7.C From the pole subtraction to the passage path: locally reacting surface 164
References 165

8 Porous frame excitation by point sources in air and by stress circular and line sources – modes of air saturated porous frames 167
8.1 Introduction 167
8.2 Prediction of the frame displacement 168
8.2.1 Excitation with a given wave number component parallel to the faces 168
8.2.2 Circular and line sources 172
8.3 Semi-infinite layer – Rayleigh wave 173
8.4 Layer of finite thickness – modified Rayleigh wave 176
8.5 Layer of finite thickness – modes and resonances 177
8.5.1 Modes and resonances for an elastic solid layer and a poroelastic layer 177
8.5.2 Excitation of the resonances by a point source in air 179
Appendix 8.A Coefficients r_{ij} and $M_{i,j}$ 182
Appendix 8.B Double Fourier transform and Hankel transform 183
Appendix 8.C Rayleigh pole contribution 185
References 185

9 Porous materials with perforated facings 187
9.1 Introduction 187
9.2 Inertial effect and flow resistance 187
9.2.1 Inertial effect 187

9.2.2 *Calculation of the added mass and the added length* 188
9.2.3 *Flow resistance* 191
9.2.4 *Apertures having a square cross-section* 192
9.3 Impedance at normal incidence of a layered porous material covered by a perforated facing – Helmoltz resonator 194
9.3.1 *Evaluation of the impedance for the case of circular holes* 194
9.3.2 *Evaluation at normal incidence of the impedance for the case of square holes* 198
9.3.3 *Examples* 199
9.3.4 *Design of stratified porous materials covered by perforated facings* 202
9.3.5 *Helmholtz resonators* 203
9.4 Impedance at oblique incidence of a layered porous material covered by a facing having circular perforations 205
9.4.1 *Evaluation of the impedance in a hole at the boundary surface between the facing and the material* 205
9.4.2 *Evaluation of the external added length at oblique incidence* 208
9.4.3 *Evaluation of the impedance of a faced porous layer at oblique incidence* 209
9.4.4 *Evaluation of the surface impedance at oblique incidence for the case of square perforations* 210
References 211

10 Transversally isotropic poroelastic media **213**
10.1 Introduction 213
10.2 Frame in vacuum 214
10.3 Transversally isotropic poroelastic layer 215
10.3.1 *Stress–strain equations* 215
10.3.2 *Wave equations* 216
10.4 Waves with a given slowness component in the symmetry plane 217
10.4.1 *General equations* 217
10.4.2 *Waves polarized in a meridian plane* 219
10.4.3 *Waves with polarization perpendicular to the meridian plane* 219
10.4.4 *Nature of the different waves* 219
10.4.5 *Illustration* 220
10.5 Sound source in air above a layer of finite thickness 222
10.5.1 *Description of the problems* 222
10.5.2 *Plane field in air* 223
10.5.3 *Decoupling of the air wave* 226
10.6 Mechanical excitation at the surface of the porous layer 227
10.7 Symmetry axis different from the normal to the surface 228
10.7.1 *Prediction of the slowness vector components of the different waves* 228
10.7.2 *Slowness vectors when the symmetry axis is parallel to the surface* 230
10.7.3 *Description of the different waves* 230

10.8 Rayleigh poles and Rayleigh waves 232
10.8.1 Example 234
10.9 Transfer matrix representation of transversally isotropic poroelastic media 236
Appendix 10.A: Coefficients T_i in Equation (10.46) 238
Appendix 10.B: Coefficients A_i in Equation (10.97) 239
References 240

11 Modelling multilayered systems with porous materials using the transfer matrix method **243**
11.1 Introduction 243
11.2 Transfer matrix method 244
11.2.1 Principle of the method 244
11.3 Matrix representation of classical media 244
11.3.1 Fluid layer 244
11.3.2 Solid layer 245
11.3.3 Poroelastic layer 247
11.3.4 Rigid and limp frame limits 251
11.3.5 Thin elastic plate 254
11.3.6 Impervious screens 255
11.3.7 Porous screens and perforated plates 256
11.3.8 Other media 256
11.4 Coupling transfer matrices 257
11.4.1 Two layers of the same nature 257
11.4.2 Interface between layers of different nature 258
11.5 Assembling the global transfer matrix 260
11.5.1 Hard wall termination condition 261
11.5.2 Semi-infinite fluid termination condition 261
11.6 Calculation of the acoustic indicators 263
11.6.1 Surface impedance, reflection and absorption coefficients 263
11.6.2 Transmission coefficient and transmission loss 263
11.6.3 Piston excitation 265
11.7 Applications 266
11.7.1 Materials with porous screens 266
11.7.2 Materials with impervious screens 271
11.7.3 Normal incidence sound transmission through a plate–porous system 274
11.7.4 Diffuse field transmission of a plate–foam system 275
Appendix 11.A The elements T_{ij} of the Transfer Matrix T] 277
References 280

12 Extensions to the transfer matrix method **283**
12.1 Introduction 283
12.2 Finite size correction for the transmission problem 283
12.2.1 Transmitted power 283
12.2.2 Transmission coefficient 287
12.3 Finite size correction for the absorption problem 288

12.3.1 Surface pressure 288
12.3.2 Absorption coefficient 289
12.3.3 Examples 291
12.4 Point load excitation 295
12.4.1 Formulation 295
12.4.2 The TMM, SEA and modal methods 297
12.4.3 Examples 298
12.5 Point source excitation 303
12.6 Other applications 304
Appendix 12.A: An algorithm to evaluate the geometrical radiation impedance 305
References 306

13 Finite element modelling of poroelastic materials **309**
13.1 Introduction 309
13.2 Displacement based formulations 310
13.3 The mixed displacement–pressure formulation 311
13.4 Coupling conditions 313
13.4.1 Poroelastic–elastic coupling condition 313
13.4.2 Poroelastic–acoustic coupling condition 314
13.4.3 Poroelastic–poroelastic coupling condition 315
13.4.4 Poroelastic–impervious screen coupling condition 315
13.4.5 Case of an imposed pressure field 316
13.4.6 Case of an imposed displacement field 317
13.4.7 Coupling with a semi-infinite waveguide 317
13.5 Other formulations in terms of mixed variables 320
13.6 Numerical implementation 320
13.7 Dissipated power within a porous medium 323
13.8 Radiation conditions 324
13.9 Examples 327
13.9.1 Normal incidence absorption and transmission loss of a foam: finite size effects 327
13.9.2 Radiation effects of a plate–foam system 329
13.9.3 Damping effects of a plate–foam system 331
13.9.4 Diffuse transmission loss of a plate–foam system 333
13.9.5 Application to the modelling of double porosity materials 335
13.9.6 Modelling of smart foams 339
13.9.7 An industrial application 343
References 347

Index **351**

Preface to the Second Edition

In the first edition, models initially developed to describe wave propagation in porous media saturated by heavy fluids are used to predict the acoustical performances of air saturated sound absorbing porous media. In this expanded and revised edition, we have retained, with slight modifications, most of the basic material of the first edition and expanded it by revisiting several original topics and adding new topics to integrate recent developments in the domain of wave propagation in porous media and practical numerical prediction methods that are widley used by researchers and engineers.

Chapters 1 to 3 dealing with sound propagation in solids and fluid and Chapter 9 dealing with the modelling of perforated facings were slightly modified. Chapters 4 to 6 were greatly revisited. A more detailed description of sound propagation in cylindrical pores is presented (Chapter 4), related to the more general presentation of new parameters and new models for sound propagation in rigid-framed porous media (Chapter 5). Also in Chapter 5 a short presentation of homogenization, with some results concerning double porosity media, is added. In Chapter 6, different formulations of the Biot theory for poroelastic media are given, with a simplified version for the case of media with a limp frame. In Chapter 11 we have revisited the original representation of the modelling of layered media (Chapter 7 of the first edition) and extended it to cover the systematic modelling of layered media using the Transfer Matrix Method (TMM). In particular, a step by step presentation of the numerical implementation of the method is given with several application examples.

Major additions include five new chapters. Chapter 7 discusses the acoustic field created by a point source above a rigid framed porous layer, with recent advances concerning the poles of the reflection coefficient and the reflected field at grazing incidence. Chapter 8 is concerned by the poroelastic layers excited by a point source in air or by a localized stress source on the free face of the layer, with a description of the Rayleigh waves and the resonances. Axisymmetrical poroelastic media are studied in Chapter 10. In Chapter 12, complements to the transfer matrix method are given. They concern mainly the effect of the finite lateral extend, and the excitation by point loads, of sound packages. Several examples illustrating the practical importance of these extensions are given (e.g. size effects on the random incidence absorption and transmission loss of porous media; airborne vs. structure borne insertion loss of sound packages). In Chapter 13, an introduction to the finite element modelling of poroelastic media is presented. Emphasis is put on the use of the mixed displacement-pressure formulation of the Biot theory, which appears in the Appendix of Chap. 6. Detailed description of coupling conditions between various domains including a waveguide are presented together with sections on the breakdown of the power dissipation mechanisms within a porous media and radiation conditions. Several applications are chosen to illustrate the practical use of the presented methods including modelling of double porosity materials and smart foams.

As in the first edition, the goal of the book remains to provide in a concrete manner a physical basis, as simple as possible, and the developments, analytical calculations and numerical methods, that will be useful in different fields where sound absorption and transmission and vibration damping by air saturated porous media are concerned.

Acknowledgments

The first authors (Prof. Allard) is grateful to Professor Walter Lauriks (Katholieke Universiteit Leuven) for his collaboration for more than twenty years which has brought a significant contribution to the book. The second author (Prof. Atalla) would like to single out for special thanks Dr Franck Sgard (Institut de Recherche Robert-Sauvé en Santé et en Sécurité du Travail), Dr Raymond Panneton (Université de Sherbrooke), Dr Mohamed Ali Hamdi (Université de Technologie de Compiègne) and Arnaud Duval (Faurecia) for their various collaborations and discussions that resulted in many beneficial improvements to the book.

Jean-Francois Allard, Le Mans, France
August 2009 Noureddine Atalla, Sherbrooke, Canada

1

Plane waves in isotropic fluids and solids

1.1 Introduction

The aim of this chapter is to introduce the stress–strain relations, the basic equations governing sound propagation which will be useful for the understanding of the Biot theory. The framework of the presentation is the linear theory of elasticity. Total derivatives with respect to time d/d*t* are systematically replaced by partial derivatives $\partial/\partial t$. The presentation is carried out with little explanation. Detailed derivation can be found in the literature (Ewing *et al.* 1957, Cagniard 1962, Miklowitz 1966, Brekhovskikh 1960, Morse and Ingard 1968, Achenbach 1973).

1.2 Notation – vector operators

A system of rectangular cartesian coordinates (x_1, x_2, x_3) will be used in the following, having unit vectors $\mathbf{i}_1$, $\mathbf{i}_2$ and $\mathbf{i}_3$. The vector operator del (or nabla) denoted by ∇ can be defined by

$$\nabla = \mathbf{i}_1 \frac{\partial}{\partial x_1} + \mathbf{i}_2 \frac{\partial}{\partial x_2} + \mathbf{i}_3 \frac{\partial}{\partial x_3} \tag{1.1}$$

When operating on a scalar field $\varphi(x_1, x_2, x_3)$ the vector operator ∇ yields the gradient of φ

$$\mathbf{grad}\ \varphi = \nabla\varphi = \mathbf{i}_1 \frac{\partial\varphi}{\partial x_1} + \mathbf{i}_2 \frac{\partial\varphi}{\partial x_2} + \mathbf{i}_3 \frac{\partial\varphi}{\partial x_3} \tag{1.2}$$

Propagation of Sound in Porous Media: Modelling Sound Absorbing Materials, Second Edition J. F. Allard and N. Atalla

When operating on a vector field $\mathbf{v}$ with components (v_1, v_2, v_3), the vector operator ∇ yields the divergence of $\mathbf{v}$

$$div\ \mathbf{v} = \nabla \cdot \mathbf{v} = \frac{\partial v_1}{\partial x_1} + \frac{\partial v_2}{\partial x_2} + \frac{\partial v_3}{\partial x_3} \tag{1.3}$$

The Laplacian of φ is:

$$\nabla \cdot \nabla\varphi = \nabla^2\varphi = \text{div } \mathbf{grad}\ \varphi = \frac{\partial^2\varphi}{\partial x_1^2} + \frac{\partial^2\varphi}{\partial x_2^2} + \frac{\partial^2\varphi}{\partial x_3^2} \tag{1.4}$$

When operating on the vector $\mathbf{v}$, the Laplacian operator yields a vector field whose components are the Laplacians of v_1, v_2 and v_3

$$(\nabla^2\mathbf{v})_i = \frac{\partial^2 v_i}{\partial \varphi_1^2} + \frac{\partial^2 v_i}{\partial \varphi_2^2} + \frac{\partial^2 v_i}{\partial \varphi_3^2} \tag{1.5}$$

The gradient of the divergence of a vector $\mathbf{v}$ is a vector of components

$$(\nabla\nabla \cdot \mathbf{v})_i = \frac{\partial}{\partial x_i}\left(\frac{\partial v_1}{\partial x_1} + \frac{\partial v_2}{\partial x_2} + \frac{\partial v_3}{\partial x_3}\right) \tag{1.6}$$

The vector **curl** is denoted by

$$\mathbf{curl}\ \mathbf{v} = \nabla \wedge \mathbf{v} \tag{1.7}$$

and is equal to

$$\mathbf{curl}\ \mathbf{v} = \mathbf{i}_1\left(\frac{\partial v_3}{\partial x_2} - \frac{\partial v_2}{\partial x_3}\right) + \mathbf{i}_2\left(\frac{\partial v_1}{\partial x_3} - \frac{\partial v_3}{\partial x_1}\right) + \mathbf{i}_3\left(\frac{\partial v_2}{\partial x_1} - \frac{\partial v_3}{\partial x_2}\right) \tag{1.8}$$

1.3 Strain in a deformable medium

Let us consider the coordinates of the two points P and Q in a deformable medium before and after deformation. The two points P and Q are represented in Figure 1.1.

The coordinates of P are (x_1, x_2, x_3) and become $(x_1 + u_1, x_2 + u_2, x_3 + u_3)$ after deformation. The quantities (u_1, u_2, u_3) are then the components of the displacement vector $\mathbf{u}$ of P. The components of the displacement vector for the neighbouring point Q, having initial coordinates $(x_1 + \Delta x_1, x_2 + \Delta x_2, x_3 + \Delta x_3)$, are to a first-order approximation

$$\begin{aligned}
u_1' &= u_1 + \frac{\partial u_1}{\partial x_1}\Delta x_1 + \frac{\partial u_1}{\partial x_2}\Delta x_2 + \frac{\partial u_1}{\partial x_3}\Delta x_3 \\
u_2' &= u_2 + \frac{\partial u_2}{\partial x_1}\Delta x_1 + \frac{\partial u_2}{\partial x_2}\Delta x_2 + \frac{\partial u_3}{\partial x_3}\Delta x_3 \\
u_3' &= u_3 + \frac{\partial u_3}{\partial x_1}\Delta x_1 + \frac{\partial u_3}{\partial x_2}\Delta x_2 + \frac{\partial u_3}{\partial x_3}\Delta x_3
\end{aligned} \tag{1.9}$$

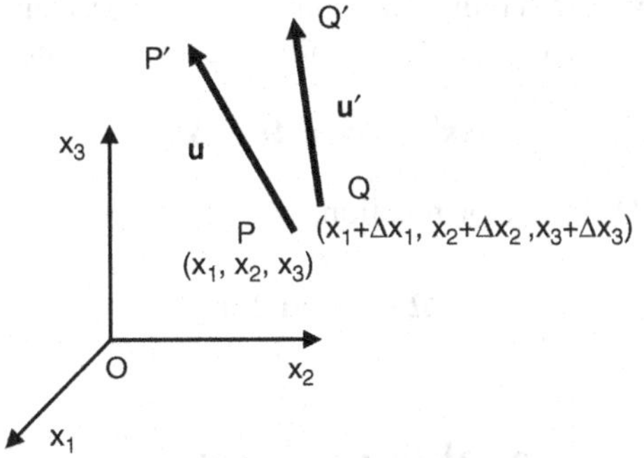

Figure 1.1 The displacement of P and Q to P' and Q' in a deformable medium.

A rotation vector $\boldsymbol{\Omega}(\Omega_1, \Omega_2, \Omega_3)$ and a 3×3 strain tensor e can be defined at P by the following equations:

$$\Omega_1 = \frac{1}{2}\left(\frac{\partial u_3}{\partial x_2} - \frac{\partial u_2}{\partial x_3}\right), \qquad \Omega_2 = \frac{1}{2}\left(\frac{\partial u_1}{\partial x_3} - \frac{\partial u_3}{\partial x_1}\right)$$
$$\Omega_3 = \frac{1}{2}\left(\frac{\partial u_2}{\partial x_1} - \frac{\partial u_1}{\partial x_2}\right) \tag{1.10}$$

$$e_{ij} = \frac{1}{2}\left(\frac{\partial u_i}{\partial x_j} + \frac{\partial u_j}{\partial x_i}\right) \tag{1.11}$$

The displacement components of Q can be rewritten as

$$\begin{aligned} u'_1 &= u_1 + (\Omega_2 \Delta x_3 - \Omega_3 \Delta x_2) + (e_{11}\Delta x_1 + e_{12}\Delta x_2 + e_{13}\Delta x_3) \\ u'_2 &= u_2 + (\Omega_3 \Delta x_1 - \Omega_1 \Delta x_3) + (e_{21}\Delta x_1 + e_{22}\Delta x_2 + e_{23}\Delta x_3) \\ u'_3 &= u_3 + (\Omega_1 \Delta x_2 - \Omega_2 \Delta x_1) + (e_{31}\Delta x_1 + e_{32}\Delta x_2 + e_{33}\Delta x_3) \end{aligned} \tag{1.12}$$

The terms in the first parenthesis of each equation are associated with rotations around P, while those in the second parenthesis are related to deformations. The three components e_{11}, e_{22} and e_{33}, which are equal to

$$e_{11} = \frac{\partial u_1}{\partial x_1}, \qquad e_{22} = \frac{\partial u_2}{\partial x_2}, \qquad e_{33} = \frac{\partial u_3}{\partial x_3} \tag{1.13}$$

are an estimation of the extensions parallel to the axes.

The cubical dilatation θ is the limit of the ratio of the change in the volume to the initial volume when the dimensions of the initial volume approach zero. Hence,

$$\theta = \lim \frac{(\Delta x_1 + e_{11}\Delta x_1)(\Delta x_2 + e_{22}\Delta x_2)(\Delta x_3 + e_{33}\Delta x_3) - \Delta x_1 \Delta x_2 \Delta x_3}{\Delta x_1 \Delta x_2 \Delta x_3} \tag{1.14}$$

and is equal to the divergence of $\boldsymbol{u}$:

$$\theta = \nabla \cdot \boldsymbol{u} = \frac{\partial u_1}{\partial x_1} + \frac{\partial u_2}{\partial x_2} + \frac{\partial u_3}{\partial x_3} = e_{11} + e_{22} + e_{33} \tag{1.15}$$

If $\mathbf{\Delta x}$ denotes the vector having components Δx_1, Δx_2 and Δx_3, after a rotation characterized by the rotation vector $\mathbf{\Omega}$, the initial vector becomes $\mathbf{\Delta x'}$ related to $\mathbf{\Delta x}$ by

$$\mathbf{\Delta x'} - \mathbf{\Delta x} = \mathbf{\Omega} \wedge \mathbf{\Delta x} \tag{1.16}$$

The rotation vector $\mathbf{\Omega}$, in vector notation, is

$$\mathbf{\Omega} = \frac{1}{2}\,\mathbf{curl\ u} \tag{1.17}$$

1.4 Stress in a deformable medium

Two kinds of forces may act on a body, body forces and surface forces. Surface forces act across the surface, including its boundary. Consider a volume V in a deformable medium as represented in Figure 1.2.

Let S be the surface limiting V and ΔS an element of S around a point P that lies on S. The side of S which is outside V is called (+) while the other is called (−). The force exerted on V across ΔS is denoted by $\mathbf{\Delta F}$. A stress vector at P is defined by

$$\mathbf{T}(P) = \lim_{\Delta S \to 0} \frac{\mathbf{\Delta F}}{\Delta S} \tag{1.18}$$

The stress vector $\mathbf{T}(P)$ depends on P and on the direction of the positive outward unit normal $\mathbf{n}$ to the surface S at P. The stress vectors can be obtained from $\mathbf{T}^1(\sigma_{11}, \sigma_{12}, \sigma_{13})$, $\mathbf{T}^2(\sigma_{21}, \sigma_{22}, \sigma_{23})$, and $\mathbf{T}^3(\sigma_{31}, \sigma_{32}, \sigma_{33})$ corresponding to surfaces with normal $\mathbf{n}$ parallel to the x_1, x_2 and x_3 axes, respectively.

The components T_1, T_2, T_3 of $\mathbf{T}$ can be expressed in the general case as

$$\begin{aligned} T_1 &= \sigma_{11}n_1 + \sigma_{21}n_2 + \sigma_{31}n_3 \\ T_2 &= \sigma_{12}n_1 + \sigma_{22}n_2 + \sigma_{32}n_3 \\ T_3 &= \sigma_{13}n_1 + \sigma_{23}n_2 + \sigma_{33}n_3 \end{aligned} \tag{1.19}$$

In these equations n_1, n_2 and n_3 are the direction cosines of the positive normal $\mathbf{n}$ to S at P. The quantities σ_{ij} are the nine components of the stress tensor at P. These components are symmetrical, i.e. $\sigma_{ij} = \sigma_{ji}$, like the components e_{ij}. An illustration is given in Figure 1.3 for a cube with faces of unit area parallel to the coordinate planes.

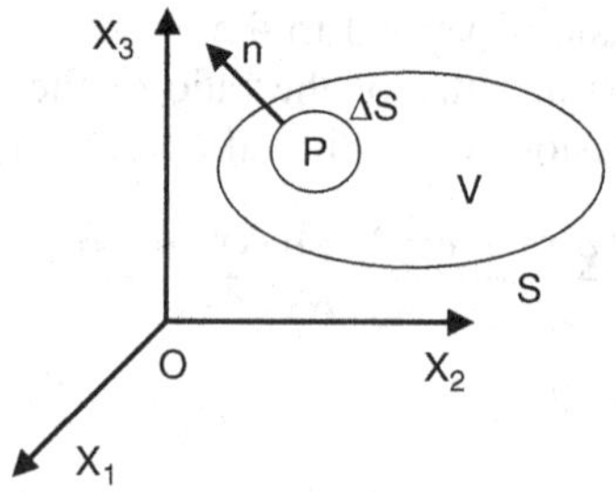

Figure 1.2 A volume V in a deformable medium, with an element ΔS belonging to the surface S limiting V.

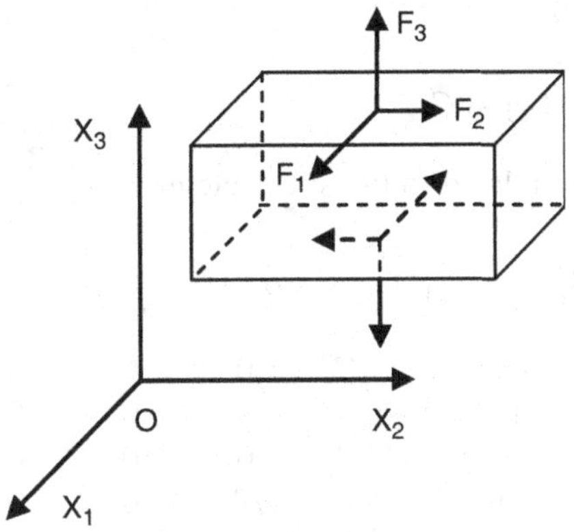

Figure 1.3 A cube with faces of unit area parallel to the coordinate planes. The three components of the forces acting on the upper and the lower faces are represented.

The variations of the components σ_{ij} are assumed to be negligible at the surface of the cube. With the components of the positive unit normal on the upper face being (0, 0, 1), Equations (1.19) reduce to

$$T_1 = \sigma_{31}, \qquad T_2 = \sigma_{32}, \qquad T_3 = \sigma_{33} \tag{1.20}$$

The force $\mathbf{F}(F_1, F_2, F_3)$ acting on the upper face is equal to $\mathbf{T}^3$. The components of the unit normal on the lower face are (0, 0, −1). The forces on the lower and the upper face are equal in magnitude and lie in opposite directions. The same property holds for the two other pairs of opposite faces. The elements σ_{ij} where $i = j$ correspond to normal forces while those with $i \neq j$ correspond to tangential forces.

1.5 Stress–strain relations for an isotropic elastic medium

The stress–strain relations for an isotropic elastic medium are as follows:

$$\sigma_{ij} = \lambda\theta\delta_{ij} + 2\mu e_{ij} \tag{1.21}$$

The quantities λ and μ are the Lamé coefficients and δ_{ij} is the Kronecker delta:

$$\begin{aligned} \delta_{ij} &= 1 \quad \textit{if} \quad i = j \\ \delta_{ij} &= 0 \quad \textit{if} \quad i \neq j \end{aligned} \tag{1.22}$$

In matrix form Equation (1.21) can be rewritten

$$\begin{pmatrix} \sigma_{11} \\ \sigma_{22} \\ \sigma_{33} \\ \sigma_{13} \\ \sigma_{23} \\ \sigma_{12} \end{pmatrix} = \begin{pmatrix} C_{11} & C_{12} & C_{12} & 0 & 0 & 0 \\ C_{12} & C_{11} & C_{12} & 0 & 0 & 0 \\ C_{12} & C_{12} & C_{11} & 0 & 0 & 0 \\ 0 & 0 & 0 & C_{44} & 0 & 0 \\ 0 & 0 & 0 & 0 & C_{44} & 0 \\ 0 & 0 & 0 & 0 & 0 & C_{44} \end{pmatrix} \begin{pmatrix} e_{11} \\ e_{22} \\ e_{33} \\ e_{13} \\ e_{23} \\ e_{12} \end{pmatrix} \tag{1.23}$$

$$\begin{aligned} C_{11} &= \lambda + 2\mu \\ C_{12} &= \lambda \\ C_{44} &= 2\mu = C_{11} - C_{12} \end{aligned} \tag{1.24}$$

The strain elements are related to the stress elements by

$$e_{ij} = -\frac{\lambda\delta_{ij}}{2\mu(3\lambda + 2\mu)}(\sigma_{11} + \sigma_{22} + \sigma_{33}) + \frac{1}{2\mu}\sigma_{ij} \tag{1.25}$$

$$\begin{pmatrix} e_{11} \\ e_{22} \\ e_{33} \\ e_{13} \\ e_{23} \\ e_{12} \end{pmatrix} = \begin{pmatrix} 1/E & -\nu/E & -\nu/E & 0 & 0 & 0 \\ -\nu/E & 1/E & -\nu/E & 0 & 0 & 0 \\ -\nu/E & -\nu/E & 1/E & 0 & 0 & 0 \\ 0 & 0 & 0 & 1/2\mu & 0 & 0 \\ 0 & 0 & 0 & 0 & 1/2\mu & 0 \\ 0 & 0 & 0 & 0 & 0 & 1/2\mu \end{pmatrix} \begin{pmatrix} \sigma_{11} \\ \sigma_{22} \\ \sigma_{33} \\ \sigma_{13} \\ \sigma_{23} \\ \sigma_{12} \end{pmatrix} \tag{1.26}$$

where E is the Young's modulus and ν is the Poisson ratio. They are related to the Lamé coefficients by

$$\begin{aligned} E &= \frac{\mu(3\lambda + 2\mu)}{\lambda + \mu} \\ \nu &= \frac{\lambda}{2(\lambda + \mu)} \end{aligned} \tag{1.27}$$

The shear modulus G is related to E and ν via

$$G = \mu = \frac{E}{2(1 + \nu)} \tag{1.28}$$

Examples

Antiplane shear

The displacement field is represented in Figure 1.4. For this case, the two components e_{ij} which differ from zero are

$$e_{32} = e_{23} = \frac{1}{2}\frac{\partial u_2}{\partial x_3} \tag{1.29}$$

The angle α is equal to

$$\alpha = \frac{\partial u_2}{\partial x_3} \tag{1.30}$$

Using Equation (1.21), one obtains two components σ_{ij} which differ from zero:

$$\sigma_{32} = \sigma_{23} = \mu\alpha \tag{1.31}$$

The coefficient μ is the shear modulus of the medium, which relates the angle of deformation and the tangential force per unit area. The three components of the rotation

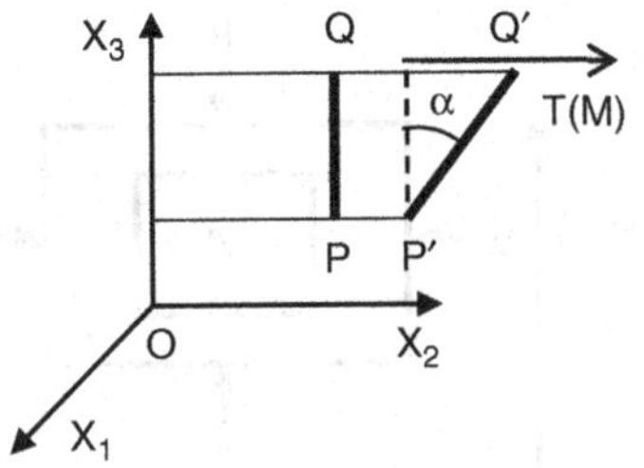

Figure 1.4 Antiplane shear in an elastic medium. A vector **PQ** initially parallel to x_3 becomes oblique with an angle α to the initial direction.

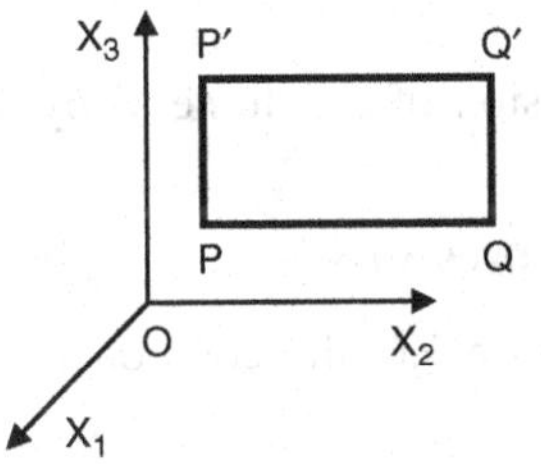

Figure 1.5 Longitudinal strain in the x_3 direction.

vector $\mathbf{\Omega}$ are

$$\Omega_1 = -\frac{1}{2}\frac{\partial u_2}{\partial x_3}, \qquad \Omega_2 = \Omega_3 = 0 \tag{1.32}$$

The deformation is equivoluminal, the dilatation θ being equal to zero, and there is a rotation around x_1.

Longitudinal strain

For this case only the component e_{33} of the strain tensor is different from zero. The vectors **PQ** and **P′Q′** are represented in Figure 1.5.

The stress tensor components that do not vanish are

$$\begin{aligned} \sigma_{33} &= (\lambda + 2\mu)e_{33} \\ \sigma_{11} &= \sigma_{22} = \lambda e_{33} \end{aligned} \tag{1.33}$$

Unidirectional stress

From Equation (1.26) the stress component σ_{33} transforms a vector **PQ** parallel to the axis x_3 into a vector **P′Q′** parallel to x_3. The ratio P′Q′/PQ is given by

$$P'Q'/PQ = \sigma_{33}/E \tag{1.34}$$

A vector **PQ** perpendicular to x_3 is transformed in a vector **P′Q′** parallel to **PQ** and the ratio P′Q′/PQ is now given by

$$P'Q'/PQ = -\nu\sigma_{33}/E \tag{1.35}$$

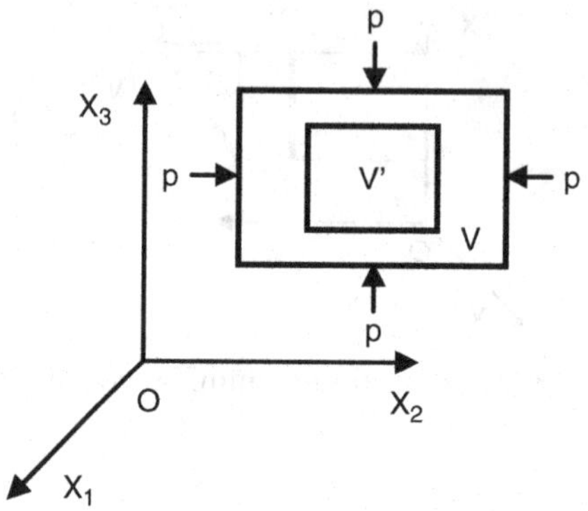

Figure 1.6 Compression of a volume V by a hydrostatic pressure.

Compression by a hydrostatic pressure

For this case, represented in Figure 1.6, the components of the stress tensor that do not vanish are

$$\sigma_{11} = \sigma_{22} = \sigma_{33} = -p \tag{1.36}$$

From Equation (1.21) it follows that the dilatation θ is related to p by

$$\theta = -p \Big/ \left(\lambda + \frac{2\mu}{3}\right) \tag{1.37}$$

The ratio $-p/\theta$ is the bulk modulus K of the material, which is equal to

$$K = \lambda + \frac{2\mu}{3} \tag{1.38}$$

Contrary to the case of simple shear, $\mathbf{\Omega} = 0$ and θ is nonzero. The deformation is irrotational, as in the case with a longitudinal strain. Note that since a hydrostatic pressure leads to a negative volume change, the bulk modulus K is positive for all materials and in consequence Poisson's ratio is less than or equal to 0.5 for all materials.

1.6 Equations of motion

The total surface force $\mathbf{F}_v$ acting on the volume V represented in Figure 1.2 is

$$\mathbf{F}_v = \iint \mathbf{T}\,\mathrm{d}S \tag{1.39}$$

The projection of the force $\mathbf{F}_v$ on to the x_i axis is

$$F_{v_i} = \iint_S (\sigma_{1i}n_1 + \sigma_{2i}n_2 + \sigma_{3i}n_3)\,\mathrm{d}S \tag{1.40}$$

By using the divergence theorem, Equation (1.40) becomes

$$F_{v_i} = \iiint\limits_V \left(\frac{\partial \sigma_{1i}}{\partial x_1} + \frac{\partial \sigma_{2i}}{\partial x_2} + \frac{\partial \sigma_{3i}}{\partial x_3}\right) \mathrm{d}V \tag{1.41}$$

Adding the component X_i of the body force per unit volume, the linearized Newton equation for V may be written as

$$\iiint\limits_V \left(\frac{\partial \sigma_{1i}}{\partial x_1} + \frac{\partial \sigma_{2i}}{\partial x_2} + \frac{\partial \sigma_{3i}}{\partial x_3} + X_i - \rho\frac{\partial^2 u_i}{\partial t^2}\right) \mathrm{d}V = 0 \tag{1.42}$$

where ρ is the mass density of the material. This equation leads to the stress equations of motion

$$\frac{\partial \sigma_{1i}}{\partial x_1} + \frac{\partial \sigma_{2i}}{\partial x_2} + \frac{\partial \sigma_{3i}}{\partial x_3} + X_i - \rho\frac{\partial^2 u_i}{\partial t^2} = 0 \qquad i = 1, 2, 3 \tag{1.43}$$

With the aid of Equation (1.21) the equations of motion become

$$\rho\frac{\partial^2 u_i}{\partial t^2} = \lambda\frac{\partial \theta}{\partial x_i} + 2\mu\frac{\partial e_{ii}}{\partial x_i} + \sum_{j\neq i} 2\mu\frac{\partial e_{ji}}{\partial x_j} + X_i \quad i = 1, 2, 3 \tag{1.44}$$

Replacing e_{ji} by $1/2(\partial u_j/\partial x_i + \partial u_i/\partial x_j)$, Equations (1.44) can be written in terms of displacement as

$$\rho\frac{\partial^2 u_i}{\partial t^2} = (\lambda + \mu)\frac{\partial \nabla.\mathbf{u}}{\partial x_i} + \mu\nabla^2 u_i + X_i \quad i = 1, 2, 3 \tag{1.45}$$

where ∇^2 is the Laplacian operator $\dfrac{\partial^2}{\partial x_1^2} + \dfrac{\partial^2}{\partial x_2^2} + \dfrac{\partial^2}{\partial x_3^2}$.

Using vector notation, Equations (1.45) can be written

$$\rho\frac{\partial^2 \mathbf{u}}{\partial t^2} = (\lambda + \mu)\nabla\nabla \cdot \mathbf{u} + \mu\nabla^2\mathbf{u} + \mathbf{X} \qquad i = 1, 2, 3 \tag{1.46}$$

In this equation, $\nabla\nabla \cdot \mathbf{u}$ is the gradient of the divergence $\nabla \cdot \mathbf{u}$ of the vector field $\mathbf{u}$, and its components are

$$\frac{\partial}{\partial x_i}\left[\frac{\partial u_1}{\partial x_1} + \frac{\partial u_2}{\partial x_2} + \frac{\partial u_3}{\partial x_3}\right] \qquad i = 1, 2, 3 \tag{1.47}$$

and the quantity $\nabla^2\mathbf{u}$ is the Laplacian of the vector field $\mathbf{u}$, having components

$$\sum_{j=1,2,3} \frac{\partial^2 u_i}{\partial x_j^2} \qquad i = 1, 2, 3 \tag{1.48}$$

as indicated in Section 1.2.

1.7 Wave equation in a fluid

In the case of an inviscid fluid, μ vanishes. The stress coefficients reduce to

$$\begin{aligned}\sigma_{11} = \sigma_{22} = \sigma_{33} = \lambda\theta \\ \sigma_{12} = \sigma_{13} = \sigma_{23} = 0\end{aligned} \tag{1.49}$$

The three nonzero stress elements are equal to $-p$, where p is the pressure. The bulk modulus K, given by Equation (1.38), becomes simply λ:

$$K = \lambda \tag{1.50}$$

The stress field (Equation 1.49) generates only irrotational deformations such as $\mathbf{\Omega} = 0$.

A representation of the displacement vector $\mathbf{u}$ in the following form can be used:

$$u_1 = \partial\varphi/\partial x_1, \qquad u_2 = \partial\varphi/\partial x_2, \qquad u_3 = \partial\varphi/\partial x_3 \tag{1.51}$$

where φ is a displacement potential.

In vector form, Equations (1.51) can be written as

$$\boldsymbol{u} = \nabla\varphi \tag{1.52}$$

Using this representation, the rotation vector $\mathbf{\Omega}$ can be rewritten

$$\mathbf{\Omega} = \frac{1}{2}\mathbf{curl}\ \nabla\varphi = 0 \tag{1.53}$$

and the displacement field is irrotational.

Substitution of this displacement representation into Equation (1.46) with $\mu = 0$ and $\mathbf{X} = 0$ yields

$$\lambda\ \nabla\nabla \cdot \nabla\varphi = \rho\frac{\partial^2}{\partial t^2}\nabla\varphi \tag{1.54}$$

Since $\nabla \cdot \nabla\varphi = \nabla^2\varphi$, Equation (1.54) reduces, with Equation (1.50), to

$$\nabla\left[K\nabla^2\varphi - \rho\frac{\partial^2}{\partial t^2}\varphi\right] = 0 \tag{1.55}$$

The displacement potential φ satisfies the equation of motion if

$$\nabla^2\varphi = \rho\frac{\partial^2\varphi}{K\partial t^2} \tag{1.56}$$

If the fluid is a perfectly elastic fluid, with no damping, K is a real number.

This displacement potential is related to pressure in a simple way. From Equations (1.49), (1.50) and (1.52), p can be written as

$$p = -K\theta = -K\nabla^2\varphi \tag{1.57}$$

By the use of Equations (1.56) and (1.57) one obtains

$$p = -\rho \frac{\partial^2 \varphi}{\partial t^2} \tag{1.58}$$

At an angular frequency ω ($\omega = 2\pi f$, where f is frequency), p can be rewritten as

$$p = \rho \omega^2 \varphi \tag{1.59}$$

As an example, a simple solution of Equation (1.56) is

$$\varphi = \frac{A}{\rho \omega^2} \exp[j(-kx_3 + \omega t) + \alpha] \tag{1.60}$$

In this equation, A and α are arbitrary constants, and k is the wave number

$$k = \omega (\rho / K)^{1/2} \tag{1.61}$$

The phase velocity is given by

$$c = \omega / \mathrm{Re}\, k \tag{1.62}$$

and Im(k) appears in the amplitude dependence on x_3, $\exp(\mathrm{Im}(k) x_3)$. In this example, u_3 is the only nonzero component of $\mathbf{u}$:

$$u_3 = \frac{\partial \varphi}{\partial x_3} = -\frac{jkA}{\rho \omega^2} \exp[j(-kx_3 + \omega t + \alpha)] \tag{1.63}$$

The pressure p is

$$p = -\rho \frac{\partial^2 \varphi}{\partial t^2} = A \exp[j(-kx_3 + \omega t + \alpha)] \tag{1.64}$$

This field of deformation corresponds to the propagation parallel to the x_3 axis of a longitudinal strain, with a phase velocity c.

1.8 Wave equations in an elastic solid

A scalar potential φ and a vector potential $\boldsymbol{\psi}(\psi_1, \psi_2, \psi_3)$ can be used to represent displacements in a solid

$$\begin{aligned} u_1 &= \frac{\partial \varphi}{\partial x_1} + \frac{\partial \psi_3}{\partial x_2} - \frac{\partial \psi_2}{\partial x_3} \\ u_2 &= \frac{\partial \varphi}{\partial x_2} + \frac{\partial \psi_1}{\partial x_3} - \frac{\partial \psi_3}{\partial x_1} \\ u_3 &= \frac{\partial \varphi}{\partial x_3} + \frac{\partial \psi_2}{\partial x_1} - \frac{\partial \psi_1}{\partial x_2} \end{aligned} \tag{1.65}$$

In vector form, Equations (1.65) reduce to

$$\mathbf{u} = \mathbf{grad}\ \varphi + \mathbf{curl}\ \boldsymbol{\psi} \tag{1.66}$$

or, using the notation ∇ for the gradient operator

$$\mathbf{u} = \nabla\varphi + \nabla \wedge \boldsymbol{\psi} \tag{1.67}$$

The rotation vector $\boldsymbol{\Omega}$ in Equation (1.17) is then equal to

$$\boldsymbol{\Omega} = \frac{1}{2}\nabla \wedge \nabla \wedge \boldsymbol{\psi} \tag{1.68}$$

Therefore, the scalar potential involves dilatation while the vector potential describes infinitesimal rotations.

In the absence of body forces, the displacement equation of motion (1.46) is

$$\rho\frac{\partial^2 \mathbf{u}}{\partial t^2} = (\lambda + \mu)\nabla\nabla \cdot \mathbf{u} + \mu\nabla^2\mathbf{u} \tag{1.69}$$

Substitution of the displacement representation given by Equation (1.67) into Equation (1.69) yields

$$\mu\nabla^2[\nabla\varphi + \nabla \wedge \boldsymbol{\psi}] + (\lambda + \mu)\nabla\nabla \cdot [\nabla\varphi + \nabla \wedge \boldsymbol{\psi}] = \rho\frac{\partial^2}{\partial t^2}[\nabla\varphi + \nabla \wedge \boldsymbol{\psi}] \tag{1.70}$$

In Equation (1.70), $\nabla \cdot \nabla\varphi$ can be replaced by $\nabla^2\varphi$, $\nabla \cdot \nabla \wedge \boldsymbol{\psi} = 0$, allowing this equation to reduce to

$$\mu\nabla^2\nabla\varphi + \lambda\nabla\nabla^2\varphi + \mu\nabla\nabla^2\varphi - \rho\frac{\partial^2}{\partial t^2}\nabla\varphi + \left(\mu\nabla^2 - \rho\frac{\partial^2}{\partial t^2}\right)\nabla \wedge \psi = \mathbf{0} \tag{1.71}$$

By using the relations $\nabla^2\nabla\varphi = \nabla\nabla^2\varphi$ and $\nabla^2\nabla \wedge \boldsymbol{\psi} = \nabla \wedge \nabla^2\boldsymbol{\psi}$, Equation (1.71) can be rewritten

$$\nabla\left[(\lambda + 2\mu)\nabla^2\varphi - \rho\frac{\partial^2\varphi}{\partial t^2}\right] + \nabla \wedge \left[\mu\nabla^2\boldsymbol{\psi} - \rho\frac{\partial^2\boldsymbol{\psi}}{\partial t^2}\right] = 0 \tag{1.72}$$

From this, we obtain two equations containing, respectively, the scalar and the vector potential

$$\nabla^2\varphi = \frac{\rho}{\lambda + 2\mu}\frac{\partial^2\varphi}{\partial t^2} \tag{1.73}$$

$$\nabla^2\boldsymbol{\psi} = \frac{\rho}{\mu}\frac{\partial^2\boldsymbol{\psi}}{\partial t^2} \tag{1.74}$$

Equation (1.73) describes the propagation of irrotational waves travelling with a wave number vector k equal to

$$k = \omega(\rho/(\lambda + 2\mu))^{1/2} \tag{1.75}$$

The phase velocity c is always related to the wave number k by Equation (1.62). The quantity K_c defined as

$$K_c = \lambda + 2\mu \tag{1.76}$$

can be substituted in Equation (1.75), resulting in

$$k = \omega(\rho_c/K_c)^{1/2} \tag{1.77}$$

while the stress–strain relations (Equations (1.21) can be rewritten as

$$\sigma_{ij} = (K_c - 2\mu)\theta\delta_{ij} + 2\mu e_{ij} \tag{1.78}$$

Equation (1.74) describes the propagation of equivoluminal (shear) waves propagating with a wave number equal to

$$k' = \omega(\rho/\mu)^{1/2} \tag{1.79}$$

As an example, a simple vector potential $\boldsymbol{\psi}$ can be used:

$$\psi_2 = \psi_3 = 0 \qquad \psi_1 = B\exp[j(-k'x_3 + \omega t)] \tag{1.80}$$

In this case, u_2 is the only component of the displacement vector which is different from zero

$$u_2 = -jBk'\exp[j(-k'x_3 + \omega t)] \tag{1.81}$$

This field of deformation corresponds to propagation, parallel to the x_3 axis, of the antiplane shear.

References

Achenbach, J.D. (1973) *Wave Propagation in Elastic Solids*. North Holland Publishing Co., New York.

Brekhovskikh, L.M. (1960) *Waves in Layered Media*. Academic Press, New York.

Cagniard, L. (1962) *Reflection and Refraction of Progressive Waves*, translated and revised by E.A. Flinn and C.H. Dix. McGraw-Hill, New York.

Ewing, W.M., Jardetzky, W.S. and Press, F. (1957) *Elastic Waves in Layered Media*. McGraw-Hill, New York.

Miklowitz, J. (1966) Elastic Wave Propagation. In *Applied Mechanics Surveys*, eds H.N. Abramson, H. Liebowitz, J.N. Crowley and R.S. Juhasz, Spartan Books, Washington, pp. 809–39.

Morse, P.M. and Ingard, K.U. (1968) *Theoretical Acoustics*. McGraw-Hill, New York.

2

Acoustic impedance at normal incidence of fluids. Substitution of a fluid layer for a porous layer

2.1 Introduction

The concept of acoustic impedance is very useful in the field of sound absorption. In this chapter, the impedance at normal incidence of one or several layers of fluid is calculated. The laws of Delany and Bazley (1970) are presented and used to replace a layer of porous material by a layer of equivalent fluid. The surface impedance at normal incidence for a layer of porous material backed by a rigid wall with and without an air gap is calculated. The main properties of both the reflection coefficient and the absorption coefficient are also discussed in this chapter.

2.2 Plane waves in unbounded fluids

2.2.1 Travelling waves

As indicated in the previous chapter, a simple displacement potential solution of the linear wave equation (1.56) in a compressible lossless fluid is

$$\varphi(x,t) = \frac{A}{\rho\omega^2}\exp[j(\omega t - kx)] \tag{2.1}$$

In this equation, ω is the angular frequency and k the wave number, given by

$$k = \omega(\rho/K)^{1/2} \tag{2.2}$$

Propagation of Sound in Porous Media: Modelling Sound Absorbing Materials, Second Edition J. F. Allard and N. Atalla

K and ρ are the bulk modulus and the density of the fluid, respectively. The quantity A is the amplitude of the acoustic pressure. From Equations (1.63) and (1.64), it follows that the acoustic pressure p and the components of the displacement vector $\boldsymbol{u}$ are respectively

$$p(x,t) = A\exp[j\omega(t - kx)] \tag{2.3}$$

and

$$u_y = u_z = 0, \qquad u_x(x,t) = \frac{-jAk}{\rho\omega^2}\exp[j\omega(t - kx)] \tag{2.4}$$

Only the x component υ_x of the velocity vector does not vanish:

$$\upsilon_x(x,t) = \frac{kA}{\rho\omega}\exp[j\omega(t - x/c)] \tag{2.5}$$

Equations (2.3) and (2.5) describe a travelling harmonic plane wave propagating along the x direction. Pressure and velocity are related by

$$\upsilon_x(x,t) = \frac{1}{Z_c}p(x,t) \tag{2.6}$$

with

$$Z_c = (\rho K)^{1/2} \tag{2.7}$$

The quantity Z_c is the characteristic impedance of the fluid.

2.2.2 Example

As an example, for air at the normal conditions of temperature T and pressure p (18 °C and 1 033 × 10^5 Pa), the density ρ_0, the adiabatic bulk modulus K_0, the characteristic impedance Z_0, and the speed of sound c_0 are as follows (Gray 1957):

$$\rho_0 = 1\cdot 213\ \mathrm{kg\,m^{-3}}$$

$$K_0 = 1\cdot 42\times 10^5\ \mathrm{Pa}$$

$$Z_0 = 415\cdot 1\ \mathrm{Pa\,m^{-1}\,s}$$

$$c_0 = 342\ \mathrm{m\,s^{-1}}$$

2.2.3 Attenuation

In a free field in air at acoustical frequencies, the damping can be neglected to a first approximation when the order of magnitude of the propagation length is 10 m or less. In the previous example, the effects of viscosity, heat conduction, and other dissipative processes have been neglected. The phenomena of viscosity and thermal conduction in fluids are a consequence of their molecular constitution. The description of sound propagation in viscothermal fluids can be found in the literature (Pierce 1981, Morse and Ingard 1986). The effects of viscosity and heat conduction on sound propagation in tubes are described in Chapter 4. Viscosity and heat conduction in tubes lead to dissipative processes, and in a macroscopic description of sound propagation, the density ρ and the bulk modulus K must be replaced by complex quantities. The wave number k and the characteristic impedance Z_c given by Equations (2.2) and (2.7) respectively, become complex:

$$\begin{aligned} k &= \mathrm{Re}(k) + j\ \mathrm{Im}(k) \\ Z_c &= \mathrm{Re}(Z_c) + j\ \mathrm{Im}(Z_c) \end{aligned} \tag{2.8}$$

If the amplitude of the waves decreases in the direction of propagation, the quantity $\mathrm{Im}(k)/\mathrm{Re}(k)$ must be negative if the time dependence is chosen as $\exp(j\omega t)$. In the alternative convention, $\exp(-j\omega t)$, $\mathrm{Im}(k)/\mathrm{Re}(k)$ must be positive (see Section 2.7).

2.2.4 Superposition of two waves propagating in opposite directions

The subscript x is removed for clarity. The pressure and the velocity, for a wave propagating toward the negative abcissa are, respectively,

$$p'(x, t) = A' \exp[j(kx + \omega t)] \tag{2.9}$$

$$v'(x, t) = -\frac{A'}{Z_c} \exp[j(kx + \omega t)] \tag{2.10}$$

If the acoustic field is a superposition of the two waves described by Equations (2.3) and (2.5) and by Equations (2.9), (2.10), the total pressure p_T and the total velocity v_T are

$$p_T(x, t) = A \exp[j(-kx + \omega t)] + A' \exp[j(kx + \omega t)] \tag{2.11}$$

$$v_T(x, t) = \frac{A}{Z_c} \exp[j(-kx + \omega t)] - \frac{A'}{Z_c} \exp[j(kx + \omega t)] \tag{2.12}$$

A superposition of several waves of the same ω and k propagating in a given direction is equivalent to one resulting wave propagating in the same direction. The acoustic field described by Equations (2.11) and (2.12) is the most general unidimensional monochromatic field. The ratio $p_T(x, t)/v_T(x, t)$ is called the impedance at x. The main properties of the impedance are studied in the following sections.

2.3 Main properties of impedance at normal incidence

2.3.1 Impedance variation along a direction of propagation

In Figure 2.1, two waves propagate in opposite directions parallel to the x axis. The impedance $Z(M_1)$ at M_1 is known. By employing Equations (2.11) and (2.12) for the pressure and the velocity, the impedance $Z(M_1)$ can be written

$$Z(M_1) = \frac{p_T(M_1)}{v_T(M_1)} = Z_c \frac{A \exp[-jkx(M_1)] + A' \exp[jkx(M_1)]}{A \exp[-jkx(M_1)] - A' \exp[jkx(M_1)]} \tag{2.13}$$

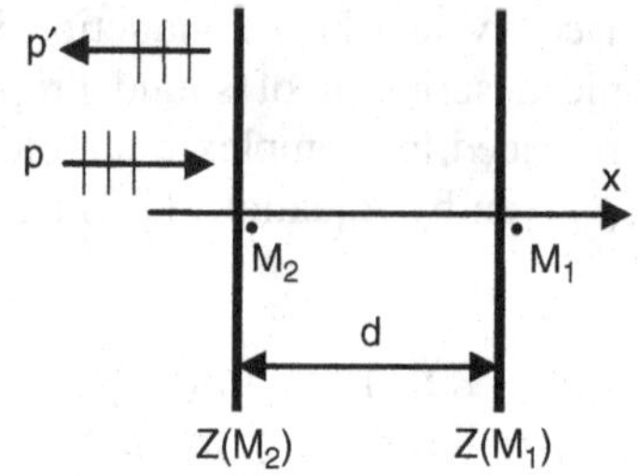

Figure 2.1 Plane waves propagate both in the x direction and in the opposite direction. The impedance at M_1 is $Z(M_1)$.

At M_2, the impedance $Z(M_2)$ is given by

$$Z(M_2) = Z_c \frac{A\exp[-jkx(M_2)] + A'\exp[jkx(M_2)]}{A\exp[-jkx(M_2)] - A'\exp[jkx(M_2)]} \tag{2.14}$$

From Equation (2.13) it follows that

$$\frac{A'}{A} = \frac{Z(M_1) - Z_c}{Z(M_1) + Z_c}\exp[-2jkx(M_1)] \tag{2.15}$$

By the use of Equations (2.14) and (2.15) we finally obtain

$$Z(M_2) = Z_c \frac{-jZ(M_1)\operatorname{cotg} kd + Z_c}{Z(M_1) - jZ_c \operatorname{cotg} kd} \tag{2.16}$$

where d is equal to $x(M_1) - x(M_2)$. Equation (2.16) is known as the impedance translation theorem.

2.3.2 Impedance at normal incidence of a layer of fluid backed by an impervious rigid wall

A layer of fluid 1 backed by a rigid impervious plane of infinite impedance at $x = 0$ is represented in Figure 2.2. Two points M_2 and M_3 are shown at the boundary of fluids 1 and 2, M_3 being in fluid 2 and M_2 in fluid 1. The impedance at M_2 at the surface of the

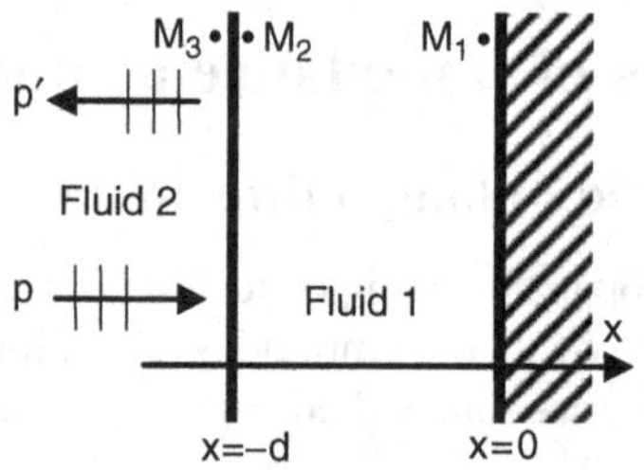

Figure 2.2 A layer of fluid of finite thickness in contact with another fluid on its front face and backed by a rigid impervious wall on its rear face.

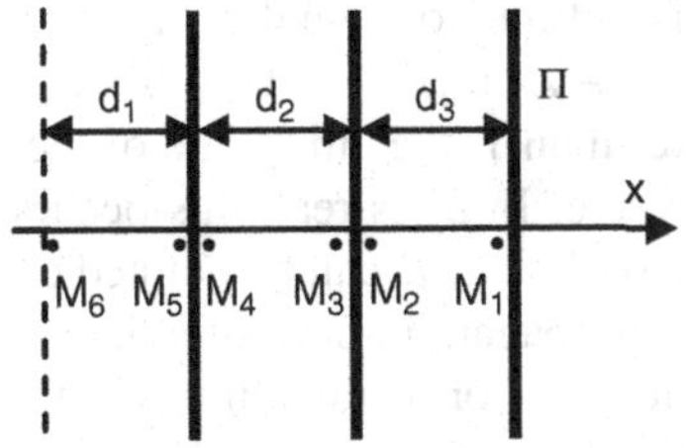

Figure 2.3 Three layers of fluid backed by an impedance plane II.

layer of fluid 1 is obtained from Equation (2.16) with $Z(M_1)$ infinite:

$$Z(M_2) = -jZ_c \text{ cotg } kd \tag{2.17}$$

where Z_c is the characteristic impedance and k the wave number in fluid 1.

The pressure and the velocity are continuous at the boundary. The impedance at M_3 is equal to the impedance at M_2, the velocities and pressures being the same on either side of the boundary:

$$Z(M_3) = Z(M_2) \tag{2.18}$$

2.3.3 Impedance at normal incidence of a multilayered fluid

A multilayered fluid is represented in Figure 2.3. If the impedance $Z(M_1)$ is known, the impedance $Z(M_2)$ inside fluid 1 can be obtained from Equation (2.16). The impedance $Z(M_3)$ is equal to the impedance at M_2. The impedance at M_4, M_5 and M_6 can be obtained successively in the same way.

2.4 Reflection coefficient and absorption coefficient at normal incidence

2.4.1 Reflection coefficient

The reflection coefficient R at the surface of a layer is the ratio of the pressures p' and p created by the outgoing and the ingoing waves at the surface of the layer. For instance, at M_3, in Figure 2.2, the reflection coefficient $R(M_3)$ is equal to

$$R(M_3) = p'(M_3, t)/p(M_3, t) \tag{2.19}$$

This coefficient does not depend on t because the numerator and the denominator have the same dependence on t. Using Equation (2.15), the reflection coefficient at M_3 can be written as

$$R(M_3) = (Z(M_3)) - Z'_c)/(Z(M_3) + Z'_c) \tag{2.20}$$

where Z'_c is the characteristic impedance in fluid 2. The ingoing and outgoing waves at M_3 have the same amplitude if

$$|R(M_3)| = 1 \tag{2.21}$$

This occurs if $|Z(M_3)|$ is infinite or equal to zero. If $|Z(M_3)|$ is finite, a more general condition is $Z^*(M_3)Z'_c + Z(M_3)Z'^*_c = 0$. If Z'_c is real, this occurs if $Z(M_3)$ is imaginary. If $|Z(M_3)|$ is greater than 1, the amplitude of the outgoing wave is larger than the amplitude of the ingoing wave. If Z'_c is real, this occurs if the real part of $Z(M_3)$ is negative. More generally, the coefficient R can be defined everywhere in a fluid where an ingoing and an outgoing wave propagate in opposite directions. For instance, it has been shown previously that the ratio p/v for a travelling wave propagating in the positive x direction is Z'_c. As indicated by Equation (2.20) there exists only an ingoing wave at M inside a fluid if the impedance at M is the characteristic impedance. The behaviour of the reflection coefficient as a function of x is much simpler than the behaviour of the impedance. Returning to Figure 2.1, Equations (2.3) and (2.9) provide the following relation between $R(M_2)$ and $R(M_1)$:

$$R(M_2) = R(M_1)\exp(-2jkd) \tag{2.22}$$

where $d = x(M_1) - x(M_2)$. Hence, the reflection coefficient describes a circle in the complex plane if k is real. If k is complex, the reflection coefficient describes a spiral.

It should be noted that the propagation of electromagnetic plane waves in a waveguide can be described with impedances and reflection coefficients in a similar way to the use of those concepts in describing sound propagation. The Smith chart, which provides a graphical representation for the propagation of electromagnetic waves, can also be used to describe the acoustic plane-wave propagation.

2.4.2 Absorption coefficient

The absorption coefficient $\alpha(M)$ is related to the reflection coefficient $R(M)$ as follows

$$\alpha(M) = 1 - |R(M)|^2 \tag{2.23}$$

The phase of $R(M)$ is removed, and the absorption coefficient does not carry as much information as the impedance or the reflection coefficient. The absorption coefficient is often used in architectural acoustics, where this simplification can be advantageous. It can be rewritten as

$$\alpha(M) = 1 - \frac{E'(M)}{E(M)} \tag{2.24}$$

where $E(M)$ and $E'(M)$ are the average energy flux through the plane $x = x(M)$ of the incident and the reflected waves, respectively.

2.5 Fluids equivalent to porous materials: the laws of Delany and Bazley

2.5.1 Porosity and flow resistivity in porous materials

Porosity

Materials such as fibreglass and plastic foam with open bubbles consist of an elastic frame which is surrounded by air. The porosity ϕ is the ratio of the air volume V_a to the

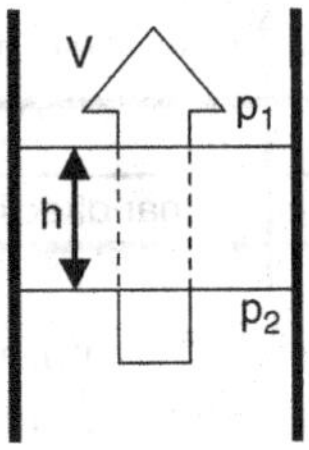

Figure 2.4 A slice of porous material is placed in a pipe. A differential pressure $p_2 - p_1$ induces a steady flow V of air per unit area of material.

total volume of porous material V_T. Thus,

$$\phi = V_a/V_T \tag{2.25}$$

Let V_b be the volume occupied by the frame in V_T. The quantities V_a, V_b and V_T are then related by

$$V_a + V_b = V_T \tag{2.26}$$

Only the volume of air which is not locked within the frame must be considered in V_a and thus in the calculation of the porosity. The latter is also known as the open porosity or the connected porosity. For instance, a closed bubble in a plastic foam is considered locked within the frame, and its volume therefore belongs to V_b. For most of the fibrous materials and plastic foams, the porosity lies very close to 1. Methods for measuring porosity are given in Zwikker and Kosten (1949) and Champoux *et al.* (1991).

Flow resistivity

One of the important parameters governing the absorption of a porous material is its flow resistance. It is defined by the ratio of the pressure differential across a sample of the material to the normal flow velocity through the material. The flow resistivity σ is the specific (unit area) flow resistance per unit thickness. A sketch of the set-up for the measurement of the flow resistivity σ is shown in Figure 2.4.

The material is placed in a pipe, and a differential pressure induces a steady flow of air. The flow resistivity σ is given by

$$\sigma = (p_2 - p_1)/Vh \tag{2.27}$$

In this equation, the quantities V and h are the mean flow of air per unit area of material and the thickness of the material, respectively. In MKSA units, σ is expressed in $\mathrm{N\,m^{-4}\,s}$. More information about the measurement of flow resistivity can be found in standards ASTM C-522, ISO 9053 (1991), Bies and Hansen (1980), and Stinson and Daigle (1988).

It should be pointed out that fibrous materials are generally anisotropic (Attenborough 1971, Burke 1983, Nicolas and Berry 1984, Allard *et al.* 1987, Tarnow 2005). A panel of fibreglass is represented in Figure 2.5. Fibres in the material generally lie in planes parallel to the surface of the material. The flow resistivity in the normal direction is different from that in the planar directions. In the former case, air flows perpendicularly

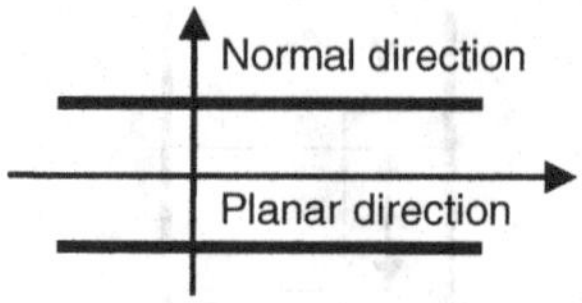

Figure 2.5 A panel of fibrous material. The normal direction is perpendicular to the surface of the panel, and the planar directions lie in planes parallel to the surface.

to the surface of the panel while in the latter case it flows parallel to the surface of the layer. The normal flow resistivity σ_N is larger than the planar flow resistivity σ_P. The flow resistivity of fibreglass and open-bubble foam generally lies between 1000 and 100 000 $\mathrm{N\,m^{-4}\,s}$.

2.5.2 Microscopic and macroscopic description of sound propagation in porous media

The quantities that are involved in sound propagation can be defined locally, on a microscopic scale, for instance in a porous material with cylindrical pores having a circular cross-section, as functions of the distance to the axis of the pores. On a microscopic scale, sound propagation in porous materials is generally difficult to study because of the complicated geometries of the frames. Only the mean values of the quantities involved are of practical interest. The averaging must be performed on a macroscopic scale, on a homogenization volume with dimensions sufficiently large for the average to be significant. At the same time, these dimensions must be much smaller than the acoustic wavelength. The description of sound propagation in porous material can be complicated by the fact that sound also excites and moves the frame of the material. If the frame is motionless, in a first step, the air inside the porous medium can be replaced on the macroscopic scale by an equivalent free fluid. This equivalent fluid has a complex effective density ρ and a complex bulk modulus K. The wave number k and the characteristic impedance Z_c of the equivalent fluid are also complex. In a second step, as shown in Chapter 5, Section 5.7, the porous layer can be replaced by a fluid layer of density ρ/ϕ and of bulk modulus K/ϕ.

2.5.3 The Laws of Delany and Bazley and flow resistivity

The complex wave number k and the characteristic impedance Z_c have been measured by Delany and Bazley (1970) for a large range of frequencies in many fibrous materials with porosity close to 1. According to these measurements, the quantities k and Z_c depend mainly on the angular frequency ω and on the flow resistivity σ of the material. A good fit of the measured values of k and Z_c has been obtained with the following expressions:

$$Z_c = \rho_o c_o[1 + 0.057X^{-0.754} - j0.087X^{-0.732}] \tag{2.28}$$

$$k = \frac{\omega}{c_o}[1 + 0.0978X^{-0.700} - j0.189X^{-0.595}] \tag{2.29}$$

where ρ_o and c_o are the density of air and the speed of sound in air (see Section 2.2.2), and X is a dimensionless parameter equal to

$$X = \rho_o f / \sigma \tag{2.30}$$

f being the frequency related to ω by $\omega = 2\pi f$.

Delany and Bazley suggest the following boundary for the validity of their laws in terms of boundaries of X, as follows:

$$0.01 < X < 1.0 \tag{2.31}$$

It may not be expected that single relations provide a perfect prediction of acoustic behaviour of all the porous materials in the frequency range defined by Equation (2.31). More elaborate models will be studied in Chapter 5. Nevertheless, the laws of Delany and Bazley are widely used and can provide reasonable orders of magnitude for Z_c and k. With fibrous materials which are anisotropic, as indicated previously, the flow resistivity must be measured in the direction of propagation for waves travelling in either the normal or the planar direction. The case of oblique incidence is more complicated and is considered in Chapter 3. It should be pointed out that after the work by Delany and Bazley, several authors suggested slightly different empirical expressions of k and Z_c for specific frequency ranges and for different materials (Mechel 1976, Dunn and Davern 1986, Miki 1990).

2.6 Examples

As a first example, the impedance at the surface of a layer of fibrous material of thickness d equal to 10 cm, and of normal flow resistivity equal to 10 000 Nm^{-4} s, fixed on a rigid impervious wall (Figure 2.6), has been calculated with the use of Equations (2.17), (2.28) and (2.29).

The real and the imaginary parts of the impedance Z are shown in Figure 2.7.

As a second example, the impedance of the same layer of fibrous material with an air gap of thickness d' equal to 10 cm (Figure 2.8) has been calculated.

The general method of calculating $Z(M)$ is given in Section 2.3.3. In the example considered, Equation (2.16) can be used, $Z(M_1)$ being the impedance of the air gap. The values of Z_o and c_o are used to calculate $Z(M_1)$ with Equation (2.17). Expressions (2.28) and (2.29) for Z_c and k for the fibrous material have been used. The impedance is

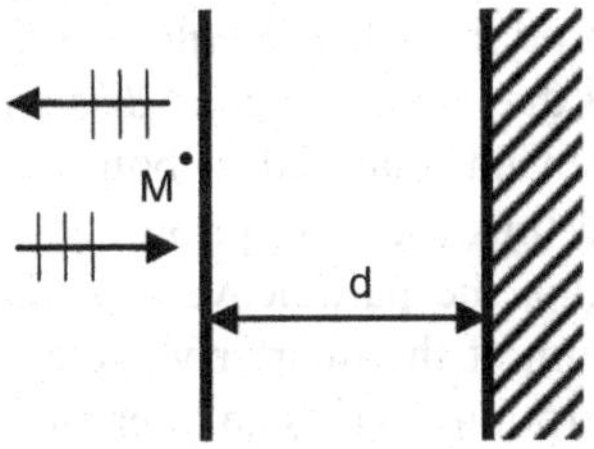

Figure 2.6 A layer of porous material fixed on a rigid impervious wall.

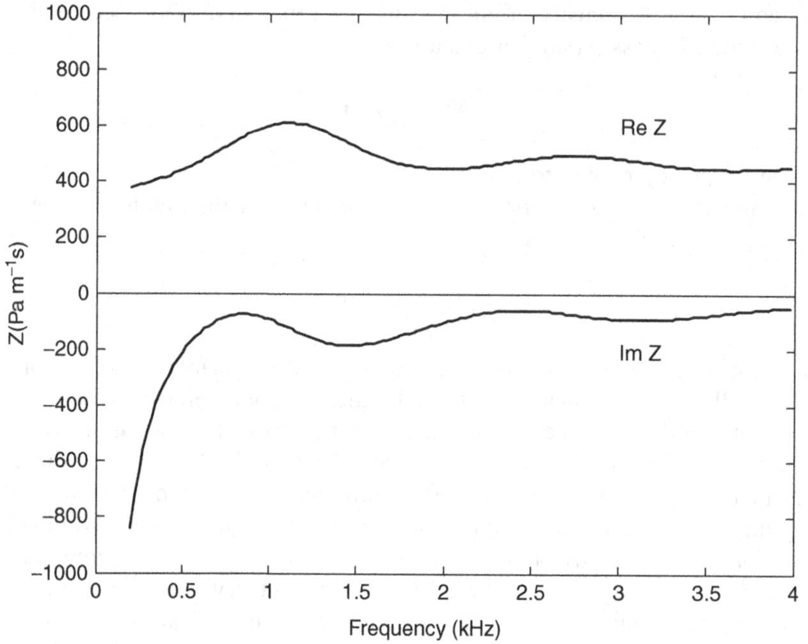

Figure 2.7 The impedance Z at normal incidence of a layer of fibrous material of thickness $d = 10$ cm, of normal flow resistivity $\sigma = 10\,000$ N m^{-4} s, calculated according to the laws of Delany and Bazley.

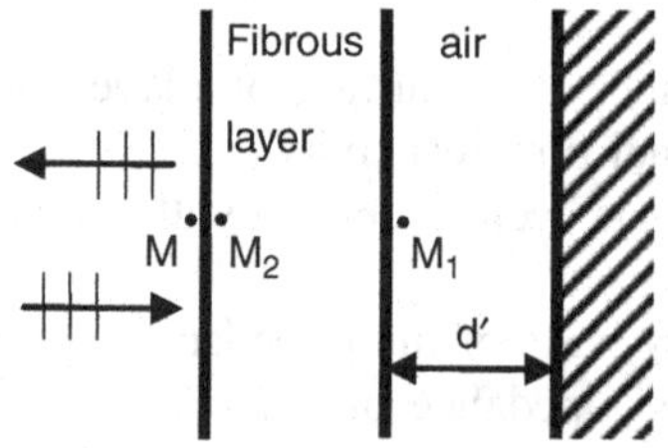

Figure 2.8 A layer of fibrous material with an air gap between the material and the rigid wall.

represented in Figure 2.9, and the absorption coefficient for the material with and without an air gap is shown in Figure 2.10.

The interesting effect of the air gap appears clearly in Figure 2.10. The air gap increases significantly the absorption at low frequencies. This is explained by the fact that sound absorption is mainly due to the viscous dissipation, related to the velocity of air in the porous medium. When the material is bonded onto a hard wall, the particle velocity at the wall is zero, and thus the absorption deteriorates rapidly at low frequencies. When backed by an air gap, the particle velocity at the rear face of the material oscillates and reaches a maximum at the quarter-wavelength of the lowest frequency of interest, thus increasing the absorption. This is an alternative to an increase of the material thickness.

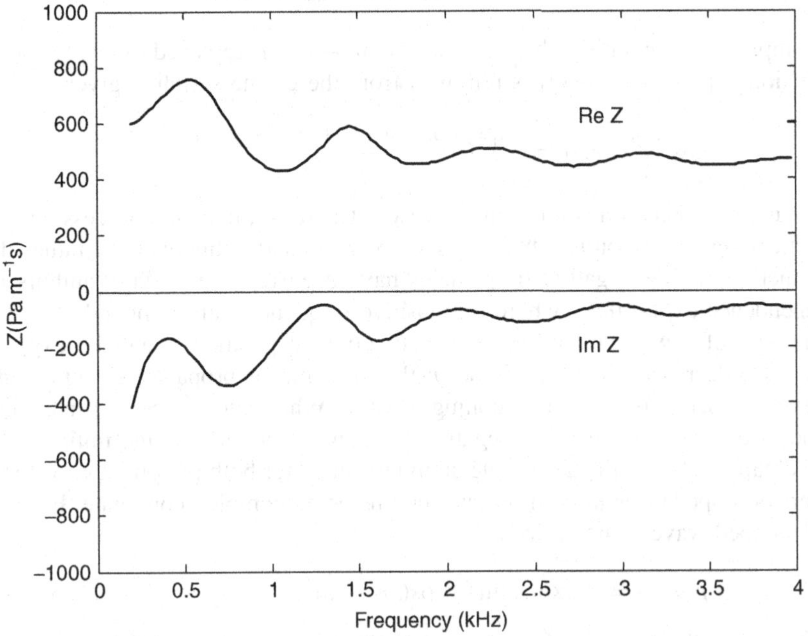

Figure 2.9 The impedance Z at normal incidence of a layer of fibrous material of thickness $d = 10$ cm, and flow resistivity $\sigma = 10\,000\ \text{N}\,\text{m}^{-4}\,\text{s}$, with an air gap of thickness $d' = 10$ cm.

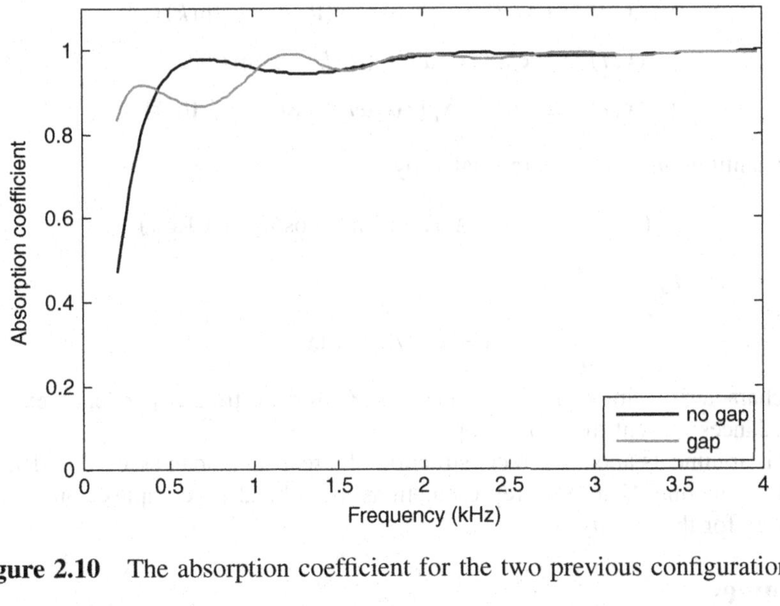

Figure 2.10 The absorption coefficient for the two previous configurations.

2.7 The complex exponential representation

In the complex representation, the function $\cos(\omega t - kx)$ is replaced by $\exp[j(\omega t - kx)]$. The function $\exp[-j(\omega t - kx)]$ is removed from the cosine which is given by

$$\cos(\omega t - kx) = \frac{\exp[j(\omega t - kx)] + \exp[-j(\omega t - kx)]}{2} \tag{2.32}$$

and the remaining function is multiplied by two. The result, in signal processing, is called an analytical signal (Papoulis 1984). It is simpler to handle this quantity rather than the initial function, as the negative frequencies have been removed. Many authors use the time dependence $\exp(-j\omega t)$, where the positive frequencies are removed. The complex representation of a wave travelling in the direction of positive x will be $\exp[j(k^*x - \omega t)]$, because decreasing the amplitude in the direction of propagation implies that the new wave number is the complex conjugate of k. When adding the positive frequency components of a real signal to its negative frequency components, the initial real signal must be obtained. This will be possible simultaneously for both pressure and velocity, the characteristic impedances in both representations being complex conjugate. For instance, the real damped wave characterized by

$$p(x, t) = A \exp(x \operatorname{Im} k) \cos(\omega t - x \operatorname{Re} k) \tag{2.33}$$

$$v(x, t) = A/|Z_c| \exp(x \operatorname{Im} k) \cos(\omega t - x \operatorname{Re} k - \operatorname{Arg} Z_c) \tag{2.34}$$

has the following two representations:

$$p_+(x, t) = A \exp[j(\omega t - (\operatorname{Re} k + j \operatorname{Im} k)x)] \tag{2.35}$$

$$v_+(x, t) = (A/Z_c) \exp[j(\omega t - (\operatorname{Re} k + j \operatorname{Im} k)x)] \tag{2.36}$$

$$p_-(x, t) = A \exp[j(-\omega t + (\operatorname{Re} k - j \operatorname{Im} k)x)] \tag{2.37}$$

$$v_-(x, t) = (A/Z_c^*) \exp[j(-\omega t + (\operatorname{Re} k - j \operatorname{Im} k)x)] \tag{2.38}$$

The quantities p_- and p_+ are related by

$$(p_+ + p_-)/2 = A \exp(x \operatorname{Im} k) \cos(\omega t - x \operatorname{Re} k) \tag{2.39}$$

In the same way

$$(v_+ + v_-)/2 = v(x, t) \tag{2.40}$$

The characteristic impedance Z_c becomes Z_c^* for the time dependence $\exp(-j\omega t)$. The impedances present the same property.

Similar arguments about the reconstruction of a real signal can be used to demonstrate that the bulk moduli K in both representations are related by complex conjugation; the same is true for the density ρ.

References

Allard, J. F., Bourdier, R. and L'Espérance, A. (1987) Anisotropy effect in glass wool on normal impedance at oblique incidence. *J. Sound Vib.*, **114**, 233–8.

Attenborough, K. (1971) The prediction of oblique-incidence behaviour of fibrous absorbents. *J. Sound Vib.*, **14**, 183–91.

Bies, D. A. and Hansen, C.H. (1980) Flow resistance information for acoustical design. *Applied Acoustics*, **13**, 357–91.

Burke, S. (1983) The absorption of sound by anisotropic porous layers. *Paper presented at 106th Meeting of the ASA*, San Diego, CA.

Champoux, Y, Stinson, M. R. and Daigle, G. A. (1991) Air-based system for the measurement of porosity, *J. Acoust. Soc. Amer.*, **89**, 910–6.

Delany, M. E. and Bazley, E. N. (1970) Acoustical properties of fibrous materials. *Applied Acoustics*, **3**, 105–16.

Dunn, I. P. and Davern, W. A. (1986) Calculation of acoustic impedance of multilayer absorbers. *Applied Acoustics*, **19**, 321–34.

Gray, D. E., ed., (1957) *American Institute of Physics Handbook*. McGraw-Hill, New York.

ISO 9053: (1991) Acoustics-Materials for acoustical applications-Determination of airflow resistance.

Mechel, F. P. (1976) Ausweitung der Absorberformel von Delany and Bazley zu tiefen Frequenzen. *Acustica*, **35**, 210–13.

Miki, Y. (1990) Acoustical properties of porous materials – Modifications of Delany–Bazley models. *J. Acoust. Soc. Japan*, **11**, 19–24.

Morse, M. K., and Ingard K. U. (1986) *Theoretical Acoustics*. Princeton University Press, Princeton.

Nicolas, J. and Berry, J. L. (1984) Propagation du son et effet de sol. *Revue d'Acoustique*, **71**, 191–200.

Papoulis, A. (1984) *Signal Analysis*. McGraw-Hill, Singapore.

Pierce, A. D. (1981) *Acoustics: An Introduction to its Physical Principles and Applications*. McGraw-Hill, New York.

Stinson, M. R. and Daigle, G. A. (1988) Electronic system for the measurement of flow resistance. *J. Acoust. Soc. Amer.* **83**, 2422–2428

Tarnow, V. (2005) Dynamic measurement of the elastic constants of glass wool. *J. Acoust. Soc. Amer.* **118**, 3672–3678.

Zwikker, C. and Kosten, C. W. (1949) *Sound Absorbing Materials*. Elsevier, New York.

3

Acoustic impedance at oblique incidence in fluids. Substitution of a fluid layer for a porous layer

3.1 Introduction

The dilatational plane waves studied previously have equiphase planes and equiamplitude planes which are both perpendicular to the direction of propagation. These waves are called homogeneous plane waves. They have been used to represent the acoustic field in layered fluids having plane boundaries perpendicular to the direction of propagation. Transmission and reflection at oblique incidence may be described by inhomogeneous waves, with nonparallel equiphase and equiamplitude planes. A short description of these waves is presented in this chapter, both in isotropic and in anisotropic fluids. Next, these waves are used to calculate the surface impedance and the absorption coefficient of both isotropic and anisotropic highly porous materials according to the laws of Delany and Bazley (1970).

3.2 Inhomogeneous plane waves in isotropic fluids

In Chapter 2, waves propagating parallel to a coordinate axis have been considered. The same plane waves can propagate in any direction defined by a unit vector **n** if the fluid is isotropic. Let n_1, n_2 and n_3 be the three components of **n**. The wave propagating in the direction **n** is then described by the following two equations:

$$p(x_1, x_2, x_3, t) = A\exp[j(-k_1x_1 - k_2x_2 - k_3x_3 + \omega t)] \tag{3.1}$$

$$v(x_1, x_2, x_3, t) = \frac{A\mathbf{n}}{Z_c}\exp[j(-k_1x_1 - k_2x_2 - k_3x_3 + \omega t)] \tag{3.2}$$

Propagation of Sound in Porous Media: Modelling Sound Absorbing Materials, Second Edition J. F. Allard and N. Atalla

In these equations k_1, k_2 and k_3 are the components of the wave number vector $\boldsymbol{k}$

$$k_1 = n_1 k_1 \qquad k_2 = n_2 k_2 \qquad k_3 = n_3 k_3 \tag{3.3}$$

Using the wave equation (1.56), the components of the wave vector satisfy the equation

$$k_1^2 + k_2^2 + k_2^3 = k^2 = \frac{\rho}{K}\omega^2 \tag{3.4}$$

Equation 3.2 can be rewritten as

$$\begin{aligned}
v_1(x_1, x_2, x_3, t) &= \frac{A}{Z_c}\frac{k_1}{k}\exp[j(-k_1x_1 - k_2x_2 - k_3x_3 + \omega t)] \\
v_2(x_1, x_2, x_3, t) &= \frac{A}{Z_c}\frac{k_2}{k}\exp[j(-k_1x_1 - k_2x_2 - k_3x_3 + \omega t)] \\
v_3(x_1, x_2, x_3, t) &= \frac{a}{Z_c}\frac{k_3}{k}\exp[j(-k_1x_1 - k_2x_2 - k_3x_3 + \omega t)]
\end{aligned} \tag{3.5}$$

For the plane wave described by Equations (3.1)–(3.3), $\boldsymbol{k}$ is perpendicular to the equiphase and equiamplitude planes. A more significant generalization can be achieved by discarding Equation (3.3) and using only Equation (3.4) to define k_1, k_2 and k_3. A first simple example is considered where k^2 is real in Equation (3.4).

The components of the wave number vector $\boldsymbol{k}$ are

$$k_1 = l, \qquad k_2 = -jm, \qquad k_3 = 0 \tag{3.6}$$

where the quantities m and l are real, positive and are linked by Equation (3.4), which can be rewritten as

$$l^2 - m^2 = \frac{\rho}{K}\omega^2 \tag{3.7}$$

From Equations (3.1) and (3.5), it follows that

$$p(x_1, x_2, x_3, t) = A\exp[j(-lx_1 + \omega t) - mx_2] \tag{3.8}$$

$$v_1 = \frac{Al}{Z_c k}\exp[j(-lx_1 + \omega t) - mx_2] \tag{3.9}$$

$$v_2 = -j\frac{Am}{Z_c k}\exp[j(-lx_1 + \omega t) - mx_2] \tag{3.10}$$

$$v_3 = 0 \tag{3.11}$$

From the above equations it is seen that the equiamplitude planes are parallel to the $x_2 = 0$ plane, and the equiphase planes are parallel to the $x_1 = 0$ plane. The factor $-j$ in v_2 indicates a phase difference of $\pi/2$ between v_1 and v_2. These two quantities cannot be the geometrical projections of a vector on the x_1 and x_2 axes unless complex angles are used to take into account the phase difference between the components.

Plane waves with distinct equiphase and equiamplitude planes are called inhomo geneous plane waves. Equations (3.3) can always be used to define the three complex

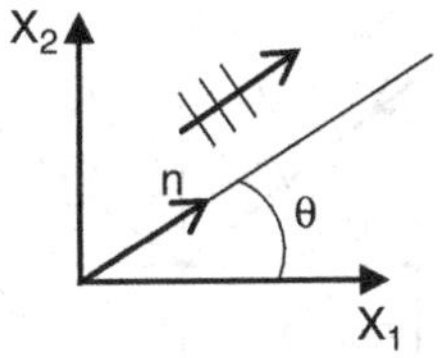

Figure 3.1 Symbolic representation of an inhomogeneous plane wave with $\boldsymbol{k}$ parallel to the plane $x_3 = 0$.

components n_1, n_2 and n_3 of a unit vector **n** such that $n_1^2 + n_2^2 + n_2^3 = 1$. In the following, we will be concerned only with waves with a vector velocity parallel to a coordinate plane, for instance the plane $x_3 = 0$ if $k_3 = 0$. A complex angle θ can be used to indicate the direction of propagation. Equation (3.4) then reduces to

$$k_1 = k(1 - \sin^2\theta)^{1/2}, \qquad k_2 = k\sin\theta, \qquad k_3 = 0 \tag{3.12}$$

The quantity $\sin\theta$ can be any complex number, associated with a complex angle such that

$$\sin\theta = (e^{j\theta} - e^{-j\theta})/(2\mathrm{j}) \tag{3.13}$$

A symbolic representation is given in Figure 3.1.

The components of the unit vector **n** in Equations (3.3) are

$$n_1 = k_1/k = (1 - \sin^2\theta)^{1/2} = \cos\theta, \quad n_2 = k_2/k = \sin\theta, \quad n_3 = k_3/k = 0 \tag{3.14}$$

In the previous example

$$n_1 = l/k, \qquad n_2 = -jm/k, \qquad n_3 = 0$$

The attenuation in the direction x_2 is not due to the dissipation in the fluid, but to the fact that l is larger than k. The equiphase and equiamplitude planes are perpendicular. It is easy to show that this property is always valid for media with k^2 real. The vector $\mathbf{k} = \mathbf{n}k$ with components k_1, k_2 and k_3 will be used in the following, even if its components are complex.

3.3 Reflection and refraction at oblique incidence

Two fluids with a plane boundary are represented in Figure 3.2.

Let k and k' be the complex wave numbers in fluid 1 and fluid 2. Let $\boldsymbol{k}_\mathrm{i}(k_{i1}, k_{i2}, k_{i3})$, $\boldsymbol{k}_\mathrm{r}(k_{r1}, k_{r2}, k_{r3})$ and $\boldsymbol{k}_\mathrm{t}(k_{t1}, k_{t2}, k_{t3})$ be the wave number vectors of the incident, reflected and refracted waves, respectively. The three components of $\boldsymbol{k}_\mathrm{i}$ are

$$k_{i1} = k\sin\theta_i, \qquad k_{i2} = 0, \qquad k_{i3} = k(1 - \sin^2\theta_i)^{1/2} = k\cos\theta_i \tag{3.15}$$

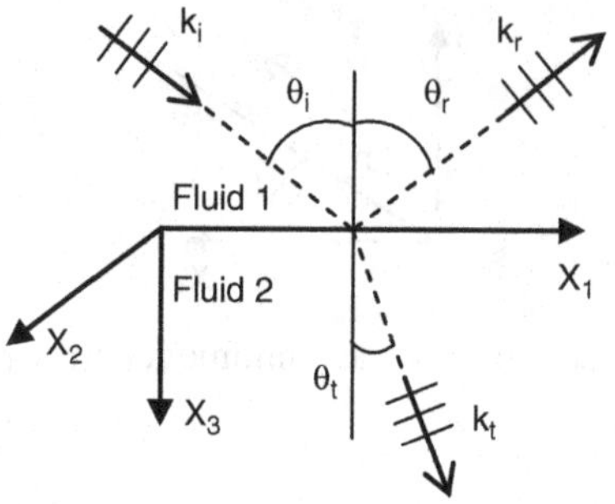

Figure 3.2 Plane wave reflection and refraction at a plane interface between two fluids.

The angle of incidence θ_i can be either real or complex. The incident wave is represented by the following equations which are similar to Equations (3.1) and (3.5):

$$p_i = A_i \exp[j(-k(\sin\theta_i x_1 + \cos\theta_i x_3) + \omega t)] \tag{3.16}$$

$$v_{i1} = \frac{A_i}{Z_c} \sin\theta_i \exp[j(-k(\sin\theta_i x_1 + \cos\theta_i x_3) + \omega t)] \tag{3.17}$$

$$v_{i3} = \frac{A_i}{Z_c} \sin\theta_i \exp[j(-k(\sin\theta_i x_1 + \cos\theta_i x_3) + \omega t)] \tag{3.18}$$

$$v_{i2} = 0 \tag{3.19}$$

Two similar sets of equations describe the reflected and the transmitted waves. At the boundary, velocities and pressures are the same in both media. From these conditions, it can be shown that reflection and refraction are described by the laws of Snell–Descartes.

The vectors $\mathbf{k_i}$, $\mathbf{k_r}$ and $\mathbf{k_t}$ are in the same plane. In the given example, the x_2 component of the three wave number vectors vanishes:

$$k_{r2} = k_{t2} = k_{i2} = 0 \tag{3.20}$$

Continuity of the velocity at the interface shows that the angles of incidence and reflection, θ_i and θ_r, are equal:

$$\theta_i = \theta_r \tag{3.21}$$

and that the angles of incidence θ_i and of transmission θ_t are related by the equation

$$k \sin\theta_i = k' \sin\theta_t \tag{3.22}$$

It follows that

$$k_{i1} = k \sin\theta_i = k \sin\theta_r = k_{r1} \tag{3.23}$$

$$k_{i3} = k \cos\theta_i = k \cos\theta_r = -k_{r3} \tag{3.24}$$

The component k_{t3} is given by

$$k_{t3} = k' \cos\theta_t = k'(1 - \sin\theta_t^2)^{1/2} \tag{3.25}$$

From Equation (3.22) it follows that

$$k_{i1} = k_{t1} \tag{3.26}$$

and

$$k_{t3} = (k'^2 - k^2 \sin^2 \theta_i)^{1/2} \tag{3.27}$$

The appropriate choice of the square root to avoid the amplitude of the refracted wave tending to infinity with x_3 is the one with $\mathrm{Im}(k_{t3}) < 0$.

In terms of the wave speeds c and c' in the two fluids, if the wave numbers are real, Equation (3.26) reads

$$\frac{\sin \theta_i}{c} = \frac{\sin \theta_t}{c'} \tag{3.28}$$

This is the classical form of Snell's law of refraction.

3.4 Impedance at oblique incidence in isotropic fluids

3.4.1 Impedance variation along a direction perpendicular to an impedance plane

An ingoing and an outgoing plane wave are represented in Figure 3.3. Let **n** and **n′** be the unit vectors associated with the incoming and outgoing wave, respectively. Both are parallel to the plane $x_2 = 0$. The $0x_2$ axis is perpendicular to the plane of the figure. Let θ be the angle between the x_3 axis and **n**, and between the x_3 axis and **n′**. This angle can be real or complex. First, it will be shown that with this geometry, the ratio of the pressure p to υ_3 (the component υ_3 of velocity on the x_3 axis) is constant on planes parallel to $x_3 = 0$. These planes are impedance planes on which the impedance Z in the normal direction

$$Z = p/\upsilon_3 \tag{3.29}$$

is constant, this impedance depending only on x_3.

By employing Equations (3.1) and (3.4), the pressure and the velocity component on the x_3 axis can be written for the two waves respectively

$$p(x_1, x_3) = A \exp(j(-k_1 x_1 - k_3 x_3 + \omega t)) \tag{3.30}$$

$$\upsilon_3(x_1, x_3) = \frac{A}{Z_c}\frac{k_3}{k} \exp(j(-k_1 x - k_3 x_3 + \omega t)) \tag{3.31}$$

$$p'(x_1, x_3) = A' \exp(j(-k_1 x_1 + k_3 x_3 + \omega t)) \tag{3.32}$$

$$\upsilon'_3(x_1, x_3) = -\frac{A'}{Z_c}\frac{k_3}{k} \exp(j(-k_1 x_1 + k_3 x_3 + \omega t)) \tag{3.33}$$

On the plane $x_3 = x_3(M_1)$, the impedance in the direction x_3 is equal to

$$Z(M_1) = \frac{p(x_1, x_3) + p'(x_1, x_3)}{\upsilon_3(x_1, x_3) + \upsilon'_3(x_1, x_3)} \tag{3.34}$$

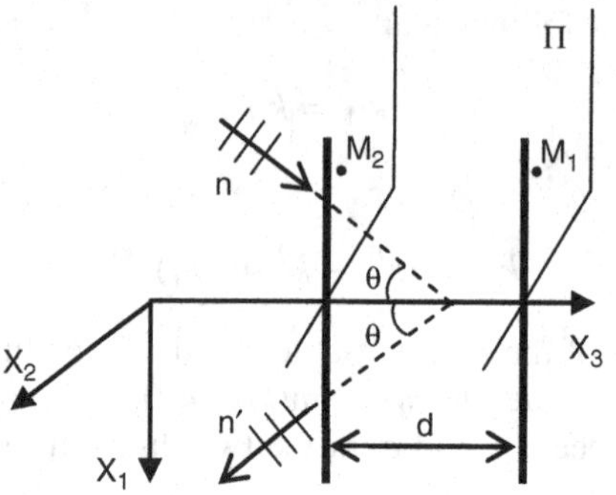

Figure 3.3 The ingoing and the outgoing plane waves associated with an impedance plane II.

Substitution of Equations (3.30)–(3.33) into Equation (3.34) yields

$$Z(M_1) = \frac{A\exp(-j\,k_3x_3) + A'\exp(j\,k_3x_3)}{\dfrac{A}{Z_c}\dfrac{k_3}{k}\exp(-j\,k_3x_3) - \dfrac{A'}{Z_c}\dfrac{k_3}{k}\exp(j\,k_3x_3)} \tag{3.35}$$

This expression depends only on x_3.

Equation (3.35) can be obtained by the substitution of Z_ck/k_3 and k_3, instead of Z_c and k, respectively, into Equation (2.14). With the same substitution, the impedance translation theorem, Equation (2.16), can be rewritten

$$Z(M_2) = \frac{Z_ck}{k_3}\left[\frac{-jZ(M_1)\operatorname{cotg} k_3d + Z_c\dfrac{k}{k_3}}{Z(M_1) - jZ_c\dfrac{k}{k_3}\operatorname{cotg} k_3d}\right] \tag{3.36}$$

where d is equal to $x(M_1) - x(M_2)$.

3.4.2 Impedance at oblique incidence for a layer of finite thickness backed by an impervious rigid wall

The layer is represented in Figure 3.4 in the incidence plane with the incident and the reflected waves. The angles are real or complex. For instance, fluid 2 can be a nondissipative medium with a real wave number k' and fluid 1 a dissipative medium with a complex wave number k. If θ' in fluid 2 is real, θ in fluid 1 is complex and defined by Equation (3.22).

$$k_1 = k'_1 = k'\sin\theta' = k\sin\theta \tag{3.37}$$

Equation (3.4) yields

$$k_3 = (k^2 - k_1'^2)^{1/2} \tag{3.38}$$

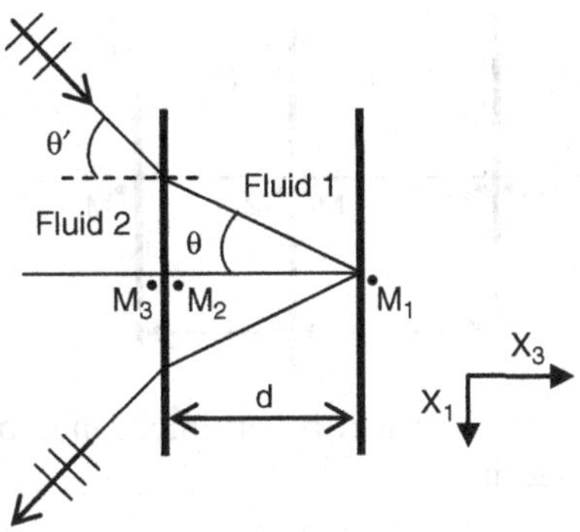

Figure 3.4 A layer of fluid of thickness d backed by an impervious rigid wall.

Both determinations of the square root can be used if d is finite. At the impervious rigid wall $Z(M_1)$ is infinite and Equation (3.36) becomes

$$Z(M_2) = -Z_c \frac{k}{k_3} j \text{ cotg } k_3 d \tag{3.39}$$

where Z_c is the characteristic impedance in fluid 1. The impedance $Z(M_2)$ at the boundary in fluid 1 is equal to the impedance $Z(M_3)$ at the boundary in fluid 2, as in the case for normal incidence. If the velocity in fluid 1 is much smaller than that in fluid 2, it follows from Equation (3.22) that $\sin\theta$ is close to zero and k_3 is close to k for large range of angles θ' around normal incidence. Therefore, $Z(M_2)$ as given by Equation (3.39) is close to its value at $\theta' = 0$, and weakly depends on the angle of incidence. Medium 1 is called a locally reacting medium (Cremer and Muller 1982). The locally reacting property means that the material response at a given point on the surface is independent on the behaviour at other points of the surface. It is a good approximation for several engineered materials such as honeycombs and acceptable for a number of thin porous materials. This approximation has been and is still widely used in finite element and boundary element vibro-acoustic simulations. In these applications, the sound package attenuation is accounted for in terms of a surface admittance, the latter is measured or simulated in a plane wave incidence context. Chapter 13 replaces this approximation with detailed modelling using poroelastic finite elements.

3.4.3 Impedance at oblique incidence in a multilayered fluid

A multilayered fluid is represented in Figure 3.5. Let k, k' and k'' be the wave numbers in fluids 1, 2 and 3, respectively. The angle of incidence is θ''. The $0x_1$ component of the wave number vector is the same in the different media and is equal to

$$k_1 = k_1' = k_1'' = k'' \sin\theta'' \tag{3.40}$$

In fluid 1, the component k_3 of $\boldsymbol{k}$ is

$$k_3 = (k^2 - k_1^2)^{1/2} \tag{3.41}$$

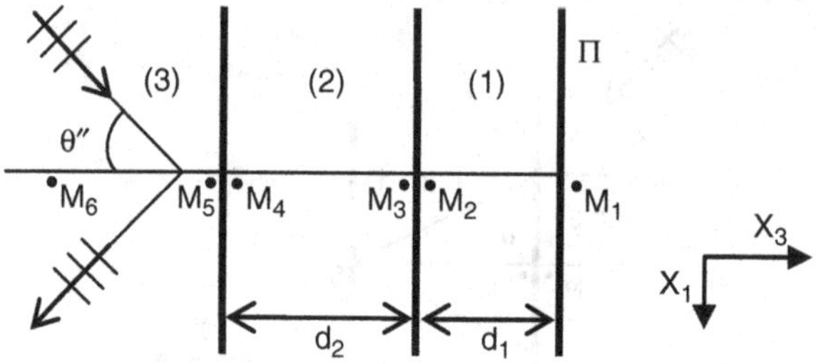

Figure 3.5 Three layers of fluid backed by an impedance plane Π. The incident and reflected waves are represented in fluid 3.

The impedance $Z(M_2)$ can be calculated by Equation (3.36), d being replaced by d_1. In the same way, $Z(M_4)$ and $Z(M_6)$ can be calculated successively. This approach can be used for any number of layers.

3.5 Reflection coefficient and absorption coefficient at oblique incidence

The reflection coefficient at the surface of a layer is the ratio of the pressures p' and p created by the outgoing and the incoming wave, as in the case of normal incidence. More generally, this coefficient can be defined everywhere in a fluid backed by an impedance plane. At any point $M(x_1, x_2, x_3)$ in the fluid represented in Figure 3.3, the reflection coefficient is given by

$$R(M) = p'(M)/p(M) \tag{3.42}$$

where the pressures $p(M)$ and $p'(M)$ are given by Equations (3.30) and (3.32). Substitution of Equations (3.30) and (3.32) in Equation (3.42) yields

$$R(x_3) = \frac{A'}{A}\exp(2jk_3x_3) \tag{3.43}$$

and, like the impedance, R is a function of x_3 only. If k_3 is real, $R(x_3)$ describes a circle in the complex plane as x_3 varies, while if k_3 is complex, $R(x_3)$ describes a spiral. With the aid of Equation (3.35), the reflection coefficient can be written as

$$R(x_3) = \frac{Z(x_3) - Z_c\dfrac{k}{k_3}}{Z(x_3) + Z_c\dfrac{k}{k_3}} \tag{3.44}$$

where k/k_3 can be replaced by $1/\cos\theta$.

An absorption coefficient α at oblique incidence can also be defined by Equation (2.23):

$$\alpha(M) = 1 - |R(M)|^2 \tag{3.45}$$

This coefficient has the same interpretation as the absorption coefficient at normal incidence.

3.6 Examples

In Chapter 2, the surface impedance and the absorption coefficient for a layer of fibrous material with and without an air gap were calculated at normal incidence. Since fibrous materials are not isotropic, the formalism which has been worked out in the present chapter cannot be used for these materials. The sound propagation in porous foams with a flow resistivity of about 10 000 N m^{-4} s or less can be described approximately by the laws of Delany and Bazley. As these materials are generally not noticeably anisotropic, the formalism previously worked out can be used.

The surface impedance of an isotropic porous material of porosity close to 1, of flow resistivity σ equal to 10 000 N m^{-4} s, and of thickness d equal to 10 cm, has been calculated for the two configurations indicated in Figure 3.6.

In the first configuration, the material is fixed to a rigid impervious wall. The front face of the material is in contact with air. The angle of incidence is real and equal to $\pi/4$ radians. In the second configuration, there is an air gap of thickness $d' = 10$ cm between the porous material and the wall. The angle of incidence θ is the same in the air gap as in the air in front of the porous material.

It follows from Equation (3.27) that, in both configurations, the component k_3 of $\mathbf{k}$ in the porous material is

$$k_3 = \left(k^2 - k_o^2 \sin^2 \frac{\pi}{4}\right)^{1/2} \tag{3.46}$$

where k_o and k are the wave number in air and in the porous material, respectively. The wave number k is calculated by means of Equation (2.29). The surface impedance for the first configuration is evaluated by Equation (3.39), Z_c being given by Equation (2.28). In the case of the second configuration, Equation (3.36) may be used. The impedance

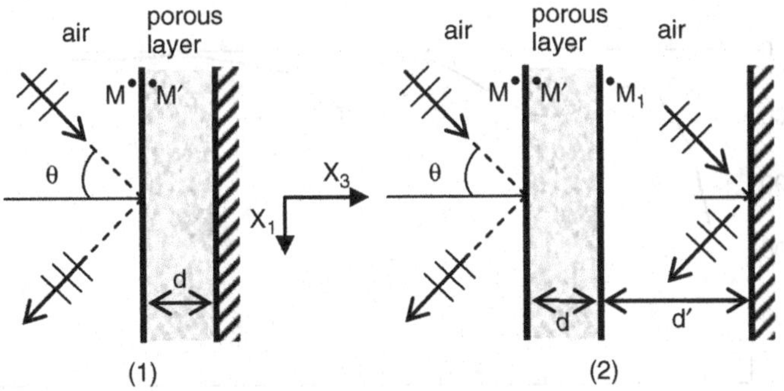

Figure 3.6 First configuration: the material is fixed to an impervious rigid wall. Second configuration: there is a gap of air of thickness d′ between the material and the rigid wall. In both cases, the front face of the porous material is in contact with air.

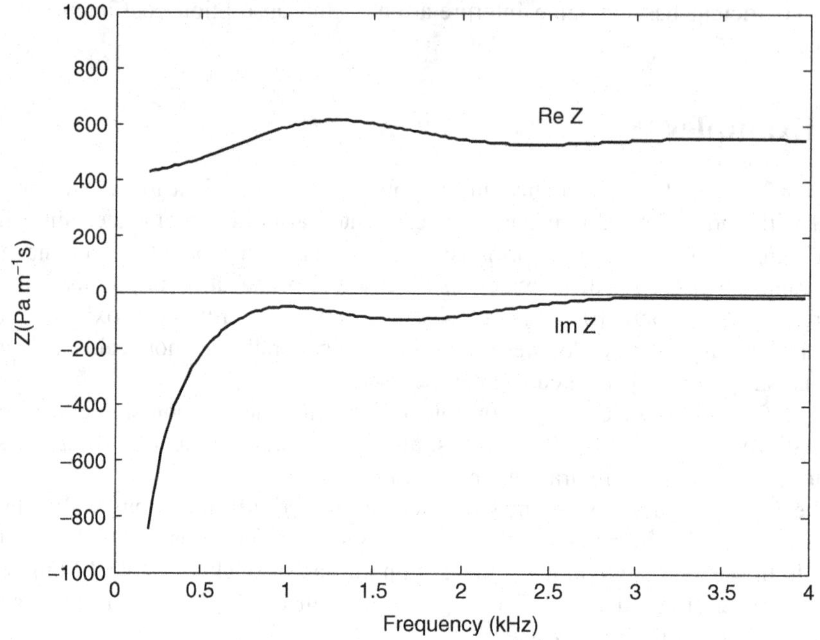

Figure 3.7 Impedance Z at oblique incidence ($\theta = \pi/4$) for a layer of thickness $d = 10$ cm, and flow resistivity $\sigma = 10\,000$ N m^{-4} s (see Figure 3.6; configuration 1).

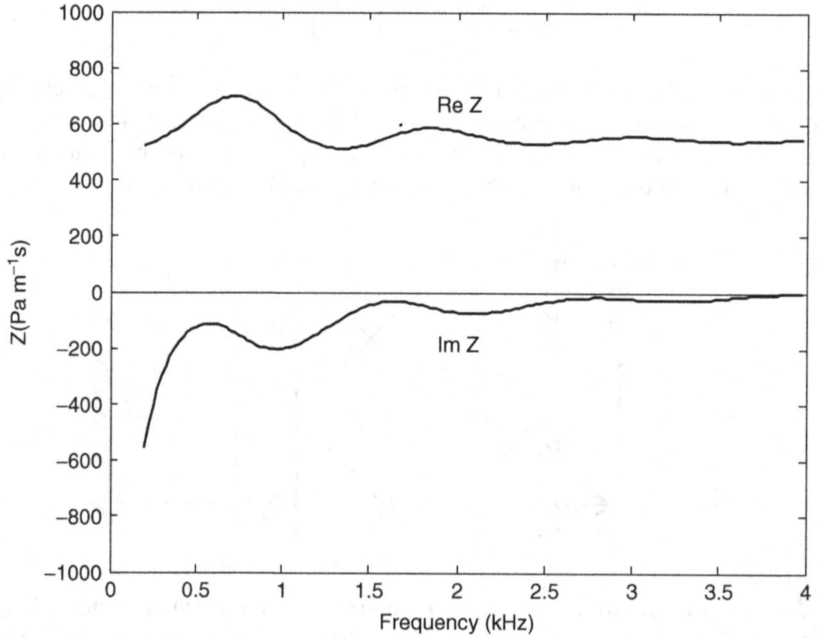

Figure 3.8 The impedance Z at oblique incidence ($\theta = \pi/4$) on the same layer of porous material as in Figure 3.7, with an air gap of thickness $d' = 10$ cm (see Figure 3.6; configuration 2).

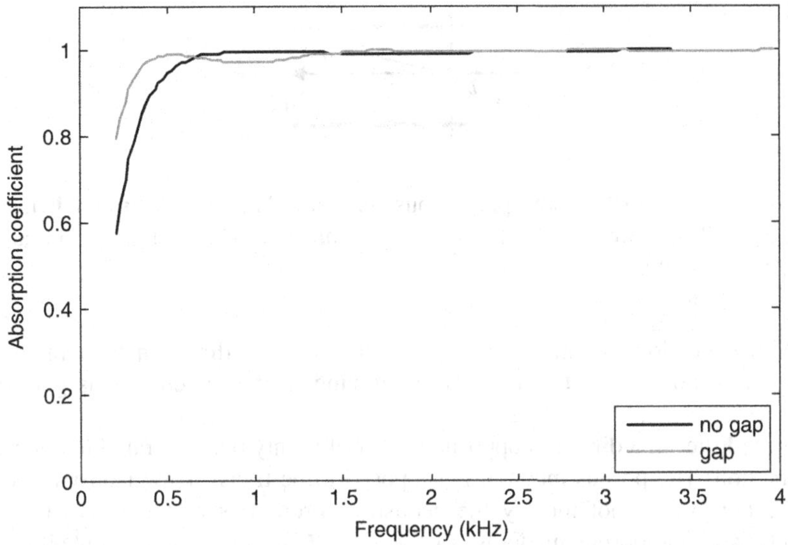

Figure 3.9 The absorption coefficient at oblique incident ($\theta = \pi/4$) for both configurations of Figure 3.6.

$Z(M_1)$ is the impedance of the air gap. This impedance may be evaluated by the use of Equation (3.39), where k, k_3 and Z_c are given by

$$\begin{aligned} k &= k_o \\ k_3 &= k_o\left[\left(1 - \sin^2\frac{\pi}{4}\right)\right]^{1/2} = k_o/\sqrt{2} \\ Z_c &= Z_o \end{aligned} \tag{3.47}$$

where Z_c is the characteristic impedance in air.

The impedance is represented in Figure 3.7 for the first configuration while Figure 3.8 shows the result for the second configuration.

The absorption coefficients in air for both configurations of Figure 3.6, evaluated from the calculated impedances by using Equations (3.44) and (3.45), are represented in Figure 3.9. It may be noted that as in the case for normal incidence, pressures, velocities and impedances are individually equal at either side of the boundary between air and the fluid equivalent to the porous material; however, due to the variation of the characteristic impedance, the reflection and absorption coefficients are different.

3.7 Plane waves in fluids equivalent to transversely isotropic porous media

As indicated in Section 2.5, fibrous materials such as fibreglass are anisotropic (Nicolas and Berry 1984, Allard *et al.* 1987, Tarnow 2005) because generally most of the fibres lie in planes parallel to the surface. A panel of fibrous material is represented in Figure 3.10. The normal direction is parallel to x_3 and the planar directions are parallel to the plane

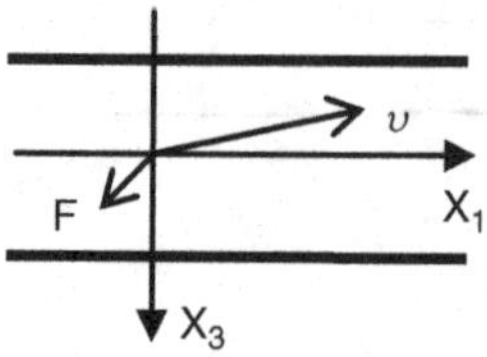

Figure 3.10 A panel of anisotropic fibrous material. The vectors $\boldsymbol{\upsilon}$ and $\mathbf{F}$ indicate in the plane $x_2 = 0$ the direction of the flow of air, and the force acting on the air due to viscosity, respectively.

$x_3 = 0$. Since the flow resistivity is larger in the normal direction than in the planar direction, the viscous force $\mathbf{F}$ acting on air flowing in the direction $\boldsymbol{\upsilon}$ is generally not parallel to $\boldsymbol{\upsilon}$.

The force $\mathbf{F}$ lies in a direction opposite to that of $\mathbf{v}$ only if $\mathbf{v}$ is parallel to a normal or a planar direction. The porous medium is supposed to be transversely isotropic. Rotations around the axis x_3 do not modify the acoustical properties of the layer. The case of transversely isotropic porous media is considered in Chapter 10 in the context of the full Biot theory. If the frame is motionless, the material can be replaced by a transversely isotropic fluid. In the transversely isotropic fluid, the bulk modulus is a scalar quantity. In the system x_1, x_2, x_3 the density is a diagonal tensor with two different components $\rho_3 = \rho_N$, $\rho_1 = \rho_2 = \rho_P$. Let $\boldsymbol{u}$ be the displacement vector. The Newton equations (1.43) and the stress-strain equations (1.57) can be written, respectively

$$-\frac{\partial p}{\partial x_1} = \rho_P \frac{\partial^2 u_1}{\partial t^2}, \qquad -\frac{\partial p}{\partial x_2} = \rho_P \frac{\partial^2 u_2}{\partial t^2}, \qquad -\frac{\partial p}{\partial x_3} = \rho_N \frac{\partial^2 u_3}{\partial t^2}, \tag{3.48}$$

$$-p = K\left(\frac{\partial u_1}{\partial x_1} + \frac{\partial u_2}{\partial x_2} + \frac{\partial u_3}{\partial x_3}\right) \tag{3.49}$$

The wave equation can be written

$$\frac{\partial^2 p}{\partial t^2} = \frac{K}{\rho_P}\left(\frac{\partial^2 p}{\partial x_1^2} + \frac{\partial^2 p}{\partial x_2^2}\right) + \frac{K}{\rho_N}\frac{\partial^2 p}{\partial x_3^2} \tag{3.50}$$

With the $\exp(j\omega t)$ time dependence, this equation becomes

$$\frac{K}{\rho_P}\left(\frac{\partial^2 p}{\partial x_1^2} + \frac{\partial^2 p}{\partial x_2^2}\right) + \frac{K}{\rho_N}\frac{\partial^2 p}{\partial x_3^2} + \omega^2 p = 0 \tag{3.51}$$

The quantities $\omega(\rho_P/K)^{1/2}$ and $\omega(\rho_N/K)^{1/2}$ will be denoted by k_P and k_N

$$k_P = \omega\left(\frac{\rho_P}{K}\right)^{1/2}, \qquad k_N = \omega\left(\frac{\rho_N}{K}\right)^{1/2} \tag{3.52}$$

A solution of Equation (3.51) is

$$p(x_1, x_2, x_3, t) = A\exp[j(-k_1 x_1 - k_2 x_2 - k_3 x_3 + \omega t)] \tag{3.53}$$

The components k_1, k_2 and k_3 of the vector **k** are related by

$$\frac{k_1^2}{k_P^2} + \frac{k_2^2}{k_P^2} + \frac{k_3^2}{k_N^2} = 1 \tag{3.54}$$

and A is an arbitrary constant.

Substitution of Equation (3.53) into Equation (3.48) yields the three components of the velocity $\boldsymbol{v}$

$$\begin{aligned} v_3(x_1, x_2, x_3, t) &= \frac{Ak_3/k_N}{(\rho_N K)^{1/2}} \exp[j(-k_1x_1 - k_2x_2 - k_3x_3 + \omega t)] \\ v_i(x_1, x_2, x_3, t) &= \frac{Ak_i/k_P}{(\rho_P K)^{1/2}} \exp[j(-k_1x_1 - k_2x_2 - k_3x_3 + \omega t)] i = 1, 2 \end{aligned} \tag{3.55}$$

Planes waves in fluids equivalent to fibrous materials can be described by Equations (3.53)–(3.55).

3.8 Impedance at oblique incidence at the surface of a fluid equivalent to an anisotropic porous material

The layer of a fibrous material is fixed to a rigid, impervious wall, on its rear face, and is in contact with air on its front face, as shown in Figure 3.11.

The acoustic field in the air is homogeneous and the angle of incidence θ is real. The formalism is the same if θ is complex. The Snell–Descartes law for refraction may be written

$$k_o \sin\theta = k_1 \tag{3.56}$$

where k_o is the wave number in air, and k_1 the component on x_1 of the wave number vector $\boldsymbol{k}$ in the fibrous material.

The component k_3 of $\boldsymbol{k}$, obtained from Equation (3.54) is

$$k_3 = k_N \left(1 - \frac{k_1^2}{k_P^2}\right)^{1/2} \tag{3.57}$$

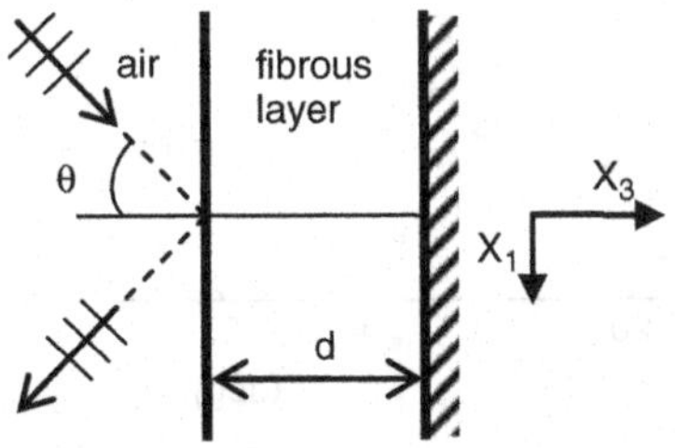

Figure 3.11 A layer of fibrous material backed by an impervious rigid wall and in contact with air on its front face.

The pressure p in the fibrous material has the form

$$p(x_1, x_2, x_3, t) = A\exp[j(-k_1x_1 - k_3x_3 + \omega t)] + A'\exp[j - (k_1x_1 + k_3x_3 + \omega t)] \quad (3.58)$$

The component of velocity in the x_3 direction obtained from Eq. (3.55) is

$$v_3(x_1, x_2, x_3, t) = \frac{1}{(\rho_N K)^{1/2}}\frac{k_3}{k_N}[A\exp[j(-k_1x_1 - k_3x_3 + \omega t)] - A'\exp[j(-k_1x_1 + k_3x_3 + \omega t)]] \quad (3.59)$$

At $x_3 = 0$, $v_3 = 0$, and $A = A'$. At $x_3 = -d$, the impedance Z is equal to

$$Z = (\rho_N K)^{1/2}\frac{k_N}{k_3}(-j \ \text{cotg}\ k_3 d) \quad (3.60)$$

The laws of Delany and Bazley, Equations (2.28) and (2.29), can be used to evaluate the characteristic impedance $(\rho_N K)^{1/2}$ and k_N from the normal flow resistivity, as well as k_P, that is used in Equation (3.57), from the planar flow resistivity. The evaluation of the reflection coefficient and the absorption coefficient is the same as in Section 3.5.

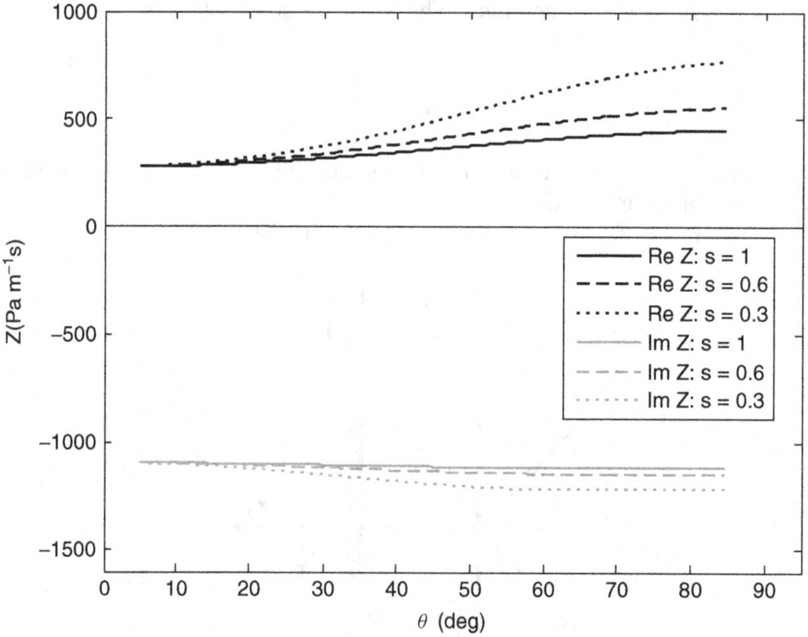

Figure 3.12 Predicted impedance Z in Pa m^{-1} s versus the angle of incidence θ in degrees for three values of the anisotropy factor $s = \sigma_P/\sigma_N$.

3.9 Example

The impedance Z of a layer of anisotropic fibrous material is represented in Figure 3.12. The normal flow resistivity σ_N and the thickness d of the material are respectively $\sigma_N = 25\,000$ N m^{-4} s and $d = 2.5$ cm.

The impedance at 630 Hz is represented as a function of the angle of incidence θ for three values of the anisotropy factor $s = \sigma_P/\sigma_N = 1$, 0.6, 0.3. A value of s close to 0·6 was obtained in measurements performed on several glass wools (Nicolas and Berry 1984, Allard *et al.* 1987, Tarnow 2005).

References

Allard, J.F., Bourdier, R. & L'Espérance, A. (1987) Anisotropy effect in glass wool on normal impedance in oblique incidence. *J. Sound Vib.*, **114**, 233–8.

Cremer, L. and Muller, A. (1982) *Principles and Applications of Room Acoustics*. Elsevier Applied Science Publishers, London.

Delany, M.E. and Bazley, E.N. (1970) Acoustical properties of fibrous materials. *Applied Acoustics*, **3**, 105–16.

Nicolas, J. and Berry, J.L. (1984) Propagation du son et effet de sol. *Revue d'Acoustique*, **71**, 191–200.

Tarnow, V. (2005) Dynamic measurement of the elastic constants of glass wool. *J. Acoust. Soc. Amer.* **118**, 3672–3678.

4

Sound propagation in cylindrical tubes and porous materials having cylindrical pores

4.1 Introduction

The geometry of the pores in ordinary porous materials is not simple, and a direct calculation of the viscous and thermal interaction between the air and these materials is generally impossible to perform. Useful information can be obtained from the simple case of porous materials with cylindrical pores. In this chapter, a simple modelling of sound propagation in cylindrical tubes of various cross-sectional shapes is presented, and this modelling is used to predict the acoustical properties of porous materials with cylindrical pores and to define important concepts such as tortuosity.

4.2 Viscosity effects

The Kirchhoff theory (Kirchhoff 1868) of sound propagation in cylindrical tubes provides a general description of viscous and thermal effects, but this description is unnecessarily complicated for many applications. Moreover, the fundamental equations of acoustics that are used in the Kirchhoff theory can be very difficult to solve in the case of a noncircular cross-section. A simplified model in which thermal and viscous effects are treated separately has been worked out by Zwikker and Kosten (1949) for the case of circular cross-sections. The validity of this model is justified later (Tijdeman 1975, Kergomard 1981, Stinson, 1991) for the range of radii from 10^{-3} cm to several centimetres at acoustical frequencies. This model will be used to describe viscosity effects in slits and cylindrical tubes having a circular cross-section. The thermal exchange effects will

Propagation of Sound in Porous Media: Modelling Sound Absorbing Materials, Second Edition J. F. Allard and N. Atalla

be related to the viscous effects by a model worked out by Stinson (1991). The simplified linear equations that satisfy the velocity $\boldsymbol{v}$, the pressure p, and the acoustic temperature τ of the fluid in a pore, i.e. the variation of the temperature in an acoustic field, are

$$\rho_0 \frac{\partial \boldsymbol{v}}{\partial t} = -\nabla p + \eta \Delta \boldsymbol{v} \tag{4.1}$$

$$\rho_0 c_p \frac{\partial \tau}{\partial t} = \kappa \nabla^2 \tau + \frac{\partial p}{\partial t} \tag{4.2}$$

where η is the shear viscosity (the volume viscosity is neglected), κ is the thermal conductivity, c_p is the specific heat per unit mass at constant pressure. For air in standard conditions, $\eta = 1.84\ 10^{-5}\mathrm{kg\,m^{-1}\,s^{-1}}$, $\kappa = 2.6\ 10^{-2}\mathrm{w\,m^{-1}\,k^{-1}}$. The pressure is considered constant on a cross- section of the tube. The boundary conditions at the air–frame interface are

$$\boldsymbol{v} = \mathbf{0} \tag{4.3}$$

$$\tau = 0 \tag{4.4}$$

A volume of air is represented in Figure 4.1, with a velocity $\boldsymbol{v}(0, 0, v_3)$ parallel to the x_3 axis and whose magnitude only depends on x_1. The variation of this velocity with x_1 produces viscous stress. Due to the viscosity, the air is subjected to a shear force parallel to the x_3 axis, and proportional to $\partial v_3/\partial x_1$. More precisely, the right-hand side of plane Π is subjected from the left-hand side of Π to a stress $\boldsymbol{T}$ parallel to X_3. The projection of $\boldsymbol{T}$ on the x_3 axis is

$$T_3(x_1) = -\eta \frac{\partial v_3(x_1)}{\partial x_1} \tag{4.5}$$

The resulting force due to the stresses at Π and Π' for a layer of unit lateral area is $\mathbf{d}\boldsymbol{F}$ parallel to the x_3 axis such that

$$\mathrm{d}F_3 = -\eta \frac{\partial v_3(x_1)}{\partial x_1} + \eta \frac{\partial v_3(x_1 + \Delta x_1)}{\partial x_1} \tag{4.6}$$

Per unit volume of air, this force is equal to

$$X_3 = \eta \frac{\partial^2 v_3}{\partial x_1^2} \tag{4.7}$$

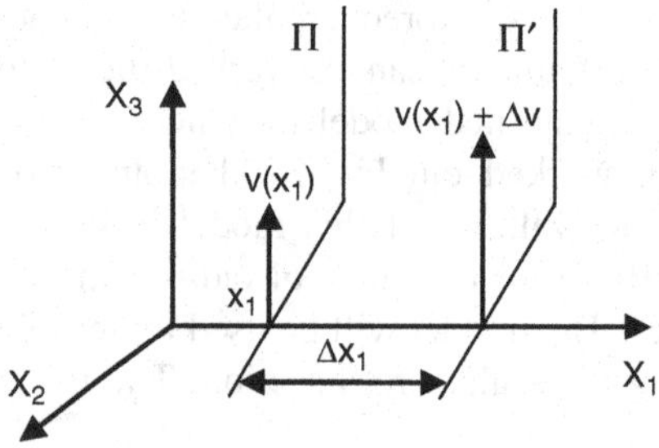

Figure 4.1 A simple velocity field in a viscous fluid.

If υ_3 depends also on x_2, X_3 is equal to

$$X_3 = \eta \left(\frac{\partial^2 \upsilon_3}{\partial x_1^2} + \frac{\partial^2 \upsilon_3}{\partial x_2^2} \right) \tag{4.8}$$

This simple description will be used to evaluate the effect of viscosity in cylindrical pores, the radial velocity components being neglected, and the pressure depending only on the x_3 direction of the pores.

Cylindrical tubes having a circular cross-section

A cylindrical tube having a circular cross-section is represented in Figure 4.2. The axis of the cylinder is $0x_3$. By using Equation (4.8), Newton's law reduces to

$$j\omega\rho_o\upsilon_3 = -\frac{\partial p}{\partial x_3} + \eta \left[\frac{\partial^2 \upsilon_3}{\partial x_1^2} + \frac{\partial^2 \upsilon_3}{\partial x_2^2} \right] \tag{4.9}$$

ρ_o being the density of air, and p the pressure.

The geometry of the problem is axisymmetrical around $0x_3$, and Equation (4.9) can be rewritten

$$j\omega\rho_o\upsilon_3 = -\frac{\partial p}{\partial x_3} + \frac{\eta}{r}\frac{\partial}{\partial r}\left(r\frac{\partial \upsilon_3}{\partial r} \right) \tag{4.10}$$

The velocity υ must vanish at the surface of the cylinder, where the air is in contact with the motionless frame. The solution of Equation (4.10), where the velocity vanishes at the surface $r = R$ of the cylinder is

$$\upsilon_3 = -\frac{1}{j\omega\rho_o}\frac{\partial p}{\partial x_3}\left(1 - \frac{J_o(lr)}{J_o(lR)} \right) \tag{4.11}$$

In this equation, l is equal to

$$l = (-j\omega\rho_o/\eta)^{1/2} \tag{4.12}$$

and J_o is the Bessel function of zero order.

Both determinations of the square root in Equation (4.12) give identical results, because the Bessel function J_0 is even. The mean velocity $\bar{\upsilon}_3$ over the cross-section is equal to

$$\bar{\upsilon}_3 = \frac{\int_o^R \upsilon_3 2\pi r\, \mathrm{d}r}{\pi R^2} \tag{4.13}$$

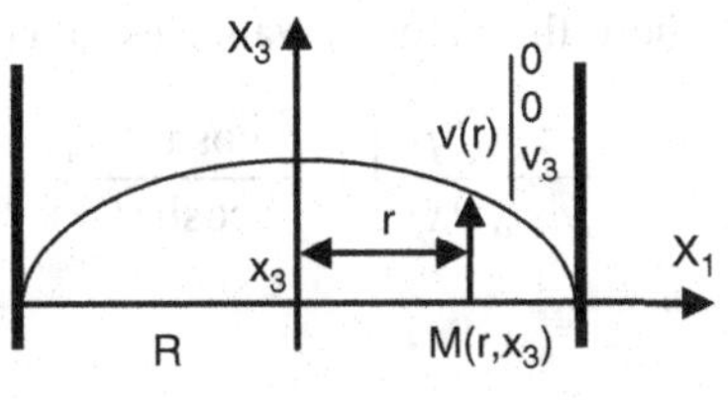

Figure 4.2 A cylindrical tube having a circular cross-section of radius R.

Making use of

$$\int_0^a r J_o(r)\,\mathrm{d}r = a J_1(a) \tag{4.14}$$

and substitution of Equation (4.11) into Equation (4.13) yields

$$\bar{v}_3 = -\frac{1}{j\omega\rho_o}\frac{\partial p}{\partial x_3}\left[1 - \frac{2}{s\sqrt{-j}}\frac{J_1(s\sqrt{-j})}{J_o(s\sqrt{-j})}\right] \tag{4.15}$$

In this equation, s is equal to

$$s = \left(\frac{\omega\rho_o R^2}{\eta}\right)^{1/2} \tag{4.16}$$

The effective density ρ of the air in the tube is defined by rewriting Equation (4.15) in a more compact form as

$$-\frac{\partial p}{\partial x_3} = j\omega\rho\bar{v}_3 \tag{4.17}$$

with

$$\rho = \rho_o \Big/ \left[1 - \frac{2}{s\sqrt{-j}}\frac{J_1(s\sqrt{-j})}{J_o(s\sqrt{-j})}\right] \tag{4.18}$$

On the other hand, the true density ρ_o can be used in the Newton equation (4.15)

$$-\frac{\partial p}{\partial x_3} = j\omega\rho_o\bar{v}_3 + \frac{2}{s\sqrt{-j}}\frac{J_1(s\sqrt{-j})}{J_o(s\sqrt{-j})}\frac{\rho_o j\omega\bar{v}_3}{\left[1 - \dfrac{2}{s\sqrt{-j}}\dfrac{J_1(s\sqrt{-j})}{J_o(s\sqrt{-j})}\right]} \tag{4.19}$$

Slit

The geometry of the problem is represented in Figure 4.3. With the same simplifications as in the case of cylindrical tubes having a circular cross-section, the Newton equation (4.9) reduces to

$$\rho_o\frac{\partial v_3}{\partial t} = -\frac{\partial p}{\partial x_3} + \eta\frac{\partial^2 v_3}{\partial x_1^2} \tag{4.20}$$

where v_3 is not dependent on x_2. The velocity must vanish at the surface of the slit. The solution of Equation (4.20), where the velocity vanishes at $x_1 = \pm a$, is

$$v_3 = -\frac{1}{j\omega\rho_o}\frac{\partial p}{\partial x_3}\left[1 - \frac{\cosh(l'x_1)}{\cosh(l'a)}\right] \tag{4.21}$$

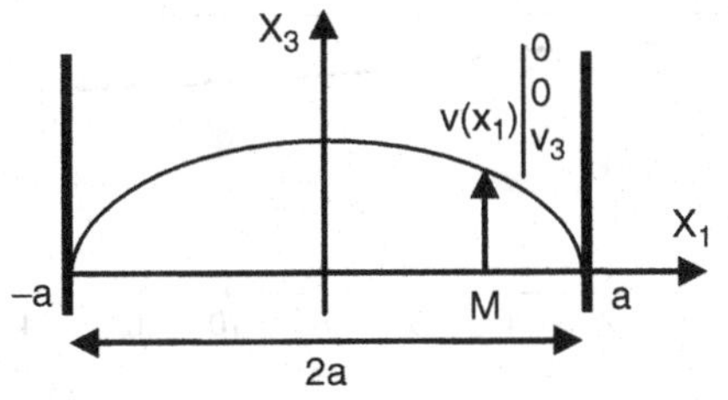

Figure 4.3 Flow in a slit limited by the planes $x_1 = $ a and $x_1 = -a$.

In this equation, l' is equal to

$$l' = (j\,\omega \rho_o/\eta)^{1/2} \tag{4.22}$$

As with the case of l, both determinations of the square root in Equation (4.22) give identical results. The mean velocity is equal to

$$\bar{\upsilon}_3 = \frac{\int_{-a}^{a} \upsilon_3 \, \mathrm{d}x_1}{2a} \tag{4.23}$$

Upon integrating, we find

$$\bar{\upsilon}_3 = -\frac{1}{j\,\omega \rho_o}\frac{\partial p}{\partial x_3}\left[1 - \frac{\tanh(s' \mathrm{j}^{1/2})}{s' \mathrm{j}^{1/2}}\right] \tag{4.24}$$

In this equation, s' is equal to

$$s' = \left(\frac{\omega \rho_o a^2}{\eta}\right)^{1/2} \tag{4.25}$$

The effective density of the air in the slit is defined by rewriting Equation (4.24) as

$$j\,\omega \rho \bar{\upsilon}_3 = -\frac{\partial p}{\partial x_3} \tag{4.26}$$

where

$$\rho = \rho_o \Big/ \left[1 - \frac{\tanh(s'\, j^{1/2})}{s'\, j^{1/2}}\right] \tag{4.27}$$

On the other hand, the true density ρ_o can be used in the Newton equation

$$-\frac{\partial p}{\partial x_3} = j\,\omega \rho_o \bar{\upsilon}_3 + j\,\omega \rho_o \bar{\upsilon}_3 \frac{\tanh(s'\, j^{1/2})(s'\, j^{1/2})}{1 - [\tanh(s'\, j^{1/2})(s'\, j^{1/2})]} \tag{4.28}$$

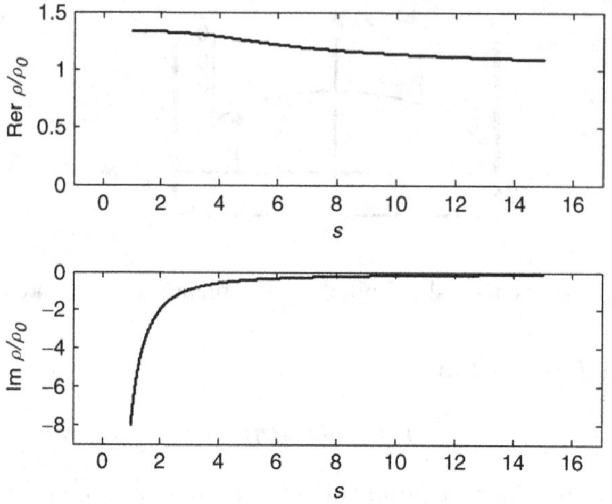

Figure 4.4 The ratio ρ/ρ_o, where ρ is the effective density of a fluid of density ρ_0 in a cylindrical tube having a circular cross-section, as a function of s.

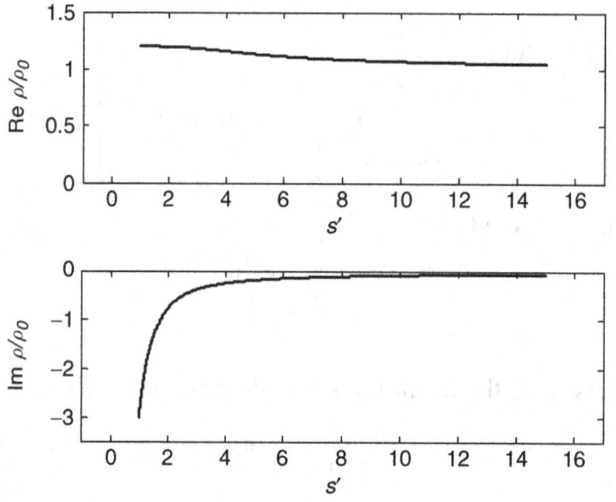

Figure 4.5 The ratio ρ/ρ_o, where ρ is the effective density of a fluid of density ρ_0, in a slit, as a function of s'.

This formulation shows the contribution added to the inertial term by the interactions with the tube. The effective density ρ is represented in Figures 4.4 and 4.5 in the case of the circular cross-section, and the slit, respectively.

4.3 Thermal effects

The equation of state of air, considered to be an ideal gas, is

$$p = \frac{P_o}{\rho_o T_o}[\rho_o \tau + T_o \xi] \tag{4.29}$$

In this equation ξ is the acoustical density, P_o and T_o are the ambient mean pressure and mean temperature. The variations of τ in the x_3 direction parallel to the tube are smaller than in the radial directions x_1 and x_2, and Equation (4.2) reduces to

$$\frac{\partial^2 \tau}{\partial x_1^2} + \frac{\partial^2 \tau}{\partial x_2^2} - j\omega\frac{\tau}{\nu'} = -j\frac{\omega}{\kappa}p \tag{4.30}$$

where ν' is given by

$$\nu' = \frac{\kappa}{\rho_0 c_p} \tag{4.31}$$

This equation has the same form as Equation (4.9) for υ_3. Moreover, the boundary conditions are the same, τ being equal to zero at the surface of the tube.

An angular frequency ω' defined by

$$\omega' = \omega\frac{\eta}{\rho_o \nu'} \tag{4.32}$$

will be used instead of ω in Equation (4.30) which then becomes

$$\frac{\partial^2 \tau}{\partial x_1^2} + \frac{\partial^2 \tau}{\partial x_2^2} - j\,\omega'\frac{\rho_o \tau}{\eta} = -\frac{j\,\omega'\upsilon'\rho_o p}{\kappa\eta} \tag{4.33}$$

The quantity $\eta/(\rho_0\nu')$ is the Prandtl number, and will be denoted by B^2. We rewrite Equation (4.9) as

$$\frac{\partial^2 \upsilon_3}{\partial x_1^2} + \frac{\partial^2 \upsilon_3}{\partial x_2^2} - \frac{j\,\omega\rho_o}{\eta}\upsilon_3 = \frac{1}{\eta}\frac{\partial p}{\partial x_3} \tag{4.34}$$

By comparing Equations (4.33) and (4.34), it follows that τ and υ_3 are equal to

$$\tau = \frac{p\nu'}{\kappa}\psi(x_1,\, x_2,\, B^2\omega) \tag{4.35}$$

$$\upsilon_3 = -\frac{\partial p}{\partial x_3}\psi(x_1, x_2, \omega)\,\frac{1}{j\,\omega\rho_o} \tag{4.36}$$

where $\psi(x_1,\, x_2,\, \omega)$ is a solution of the equation

$$\frac{\partial^2 \psi}{\partial x_1^2} + \frac{\partial^2 \psi}{\partial x_2^2} - j\,\omega\frac{\rho_o}{\eta}\psi = -\frac{j\,\omega\rho_o}{\eta} \tag{4.37}$$

with the boundary condition that ψ vanishes at the surface of the tube.

The effective density is calculated from the average value $\bar{\upsilon}_3$ of υ_3 over the cross-section of the tube (Equation 4.17)

$$\rho = -\frac{\partial p}{\partial x_3}\frac{1}{j\,\omega\bar{\upsilon}_3} = \frac{\rho_0}{\bar{\psi}(x_1, x_2, \omega)} \tag{4.38}$$

Using Equation (1.57) and the linearized continuity equation, the bulk modulus is given by

$$K = -p/\bar{\nabla}\boldsymbol{u} = \rho_0 p/\bar{\xi} \tag{4.39}$$

where the mean density $\bar{\xi}$ is the average of ξ, given by Equation (4.29), on the cross-section:

$$\bar{\xi} = \frac{\rho_0}{P_0} p - \frac{\rho_0}{T_0} \bar{\tau} \tag{4.40}$$

The expression for K can be rewritten with the use of Equation (4.35) as

$$K = \frac{P_0}{1 - \dfrac{P_0}{T_0} \dfrac{\nu'}{\kappa} \bar{\psi}(x_1, x_2, B^2\omega)} \tag{4.41}$$

Let c_v be the specific heat at constant volume. Making use of $\rho_0(c_p - c_v) = P_0/T_0$ which is valid for ideal gases, Equation (4.41) can be rewritten as

$$K = \frac{\gamma P_0}{\gamma - (\gamma - 1)\bar{\psi}(x_1, x_2, B^2\omega)} \tag{4.42}$$

where $\gamma = c_p/c_v$. Denoting $\bar{\psi}(x_1, x_2, \omega)$ by $F(\omega)$

$$\bar{\psi}(x_1, x_2, \omega) = F(\omega) \tag{4.43}$$

$\bar{\psi}(x_1, x_2, B^2\omega)$ can be written as

$$\bar{\psi}(x_1, x_2, B^2\omega) = F(B^2\omega) \tag{4.44}$$

Equations (4.38) and (4.42) then become

$$\rho = \rho_o/F(\omega) \tag{4.45}$$

$$K = \gamma P_o/[\gamma - (\gamma - 1)F(B^2\omega)] \tag{4.46}$$

For air at 18 °C with $P_o = 1{\cdot}0132 \times 10^5$ Pa, $\gamma = 1{\cdot}4$ and $B^2 = 0{\cdot}71$ (Gray 1956).

For cylindrical tubes having circular cross-section, F is obtained from Equations (4.18) and (4.45):

$$F(\omega) = \left[1 - \frac{2}{s\sqrt{-j}} \frac{J_1(s\sqrt{-j})}{J_o(s\sqrt{-j})}\right] \tag{4.47}$$

The dependence of F on ω is given via Equation (4.16).

The bulk modulus K is obtained from Equation (4.46). It is equal to

$$K = \gamma P_o \Big/ \left[1 + (\gamma - 1)\frac{2}{Bs\sqrt{-j}} \frac{J_1(Bs\sqrt{-j})}{J_o(Bs\sqrt{-j})}\right] \tag{4.48}$$

The bulk modulus for slits is obtained in the same way

$$K = \gamma P_o \Big/ \left[1 + (\gamma - 1)\frac{\tanh(Bs'\sqrt{j})}{Bs'\sqrt{j}}\right] \tag{4.49}$$

This method of determining K holds for all cylindrical tubes. The bulk modulus K, for cylindrical tubes having a circular cross-section, and slits, is represented in Figures 4.6 and 4.7 as a function of s and s', respectively. It is seen that its real part is equal to P_0 at low frequencies (isothermal limit) and to γP_0 at high frequencies (adiabatic limit). These limits are detailed in Sections 4.5 and 4.6.

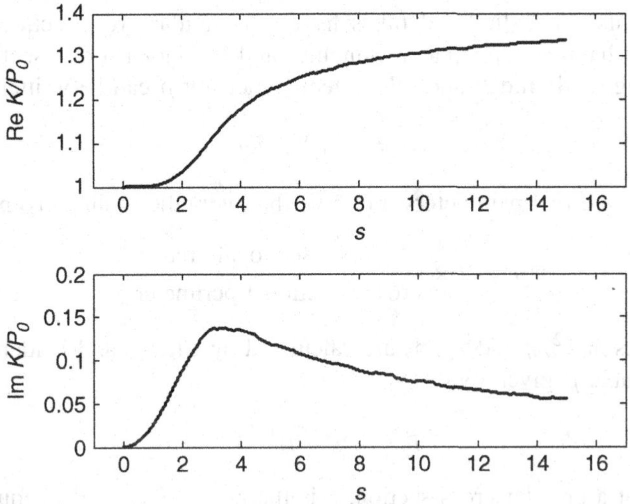

Figure 4.6 The bulk modulus K of air in a cylindrical tube having a circular cross-section, as a function of s. The unit is the atmospheric pressure P_0.

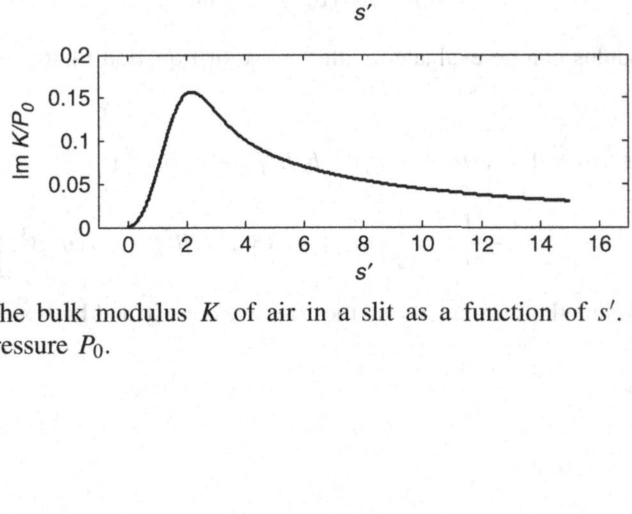

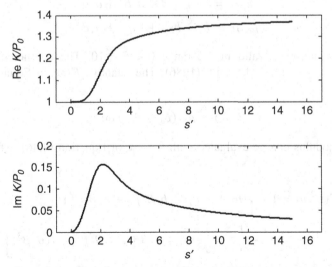

Figure 4.7 The bulk modulus K of air in a slit as a function of s'. The unit is the atmospheric pressure P_0.

4.4 Effective density and bulk modulus for cylindrical tubes having triangular, rectangular and hexagonal cross-sections

The effective density has been calculated by Craggs and Hildebrandt (1984, 1986) using the finite element method (Hubner 1974, Zienkiewicz 1971). The calculation has been performed for slits, for cylindrical tubes having a circular cross-section, and also for cylindrical tubes having triangular, rectangular and hexagonal cross-sections. With the notation of Craggs and Hildebrandt, the effective density ρ can be written

$$\rho = \rho_e + \Re_e / j\,\omega \tag{4.50}$$

where ρ_e and $\Re_e$ are real parameters. Let $\bar{r}$ be the hydraulic radius, given by

$$\bar{r} = 2 \times \frac{\text{cross - sectional area}}{\text{cross - sectional perimeter}} \tag{4.51}$$

The quantities $\Re_e \bar{r}^2/\eta$ and ρ_e/ρ_o are calculated by Craggs and Hildebrandt as functions of the variable β given by

$$\beta = (\rho_o \omega/\eta)^{1/2}\,\bar{r} \tag{4.52}$$

In the case of a circular cross-section, $\bar{r}$ is the radius R, and β is equal to s. In the case of a slit, $\bar{r}$ is equal to the width $2a$, and β is equal to $2s'$.

The quantities $\Re_e \bar{r}^2/\eta$ and ρ_e/ρ_o have been fitted by Craggs and Hildebrandt (1986) to polynomials in the case of slits, circular, triangular and square cross-sections,

$$\Re_e \bar{r}^2/\eta = a_1 + a_2\beta + a_3\beta^2 + a_4\beta^3 \tag{4.53}$$

$$\rho_e/\rho_o = b_1 + b_2\beta + b_3\beta^2 + b_4\beta^3 \tag{4.54}$$

This representation is valid in the range $0 < \beta < 10$. The coefficients a_i and b_i are given by Craggs and Hildebrandt (1986). The quantity $F(\omega)$ in Equation (4.45) is equal to

$$F(\omega) = \rho_o / (\rho_e + \Re_e / j\,\omega) \tag{4.55}$$

and the bulk modulus can be evaluated with the use of Equation (4.46), where $F(B^2\omega)$ is given by

$$F(B^2\omega) = 1 \bigg/ \bigg[(b_1 + b_2 B\beta + b_3 B^2\beta^2 + b_4 B^3\beta^3) + \frac{1}{j\,\omega B^2}\frac{\eta}{\rho_o \bar{r}^2}(a_1 + a_2 B\beta + a_3 B^2\beta^2 + a_4 B^3\beta^3) \bigg] \tag{4.56}$$

A more general method for predicting the effective density and bulk modulus is given in Section 4.7.

4.5 High- and low-frequency approximation

Tubes having a circular cross-section

For small and large values of s, asymptotic expressions of J_1/J_o can be used in Equation (4.19). With the Prandtl number close to 1, in the same ranges of frequencies, asymptotic expressions can be used simultaneously in Equation (4.48) to evaluate the bulk modulus K. The quantity s is large if

$$\left(\frac{\eta}{\omega\rho_o}\right)^{1/2} \ll R \tag{4.57}$$

The range of frequencies where Equation (4.57) is valid is called the high-frequency range. The length δ

$$\delta = \left(\frac{2\eta}{\omega\rho_o}\right)^{1/2} \tag{4.58}$$

is called the viscous skin depth. This length is approximately equal to the thickness of the layer of air close to the surface of the tube where the velocity distribution is considerably perturbed by the viscous forces generated by the motionless frame. If Equation (4.57) is valid, the effect of the viscous forces is negligible in a large part of the tube centred around the axis of symmetry and a flat central core appears in the velocity distribution over the cross-section of the tube. Similarly, a thermal skin depth δ' can be defined by the relation

$$\delta' = \left(\frac{2\eta}{\omega B^2\rho_0}\right)^{1/2} = \left(\frac{2\kappa}{\omega\rho_0 C_p}\right)^{1/2} \tag{4.59}$$

At high frequencies, choosing

$$(-j)^{1/2} = (-1+j)/\sqrt{2} \tag{4.60}$$

J_1/J_o is equal to j and Equation (4.18) approximates to

$$\frac{\rho}{\rho_0} = 1\Bigg/\left[1 - \frac{2}{s\sqrt{-j}}\frac{\mathrm{J}_1(s\sqrt{-j})}{\mathrm{J}_0(s\sqrt{-j})}\right] = 1 + \frac{\sqrt{2}(1+j)}{js} \tag{4.61}$$

The Newton equation (4.19) becomes

$$-\frac{\partial p}{\partial x} = j\omega\rho_o\bar{v}_3 + (1+j)\bar{v}_3\left(\frac{2\eta}{R^2}\rho_o\omega\right)^{1/2} \tag{4.62}$$

and the bulk modulus (Equation 4.48) can be written

$$K = \gamma P_o[1 + \sqrt{2}(-1+j)(\gamma-1)/(Bs)] \tag{4.63}$$

Conversely, at sufficiently low frequencies, the quantity s is less than 1. Here the effect of the viscous forces is important everywhere in the pore. The range of frequencies such that

$$\left(\frac{\eta}{\omega\rho_o}\right)^{1/2} \gg R \tag{4.64}$$

is called the low-frequency range. At low frequencies, the following approximations can be used:

$$\frac{2}{s\sqrt{-j}}\frac{J_1(s\sqrt{-j})}{J_o(s\sqrt{-j})} = 1 - \frac{js^2}{8} - \frac{4}{192}s^4 \tag{4.65}$$

and

$$\frac{2}{s\sqrt{-j}}\frac{J_1(s\sqrt{-j})}{J_o(\sqrt{-j})\left\{1 - \dfrac{2}{s\sqrt{-j}}\dfrac{J_1(s\sqrt{-j})}{J_o(s\sqrt{-j})}\right\}} = -\frac{8j}{s^2} + \frac{1}{3} \tag{4.66}$$

Using this last equation, Equation (4.19) can be rewritten

$$-\frac{\partial p}{\partial x_3} = \frac{4}{3}\, j\,\omega\rho_o\bar{v}_3 + \frac{8\eta}{R^2}\,\bar{v}_3 \tag{4.67}$$

It should be noted that, at low frequencies

$$\frac{\eta}{R^2} \gg \rho_o\omega \tag{4.68}$$

Hence, in the Newton equation (4.67), the magnitude of the term $(4/3)\,j\omega\rho_o\bar{v}_3$ is much smaller than that of $(8\eta/R^2)\bar{v}_3$. The bulk modulus K given by Equation (4.48) is

$$K = P_o\left[1 + \frac{1}{8}j\frac{\gamma - 1}{\gamma}(B^2s^2)\right] \tag{4.69}$$

Slit

The high- and low-frequency ranges are defined by

$$\left(\frac{\eta}{\omega\rho_0}\right)^{1/2} \ll a,\ s' = \left(\frac{\omega\rho_o a^2}{\eta}\right)^{1/2} \gg 1 \text{ High - frequency range}$$

$$\left(\frac{\eta}{\omega\rho_0}\right)^{1/2} \gg a,\ s' = \left(\frac{\omega\rho_o a^2}{\eta}\right)^{1/2} \ll 1 \text{ Low - frequency range}$$

At high frequencies, with

$$j^{1/2} = \frac{1 + j}{\sqrt{2}} \tag{4.70}$$

the term $\tanh(s'j^{1/2})$ is equal to 1. The Newton equation (4.28) becomes

$$-\frac{\partial p}{\partial x_3} = j\omega\rho_o\bar{v}_3 + \frac{(1+j)}{\sqrt{2}}\bar{v}_3\left(\frac{\eta}{a^2}\rho_o\omega\right)^{1/2} \tag{4.71}$$

and the bulk modulus K is

$$K = \gamma P_o\left[1 + \frac{\sqrt{2}}{2}(-1+j)(\gamma-1)/(Bs')\right] \tag{4.72}$$

At low frequencies, the following approximation can be used:

$$\frac{1}{s'j^{1/2}}\tanh(s'j^{1/2}) \Big/ \left(1 - \frac{1}{s'j^{1/2}}\tanh(s'j^{1/2})\right) = \frac{3}{(s'j^{1/2})^2} + \frac{1}{5} \tag{4.73}$$

The Newton equation (4.28) becomes

$$-\frac{\partial p}{\partial x_3} = \frac{6}{5}j\omega\rho_o\bar{v}_3 + \frac{3\eta}{a^2}\bar{v}_3 \tag{4.74}$$

and the bulk modulus is

$$K = P_o\left[1 + \frac{1}{3}j\frac{\gamma-1}{\gamma}(Bs')^2\right] \tag{4.75}$$

4.6 Evaluation of the effective density and the bulk modulus of the air in layers of porous materials with identical pores perpendicular to the surface

First, the cylindrical pore having a circular cross-section and the slit are considered. Next, general models with adjustable parameters, used for the evaluation of the effective density and the bulk modulus of the air in pores with other cross-sectional shapes, are presented. The flow resistivity is used as an acoustical parameter.

4.6.1 Effective density and bulk modulus in cylindrical pores having a circular cross-section

The sample of porous material represented in Figure 4.8 has n pores of radius R per unit area of cross-section. The flow resistivity σ given by Equation (2.27) is equal to

$$\sigma = \frac{p_2 - p_1}{\bar{v}n\pi R^2 e} \tag{4.76}$$

The air in the pores is submitted to two opposite forces due to the viscosity and the pressure gradient. Using Equation (4.67) at $\omega = 0$ gives

$$\bar{v} = \frac{R^2}{8\eta}\left(-\frac{\partial p}{\partial x}\right) \tag{4.77}$$

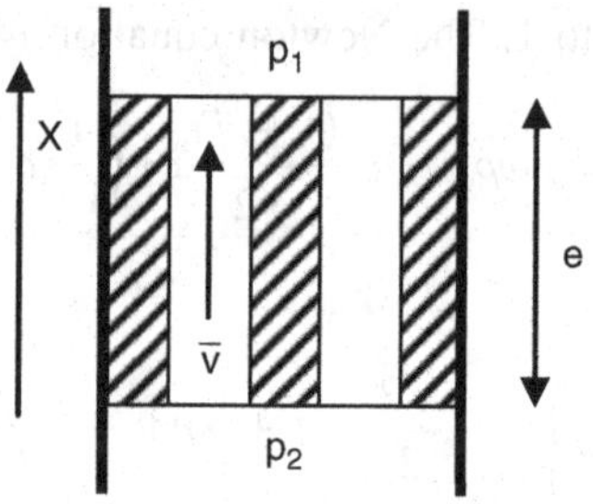

Figure 4.8 A sample of porous material of thickness e and of unit cross-section, submitted to a constant differential pressure $p_2 - p_1$. The mean molecular velocity in the pores is $\bar{v}$.

The flow resistivity σ is given by

$$\sigma = \frac{8\eta}{R^2(n\pi R^2)} \tag{4.78}$$

The quantity $n\pi R^2$ is the porosity ϕ of the material, and Equation (4.78) may be rewritten

$$\sigma = \frac{8\eta}{R^2\phi} \tag{4.79}$$

Using Equation (4.79), Equation (4.16) can be rewritten

$$s = \left(\frac{8\omega\rho_o}{\sigma\phi}\right)^{1/2} \tag{4.80}$$

Equations (4.18), (4.48) and (4.80) can be used to calculate ρ and K at a given angular frequency ω in the porous material represented in Figure 4.8. The description of the viscous force in the Newton equation (4.19) can be modified in the following way. The quantity $\rho_o j\omega/s$ is given by

$$\rho_o j\omega/s = -\sigma\phi s(\sqrt{-j})^2/8 \tag{4.81}$$

Substituting the right-hand side of Equation (4.81) for $\rho_o j\omega/s$ in Equation (4.19) gives

$$-\frac{\partial p}{\partial x} = j\omega\rho_o\bar{v} + \sigma\phi\bar{v}G_c(s) \tag{4.82}$$

where $G_c(s)$ is given by

$$G_c(s) = -\frac{s}{4}\sqrt{-j}\frac{J_1(s\sqrt{-j})}{J_o(s\sqrt{-j})}\left[1 - \frac{2}{s\sqrt{-j}}\frac{J_1(s\sqrt{-j})}{J_o(s\sqrt{-j})}\right] \tag{4.83}$$

The limit at low frequencies of $G_c(s)$ is 1 at $\omega = 0$, and Eq. (4.82) becomes

$$-\frac{\partial p}{\partial x} = \sigma\phi\bar{v} \tag{4.84}$$

This last equation describes the static measurement of flow resistivity as represented in Figure 2.4.

At the first-order approximation in $1/\omega$, Equations (4.67) and (4.69) yield

$$\rho = \frac{4}{3}\rho_0 + \frac{\sigma\phi}{j\omega} \tag{4.85}$$

$$K = \frac{\gamma P_0}{\gamma - (\gamma - 1)/\left(\frac{4}{3} + \frac{\sigma\phi}{jB^2\omega\rho_0}\right)} \tag{4.86}$$

In the high-frequency range, Equations (4.61) and (4.63) yield

$$\rho = \rho_0\left(1 + \sqrt{2/j}\frac{\delta}{R}\right) \tag{4.87}$$

$$K = \frac{\gamma P_0}{\gamma - (\gamma - 1)/\left(1 + \sqrt{2/j}\frac{\delta}{BR}\right)} \tag{4.88}$$

where $\sqrt{2/\mathrm{j}} = 1 - \mathrm{j}$ and δ is the viscous skin depth given by Equation (4.58).

4.6.2 Effective density and bulk modulus in slits

For the material represented in Figure 4.8, the circular pores are now replaced by slits. Equation (4.76) becomes

$$\sigma = \frac{p_2 - p_1}{2na\bar{v}e} = \frac{p_2 - p_1}{\phi\bar{v}e} \tag{4.89}$$

Using Equation (4.74) at $\omega = 0$ gives

$$\bar{v} = \frac{a^2}{3\eta}\left(-\frac{\partial p}{\partial x}\right) \tag{4.90}$$

The flow resistivity σ is given by

$$\sigma = \frac{3\eta}{\phi a^2} \tag{4.91}$$

Using Equation (4.91), Equation (4.25) can be rewritten as

$$s' = \left(\frac{3\omega\rho_o}{\phi\sigma}\right)^{1/2} \tag{4.92}$$

Equations (4.27) and (4.92) can be used to calculate ρ and K. The description of the viscous force in the Newton equation (4.28) can be modified in the following way; the quantity $\rho_o\omega/s'$ is given by

$$\rho_o\omega/s' = \sigma\phi s'/3 \tag{4.93}$$

Substituting the right-hand side of Equation (4.93) for $\rho_o\omega/s'$ in Equation (4.28) yields

$$-\frac{\partial p}{\partial x_3} = j\,\omega\rho_o\bar{v} + \phi\bar{v}\sigma G_s(s') \tag{4.94}$$

where

$$G_s(s') = \frac{j^{1/2}s'\tanh(s'\,j^{1/2})}{3\left[1 - \dfrac{\tanh(s'\,j^{1/2})}{s'\,j^{1/2}}\right]} \tag{4.95}$$

The limiting value of $G_s(s')$ at $\omega = 0$ is 1 as in the previous case, and Equation (4.94) can be rewritten

$$-\frac{\partial p}{\partial x} = \sigma\phi\bar{v} \tag{4.96}$$

At the first-order approximation in $1/\omega$, Equations (4.74) and (4.75) yield

$$\rho = \frac{6}{5}\rho_0 + \frac{\sigma\phi}{j\,\omega} \tag{4.97}$$

$$K = \frac{\gamma P_0}{\gamma - (\gamma - 1)/\left(\dfrac{6}{5} + \dfrac{\sigma\phi}{jB^2\omega\rho_0}\right)} \tag{4.98}$$

In the high-frequency range, Equations (4.71) and (4.72) yield

$$\rho = \rho_0\left(1 + \sqrt{2/j}\,\frac{\delta}{2a}\right) \tag{4.99}$$

$$K = \frac{\gamma P_0}{\gamma - (\gamma - 1)/\left(1 + \sqrt{2/j}\,\dfrac{\delta}{2Ba}\right)} \tag{4.100}$$

4.6.3 High- and low-frequency limits of the effective density and the bulk modulus for pores of arbitrary cross-sectional shape

At low frequencies the limit of the effective density is at the 0 order approximation in ω, from the definition of flow resistivity, Equation (4.84)

$$\rho = \frac{\phi\sigma}{j\omega} \tag{4.101}$$

From Equation (4.46), the limit for the bulk modulus at the first-order approximation in ω is

$$K = \frac{\gamma P_0}{\gamma - j(\gamma - 1)\omega B^2\rho_0/(\phi\sigma)} \tag{4.102}$$

Both parameters ρ and K only depend on the flow resistivity in the previous equations. At high frequencies, when ω tends to infinity, the viscous skin depth tends to zero and the velocity is the velocity without viscosity, except in a small layer close to the surface of the cylinders. When the viscous skin depth is very thin compared with the smallest lateral dimension of the pores, the velocity distribution close to the surface of the pores is the same as for a plane surface. Le q be the distance from the surface to a point close to the surface. In Figure 4.3 $q = a - x_1$. The mean velocity component in the direction x_3 is given by

$$j\omega\rho_0\overline{v_3} = -\frac{\partial p}{\partial x_3}\frac{\int_S\{1-\exp[-q(1+j)/\delta]\}\,\mathrm{d}S}{S} \tag{4.103}$$

where $\mathrm{d}S$ is the infinitesimal area related to $\mathrm{d}q$, and S is the area of the cross-section. This expression is valid close to the surface, where the plane approximation is valid and $\mathrm{d}S = l\mathrm{d}q$, l being the perimeter of the pore, and far from the surface, where the exponential is negligible. The contribution of the exponential function to the integral is

$$-\int_S \exp[-q(1+j)/\delta]\,\mathrm{d}S = \frac{\delta}{1+j}l \tag{4.104}$$

where l is the perimeter of the pore. This leads to the following approximation for the effective density

$$\rho = \rho_0\left(1 + \delta\sqrt{2/j}\frac{l}{2S}\right) \tag{4.105}$$

which can be written in terms of the hydraulic radius $\bar{r} = 2S/l$

$$\rho = \rho_0\left(1 + \sqrt{2/j}\,\frac{\delta}{\bar{r}}\right) \tag{4.106}$$

From Equation (4.46), the related bulk modulus is given by

$$K = \frac{\gamma P_0}{\gamma - (\gamma-1)/\left(1 + \sqrt{2/j}\,\frac{\delta l}{2SB}\right)} \tag{4.107}$$

which can be rewritten

$$K = \gamma P_0\left[1 + (\gamma-1)\sqrt{2/j}\frac{\delta}{\bar{r}B}\right] \tag{4.108}$$

Both the high-frequency limits of the effective density and the bulk modulus depend only on the hydraulic radius.

4.7 The Biot model for rigid framed materials

4.7.1 Similarity between G_c and G_s

Biot (1956) has pointed out that $G_s(s')$ is very similar to $G_c(s)$ if s is taken as

$$s = \tfrac{4}{3}s' \tag{4.109}$$

This is easy to show at high frequencies, the quantities G_c and G_s being close to

$$G_c(s) = s\sqrt{j}/4 \tag{4.110}$$

$$G_s(s') = s'\sqrt{j}/3 \tag{4.111}$$

and at low frequencies, where s tends to zero and

$$G_c(s) = 1 + js^2/24 \tag{4.112}$$

$$G_s(s') = 1 + js'^2/15 \tag{4.113}$$

Figure 4.9 compares $G_c(s)$ and $G_s(s')$ when $s' = \frac{3}{4}s$. It appears from Figure 4.9 that this property is valid for the entire range of variation of s and s'. Substituting for s' in Equation (4.109) the expression of s' given by Equation (4.92) yields

$$s = \sqrt{\frac{2}{3}}\left(\frac{8\omega\rho_o}{\sigma\phi}\right)^{1/2} \tag{4.114}$$

σ being the flow resistivity of the porous material having rectangular slits. The quantity $G_s(s')$ can be replaced by $G_c(s)$, s being given by Equation (4.114).

For a given value of σ and ϕ, Equation (4.82), which is valid for circular pores, can also be used in the case of slits, but Equation (4.80) needs to be modified, since s must be expressed as

$$s = c\left(\frac{8\omega\rho_o}{\sigma\phi}\right)^{1/2} \tag{4.115}$$

where

$$c = (\tfrac{2}{3})^{1/2} \tag{4.116}$$

4.7.2 Bulk modulus of the air in slits

Biot did not proceed to obtain a prediction for the frequency dependence of the bulk modulus. It is now easy to complete his model by using Equations (4.45) and (4.46).

The bulk modulus can be calculated, for the case of slits, in the following way. From Equation (4.82) the effective density ρ is

$$\rho = \rho_o\left[1 + \frac{\sigma\phi}{j\,\omega\rho_o}G_c(s)\right] \tag{4.117}$$

s being given by Equation (4.115) with $c = (\frac{2}{3})^{1/2}$. Using Equation (4.115), ρ can be rewritten

$$\rho = \rho_o\left[1 + \frac{8c^2}{js^2}G_c(s)\right] \tag{4.118}$$

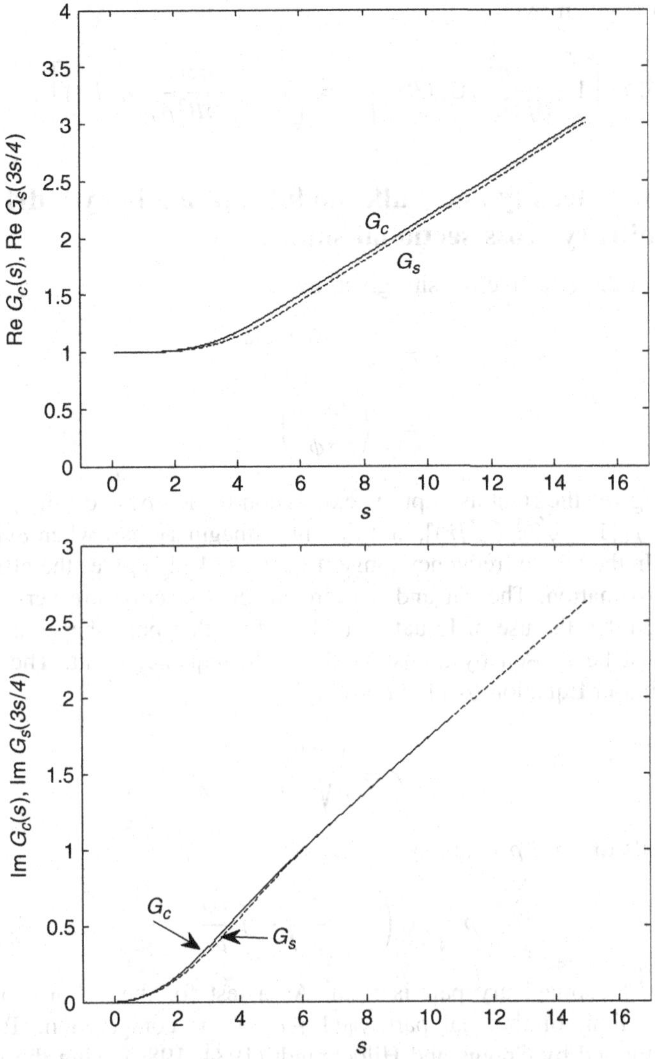

Figure 4.9 The form factors $G_c(s)$ and $G_s(s')$ for $s' = 3s/4$.

The function F in Equation (4.45) is equal to ρ_o/ρ. It follows from the two previous equations that

$$F(\omega) = \left[1 + \frac{8c^2}{js^2} G_c(s)\right]^{-1} \tag{4.119}$$

From Equation (4.46), the bulk modulus K is

$$K = \gamma P_o / (\gamma - (\gamma - 1) F(B^2 \omega)) \tag{4.120}$$

where $F(B^2\omega)$ is given by

$$F(B^2\omega) = \left[1 + \frac{8c^2}{js^2B^2}G_c(Bs)\right]^{-1} = \left[1 + \frac{\sigma\phi}{j\,\omega B^2\rho_o}G_c(Bs)\right]^{-1} \tag{4.121}$$

4.7.3 Effective density and bulk modulus of air in cylindrical pores of arbitrary cross-sectional shape

Using for the slit the effective density given by

$$\begin{aligned} \rho &= \rho_0 + \frac{\phi\sigma G_c(s)}{j\omega} \\ s &= c\left(\frac{8\omega\rho_0}{\sigma\phi}\right)^{1/2} \end{aligned} \tag{4.122}$$

with $c = \sqrt{2/3}$ gives the right asymptotic expression for $\mathrm{Re}(\rho)$ and $\mathrm{Im}(\rho)$ when ω tends to infinity, $\rho = \rho_0[1 + \sqrt{2}\delta/(\sqrt{j}\bar{r})]$, and the right imaginary part when ω tends to zero, $\rho = \phi\sigma/(j\omega)$. In the whole frequency range, Equation (4.117) gives the effective density to a good approximation. The slit and the circular cross-section are very different, and it can be guessed that the use of Equation (4.117) for other pore shapes is possible. The parameter c must be chosen by adjusting the high-frequency limit. The right limit is obtained by using in Equation (4.115) c satisfying

$$\bar{r} = \frac{1}{c}\sqrt{\frac{8\eta}{\sigma\phi}} \tag{4.123}$$

When ω tends to zero, ρ tends to

$$\rho = \rho_0\left(1 + \frac{c^2}{3}\right) - j\frac{\sigma\phi}{\omega} \tag{4.124}$$

The limit of the imaginary part is right. As a test for the validity of the general formulation, the limit of the real part, $\rho_0(1 + c^2/3)$, is compared in Table 4.1 with evaluations performed by Craggs and Hildebrandt (1984, 1986) using the finite element method.

It appears that Equations (4.123)–(4.124) can be used to predict the limit $\mathrm{Re}(\rho/\rho_0)$ when $\omega \to 0$ with a good precision for the cross-sectional shapes considered.

Table 4.1 Flow resistivity as a function of the hydraulic radius, c obtained from Equation (4.123), $(1 + c^2/3)$, and $\mathrm{Re}(\rho/\rho_0)$ obtained with the finite element method.

Cross-sectional shape	$\sigma\phi$	c	$1 + c^2/3$	$\mathrm{Re}(\rho/\rho_0)$
Circle	$8\eta/\bar{r}^2$	1	1.33	1.33
Square	$7\eta/\bar{r}^2$	1.07	1.38	1.38
Equilateral triangle	$6.5\eta/\bar{r}^2$	1.11	1.41	1.44
Rectangular slit	$12\eta/\bar{r}^2$	0.81	1.22	1.2

In conclusion, for a porous material with porosity ϕ and flow resistivity σ, where pores are identical parallel holes normal to the surface, the following expressions for the effective mass, and the compressibility of air in the material, can be used:

$$\rho = \rho_o\left(1 + \frac{\sigma\phi}{j\,\omega\rho_o}G_c(s)\right) \tag{4.125}$$

$$K = \gamma P_o/[\gamma - [(\gamma - 1)F(B^2\omega)]] \tag{4.126}$$

where

$$F(B^2\omega) = 1\bigg/\left[1 + \frac{\sigma\phi}{jB^2\omega\rho_o}G_c(Bs)\right] \tag{4.127}$$

and

$$G_c(s) = -\frac{s}{4}\sqrt{-j}\,\frac{J_1(s\sqrt{-j})}{J_o(s\sqrt{-j})}\bigg/\left[1 - \frac{2}{s\sqrt{-j}}\frac{J_1(s\sqrt{-j})}{J_o(s\sqrt{-j})}\right] \tag{4.128}$$

In Equation (4.128), s is equal to

$$s = c\left(\frac{8\omega\rho_o}{\sigma\phi}\right)^{1/2} \tag{4.129}$$

and c depends on the hydraulic radius, $c = (8\eta/(\sigma\phi))^{1/2}/\bar{r}$. Table 4.1 gives the values of c for various cross-sections.

4.8 Impedance of a layer with identical pores perpendicular to the surface

4.8.1 Normal incidence

The layer of porous material represented in Figure 4.10 is placed on an impervious rigid floor and is in contact with air in which a normal plane acoustic field is present. The acoustic field in the air is plane up to a small distance e from the material and identical waves propagate in each pore. The distance e is smaller than the distance between two pores, and can be neglected for the more common acoustic materials. The porous material of Figure 4.10 is represented on a macroscopic scale in Figure 4.11.

Two points M_1 and M_2 are selected at the surface of the material, M_2 in free air and M_1 in the porous material. Let $v(M_1)$ be the mean velocity of the air in a pore close to the surface and $v(M_2)$ the velocity in free air in the plane field, the distance e being neglected. The pressures $p(M_2)$ and $p(M_1)$ are the pressures in free air and in a pore, respectively. The continuity of air flow and pressure at the surface of the porous material implies the following two equations:

$$p(M_2) = p(M_1) \tag{4.130}$$

$$v(M_2) = v(M_1)\phi \tag{4.131}$$

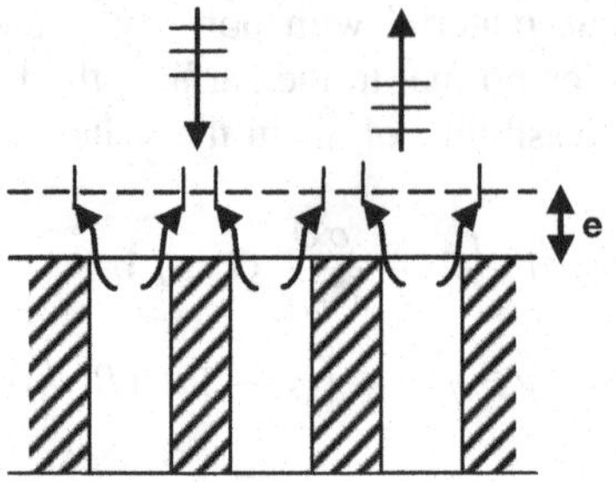

Figure 4.10 A layer of porous material with identical parallel pores perpendicular to the surface in a normal plane acoustic field. Arrows represent the trajectories of air molecules.

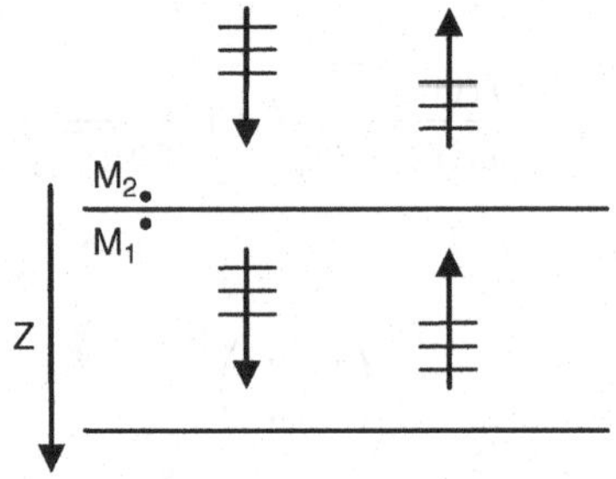

Figure 4.11 A simplified representation of the porous material of Figure 4.10.

Two impedances, $Z(M_2)$ in free air and $Z(M_1)$ in a pore, can be defined at the surface

$$Z(M_2) = p(M_2)/v(M_2), \qquad Z(M_1) = p(M_1)/v(M_1) \tag{4.132}$$

related by

$$\phi Z(M_2) = Z(M_1) \tag{4.133}$$

The wave equation in a pore is

$$K\frac{\partial^2 \bar{v}}{\partial z^2} = \rho \frac{\partial^2 \bar{v}}{\partial t^2} \tag{4.134}$$

The bulk modulus K and density ρ are given by Eqs. (4.125)–(4.129). The characteristic impedance Z_c and the complex wave number k in a pore can be calculated by the equation

$$Z_c = (K\rho)^{1/2}, \qquad k = \omega(\rho/K)^{1/2} \tag{4.135}$$

The impedance $Z(M_1)$ may be calculated by Equation (2.17)

$$Z(M_1) = -jZ_c \operatorname{cotg} kd \tag{4.136}$$

and $Z(M_2)$ is equal to

$$Z(M_2) = -j\frac{Z_c}{\phi}\,\mathrm{cotg}\,kd \tag{4.137}$$

where in both equations d is the thickness of the material.

4.8.2 Oblique incidence – locally reacting materials

Equations (4.130)–(4.137) are still valid for the case of oblique incidence. The impedance $Z(M_2)$ is given by Equation (4.137) and does not depend on the angle of incidence. Sound propagation in each pore depends only on the pressure of air above the pore, and the material is a locally reacting material. If the pores were interconnected, the surface impedance could depend on the angle of incidence, because of the interference between waves out of phase inside the material. An isotropic porous material, where pores are isotropically distributed and connected, may be represented by an equivalent isotropic fluid. It has been pointed out previously, in Section 3.4.2, that the impedance of a layer of fluid depends weakly on the angle of incidence if the velocity in the porous material is much smaller than that in air, and the transmitted wave propagates perpendicular to the surface. In such a case, the material is also called a locally reacting material.

4.9 Tortuosity and flow resistivity in a simple anisotropic material

In this section, the concept of tortuosity will be presented via a simple example. A porous layer having pores of radius R lying in two directions symmetrical with respect to the normal of the surface is represented in Figure 4.12. The thickness of the layer is d, the length of the pores is l, and the differential pressure is $p_2 - p_1$. The velocities in the two pores, averaged over the cross-sections, are $\bar{v}_1$ and $\bar{v}_2$.

With n pores per unit area of surface, the porosity ϕ is given by

$$\phi = \frac{n\pi R^2}{\cos\varphi} \tag{4.138}$$

where φ is the angle between the axes of the pores and the surface normal. The pressure gradient in the pores is

$$\frac{p_2 - p_1}{l} = \frac{p_2 - p_1}{d}\cos\varphi \tag{4.139}$$

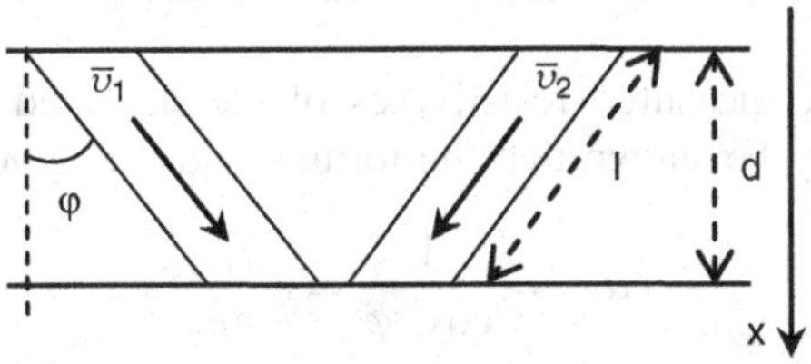

Figure 4.12 A porous material with pores of constant and equal radius, and oriented in two directions, symmetrical with respect to the normal of the surface.

Using Equation (4.77), the flow resistivity σ in the x direction is

$$\sigma = \frac{p_2 - p_1}{n(\bar{v}\pi R^2)\,\mathrm{d}} = \frac{8\eta}{n\pi R^2 \cos\varphi} \tag{4.140}$$

where $\bar{v}$ is the modulus of the average velocities $\bar{v}_1$ and $\bar{v}_2$. From Equation (4.138), σ can be rewritten

$$\sigma = \frac{8\eta}{\phi R^2 \cos^2\varphi} \tag{4.141}$$

The parameter s in Equation (4.128) is given by Equation (4.129). Eliminating η/R^2 between Equations (4.16) and (4.141) we find

$$s = \left(\frac{8\omega\rho_o}{\sigma\phi\cos^2\varphi}\right)^{1/2} \tag{4.142}$$

The quantity $1/\cos^2\varphi$ is the tortuosity of the material. The concept of tortuosity is not recent, and it appears with different notation and meanings in previous works. For Carman (1956) the tortuosity is related to $(\cos\varphi)^{-1}$ and in the book by Zwikker and Kosten (1949), tortuosity is denoted by k_s and is called the structure form factor. In latter works, tortuosity is denoted as α_∞

$$\alpha_\infty = \frac{1}{\cos^2\varphi} \tag{4.143}$$

Using α_∞ instead of $(\cos\varphi)^{-2}$, σ and s can be rewritten

$$\sigma = \frac{8\eta\alpha_\infty}{\phi R^2} \tag{4.144}$$

$$s = \left(\frac{8\omega\rho_o\alpha_\infty}{\sigma\phi}\right)^{1/2} \tag{4.145}$$

In the formalism of the Biot theory (Biot 1956) that will be presented in Chapter 6, the tortuosity is related to an inertial coupling term ρ_a by

$$\rho_a = \rho_o\phi(\alpha_\infty - 1) \tag{4.146}$$

A method for measuring tortuosity exists, which, however, can only be used if the frame does not conduct electricity. The porous material is saturated with a conducting fluid, and the resistivity of the saturated material is then measured, as indicated in Figure 4.13.

Let r_c and r_f be the measured resistivities of the saturated material and the fluid, respectively. It can easily be shown that the tortuosity α_∞ is given by

$$\alpha_\infty = \frac{1}{\cos^2\varphi} = \phi\frac{r_c}{r_f} \tag{4.147}$$

which is independent of the shape of the cross-sections of the pores. As shown by Brown (1980) the relation $\alpha_\infty = \phi r_c/r_f$ can be generalized to any porous medium. Another

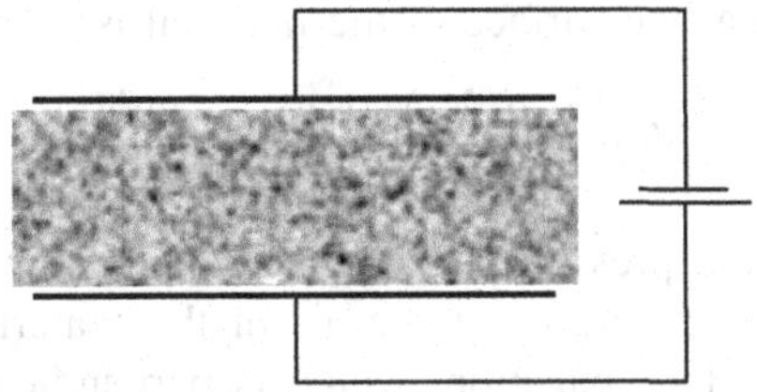

Figure 4.13 Measurement of tortuosity. The material is saturated with a conducting fluid and the resistivity of the material is measured between the two electrodes.

method of measuring the tortuosity uses the time of flight of an ultrasound pulse across the material. This method will be discussed in Chapter 5. The concept of tortuosity is generalized in Section 5.3 for the case of materials having non-cylindrical pores.

4.10 Impedance at normal incidence and sound propagation in oblique pores

4.10.1 Effective density

The material represented in Figure 4.14 is placed on an impervious rigid wall. The acoustic field in the air above the material is plane and normal to the surface.

As shown in Figure 4.14, there are two microscopic directions of propagation parallel to the two symmetrical directions of the pores. The macroscopic direction of propagation is x and only the x components $v(x)$ of the microscopic velocities $\bar{v}_1$ and $\bar{v}_2$ must be taken into consideration at the macroscopic level:

$$v(x) = \bar{v}_1(x)\cos\varphi = \bar{v}_2(x)\cos\varphi \tag{4.148}$$

The flux of air V through a unit surface area perpendicular to the x axis is given by

$$V(x) = v(x)n\pi R^2/\cos\varphi \tag{4.149}$$

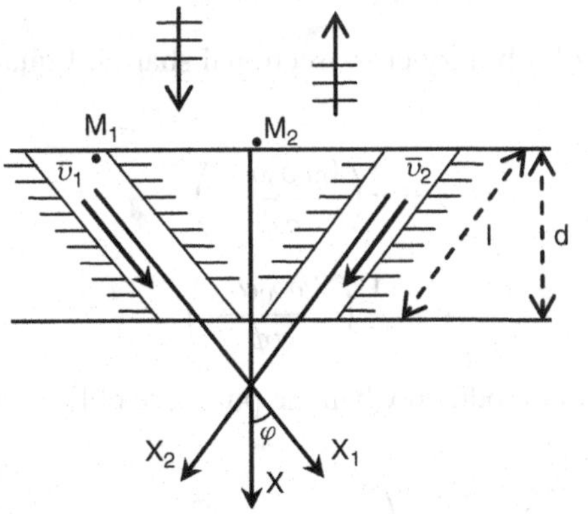

Figure 4.14 Porous material with oblique pores in a normal plane acoustic field and set upon an impervious rigid wall.

The impedance $Z(M_2)$ at the surface of the material is given by

$$Z(M_2) = \frac{p(M_1)\cos\varphi}{\upsilon(M_1)n\pi R^2} = \frac{p(M_1)}{\phi\upsilon(M_1)} \tag{4.150}$$

$p(M_1)$ and $\upsilon(M_1)$ being the pressure and the velocity of air in a pore at the surface of the material. The effective density of the air in the material must be calculated in the macroscopic direction of propagation, which is perpendicular to the surface of the layer. In the case of circular cross-sectional shape, the Newton equation (4.19) in the microscopic directions of propagation x_1 and x_2 can be rewritten, with Equation (4.128)

$$-\frac{\partial p(x)}{\partial x_i} = j\omega\rho_o\bar{\upsilon}_i(x) + \frac{8\rho_0\omega}{s^2}G_c(s)\bar{\upsilon}_i(x) \qquad i = 1, 2 \tag{4.151}$$

where G_c is given by Equation (4.128) and s by Equation (4.145).

In the x direction, by the use of Equations (4.141), (4.145) and

$$-\frac{\partial p(x)}{\partial x} = -\frac{1}{\cos\varphi}\frac{\partial p(x)}{\partial x_i} \tag{4.152}$$

Equation (4.151) can be rewritten in the macroscopic direction of propagation

$$-\frac{\partial p(x)}{\partial x} = j\omega\rho_o\alpha_\infty\upsilon(x) + \sigma\phi G_c(s)\upsilon(x) \tag{4.153}$$

where $\upsilon(x)$ is the velocity component in a pore in the macroscopic direction of propagation. The effective density can be written

$$\rho = \alpha_\infty\rho_0\left(1 + \frac{\sigma\phi G_c(s)}{j\omega\alpha_\infty\rho_0}\right) \tag{4.154}$$

The high-frequency limit of ρ is now

$$\rho = \alpha_\infty\rho_0(1 + \sqrt{2/j}\delta/R) \tag{4.155}$$

It may be shown that Eq. (4.155) is valid for all cross-sectional shapes if R is replaced by the hydaulic radius

$$\rho = \alpha_\infty\rho_0(1 + \sqrt{2/j}\delta/\bar{r}) \tag{4.156}$$

For the general case of arbitrary cross-sectional shapes, Equation (4.151) can be used with s given by

$$s = c\left(\frac{8\omega\rho_o\alpha_\infty}{\sigma\phi}\right)^{1/2} \tag{4.157}$$

$$c = \frac{1}{\bar{r}}\left(\frac{8\omega\rho_0\alpha_\infty}{\sigma\phi}\right)^{1/2} \tag{4.158}$$

The bulk modulus is not modified when the pores are oblique and is given by Equation (4.126) with

$$F(B^2\omega) = 1\bigg/\left[1 + \frac{\sigma\phi}{jB^2\omega\rho_o\alpha_\infty}G_c(Bs)\right] \tag{4.159}$$

4.10.2 Impedance

The quantities K and ρ having been evaluated, the impedance $Z(M_2)$ at normal incidence can be calculated in the same way as in the case where the pores are perpendicular to the surface. This impedance is given by

$$Z(M_2) = -j\frac{Z_c}{\phi}\operatorname{cotg} kd \tag{4.160}$$

Z_c and k being respectively

$$Z_c = (K\rho)^{1/2} \tag{4.161}$$

$$k = \omega(\rho/K)^{1/2} \tag{4.162}$$

Appendix 4.A Important Expressions

Description on the microscopic scale

Relations between, pressure, velocity, temperature, Eqs. (4.1)–(4.4)

$\rho_0\frac{\partial \boldsymbol{v}}{\partial t} = -\nabla p + \eta\Delta\boldsymbol{v}$, with the boundary condition at the air-frame interface $\boldsymbol{v} = \boldsymbol{0}$

$\rho_0 c_p\frac{\partial \tau}{\partial t} = \kappa\nabla^2\tau + \frac{\partial p}{\partial t}$, with the boundary condition at the air-frame interface $\tau = 0$.

State equation, Eq. (4.29)

$$p = \frac{P_o}{\rho_o T_o}[\rho_o\tau + T_o\xi]$$

Effective density and bulk modulus

Effective density,

$$\rho = -\frac{\partial p}{\partial x_3}\frac{1}{j\omega\bar{v}_3}, \quad \text{Equation (4.38)}$$

Bulk modulus

$$K = -p/\bar{\nabla}\boldsymbol{u}, \quad \text{Equation (4.39)}$$

Connection between the bulk modulus and the effective density

$$\rho = \rho_o/F(\omega), \quad \text{Equation (4.45)}$$

$$K = \gamma P_o/[\gamma - (\gamma - 1)F(B^2\omega)], \quad \text{Equation (4.46)}$$

Low-frequency limit:

$$\rho = \frac{\phi\sigma}{j\omega} + cte, \quad \text{Equation (4.85), (4.97)}$$

$$K = \frac{\gamma P_0}{\gamma - j(\gamma - 1)\omega B^2\rho_0/(\phi\sigma)}, \quad \text{Equation (4.102)}$$

High-frequency limit:

$$\rho = \alpha_\infty \rho_0 \left(1 + \sqrt{2/j}\frac{\delta}{\bar{r}}\right), \text{ Equation (4.106)}$$

$$K = \gamma P_0 \left[1 + (\gamma - 1)\sqrt{2/j}\frac{\delta}{\bar{r}B}\right], \text{ Equation (4.108)}$$

References

Biot, M.A., (1956) Theory of propagation of elastic waves in a fluid-saturated porous solid. I. Low frequency range, II. Higher frequency range. *J. Acoust. Soc. Amer.*, **28**, 168–91.

Brown, R.J.S., (1980) Connection between formation factor for electrical resistivity and fluid–solid coupling factor in Biot's equations for acoustic waves in fluid-filled porous media. *Geophysics*, **45**, 1269–75.

Carman, P.C., (1956) *Flow of Gases Through Porous Media*. Butterworths, London, 1956.

Craggs, A. and Hildebrandt, J.G. (1984) Effective densities and resistivities for acoustic propagation in narrow tubes. *J. Sound Vib.*, **92**, 321–31.

Craggs, A. and Hildebrandt, J.G. (1986) The normal incidence absorption coefficient of a matrix of narrow tubes with constant cross-section. *J. Sound Vib.*, **105**, 101–7.

Gray, D.E., ed., (1957) *American Institute of Physics Handbook*, McGraw-Hill, New York.

Hubner, K.H., (1974) *The Finite Element Method for Engineers*. Wiley-Interscience, New York.

Kergomard, J., (1981) Champ interne et champ externe des instruments à vent, *Thèse* Université de Paris VI.

Kirchhoff, G., (1868) Uber der Einfluss der Wärmeleitung in einem Gase auf die Schallbewegung. *Annalen der Physik and Chemie*, **134**, 177–93.

Stinson, M.R., (1991) The propagation of plane sound waves in narrow and wide circular tubes, and generalization to uniform tubes of arbitrary cross-sectional shape. *J. Acoust. Soc. Amer.*, **89**, 550–8.

Tijdeman, H., (1975) On the propagation of sound waves in cylindrical tubes. *J. Sound Vib.*, **39**, 1–33.

Zienkiewicz, O.C., (1971) *The Finite Element Method in Engineering Science*. McGraw-Hill, London.

Zwikker, C. & Kosten, C.W. (1949) *Sound Absorbing Materials*. Elsevier, New York.

5

Sound propagation in porous materials having a rigid frame

5.1 Introduction

In Chapter 4, porous materials with cylindrical pores were considered. In the case of common porous materials, a similar analytical description of sound propagation that takes into account the complete geometry of the microstructure is not possible. This explains why the models of sound propagation in these materials are mostly phenomenological and provide a description only on a large scale. A review of the models worked out before 1980 can be found in the work by Attenborough (1982). Many models have been presented since 1980. Moreover direct time-domain analysis has brought new tools for the modelling and the measurements of these materials (Carcione and Quiroga-Goode 1996, Fellah *et al.* 2003). In order to give a physical basis to the description of sound propagation in porous media, we have selected in the frequency domain a series of semi-phenomenological models involving several physical parameters. A brief description of methods used to measure these parameters is given to clarify their physical nature. As for the case of cylindrical pores, the air in the porous frame is replaced by an equivalent fluid that presents the same bulk modulus K as the saturating air and a complex density ρ that takes into account the viscous and the inertial interaction with the frame. As in Chapter 4, the wave number $k = \omega(\rho/K)^{1/2}$ and the characteristic impedance $Z_c = (\rho K)^{1/2}$ can describe the acoustical properties of the medium. A detailed description of the conditions where an equivalent fluid can be used has been given by Lafarge (2006). The main condition is the long-wavelength condition. The wavelength is much larger than the characteristic dimensions of the pores, and the saturating fluid can behave as an incompressible fluid at the microscopic scale. At the end of the chapter, the homogenization method for periodic structures introduced by Sanchez-Palencia (1974, 1980), Keller (1977), and Bensoussan *et al.* (1978), is presented with the dimensionless analysis developed by Auriault (1991). Real porous media are generally not periodic. However,

Propagation of Sound in Porous Media: Modelling Sound Absorbing Materials, Second Edition J. F. Allard and N. Atalla

as indicated by Auriault (2005), a random medium and a periodic medium built with a characteristic cell of the random medium present similar properties at the macroscopic scale. Under the long-wavelength condition, different steps which justify the use of an equivalent fluid are presented. Some properties of the double porosity media obtained with the homogenization method by Olny and Boutin (2003), Boutin *et al.* (1998) and Auriault and Boutin (1994) are summarized.

5.2 Viscous and thermal dynamic and static permeability

5.2.1 Definitions

Viscous permeability

More rigorously the viscous permeability should be defined as the visco-inertial permeability. The viscous dynamic permeability q has been defined by Johnson *et al.* (1987). The viscous dynamic permeability is a complex parameter that relates the pressure gradient and the fluid velocity $\boldsymbol{v}$ in an isotropic porous medium by

$$-q(\omega)\nabla p = \eta\phi < \boldsymbol{v} > \tag{5.1}$$

where η is the viscosity and ϕ is the porosity. The symbol $< >$ denotes an average over the fluid part Ω_f of a representative elementary volume Ω. This leads to (see Equation 4.38)

$$q(\omega) = \frac{\eta\phi}{j\omega\rho(\omega)}, \tag{5.2}$$

where ρ is the effective density. From the definition of the flow resisivity σ, the limit q_0 of q when ω tends to zero is

$$q_0 = \frac{\eta}{\sigma} \tag{5.3}$$

The static viscous permeability q_0 is a intrinsic parameter depending only on the microgeometry of the porous frame.

Thermal permeability

The dynamic thermal permeability q' has been defined by Lafarge (1993). The thermal permeability is a complex parameter that relates the pressure time derivative to the mean temperature by

$$q'(\omega)j\omega p = \phi\kappa < \tau > \tag{5.4}$$

where κ is the thermal conductivity. The use of the same denomination for the thermal and the viscous parameters is justified by the analogy which appears at the microscopic scale between Equations (4.1), (4.3) and Equations (4.2), (4.4) which are valid under the long-wavelength condition. In Equation (4.1) the source term is $-\nabla p$ and in Equation (4.2) the source term is $\partial p/\partial t$. In Equation (4.2) the thermal inertia $\rho_0 c_p$ replaces the density ρ_0 in Equation (4.1). Moreover, v and τ are equal to 0 at the air–frame contact surface. The condition $\tau = 0$ at the air–frame contact surface is due to the fact that the density of the porous media used for sound absorption and damping is generally much heavier than air, and the thermal exchanges with the saturating air do not modify the temperature of the frame. This condition must be fulfilled for the previous description of the bulk modulus to be valid. The bulk modulus of the saturating air depends on the thermal permeability q' via the averaged density. From Equation (4.40), the averaged density $<\xi>$ is given by

$$<\xi> = \frac{\rho_0}{P_0}p - \frac{\rho_0}{T_0}\frac{q'(\omega)j\omega p}{\phi\kappa} \tag{5.5}$$

From Equation (4.39), and using $P_0/T_0 = \rho_0(c_p - c_v)$, the bulk modulus of the saturating air can be written

$$K(\omega) = P_0 \Big/ \left[1 - \frac{\gamma - 1}{\gamma}\frac{j\mathrm{B}^2\omega\rho_0 q'(\omega)}{\phi\eta}\right] \tag{5.6}$$

where B^2 is the Prandtl number. When ω tends to zero, q' tends to the static thermal permeability q'_0. It has been shown by Torquato (1990) that $q'_0 \geq q_0$. A comparison of Equations (5.6) and (4.102) shows that both parameters are equal for identical cylindrical pores parallel to the direction of propagation. From Equations (4.45)–(4.46) and Equation (5.6) it can be shown that $q'(\omega) = q(B^2\omega)$ for identical cylindrical pores. When the density of the frame has the same order of magnitude as the air, the isothermal limit for the bulk modulus cannot be reached (Lafarge *et al.* 1997), and the previous description must be modified.

Static permeabilities and low frequency limits of ρ and K

The limit of the ratio $\rho/[\phi\sigma/(j\omega)]$ is 1 when $\omega \to 0$ as shown in Chapter 4 for cylindrical pores. Norris (1986) has shown that this relation was always valid. More precisely, the limit when $\omega \to 0$ can be written

$$\rho(\omega) = \frac{\eta\phi}{j\omega q_0} + cte \tag{5.7}$$

The nature of the constant is shown explicitly in Section (5.3.6).

The limit of the bulk modulus at the first order approximation in ω is obtained from Equation (5.6) which can be rewritten

$$K(\omega) = P_0\left[1 + \frac{\gamma - 1}{\gamma}\frac{jB^2\omega\rho_0 q'_0}{\phi\eta}\right] \tag{5.8}$$

This equation is the generalization of Equation (4.102) obtained for the case of parallel identical cylindrical pores. The static thermal permeability q'_0 is the limit when ω tends to zero of $C\,\mathrm{Im}K(\omega)$ where C is given by

$$C(\omega) = \frac{\gamma\phi\eta}{P_0(\gamma - 1)B^2\rho_0\omega} \tag{5.9}$$

It has also be shown by Lafarge (1993) that a physical intrinsic parameter, the trapping constant Γ of the porous structure which only depends on the geometry of the frame, is related to q_0' by

$$\Gamma = 1/q_0' \tag{5.10}$$

The evaluation of the dynamic thermal permeability of a porous frame can be performed by solving the diffusion-controlled trapping problem (Lafarge 2002, Perrot *et al.*, 2007). This is beyond the scope of this book.

A justification of the existence of the dynamic thermal and viscous permeability under the long-wavelength condition, and therefore of the use of an equivalent fluid, can be obtained by the homogenization method.

5.2.2 Direct measurement of the static permeabilities

The viscous static permeability can be evaluated from the flow resistivity measured with the techniques previously described. The thermal static permeability can be evaluated from measurements of the bulk modulus K at sufficiently low frequencies. At these frequencies the use of a Kundt tube generally does not provide precise results. A method developed by Tarnow to measure the compressibility of air in glass wools at low frequencies is described in Lafarge *et al.* (1997).

A short description of the method with simplified calculations is given in what follows. As shown in Figure 5.1, the sample is set in a long tube where a loudspeaker creates a plane field. A microphone measures the pressure around the $\lambda/4$ resonance, where the pressure is close to zero. Frequencies as low as 25 Hz can be reached with a tube longer than 3.5 m. The surface impedance Z_s is given by Equation (4.137),

$$Z_s = -\frac{jZ_c}{\phi}\cot(k_1 l).$$

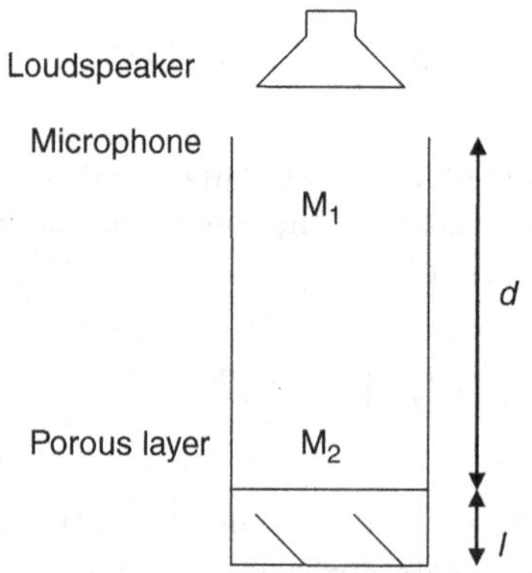

Figure 5.1 The experimental setup. The distance from the porous layer to the microphone is d and the thickness of the layer is l.

At these frequencies, for samples of thickness l around 3 cm or less, $\cot kl$ can be developed to first order in kl, and the impedance $Z(M_2)$ is given by

$$Z(M_2) = -\frac{jK}{\phi\omega l} \tag{5.11}$$

The reflection coefficient at M_2 is given by

$$R(M_2) = \frac{\mathrm{Re}Z(M_2) - Z_c + j\mathrm{Im}Z(M_2)}{\mathrm{Re}Z(M_2) + Z_c + j\mathrm{Im}Z(M_2)} \tag{5.12}$$

where Z_C is the characteristic impedance of the free air. This equation can be rewritten, under the conditions $|\mathrm{Im}Z(M_2)| \gg Z_c$ and $|\mathrm{Im}Z(M_2)| \gg \mathrm{Re}Z(M_2)$, which are fulfilled at sufficiently low frequency,

$$R(M_2) = \left[1 - 2\frac{Z_c\mathrm{Re}Z(M_2)}{\mathrm{Im}^2 Z(M_2)}\right]\exp(-j\varphi) \tag{5.13}$$

where φ is a small real angle, $\varphi = -2Z_c/\mathrm{Im}Z(M_2)$. The function in the square brackets in Equation (5.13) is a development at the first-order approximation of the modulus of R. If the losses out of the porous sample are neglected, an incident pressure amplitude p at M_1 corresponds to a total pressure p_T given by

$$p_T = p[1 + R(M_2)\exp(-2j\omega d/c)] \tag{5.14}$$

where c is the speed of sound. This equation can be rewritten

$$p_T = p\left\{1 + \left[1 - 2\frac{Z_c\mathrm{Re}Z(M_2)}{\mathrm{Im}^2 Z(M_2)}\right]\left[\cos\left(\varphi + 2\omega\frac{d}{c}\right) - j\sin\left(\varphi + 2\omega\frac{d}{c}\right)\right]\right\} \tag{5.15}$$

The minimum value for $|p_T|$ is given by

$$\min|p_T| = p\left|2\frac{Z_c\mathrm{Re}Z(M_2)}{\mathrm{Im}^2 Z(M_2)}\right| \tag{5.16}$$

and is obtained for ω satisfying the relation $\varphi + 2\omega d/c = \pi$. For small variation of ω around this value, $\cos(\varphi + 2\omega d/c)$ is stationary. The variation $\Delta\omega$ related to an increase of the amplitude $\min|p_T| \to \sqrt{2}\min|p_T|$ is given by

$$\Delta\omega\frac{2d}{c} = \left|2\frac{Z_c\mathrm{Re}Z(M_2)}{\mathrm{Im}^2 Z(M_2)}\right| \tag{5.17}$$

At sufficiently low frequency, $\mathrm{Re}K$ can be replaced by P_0 in Equation (5.11), and $\mathrm{Im}K$ is given by

$$\mathrm{Im}K = \frac{\Delta\omega}{\omega}\frac{dP_0}{2\phi l\gamma} \tag{5.18}$$

An example is presented in Figure 5.2 for layers of steel beads of thickness l ranging from 2 to 19 cm. The mean diameter of the beads is 1.5 mm, and the viscous static perme-

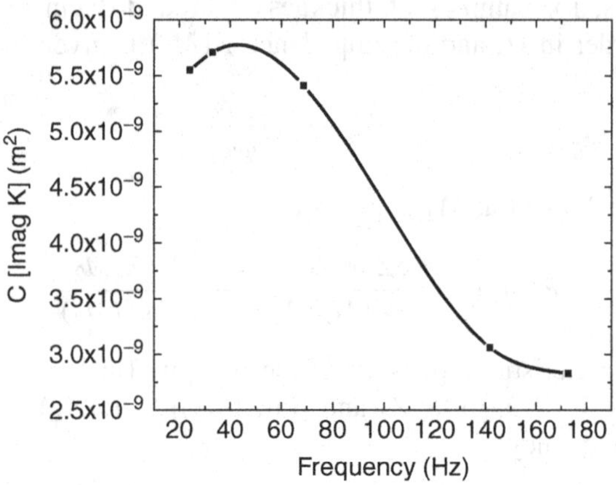

Figure 5.2 The measured quantity $C\text{Im}K$ which tends to the static thermal permeability when ω tends to 0 (Debray *et al.* 1997).

ability $q_0 = 1.510^{-9}\text{m}^2$ (the flow resistivity $\sigma = 12\,000$ N m^{-4} s). The predicted thermal permeability calculated from Figure 4 in Straley *et al.* (1987) for a similar medium is equal to $4.8\times10^{-9}\text{m}^2$.

5.3 Classical tortuosity, characteristic dimensions, quasi-static tortuosity

5.3.1 Classical tortuosity

Tortuosity, denoted as α_∞, has been precisely defined by Johnson *et al.* (1987). When a porous frame is saturated by an ideal nonviscous fluid, the effective density of the fluid is given by

$$\rho = \alpha_\infty \rho_0 \tag{5.19}$$

The apparent increase of the density can be explained in the following way. In a flow of nonviscous fluid, let us denote by $\boldsymbol{v}_\mathbf{m}(M)$ the microscopic velocity at M. The macroscopic velocity $\boldsymbol{v}(M_\mathbf{o})$ is obtained by averaging $\boldsymbol{v}_\mathbf{m}(M)$ over a representative elementary volume V around M_0

$$\boldsymbol{v}(M_0) = \langle \boldsymbol{v}_\mathbf{m}(M) \rangle_v \tag{5.20}$$

The tortuosity is defined by the relation

$$\alpha_\infty = \left\langle v_m^2(M) \right\rangle_v / v^2(M_o) \tag{5.21}$$

Per unit volume of saturating fluid, the kinetic energy E_c is given by

$$E_c = \frac{1}{2}\alpha_\infty \rho_0 v^2(M_0) \tag{5.22}$$

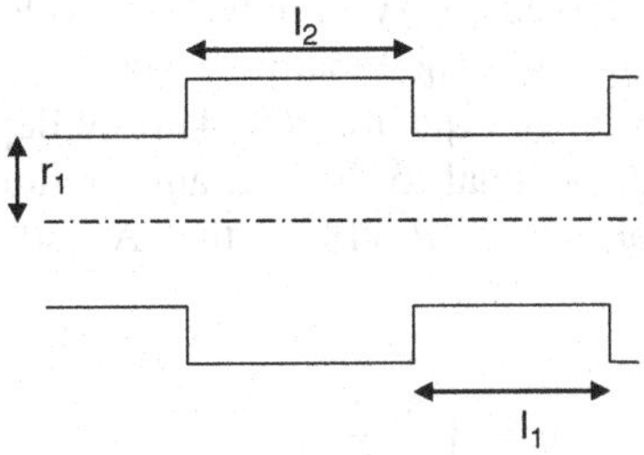

Figure 5.3 A pore made up of an alternating sequence of cylinders.

As far as the macroscopic velocity is considered, the nonviscous fluid must be replaced by a fluid of density $\rho_0\alpha_\infty$. The value of the tortuosity is an intrinsic property of the porous frame that depends on the micro-geometry. When the saturating fluid is viscous the effective density must tend to $\alpha_\infty\rho_0$ when the viscous skin depth tend to zero and the viscosity effects become negligible. For the material of Sections 4.9 and 4.10, $\alpha_\infty = 1/\cos^2\varphi$ in the direction perpendicular to the surface of the layer. A tortuosity larger than 1 is due to the dispersion of the microscopic velocity in Equation (5.21). This dispersion can be created by variations of the diameter of the pores. Let us consider, for instance, a material with identical pores parallel to the direction of propagation, made up of alternating cylinders represented in Figure 5.3, with lengths l_1 and l_2 and cross-sections S_1 and S_2, respectively. Even if the fluid is nonviscous, the description of the inertial forces and the evaluation of α_∞ by Equation (5.21) are very complicated at the junction of the two cylinders. A simple approximation is obtained by assuming constant velocities in each cylinder. It is shown in Appendix 5.A that with this approximation, α_∞ is given by

$$\alpha_\infty = \frac{(l_1S_2 + l_2S_1)(l_1S_1 + l_2S_2)}{(l_1 + l_2)^2 S_1S_2} \tag{5.23}$$

The evaluation of tortuosity can be performed from resistivity measurements, as indicated in Section 4.9. Another method simultaneously allowing the measurement of other high-frequency parameters is described at the end of Section 5.3.5.

5.3.2 Viscous characteristic length

For materials with cylindrical pores, as indicated by Equations (4.107) and (4.155), the high-frequency behaviour of ρ and K depends on tortuosity and on the hydraulic radius. The viscous characteristic length defined by Johnson *et al.* (1986) replaces the hydraulic radius for more general micro-geometries. Johnson *et al.* have defined the characteristic dimension Λ by

$$\frac{2}{\Lambda} = \frac{\int_A v_i^2(\boldsymbol{r}_w)\,\mathrm{d}A}{\int_V v_i^2(\boldsymbol{r})\,\mathrm{d}V} \tag{5.24}$$

For a static flow of nonviscous fluid in the porous structure, $v_i(\boldsymbol{r}_w)$ is the velocity of the fluid on the pore surface and the integral in the numerator is performed over the pore surfaces A in the representative elementary volume. The velocity $v_i(\mathbf{r})$ is the velocity inside the pores, the integral in the denominator is performed over the volume V of the pore. The parameter Λ given by Equation. (5.24) only depends on the geometry of the frame. With the factor 2, Λ is equal to the hydraulic radius for identical cylindrical pores. It has been noted by Johnson *et al.* (1986) that Λ and the flow resistivity σ are related by

$$\Lambda = \left(\frac{8\eta\alpha_\infty}{\sigma\phi}\right)^{1/2}\frac{1}{c} \tag{5.25}$$

with c close to 1. It has been shown by Johnson *et al.* (1987) that the effective density can be written at high frequencies at first-order approximation in $1/\sqrt{\omega}$

$$\rho = \alpha_\infty\rho_0\left[1 + (1-j)\frac{\delta}{\Lambda}\right] \tag{5.26}$$

A previous expression of the dynamic viscous permeability obtained by Auriault *et al.* (1985) for anisotropic media leads to a similar result where ρ is the sum of a real inertial term and a correction proportional to $1/\sqrt{\omega}$ due to viscosity.

5.3.3 Thermal characteristic length

It has been shown by Champoux and Allard (1991) that the high-frequency behaviour of the bulk modulus K can be characterized by a second length denoted as Λ' and given by

$$\frac{2}{\Lambda'} = \frac{\int_A \mathrm{d}A}{\int_V \mathrm{d}V} = \frac{A}{S} \tag{5.27}$$

The integral in the numerator is performed over the pore surfaces A in the elementary representative volume and the integral in the denominator is performed over the volume V of the pore, as for Equation (5.24), but there is no weighting by the squared velocity. As shown in Appendix 5.B, the bulk modulus K is given at high frequencies to first–order approximation in $1/\sqrt{\omega}$ by

$$K = \frac{\gamma P_o}{\gamma - (\gamma - 1)\left[1 - (1-j)\dfrac{\delta}{\Lambda' B}\right]} \tag{5.28}$$

For identical cylindrical pores, $\Lambda' = \Lambda = \bar{r}$. This is a direct consequence of the definition of these quantities and this can be verified by comparing Equation (4.156) with Equation (5.26), and Equation (4.108) with Equation (5.28).

5.3.4 Characteristic lengths for fibrous materials

As indicated in Chapter 2, fibres in layers of fibreglass generally lie in planes parallel to the surface of the layers. At normal incidence, the macroscopic air velocity is perpendicular to the direction of the fibres. The characteristic dimension Λ at normal incidence is calculated in Appendix 5.C. The fibres are modelled as infinitely long cylinders having a circular cross-section with radius R. In the case of materials with porosity close to 1, Λ is given by

$$\Lambda = \frac{1}{2\pi LR} \tag{5.29}$$

where L is the total length of fibres per unit volume of material. The characteristic thermal dimension, evaluated from Equation (5.27) is given by

$$\Lambda' = \frac{1}{\pi LR} = 2\Lambda \tag{5.30}$$

5.3.5 Direct measurement of the high-frequency parameters, classical tortuosity and characteristic lengths

In this present subsection, let n^2 be the squared ratio of the velocity in a free fluid to the velocity when it saturates a porous structure. In the high-frequency range, using Equation (5.26) for the effective density and Equation (5.28) for the bulk modulus, n^2 is given, to first-order approximation in $1/\sqrt{\omega}$ by

$$n^2 = \alpha_\infty \left[1 + \delta \left(\frac{1}{\Lambda} + \frac{\gamma - 1}{B\Lambda'} \right) \right] \tag{5.31}$$

A sketch of the experimental set-up for the measurement of n is represented in Figure 5.4.

The phase velocity is obtained by comparing the phase spectra at the receiver with and without the porous layer. In the domain of validity of Equation (5.31), n^2 is linearly dependent on $1/\sqrt{f}$. In Figure 5.5 the squared velocity ratio is represented as a function of the square root of the inverse of frequency.

The porous layer has been successively saturated with air and with helium. For helium, $(\gamma - 1)/B$ is close to 0.81 and for air it is close to 0.48. A comparison of both slopes in

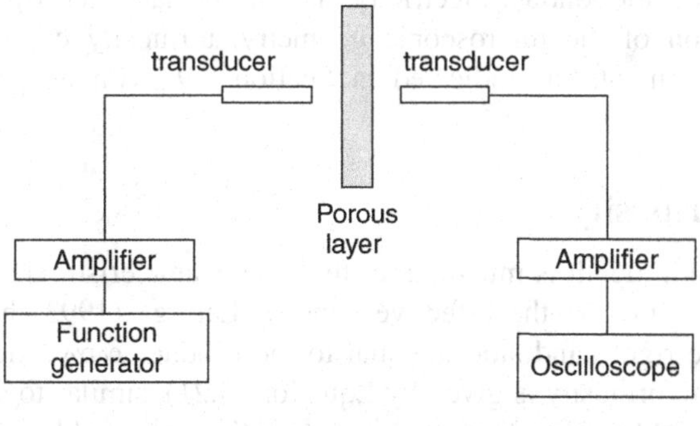

Figure 5.4 Measurement of the refraction index.

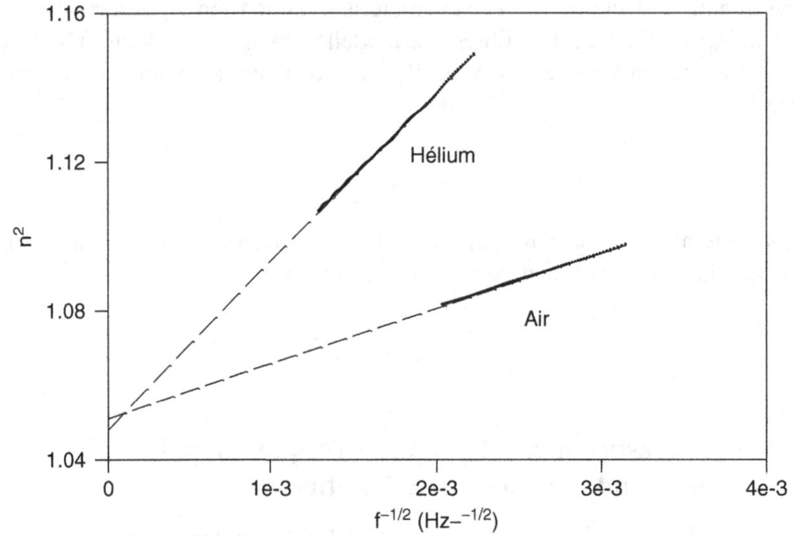

Figure 5.5 Measured n^2 for a foam of porosity $\phi = 0.98$ and viscous static permeability $q_0 = 3.08 \times 10^{-9}$ m^2 (Leclaire *et al.*, 1996). Reprinted with permission from Leclaire, Ph., Kelders, L., Lauriks, W., Melon, M., Brown, N. & Castagnède, B. Determination of the viscous and thermal characteristic lengths of plastic foams by ultrasonic measurements in helium and air. J. Appl. Phys. 80, 2009-2012. Copyright 1996, American Institute of Physics.

Figure 5.5 gives $\Lambda = 202$ μm and $\Lambda' = 367$ μm. The intercepts of the straight lines that extrapolate the measurements toward increasing frequencies give a tortuosity $\alpha_\infty = 1.05$.

These measurements can easily be performed on foams of low flow resistivity. For granular media, diffusion in the high-frequency regime can modify wave propagation. More generally, it has been shown by Ayrault *et al.* (1999) that measurements can be improved by increasing the static pressure. The coupling of air with the transducers is improved. The viscous skin depth decreases when the static pressure increases, and the high-frequency regime appears at lower frequencies.

If the frame does not conduct electricity and can be saturated by a conducting fluid with no modification of the microscopic geometry, tortuosity can be evaluated from conductivity measurements, as indicated in Section 4.9, with no problem concerning diffusion.

5.3.6 Static tortuosity

When the viscous skin depth is much larger than the characteristic dimensions, Equation (5.7) can be used to predict the effective density. Lafarge (1993) has shown that the real constant on the right-hand side is equal to the product $\rho_0\alpha_0$ where α_0 is the static tortuosity. The static tortuosity is given by Equation (5.21), similar to α_∞. The difference is that the velocity field in Equation. (5.21) is now the static field at $\omega = 0$ modified by viscosity. It was shown by Lafarge (2006) that $\alpha_0 \geq \alpha_\infty$

5.4 Models for the effective density and the bulk modulus of the saturating fluid

5.4.1 Pride *et al.* model for the effective density

A model that can be used for identical cylindrical pores with different cross-sectional shapes has been described in Chapter 4. This model predicts the right asymptotic behaviour at high and low frequencies, and gives good predictions in the intermediate frequency range, at least for slits and circular cross-sectional shaped pores. More general models have been suggested by Johnson *et al.* (1987), and Pride *et al.* (1993). The effective density suggested by Johnson *et al.* (1987) has the simplest expression for the high-frequency limit and the low-frequency limit of the imaginary part previously indicated in this chapter, which satisfies the physical constraint due to causality concerning the singularities which must be located on the positive imaginary frequency axis. The model has been modified by Pride *et al.* (1993) to adjust the low-frequency limit of the real part of the effective density with a parameter denoted as b in what follows. The ratio ρ/ρ_0 defined as the dynamic tortuosity is given by

$$\alpha(\omega) = \frac{\nu\phi}{j\omega q_0}\left\{1 - b + b\left[1 + \left(\frac{2\alpha_\infty q_0}{b\phi\Lambda}\right)^2 \frac{j\omega}{\nu}\right]^{1/2}\right\} + \alpha_\infty \tag{5.32}$$

where $\nu = \eta/\rho_0 = B^2\nu'$, B^2 being the Prandtl number. The limit of the real part of the effective density when ω tends to zero is $\rho_0[\alpha_\infty + 2\alpha_\infty^2 q_0/(b\phi\Lambda^2)]$. The right low-frequency limit α_0 for the real part of ρ is obtained by Lafarge (2006) for b, given by

$$b = \frac{2q_0\alpha_\infty^2}{\phi\Lambda^2(\alpha_0 - \alpha_\infty)} \tag{5.33}$$

The limit for the circular pores is obtained for $b = 3/4$. Simulations on simple geometries performed by Perrot (2006), and experiments with air-saturated porous media show that Equation (5.32) can provide very precise predictions of the effective density. Nevertheless, there is a limit to the applicability of Equations (5.32) and (5.33). Simulations performed by Cortis *et al.* (2003) show that the general formulations of Pride *et al.* (1993) with Λ given by Equation (5.24) become inadequate for porous structures with sharp edges.

5.4.2 Simplified Lafarge model for the bulk modulus

In Johnson *et al.* (1987) the dynamic tortuosity $\alpha(\omega)$ is the elementary function used to express the dynamic viscous permeability, and in the present work, the effective density. Similar functions exist for the description of the thermal exchanges and the incompressibility. The function denoted as $\alpha'(\omega)$, related to the bulk modulus K by

$$K = P_0 / \left(1 - \frac{\gamma - 1}{\gamma\alpha'(\omega)}\right) \tag{5.34}$$

was selected by Lafarge *et al.* (1997) as the homologue of $\alpha(\omega)$. From Equation (5.6) the thermal permeability is related to $\alpha'(\omega)$ by $q'(\omega) = \nu'\phi/(j\omega\alpha'(\omega))$. For the case of identical parallel cylindrical pores, it is seen from Equations (4.45)–(4.46) that $\alpha(\omega)$ can be identified with $1/F(\omega)$ and $\alpha'(\omega)$ with $1/F(B^2\omega)$. With the simplified Lafarge model, which gives for K the same high-frequency limit as Equation (5.28), the same low-frequency limit as Equation (5.8), and satisfies the causality condition, α' can be written

$$\alpha'(\omega) = \frac{\nu'\phi}{j\omega q_0'}\left[1 + \left(\frac{2q_0'}{\phi\Lambda'}\right)^2 \frac{j\omega}{\nu'}\right]^{1/2} + 1 \tag{5.35}$$

An additional parameter p' is present in the complete expression of $\alpha'(\omega)$ given by Lafarge (2006). This parameter can provide minor modifications of the bulk modulus in the low- and the medium-frequency range, but does not seem necessary in the description of the bulk modulus of plastic foams and fibrous materials. This parameter is equal to 1 in Equation (5.35).

5.5 Simpler models

5.5.1 The Johnson *et al.* model

The dynamic tortuosity in the work by Johnson *et al.* (1987) is given by

$$\alpha(\omega) = \frac{\nu\phi}{j\omega q_0}\left[1 + \left(\frac{2\alpha_\infty q_0}{\phi\Lambda}\right)^2 \frac{j\omega}{\nu}\right]^{1/2} + \alpha_\infty \tag{5.36}$$

The use of causality and of the asymptotic behaviour to justify the use of this expression was an important step in the description of sound propagation in porous media. The same expression is obtained by setting $b = 1$ in Equation (5.32) that was carried out later by Pride *et al.* (1993). The effective density $\rho = \alpha(\omega)\rho_0$ has the right limit to first-order approximation in $1/\sqrt{\omega}$ for large ω given by Equation (5.26) and for small ω the limit is given by

$$\rho(\omega) = \rho_0\alpha_\infty\left(1 + \frac{2\alpha_\infty q_0}{\Lambda^2\phi}\right) + \frac{\eta\phi}{j\omega q_0} \tag{5.37}$$

The limit of the imaginary part is given by Equation (5.7), $j\mathrm{Im}\rho = \eta\phi/(j\omega q_0) = \phi\sigma/(j\omega)$. As an example, for identical circular cross-sectional shaped pores, with $\Lambda = R$ and $q_0 = \eta/\sigma = R^2\phi/8$, the limit of the real part is 1.25 ρ_0. The true limit obtained in Chapter 4 is 1.33 ρ_0. In spite of this small difference for the limit of Re ρ when ω tends to zero, Equation. (5.36) and the 'exact' model give similar predictions.

5.5.2 The Champoux–Allard model

The direct measurement of the static thermal permeability is not easy. The simplified Lafarge model has been used with q_0' replaced in Equation (5.35) by the permeability

$q_0' = \phi\Lambda'^2/8$ of a porous medium with circular cylindrical pores having a radius $R = \Lambda'$ leading to

$$\alpha'(\omega) = \frac{8\nu'}{j\omega\Lambda'^2}\left[1 + \left(\frac{\Lambda'}{4}\right)^2 \frac{j\omega}{\nu'}\right]^{1/2} + 1 \tag{5.38}$$

It will be shown in Section 5.6 that this arbitrary choice for q_0' can lead to a large error in the localization of the transition frequency where the imaginary part of the bulk modulus reaches its maximum. This does not necessarily lead to a large error in the evaluation of a surface impedance because the damping is mainly created by the viscosity via the effective density.

5.5.3 The Wilson model

In the model due to Wilson (1993), the effective density and the bulk modulus are given by

$$\rho(\omega) = \phi\rho_\infty \frac{(1 + j\omega\tau_{vor})^{1/2}}{(1 + j\omega\tau_{vor})^{1/2} - 1} \tag{5.39}$$

$$K(\omega) = \phi K_\infty \frac{(1 + j\omega\tau_{ent})^{1/2}}{(1 + j\omega\tau_{ent})^{1/2} + \gamma - 1} \tag{5.40}$$

The parameters τ_{vor}, and τ_{ent}, are the vorticity-mode relaxation time, and the entropy-mode relaxation time, respectively. The model is intended to match the middle-frequency behaviour, and not to fit the asymptotic behaviour at high and low frequencies. Therefore $\phi\rho_\infty$ can be different from the effective density ρ when $\omega \to \infty$, and ϕK_∞ can be different from the bulk modulus K when $\omega \to \infty$, due to the fact that the adjustment does not concern the high- and low-frequency asymptotic expressions.

5.5.4 Prediction of the effective density with the Pride *et al.* model and the model by Johnson *et al.*

In Figure 5.6, the effective density ρ is successively predicted with Equation (5.32) for $b = 0.6$ and $b = 1$. The other parameters used for the prediction are $q_0 = 1.23\times10^{-10}$ m^2, $q_0' = 5\times10^{-10}$ m^2, $\phi = 0.37$, $\Lambda = 31$ μm, $\Lambda' = 90$ μm, and $\alpha_\infty = 1.37$. These parameters have been measured for a washed quarry sand (Tizianel *et al.* 1999). A noticeable difference exists for both evaluations of Reρ.

5.5.5 Prediction of the bulk modulus with the simplified Lafarge model and the Champoux-Allard model

Measurements of the static thermal permeability are not easy and with the Champoux–Allard model, the thermal permeability is set equal to that of the porous material having identical circular cross-sectional shaped pores with a radius $R = \Lambda'$. For a porosity $\phi = 0.95$, and $\Lambda' = 610$ μm, the static thermal permeability of this porous medium $q_0' = 4.4\times10^{-8}$ m^2. The bulk modulus of the air saturating the medium is represented in Figure 5.7 for $q_0' = 1.3\times10^{-8}$ m^2 with the simplified Lafarge model

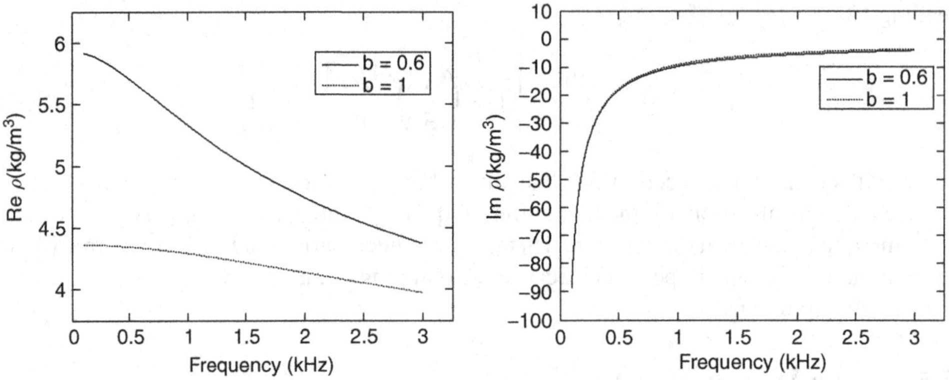

Figure 5.6 The real and the imaginary part of the effective density predicted from Equation (5.32) with the parameters $q_0 = 1.23 \times 10^{-10}$ m^2, $\phi = 0.37$, $\Lambda = 31$ μm, and $\alpha_\infty = 1.37$.

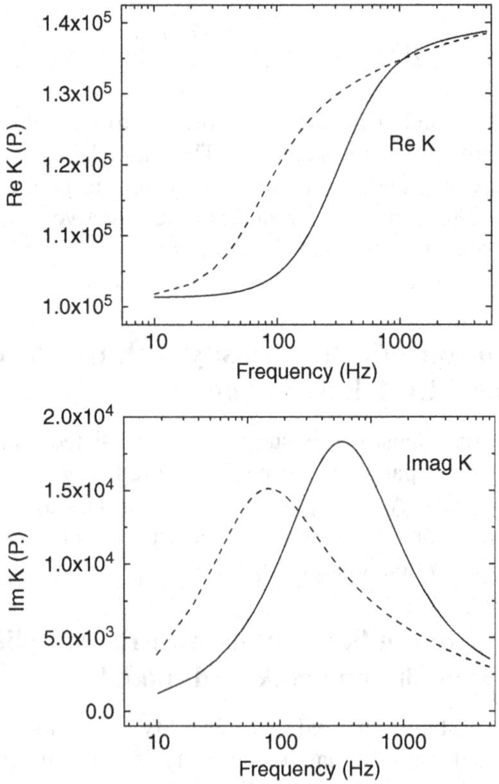

Figure 5.7 The bulk modulus predicted by Equations (5.34)–(5.35) for $\phi = 0.95$, and $\Lambda' = 610$ μm______$q_0' = 1.3 \times 10^{-8}$ m^2, - - - - - - $q_0' = 4.4 \times 10^{-8}$ m^2.

and $q_0' = 4.4 \times 10^{-8}$ m^2. The predicted bulk modulus is different for both permeabilities, the transition frequencies where Im K has a maximum are located at very different frequencies. Measurements of the bulk modulus have been performed at low frequencies with the measurement set-up described in Section (5.2.2). The porous medium was a foam with porosity, tortuosity, viscous and thermal dimensions, close to those used for the predictions. The measured bulk modulus in Lafarge *et al.* (1997) was close to the one predicted with the lowest permeability, $q_0' = 1.3 \times 10^{-8}$ m^2.

5.5.6 Prediction of the surface impedance

In Figure 5.8 the surface impedance at normal incidence is given by Equation (4.137), $Z_s = -jZ_c/(\phi \tan kl)$. All the parameters have been measured with nonacoustical methods, except the thermal static permeability q_0', and the parameter b of the Pride *et al.* model which have been chosen to adjust the predicted impedance to the measured impedance.

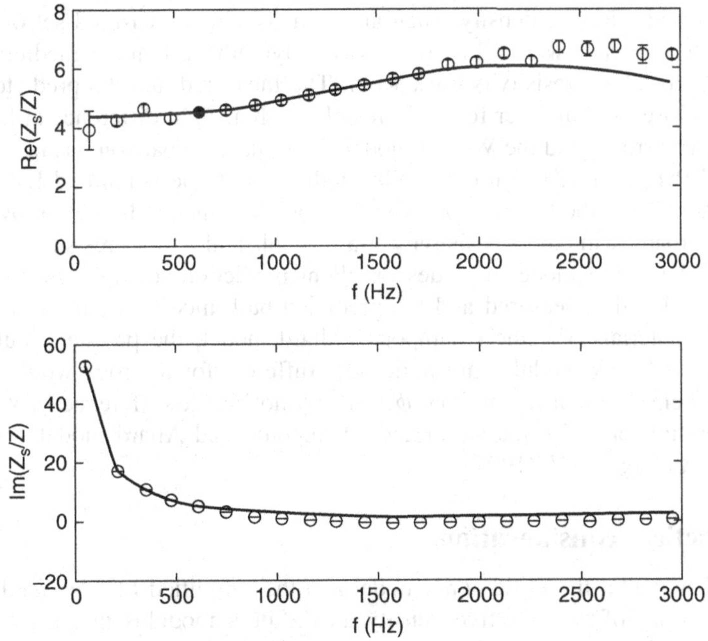

Figure 5.8 The surface impedance of a layer of sand of thickness $l = 3$ cm.——— Prediction of effective density from Equation (5.32) with the parameters $q_0 = 1.23 \times 10^{-10}$ m^2, $\phi = 0.37$, $\Lambda = 31$ μm, $\alpha_\infty = 1.37$, $b = 0.6$, prediction of bulk modulus with Equation (5.35), $\Lambda' = 90$ μm, and $q_0' = 5 \times 10^{-10}$ m^2 (Tizianel *et al.*, 1999). Reprinted with permission from Tizianel, J., Allard, J. F., Castagnède, B., Ayrault, C., Henry, M. & Gedeon, A. Transport parameters and sound propagation in an air-saturated sand. *J. Appl. Phys.* **86**, 5829–5833. Copyright 1999, American Institute of Physics.

5.6 Prediction of the effective density and the bulk modulus of open cell foams and fibrous materials with the different models

5.6.1 Comparison of the performance of different models

A systematic comparison of the performance of the Wilson model and the Johnson *et al.* model has been performed by Panneton and Olny (2006) in the following way. The parameters that characterize the effective density have been evaluated as a function of frequency from careful measurements of the effective density, with the impedance tube technique of Utsono *et al.* (1989), and the method proposed by Iwase *et al.* (1998) to minimize the vibrations of the frame which is motionless in the previous models. In a second step, the effective density has been predicted with these parameters. The first test concerns the variations of these parameters with frequency. The variations must be negligible in the range of frequencies where the model can be used and the measurements sufficiently precise. The second test concerns the agreement between the predicted and the measured effective density when an optimal unique measured set of parameters is chosen. The materials were a low-resistivity polyurethane foam, a medium-resistivity metal foam, and a high-resistivity rock wool. The measured and the predicted effective densities are close to each other for both models, with a slight difference at low frequencies for the metal foam and the Wilson model. A similar comparison has been performed by Olny and Panneton (2008) for the bulk modulus with the simplified Lafarge model, the Wilson model and the Champoux–Allard model. The materials were a low-resistivity polyurethane foam, a medium-resistivity glass wool, and a high-resistivity rock wool. The simplified Lafarge model provides excellent predictions of the bulk modulus. For the Wilson model, the measured and the predicted bulk moduli are close to each other for the three materials. For the Champoux–Allard model, the predicted bulk modulus and the measured bulk modulus are noticeably different for the rock wool, the thermal permeability being very different from $\phi\Lambda'^2/8$. A noticeable difference between prediction and measurement with the simplified Champoux–and Allard model was observed previously in Lafarge *et al.* (1997).

5.6.2 Practical considerations

Simulations show that the Pride *et al.* model and the simplified Lafarge model can give precise predictions of the effective density ad the bulk modulus in the whole audible frequency range. The use of physical parameters in these models provides a link between the acoustical and the physical domain. The improvement of the performances of sound absorbing porous media can be carried out via the modifications of these parameters. A problem for the use of the full models is that some parameters can be very difficult to measure. The use of simple phenomenological parameters, such as the Wilson model, can provide around normal temperature and pressure conditions a simple and precise representation of the measurements over a large frequency range. Moreover, natural and synthetic porous structures are to some extent nonhomogeneous, anisotropic, and not exactly reproducible in the production process. Very precise predictions can be impossible and simple approximations can be a sufficient goal. The bulk modulus appears in a square root in the characteristic impedance and the wave number, and the adiabatic and the isothermal values differ by a factor of 1.4. The thermal dissipation has a maximum around the transition frequency between the isothermal and the adiabatic regime, but this thermal dissipation is generally negligible compared to the viscous dissipation. In spite of the possible erroneous location of the transition frequency, one-parameter models such as the Champoux–Allard model can be used with confidence for the prediction of the surface impedance and the absorption coefficient.

5.7 Fluid layer equivalent to a porous layer

The surface impedance at normal incidence of a layer of isotropic porous medium, given by Equation (4.137), $Z_s = -jZ_c/(\phi \tan kl)$ is identical to the impedance of a layer of isotropic fluid with the same thickness l given by Equation (2.17) $Z_s{}' = -jZ_c{}' \cot k'l$ if

$$k' = k \tag{5.41}$$

$$Z_c' = Z_c/\phi \tag{5.42}$$

These conditions are satisfied if the density ρ' and the bulk modulus K' of the fluid are given by

$$\rho' = \rho/\phi \tag{5.43}$$

$$K' = K/\phi \tag{5.44}$$

With these conditions

$$Z_s = Z_s' \tag{5.45}$$

At oblique incidence, with an angle of incidence θ, Z_s and Z_s' become

$$Z_s = \frac{-jZ_c}{\phi \cos\theta_1} \cot kl \cos\theta_1 \tag{5.46}$$

$$Z_s' = \frac{-jZ_c'}{\cos\theta_1'} \cot k'l \cos\theta_1' \tag{5.47}$$

where θ_1 and θ_1' (with $\theta_1 = \theta_1'$) are the refraction angles defined by

$$k \sin\theta_1 = k_0 \sin\theta \tag{5.48}$$

$$k' \sin\theta' = k_0 \sin\theta \tag{5.49}$$

where k_0 is the wave number in the external medium. The porous medium can be replaced by the homogeneous fluid layer without modifying the reflected field in the external medium.

5.8 Summary of the semi-phenomenological models

The effective density ρ and the bulk modulus K can be written

$$\rho = \rho_0 \left[\alpha_\infty + \frac{\nu\phi}{j\omega q_0} G(\omega) \right] \tag{5.50}$$

$$K = \gamma P_0 / \left[\gamma - \frac{\gamma - 1}{1 + \frac{\nu'\phi}{j\omega q_0'} G'(\omega)} \right] \tag{5.51}$$

The expression for G in the Johnson *et al.* model is

$$G_j(\omega) = \left[1 + \left(\frac{2\alpha_\infty q_0}{\phi \Lambda} \right)^2 \frac{j\omega}{\nu} \right]^{1/2} \tag{5.52}$$

The expression for G' in the simplified Lafarge model is

$$G'_j(\omega) = \left[1 + \left(\frac{2q_0'}{\phi \Lambda'} \right)^2 \frac{j\omega}{\nu'} \right]^{1/2} \tag{5.53}$$

Under the hypothesis that the bulk modulus is similar to the one in circular cross-sectional shaped pores, the following expression for q_0' can be used in Equation (5.53)

$$q_0' = \phi \Lambda'^2 / 8 \tag{5.54}$$

leading to the Champoux–Allard model

$$G'_j(\omega) = \left[1 + \left(\frac{\Lambda'}{4} \right)^2 \frac{j\omega}{\nu'} \right]^{1/2} \tag{5.55}$$

With a supplementary parameter, G_j in the Pride model and G'_j in the full Lafarge model can be replaced by G_p and G'_p

$$G_p(\omega) = 1 - b + b \left[1 + \left(\frac{2\alpha_\infty q_0}{b \phi \Lambda} \right)^2 \frac{j\omega}{\nu} \right]^{1/2} \tag{5.56}$$

$$G'_p(\omega) = 1 - b' + b' \left[1 + \left(\frac{2q_0'}{b' \phi \Lambda'} \right)^2 \frac{j\omega}{\nu'} \right]^{1/2} \tag{5.57}$$

Dynamic tortuosities and permeabilities are related by

$$\begin{aligned} q(\omega) &= \nu\phi / (j\omega\alpha(\omega)) \\ q'(\omega) &= \nu'\phi / (j\omega\alpha'(\omega)) \end{aligned} \tag{5.58}$$

Effective densities and bulk moduli are related to dynamic tortuosities by

$$\begin{aligned} \rho(\omega) &= \rho_0 \alpha(\omega) \\ K(\omega) &= P_0 / [1 - (\gamma - 1)/(\gamma\alpha'(\omega))] \end{aligned} \tag{5.59}$$

5.9 Homogenization

Separation of scales

The semi-phenomenological models in the present chapter can be valid only under the long-wavelength condition. Let L be the macroscopic size defined by

$$L = O\left(\frac{|\lambda|}{2\pi}\right), \tag{5.60}$$

where λ is the complex wavelength in the porous medium. Let l be the microscopic size which characterizes the representative elementary volume (see Figure 5.9). The long-wavelength condition corresponds to $L \gg l$. The use of homogenization methods is based on the same condition. Among these methods, the homogenization method for periodic structures (HPS) has been used to describe sound propagation in ordinary porous media (Sanchez Palencia 1974, Sanchez Palencia 1980, Keller 1977; Bensoussan *et al.*1978, Auriault 1991, Auriault 2005) and in double porosity media (Auriault and Boutin 1994; Boutin *et al*, 1998; Olny and Boutin, 2003) where two networks of pores of very different characteristic sizes are interconnected. The period is defined by the representative elementary volume and the period characteristic size is l.

Porous media are generally not periodic, but for random microscopic geometries the method gives supplementary information about the parameters presented in the context of the semi-phenomenological models. In what follows, some results obtained for simple porosity media are presented. The fundamental parameter is ε given by

$$\varepsilon = l/L \tag{5.61}$$

This parameter characterizes the separation of scales and the method can be used under the condition $\varepsilon \ll 1$.

Two dimensionless space variables are used. Let X be the physical space variable. The dimensionless macroscopic space variable is $\boldsymbol{x} = \boldsymbol{X}/L$, and the dimensionless microscopic space variable is $\boldsymbol{y} = \boldsymbol{X}/l$. The dependence on $\boldsymbol{x}$ corresponds to the slow macroscopic variations and the dependence on $\boldsymbol{y}$ corresponds to the fast variations at the microscopic scale.

Governing equations describing the fluid displacements in a motionless frame

The Navier–Stokes equation in Ω_f

$$\eta\Delta\boldsymbol{v} + (\lambda + \eta)\nabla(\nabla\cdot\boldsymbol{v}) - \nabla p = \rho_0\frac{\partial\boldsymbol{v}}{\partial t} \tag{5.62}$$

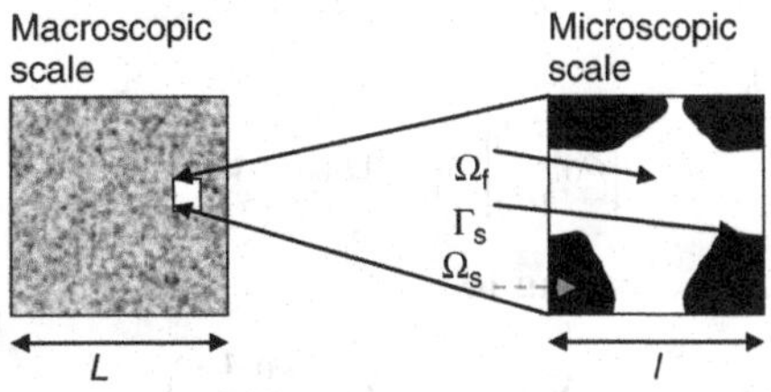

Figure 5.9 A porous medium on the macroscopic scale and on the microscopic scale.

where v is the velocity and λ is the volume viscosity.

Mass balance in Ω_f

$$\frac{d\xi}{dt} + \rho_0 \nabla \cdot \boldsymbol{v} = 0 \tag{5.63}$$

where ξ is the acoustic density.

Adherence condition on Γ_s

$$\boldsymbol{v}/\Gamma_s = 0 \tag{5.64}$$

Heat conduction equation

$$\kappa \Delta \tau = j\omega(\rho_0 c_p \tau - p) \tag{5.65}$$

Air state equation

$$p = P_0 \left(\frac{\xi}{\rho_0} + \frac{\tau}{T_0} \right), \tag{5.66}$$

Thermal boundary condition

$$\tau/\Gamma_s = 0 \tag{5.67}$$

In order to express these equations in a dimensionless form, the reference length L is chosen, and the dimensionless space variable will be $\boldsymbol{x} = \boldsymbol{X}/L$. Adequate characteristic values v_c, p_c, ... of the quantities v, p, ... are used to derive the dimensionless quantities v^*, p^*, ... obtained from $\boldsymbol{v} = v_c \boldsymbol{v}^*$ ... and similar relations for the constant parameters. The above set of equations introduces several dimensionless numbers whose orders of magnitude are related to the characteristic values by

$$Q_L = \frac{|\nabla p|}{|\eta \Delta \boldsymbol{v}|} = \frac{L p_c}{\eta_c v_c} \tag{5.68}$$

$$Rt_L = \frac{\left|\rho_0 \frac{\partial \boldsymbol{v}}{\partial t}\right|}{\eta \Delta \boldsymbol{v}} = \frac{\rho_{0c} \omega_c L^2}{\eta_c} \tag{5.69}$$

$$S_L = \frac{\left|\frac{d\xi}{dt}\right|}{|\rho_0 \nabla \boldsymbol{v}|} = \frac{\omega_c \xi_c L}{\rho_{0c} v_c} \tag{5.70}$$

The flow is forced by the macroscopic pressure gradient, and $|\nabla p| = O(p_c/L)$. The flow occurs in the pores, and the viscous forces satisfy

$$|\eta \Delta \boldsymbol{v}| = O\left(\eta_c \frac{v_c}{l^2}\right).$$

At radian frequency ω,

$$\left|\rho_0 \frac{\partial v}{\partial t}\right| = O(\rho_{0c} \omega_c v_c).$$

The wavelength $\lambda = 2\pi L$, leading to

$$|\rho_0 (\nabla \cdot \boldsymbol{v})| = O\left(\frac{\rho_{0c} v_c}{L}\right).$$

In the situation of greatest interest, the three forces in the Navier–Stokes equation have the same order of magnitude, and the characteristic values are related by

$$\rho_{0c}\omega_c v_c = O\left(\frac{p_c}{L}\right) = O\left(\eta_c \frac{v_c}{l^2}\right) \tag{5.71}$$

This leads to

$$\begin{aligned} Q_L &= O(\varepsilon^{-2}) \\ Rt_L &= O(\varepsilon^{-2}) \\ S_L &= O(1) \end{aligned} \tag{5.72}$$

The dimensionless quantities satisfy the following equations

$$\varepsilon^2 \eta^* \Delta v^* + \varepsilon^2 (\lambda^* + \eta^*) \nabla(\nabla \cdot \boldsymbol{v}^*) - \nabla p^* = j\omega^* \rho_0^* \boldsymbol{v}^* \tag{5.73}$$

$$j\omega^* \xi^* + \rho_0^* \nabla \cdot \boldsymbol{v}^* = 0 \tag{5.74}$$

$$\boldsymbol{v}^* / \Gamma_s^* = 0 \tag{5.75}$$

In Equation (5.66) the relative variations of pressure, temperature, and density, are of the same order of magnitude

$$O\left(\frac{\xi}{\rho_0}\right) = O\left(\frac{\tau}{T_0}\right) = O\left(\frac{p}{P_0}\right) \tag{5.76}$$

leading to

$$O(\rho_0 c_p \tau) = O(p) \tag{5.77}$$

In Equation (5.65) the only dimensionless number which must be estimated is

$$N_L = \frac{|j\omega\rho_0 c_p \tau|}{|\kappa \Delta \tau|} = \frac{\omega \rho_0 c_p \tau_c L^2}{\kappa \tau_c} \tag{5.78}$$

The thermal skin depth given by Equation (4.59) is of the same order of magnitude as the viscous skin depth, and is of the same order of magnitude as the pore size, leading to $\frac{\kappa}{\omega\rho_0 c_p} = O(l^2)$, and to

$$N_L = O(\varepsilon^{-2}) \tag{5.79}$$

The dimensionless quantities satisfy the following equations

$$\varepsilon^2 (\kappa^* \Delta \tau^*) = j\omega^* (\rho_0^* c_p^* \tau^* - p^*) \tag{5.80}$$

$$p^* = P_0^* \left(\frac{\xi^*}{\rho_0^*} + \frac{\tau^*}{T_0^*}\right) \tag{5.81}$$

$$\tau^* / \Gamma_s^* = 0 \tag{5.82}$$

The dimensionless acoustical pressure $p*$, velocity v^*, the acoustic air density ξ^*, and the other space-dependent quantities involved are expressed in the form of asymptotic expansions in powers of ε

$$p^*(\boldsymbol{x},\boldsymbol{y}) = p^{(0)}(\boldsymbol{x},\boldsymbol{y}) + \varepsilon p^{(1)}(\boldsymbol{x},\boldsymbol{y}) + \varepsilon^2 p^{(2)}(\boldsymbol{x},\boldsymbol{y}) + \dots \tag{5.83}$$

$$v^*(\boldsymbol{x},\boldsymbol{y}) = v^{(0)}(\boldsymbol{x},\boldsymbol{y}) + \varepsilon v^{(1)}(\boldsymbol{x},\boldsymbol{y}) + \varepsilon^2 v^{(2)}(\boldsymbol{x},\boldsymbol{y}) + \dots \tag{5.84}$$

$$\xi^*(\boldsymbol{x},\boldsymbol{y}) = \xi^{(0)}(\boldsymbol{x},\boldsymbol{y}) + \varepsilon \xi^{(1)}(\boldsymbol{x},\boldsymbol{y}) + \varepsilon^2 \xi^{(2)}(\boldsymbol{x},\boldsymbol{y}) + \dots \tag{5.85}$$

The superscript i in p^i, v^i, and ξ^i denote the different terms in the developments, not powers. The different p^i, v^i, and ξ^i are periodic with respect to $\boldsymbol{y}$ with the same periodicity as the microstructure. The gradient operator and the divergence are also dimensionless and are given by

$$\nabla = \nabla_x + \varepsilon^{-1}\nabla_y \tag{5.86}$$

$$\Delta = \Delta_x + 2\varepsilon^{-1}\nabla_x\nabla_y + \varepsilon^{-2}\Delta_y \tag{5.87}$$

Equation (5.73) gives, at the order $O(\varepsilon^{-1})$

$$\nabla_y p^{(0)} = 0, \tag{5.88}$$

And, at the order $O(\varepsilon^0)$

$$\eta^*\Delta_y \boldsymbol{v}^{(0)} + (\lambda^* + \eta^*)\nabla_y(\nabla_y \cdot \boldsymbol{v}^{(0)}) - \nabla_y p^{(1)} - \nabla_x p^{(0)} = j\omega^* \rho_0^* \boldsymbol{v}^{(0)} \tag{5.89}$$

Equation (5.74) gives, at the order $O(\varepsilon^{-1})$

$$\Delta_y.\boldsymbol{v}^{(0)} = 0 \tag{5.90}$$

And, at the order $O(\varepsilon^0)$

$$j\omega^*\xi^{(0)} + \rho_0^*\nabla_x \cdot \boldsymbol{v}^{(0)} + \rho_0^*\nabla_y \cdot \boldsymbol{v}^{(1)} = 0 \tag{5.91}$$

Equation (5.75) gives the relations

$$\boldsymbol{v}^{(0)}/\Gamma_s^* = \boldsymbol{v}^{(1)}/\Gamma_s^* = \dots = 0 \tag{5.92}$$

At the order $O(\varepsilon^0)$, Equation (5.80) gives

$$\kappa^*\Delta_y\tau^{(0)} - j\omega^*\rho_0^* c_p^* \tau^{(0)} = -j\omega^* p^{(0)} \tag{5.93}$$

At the order $O(\varepsilon^0)$, Equation (5.81) gives

$$\frac{p^{(0)}}{P_0^*} = \frac{\xi^{(0)}}{\rho_0^*} + \frac{\tau^{(0)}}{T_0^*} \tag{5.94}$$

At zeroth order the pressure does not depend on the microscopic space variable and the saturating air is not compressible at the scale of the pore. The method confirms the validity of the description of sound propagation in cylindrical pores in Chapter 4, because the period is arbitrary in the direction X_3 of the pores and at the microscopic scale $p^{(1)}$ and the other physical quantities in Equations (5.89) and (5.93) do not depend on X_3 (Auriault 1986). Therefore Equations (5.89) and (5.94) will provide, under the long-wavelength condition, a description of sound propagation in cylindrical pores similar to the one given in Chapter 4.

Moreover the justification of the semi-phenomological macroscopic description of the dynamic flow of a viscous fluid in a porous medium with a viscous permeability tensor q^*_{ij} which reduces to a dimensionless viscous permeability q^* for isotropic media has been obtained from Equations (5.89)–(5.92) by Levy (1979) and Auriault (1983)

$$\phi\langle v_i^0\rangle = -\frac{q^*_{ij}(\omega)}{\eta^*}\frac{\partial p^{(0)}}{\partial x_j} \tag{5.95}$$

A return to dimensional quantities is obtained by multiplying the right-hand side of this equation by $l^2 p_c/(L\eta_c v_c) = 1$. This leads to the linear relation

$$\begin{aligned}\phi\langle v_i\rangle &= -\frac{q_{ij}(\omega)}{\eta}\frac{\partial p}{\partial X_j}\\ q_{ij} &= q^*_{ij}/l^2\end{aligned} \tag{5.96}$$

where

$$\langle\cdot\rangle = \frac{1}{|\Omega_f|}\int_{\Omega_f}\cdot \mathrm{d}\Omega$$

(In the main papers concerning homogenization, ϕ is removed because the velocity is averaged over the total representative elementary volume). It has been shown by Sanchez-Palencia (1980) that the tensor q_{ij} is symmetrical. Similarly, Auriault (1980) has shown that Equations (5.92)–(5.93) lead to the following linear relation

$$\phi\langle\tau\rangle_\Omega = \frac{q'(\omega)}{\kappa}j\omega p \tag{5.97}$$

The use of an equivalent fluid at the macroscopic scale is justified, under the long-wavelength condition, by Equations (5.96)–(5.97). The homogenization method for periodic structures justifies the use of the semi-phenomenological models where an equivalent fluid is described from an effective density and a bulk modulus.

5.10 Double porosity media

5.10.1 Definitions

In this section, the basic definitions and some results of the work by Olny and Boutin (2003) about sound propagation in double porosity media are presented. Detailed calculations can be found in Olny and Boutin (2003), Olny (1999) and Boutin *et al.* (1998). In these media two networks of pores of very different characteristic sizes are interconnected, as for instance in an ordinary porous medium with a microporous frame. The macroscopic size characteristic of wave propagation is defined by

$$L = O\left(\left|\frac{\lambda}{2\pi}\right|\right) \tag{5.98}$$

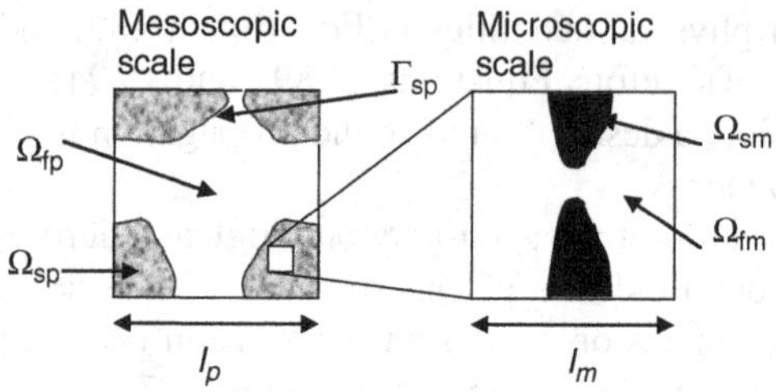

Figure 5.10 A double porosity medium on the mesoscopic scale and on the microscopic scale.

Two characteristic sizes l_p, l_m are defined for the pore and the micropore structure, related to two representative elementary volumes, REVp for the pores and REVm for the micropores (see Figure 5.10).

The subscript p is used for the pores, the subscript m for the micropores and the subscript dp for the double porosity medium. Two parameters, ε defining the (macroscopic/mesoscopic) separation of scale, and ε_0 defining the (mesoscopic/microscopic) separation are given by

$$\varepsilon = l_p/L \tag{5.99}$$

$$\varepsilon_0 = l_m/l_p \tag{5.100}$$

The homogenization method for periodic structures can be used under the condition that these parameters must be much smaller than 1. Two fictive single porosity materials are introduced, the mesoporous medium defined from the REVp with an impervious skeleton, and the microporous medium defined from the REVm without pores. Let ϕ_p be the porosity of the mesoporous medium and ϕ_m be the porosity of the microporous medium. The porosity ϕ of the actual porous medium is $\phi = \phi_p + (1 - \phi_p)\phi_m$.

5.10.2 Orders of magnitude for realistic double porosity media

Macroscopic size for a simple porosity medium

In a first step the Johnson *et al.* model is used to evaluate an order of magnitude of the wavelength and of the macroscopic characteristic size of a simple porosity medium. Two regimes characterize the viscous interaction. At low frequencies the dynamic viscous permeability is close to the static viscous permeability. The static viscous permeability q_0 is linked to the flow resistivity by $q_0 = \eta/\sigma$ (Equation 5.3), and the order of magnitude of the flow resistivity is $\sigma = O(\eta/\phi l^2)$ where l is the characteristic size of the pores (see for instance Equation 4.79, $\sigma = 8\eta/(R^2\phi)$). The order of magnitude of the static viscous permeability is

$$q_0 = O(\phi l^2) \tag{5.101}$$

At large ω, the viscous permeability has the following limit

$$q(\infty) = \frac{\eta\phi}{j\omega\rho_0\alpha_\infty} \tag{5.102}$$

A rough estimate of the transition angular frequency ω_ν between the viscous and the inertial regimes is obtained when $q_0 = q(\infty)$ and is given by

$$\omega_\nu = O\left(\frac{\eta}{l^2 \rho_0 \alpha_\infty}\right) \tag{5.103}$$

The wavelength and the macroscopic size can be evaluated from the wave number k in both regimes. The complex wave number $k = \omega(\rho/K)^{1/2} = \omega[\eta\phi/(j\omega q(\omega)K(\omega))]^{1/2}$ can be evaluated with the adiabatic bulk modulus, the order of magnitude is not modified by the factor $\sqrt{\gamma} = 1.18$. The low frequency estimation of the wave length is given by

$$\left|\frac{\lambda}{2\pi}\right| = O\left(\frac{l}{\delta_\nu}\frac{\lambda_0}{\sqrt{2}\pi}\right) \tag{5.104}$$

where λ_0 is the wave number in the free air. Tortuosity is generally close to 1 and the high-frequency estimation is given by

$$\left|\frac{\lambda}{2\pi}\right| = O\left(\frac{1}{\omega}\sqrt{\frac{\gamma P_0}{\rho_0}}\right) = O\left(\frac{\lambda_0}{2\pi}\right) \tag{5.105}$$

Realistic double porosity medium

Three conditions must be satisfied;

(i) The wavelength is much larger than the mesoscopic size l_p in the whole audible frequency range $\rightarrow l_p \leq 10^{-2}$ m.

(ii) The microporous medium must be sufficiently pervious to acoustical waves $\rightarrow l_m \geq 10^{-5}$ m.

(iii) The separation of both smaller scales must be sufficient $\rightarrow l_p/l_m > 10$.

Two different cases are selected, the first with a low contrast between static permeabilities $l_p = 10^{-3}$ m, and $l_m = 10^{-4}$ m, the second one with a high contrast, $l_p = 10^{-2}$ m, and $l_m = 10^{-5}$ m. The moduli of the wavelengths λ_p and λ_m predicted with Equations (5.104)–(5.105) are shown in Figure 5.11(a) for the low-contrast medium and in Figure 5.11(b) for the high-contrast medium.

In Figure 5.11(a) there is a domain for $\omega > \omega_{\nu m}$ where the wavelengths are similar in the porous medium and the microporous medium, and a strong coupling can be supposed between pores and micropores. In Figure 5.11(b) at $\omega = \omega_d$ the modulus of the complex wavelength in the micropores is equal to the characteristic size of the pores. There is a domain for $\omega > \omega_d$ where the wavelength in the micropores is smaller than the characteristic size of the pores. The regime in the micropores is diffusive, and fast spatial variation of pressure can occur around the pores in the microporous medium at the scale of the REVp.

5.10.3 Asymptotic development method for double porosity media

Wave propagation in double porosity media has been described with the homogenization method for periodic structures (HPS) (Olny and Boutin 2003, Boutin *et al.* 1998, Auriault and Boutin 1994). The porous medium presents a double periodicity at the mesoscopic

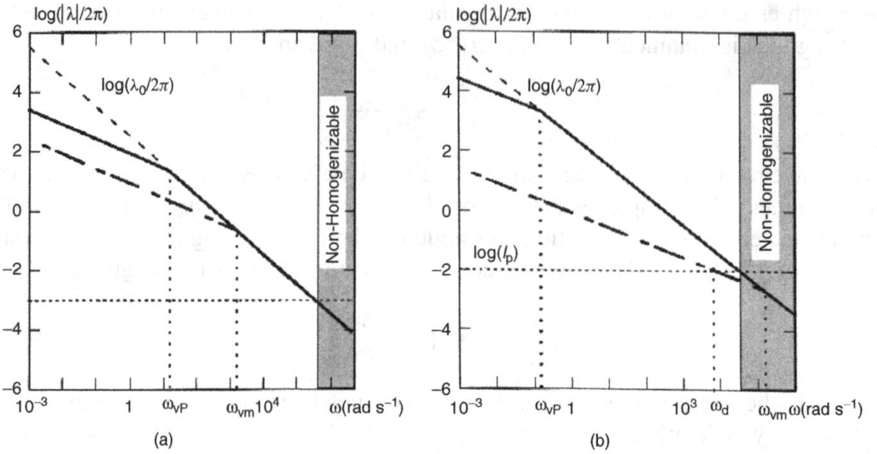

Figure 5.11 Asymptotic modulus of the wavelength in the pores (thick line) and in the micropores (broken line) compared to the wavelength in the free air: (a) low contrast, (b) high contrast (Olny and Boutin 2003). Reprinted with permission from Olny, X., & Boutin, C. Acoustic wave propagation in double porosity media. *J. Acoust. Soc. Amer.* **114**, 73–89. Copyright 2003, Acoustical Society of America.

and the microscopic scale. Three dimensionless space variables describe at each scale the pressure and the velocity fields. Let X be the ordinary space variable. The macroscopic space variable $\boldsymbol{x}$ is defined by $\boldsymbol{x} = X/L$, the mesoscopic space variable $\boldsymbol{y}$ and the microscopic space variable z are defined by $\boldsymbol{y} = X/l_p$ and $z = X/l_m$. The first separation of scales is defined by $\varepsilon = l_p/L$, the second separation of scales is defined by $\varepsilon_0 = l_m/l_p$. The parameters ε and ε_0 must satisfy the relations $\varepsilon \ll 1$ and $\varepsilon_0 \ll 1$. The acoustic pressure and velocity are described with $(\boldsymbol{x}, \boldsymbol{y}, z, \omega)$ in the micropores, and with $(\boldsymbol{x}, \boldsymbol{y}, \omega)$ in the pores except in a thin layer at the interface (Γ_{sp}) with the micropores. In Section 5.7, the porous material as a whole is replaced by an equivalent fluid of bulk modulus K' and density ρ' which occupies the full volume of the layer. This formalism is used in the next subsections because its use is easier when two different porosities are present than the initial model, where an equivalent free fluid replaces the air in the pores. Some results are summarized in what follows.

5.10.4 Low permeability contrast

For the selected example, estimating L from Equation (5.98) for $\omega = O(\omega_{vm})$ yields $\varepsilon = O(\varepsilon_0^2)$. The pressure in the micropores only varies at the macroscopic scale. In a REVp at first order, the pressure denoted as $p_{dp}^0(x)$, is uniform. At the macroscopic level, the macroscopic flow law is similar to the macroscopic flow law given by Equation (5.1) in a simple porosity medium

$$\phi\langle v_{dp}^0\rangle = -\frac{q_{dp}^*(\omega)}{\eta^*}\nabla_x p_{dp}^O$$
$$\langle.\rangle = \frac{1}{|\Omega_{fm}^* \mathrm{U} \Omega_{fp}^*|}\int_{\Omega_{fm}^* \mathrm{U}\Omega_{fp}^*} .\mathrm{d}\Omega \qquad (5.106)$$

A calculation of q_{dp} is performed in Olny and Boutin (2003) for a microporous medium drilled with parallel circular cylindrical pores. The calculation is restricted to the direction of the pores. The semi-phenomenological models cannot be used. For $\omega < \omega_{vp}$ the velocity in the pores is much larger than in the micropores and the contribution of the micropores to the flow can be neglected. From Equations (5.2) and (5.43) the density of the fluid which replaces the porous medium is given, on the macroscopic scale by

$$\rho'_{dp}(\omega) = \frac{\eta}{j\omega q_{dp}(\omega)} \tag{5.107}$$

The bulk modulus defined by Equation (5.44) is given by

$$K'_{dp}(\omega) = \left(\frac{1}{K'_p(\omega)} + (1-\phi_p)\frac{1}{K'_m(\omega)} \right)^{-1} \tag{5.108}$$

The volume of air in the pores is ϕ_p per unit volume of material and is $(1-\phi_p)\phi_m$ in the micropores. The bulk modulus of air in the pores is $\phi_p K'_p(\omega)$ and $\phi_m K'_m(\omega)$ in the micropores and

$$\frac{1}{K'_{dp}(\omega)} = \frac{\phi_p}{\phi_p K'_p(\omega)} + \frac{(1-\phi_p)\phi_m}{\phi_m K'_m(\omega)}.$$

5.10.5 High permeability contrast

For the selected medium, considering that $\omega = O(\omega_d)$ yields $\varepsilon_0 = O(\varepsilon^3)$. As in the case of low-contrast media at $\omega < \omega_{vm}$ the flow in the micropores is mainly viscous, and the bulk modulus is isothermal. The viscous permeability is given by

$$q_{dp}(\omega) = (1-\phi_p)q_m(\omega) + q_p(\omega) \tag{5.109}$$

At first-order development the pressure p_p^0 in the pores only varies on the macroscopic scale. The main difference with the low-contrast media concerns the first-order pressure p_m^0 in the micropores which can vary as a function of the mesoscopic scale variable $\mathbf{y}$, due to the poor transmission of the acoustic field in the micropores. The pressure satisfies a diffusion equation in the micropore domain Ω_{sp} (see Equation 91 in Olny and Boutin 2003). As indicated in Boutin *et al.* (1998) the description of the pressure field in the microporous domain Ω_{sp} around the pores in the double porosity medium is similar to the description of the temperature field in the fluid around the pores in a simple porosity medium. The temperature is equal to 0 at the surface of the pores in contact with air, and the pressure is equal to the pore pressurep_p at the surface of the pores in contact with the microporous medium. The thermal skin depth is $\delta' = (2\nu'/\omega)^{1/2}$ and the skin depth for the pressure is given by (see Equation 91 in Olny and Boutin 2003)

$$\delta_d = \left(\frac{2P_0 q_{0m}}{\phi_m \eta \omega} \right)^{1/2} \tag{5.110}$$

In Olny and Boutin (2003) the factor 2 on the right-hand side is removed due to the different definition of the skin depth. In order to generalize the use of the simplified

Lafarge model to pressure diffusion, in a first step, the dynamic thermal permeability with α' given by Equation (5.35) can be rewritten

$$q'(\omega) = q'_0\left[\frac{j\omega}{\omega_t} + \left(1 + j\frac{M_t}{2}\omega/\omega_t\right)^{1/2}\right] \tag{5.111}$$

$$M_t = \frac{8q'_0}{\Lambda'^2\phi} \tag{5.112}$$

$$\omega_t = \frac{\omega\phi\delta'^2}{2q'_0} \tag{5.113}$$

A function $D(\omega)$, similar to $q'(\omega)$, can be defined for the acoustic pressure by

$$D(\omega) = D(0)\left(\frac{j\omega}{\omega_d} + \left(1 + j\frac{M_d}{2}\omega/\omega_d\right)^{1/2}\right) \tag{5.114}$$

$$M_d = \frac{8D(0)}{\Lambda_d^2(1-\phi_p)} \tag{5.115}$$

$$\omega_d = \frac{(1-\phi_p)P_0 q_{0m}}{\phi_m \eta D(0)} \tag{5.116}$$

The parameters $D(0)$ and Λ_d are geometrical factors which are defined similar to q'_0 and Λ'. The contact surface Γ_{sp} is the same for both problems, but for the thermal conduction problem the volume is the volume of the pores Ω_{fp} and for the pressure diffusion it is the volume Ω_{sp} out of the pores. Therefore

$$\Lambda_d = 2\frac{\Omega_{sp}}{\Gamma_{sp}} = \frac{(1-\phi_p)}{\phi_p}\Lambda'_p \tag{5.117}$$

where $\Lambda'_p = 2\Omega_{fp}/\Gamma_{sp}$ is the thermal characteristic length of the pore network.

In a second step, the relation $\langle\tau\rangle = q'(\omega)j\omega p/(\phi\kappa)$ which relates the average temperature to the pressure for a simple porosity medium is reinterpreted. The temperature field τ can be considered as the sum of two fields τ_1 and τ_2, $\tau_1 = \nu' p/\kappa$ is the spatially constant field created for a specific mass capacity of the frame equal to 0 and a temperature of the boundary equal to τ_1. In order to satisfy the boundary condition $\tau = 0$ on Γ the second field τ_2 is the diffused temperature field related to the boundary condition $\tau = -\nu' p/\kappa$ on Γ. The average temperature τ_2 is given by

$$\begin{aligned}\langle\tau_2\rangle &= \langle\tau\rangle - \langle\tau_1\rangle \\ &= \left(1 - \frac{q'(\omega)j\omega}{q'_0\omega_t}\right)\left(-\frac{\nu' p}{\kappa}\right) \\ &= \left(1 - \frac{q'(\omega)j\omega}{q'_0\omega_t}\right)(-\tau_1)\end{aligned} \tag{5.118}$$

This expression can be transposed to express the average diffused pressure field in the air in the microporous medium Ω_{sp}

$$\langle p_m\rangle = \left(1 - j\frac{\omega}{\omega_d}\frac{D(\omega)}{D(0)}\right)p_p \tag{5.119}$$

Let F be the parameter defined by

$$F = 1 - j\frac{\omega}{\omega_d}\frac{D(\omega)}{D(0)} \tag{5.120}$$

At low frequencies the diffusion skin depth is large around the pores in the microporous medium and F is close to 1. At high frequencies, the diffusion skin depth decreases and F tends to 0. At the macroscopic scale the bulk modulus of the fluid equivalent to the porous medium is given by

$$K_{dp}{}'(\omega) = \left(\frac{1}{K'_p(\omega)} + (1-\phi_p)\frac{\phi_m F(\omega)}{P_0}\right)^{-1} \tag{5.121}$$

Illustration in the case of a slit material (from Olny and Boutin 2003)

The microporous panels of thickness $2b$ are separated with air-gaps of thickness $2a$ (see Figure 5.12). The mesoporosity $\phi_p = a/(a+b)$. The viscous permeability q of a slit material of porosity ϕ, with slits of thickness $2a$ is obtained with Equations (4.27) and (5.2)

$$q(\omega) = -j\phi\frac{\delta}{\sqrt{2}}\left(1 - \frac{\tanh j^{1/2}(a\sqrt{2}/\delta)^2}{j^{1/2}(a\sqrt{2}/\delta)^2}\right) \tag{5.122}$$

where $\delta = (2\eta/(\omega\rho_0))^{1/2}$ (see Equation 4.58) is the viscous skin depth. The thermal permeability has the same form, except that the viscous skin depth must be replaced by the thermal skin depth $\delta' = \delta/B$, B being the square root of the Prandtl number. The

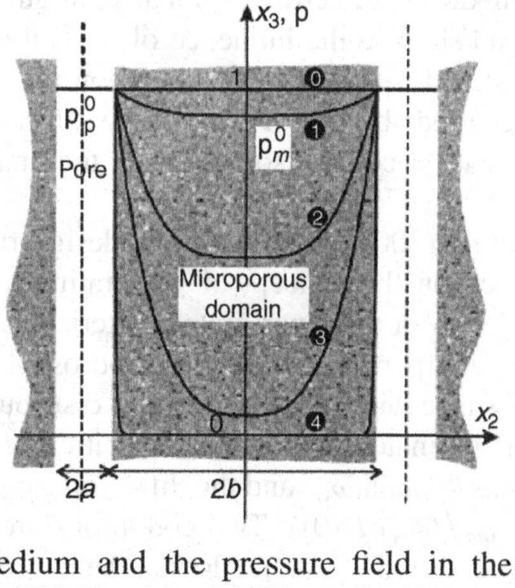

Figure 5.12 The slit medium and the pressure field in the microporous domain:(0) $b/\delta_d = 0.1$, (1)$b/\delta_d = 1$, (2)$b/\delta_d = 2$, (3)$b/\delta_d = 5$, (4)$b/\delta_d = 100$ (Olny and Boutin 2003). Reprinted with permission from Olny, X., & Boutin, C. Acoustic wave propagation in double porosity media. *J. Acoust. Soc. Amer.* **114**, 73–89. Copyright 2003, Acoustical Society of America.

volume where diffusion occurs is the microporous medium and ϕ must be replaced by $1 - \phi_p$. The skin depth is δ_d and $D(\omega)$ is given by

$$D(\omega) = -j(1-\phi_p)\frac{\delta_d}{\sqrt{2}}\left(1 - \frac{\tanh j^{1/2}(b\sqrt{2}/\delta_d)^2}{j^{1/2}(b\sqrt{2}/\delta_d)^2}\right) \tag{5.123}$$

The semi-phenomenological model could be used with $D(0) = (1-\phi_p)b^2/3$ and $\Lambda_d = 2b$. The pressure field in the microporous domain is represented in Figure 5.12. When the diffusion skin depth is much smaller than b, the pressure is negligible compared with p_p^0 except in a small volume close to the pores. Volume variations are restricted because the pressure is not transmitted in the whole volume of the microporous medium. Large variations of the macroscopic bulk modulus are created by the variations of F. Moreover, in the transition range where F varies rapidly from 1 to 0, the loss angle of the macroscopic bulk modulus can be larger than for a single porosity medium (see Figure 9 in Olny and Boutin 2003). The semi-phenomenological models for the thermal exchanges and the bulk modulus described in Sections 5.1–5.5 are not valid for double porosity media with a high permeability contrast.

5.10.6 Practical considerations

Double porosity materials may exist either in a natural state (fractured material with a porous frame) or result from a manufacturing process (recycled materials, perforated porous materials). In particular, Olny (1999) and Olny and Boutin (2003) showed both theoretically and experimentally that the absorption coefficient of highly resistive porous materials could be significantly increased in a wide frequency band by performing properly designed mesoperforations in the materials. Atalla *et al.* (2001) presented a finite-element-based numerical model accounting naturally for the assembling of air cavities and multiple porous materials, thereby alleviating the limitations of Olny's HSP-based model and extended his model to three-dimensional configurations. In particular, they confirmed Olny's results and showed the influence of several design parameters (size of holes, meso-perforation rate and distribution of holes) on the absorption coefficient. An example illustrating the increased absorption of properly selected double porous materials is given in Section 13.9.5 with a comparison between the analytical and the numerical models.

The review by Sgard *et al.* (2005) gives practical design rules to develop optimized noise control solutions based on the concept of perforating a properly selected porous material. The acoustic behaviour of these perforated materials is governed by three important parameters: the size of the perforation, the mesoporosity and a shape factor which depends on the perforation shape and on the mesopores distribution. The enhancement of the absorption coefficient is intimately linked to the position of the viscous characteristic frequency of the microporous domain ω_{vm} and the diffusion frequency defined in Equation (5.116): $\omega_d = (1-\phi_p)P_0 q_{0m} / (\phi_m \eta D(0))$. Two conditions are necessary for this potential enhancement. The first condition is that the flow in the microporous domain be viscous that is: $\omega \ll \omega_{vm}$ where $\omega_{vm} = \eta\phi_m / (\rho_0 \alpha_{\infty m} q_{om})$. The second condition imposes that the diffusion frequency ω_d be much smaller than the viscous characteristic frequency of the microporous domain ω_{vm}, that is: $\omega_d \ll \omega_{vm}$. This condition means that the wavelength in the microporous domain is of the same order as the size of the pore, so that pressure

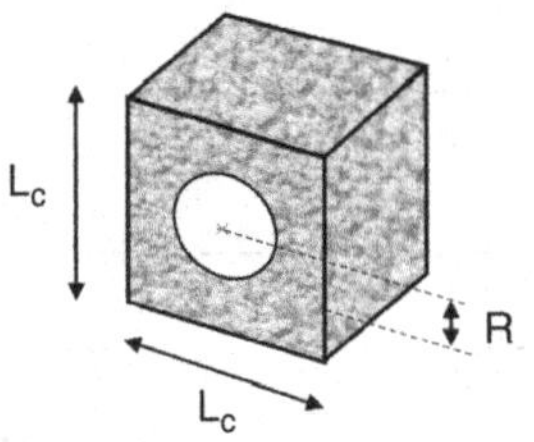

Figure 5.13 Cell geometry for a circular cross-section.

in the microporous part varies on the mesoscale. Combining these two conditions Sgard *et al.* (2005) propose the following design criterion for perforated materials:

$$P_0 \rho_0 \frac{(1-\phi_p)}{D(0)} \frac{\alpha_{\infty m} q_{0m}^2}{\phi_m^2 \eta^2} \ll 1 \tag{5.124}$$

This criterion shows that the design parameters consist of the resistivity ($\sigma_m = \eta / q_{0m}$), porosity and tortuosity of the microporous medium, and a geometrical parameter, $(1-\phi_p)/D(0)$, defined from the mesoscopic structure. Both porosity and resistivity of the substrate should be as large as possible while tortuosity as low as possible. The mesostructure parameter, $(1-\phi_p)/D(0)$, has to be evaluated for the specific geometry and should also be as small as possible. For example, in the case of circular perforations D(0) is given by (Tarnow 1996)

$$D(0) = \frac{L_c^2}{4\pi}\left(\ln\left(\frac{1}{\phi_p}\right) - \frac{3}{2} + 2\phi_p - \frac{\phi_p^2}{2}\right) \tag{5.125}$$

with $\phi_p = \pi R^2 / L_c^2$ and L_c is the size of the cell (see Figure 5.13). In consequence, a low ϕ_p with a large hole diameter must be chosen. Moreover, since the increase of absorption occurs around ω_d with a bandwidth depending on the design parameters, a small value of ω_d is needed to increase absorption at low frequencies, meaning again that a very porous, highly resistive, weakly tortuous substrate material should be selected. These conclusions are confirmed by a numerical parameter study in the case of straight holes with constant cross-section along the thickness and experimental work (Atalla *et al.* 2001, Sgard *et al.* 2005). Results show that significant enhancements of the absorption properties can be obtained over a selected frequency band by adjusting the mesopore profile. Moreover, interesting absorbing properties can be obtained when coating a double porosity medium with an impervious screen.

Appendix 5.A: Simplified calculation of the tortuosity for a porous material having pores made up of an alternating sequence of cylinders

The cylinders are represented in Figure 5.A.1. Let v_1 be the velocity in the cylinders of section S_1 and length l_1, and v_2 the velocity in the cylinders of section S_2 and length l_2.

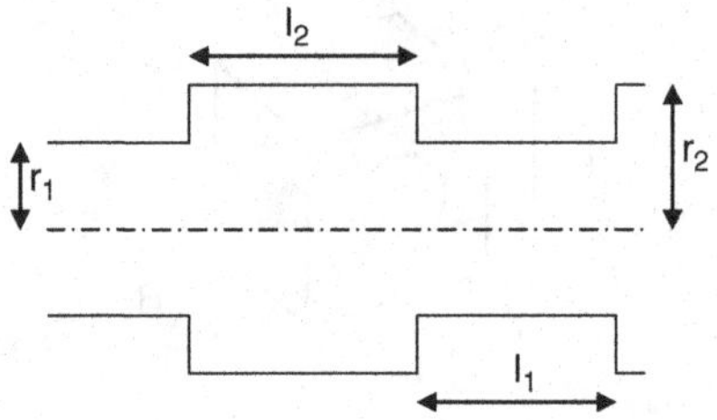

Figure 5.A.1 A pore made up of an alternating sequence of cylinders.

The velocities are supposed to be constant in each cylinder. Then, v_1 and v_2 are related by

$$\frac{v_1}{v_2} = \frac{S_2}{S_1} \tag{5.A.1}$$

If the fluid is nonviscous, the Newton equations in the cylinders are

$$-\frac{\partial p_1}{\partial x} = j\omega\rho_0 v_1 \tag{5.A.2}$$

$$-\frac{\partial p_2}{\partial x} = j\omega\rho_0 v_2 \tag{5.A.3}$$

where p_1 and p_2 are the pressures in cylinders 1 and 2 and ρ_0 is the density of the fluid.
The macroscopic pressure derivative $\partial p/\partial x$ and the macroscopic velocity v are

$$\frac{\partial p}{\partial x} = \frac{\partial p_1}{\partial x}\frac{l_1}{l_1 + l_2} + \frac{\partial p_2}{\partial x}\frac{l_2}{l_1 + l_2} \tag{5.A.4}$$

$$v = \frac{l_1 S_1 v_1}{l_1 S_1 + l_2 S_2} + \frac{l_2 S_2 v_2}{l_1 S_1 + l_2 S_2} \tag{5.A.5}$$

The quantities v and $\partial p/\partial x$ are linked by the following equation:

$$-\frac{\partial p}{\partial x} = \alpha_\infty \rho_0 j\omega v \tag{5.A.6}$$

where α_∞ is given by

$$\alpha_\infty = \frac{[l_1 S_1 + l_2 S_2][l_2 S_1 + l_1 S_2]}{(l_1 + l_2)^2 S_1 S_2} \tag{5.A.7}$$

Appendix 5.B: Calculation of the characteristic length Λ'

In a porous medium, the temperature at high frequencies is constant over the cross-section, except for a small region close to the frame–air boundary. The boundary between the frame and the air in a porous material is represented in Figure. 5.B.1.

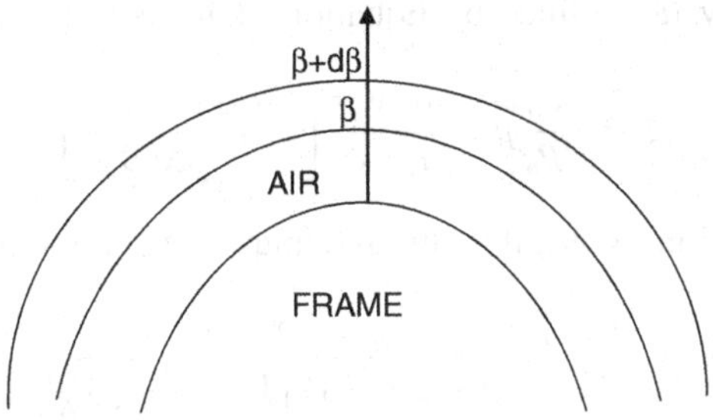

Figure 5.B.1 The frame–air boundary in a porous material.

At high frequencies, the spatial dependence of temperature close to the surface of the pore is the same as if the surface were an infinite plane. Equation (4.33) can be rewritten

$$\frac{\partial^2 \tau}{\partial \beta^2} - j\frac{\omega' \rho_o \tau}{\eta} = -j\frac{\omega' v' \rho_o p}{\kappa \eta} \tag{5.B.1}$$

The solution which vanishes at $\beta = 0$ is

$$\tau = \frac{v' p}{\kappa}(1 - \exp(-j\beta g B)) \tag{5.B.2}$$

where g is given by

$$g = \frac{1-j}{\delta} \tag{5.B.3}$$

In a volume of homogenization V, $\langle \tau \rangle$ is given by

$$\langle \tau \rangle = \frac{\int \tau \, \mathrm{d}V}{\int \mathrm{d}V} = \frac{v' p}{\kappa}\left[1 - \frac{1}{V}\int_V \exp(-j\beta g B)\, \mathrm{d}V\right] \tag{5.B.4}$$

Very close to the frame–air boundary, the volume $\mathrm{d}V$ of air related to $\mathrm{d}\beta$ is equal to $A\ \mathrm{d}\beta$, A being the area of the frame that is in contact with the air in the volume V of porous material. The quantity $\exp(-j\beta g B)$ decreases very quickly when the distance β from M to the surface increases, and the integral in the right-hand side of Equation (5.B.4) is rewritten in Champoux and Allard (1991)

$$\int_v \exp(-j\beta g B) A\ \mathrm{d}\beta = \frac{A}{jBg} = \frac{A\delta(1-j)}{2B} \tag{5.B.5}$$

and $<\tau>$ is given by

$$\langle \tau \rangle = \frac{v' p}{\kappa}\left(1 - (1-j)\frac{\delta}{B\Lambda'}\right) \tag{5.B.6}$$

where $2/\Lambda' = A/V$.

The acoustic density $\langle\xi\rangle$ defined by Equation (4.40) is

$$\langle\xi\rangle = \frac{\rho_o}{P_o}p - \frac{\rho_o}{T_o}\frac{v'p}{\kappa}\left[1 - \frac{(1-j)\delta}{\Lambda' B}\right] \tag{5.B.7}$$

and by the use of Equation (4.39), the bulk modulus of air in the material can be written

$$K = \frac{\rho_o p}{\langle\xi\rangle} = \frac{\gamma P_o}{\gamma - (\gamma - 1)\left[1 - (1-j)\frac{\delta}{\Lambda' B}\right]} \tag{5.B.8}$$

Appendix 5.C: Calculation of the characteristic length Λ for a cylinder perpendicular to the direction of propagation

The cylinder in the acoustic field is represented in Figure 5.C.1. The fluid is nonviscous, and the velocity field can be calculated with the aid of the conformal representation (Joos 1950).

The displacement potential is equal to

$$\varphi = \frac{v_0}{j\omega}x_3\left(1 + \frac{R^2}{x_1^2 + x_3^2}\right) \tag{5.C.1}$$

v_o being the modulus of the velocity far from the cylinder. The two components of the velocity are

$$v_3 = \frac{\partial\varphi}{\partial x_3} = v_o\left(1 + \frac{R^2}{x_1^2 + x_3^2} - \frac{2R^2x_3^2}{(x_1^2 + x_3^2)^2}\right) \tag{5.C.2}$$

$$v_1 = \frac{\partial\varphi}{\partial x_1} = -2v_o x_1 x_3 \frac{R^2}{(x_1^2 + x_3^2)^2} \tag{5.C.3}$$

The squared velocity v^2 at the surface of the cylinder is

$$v^2 = \left[4 - \frac{4x_3^2}{R^2}\right]\cdot v_o^2 \tag{5.C.4}$$

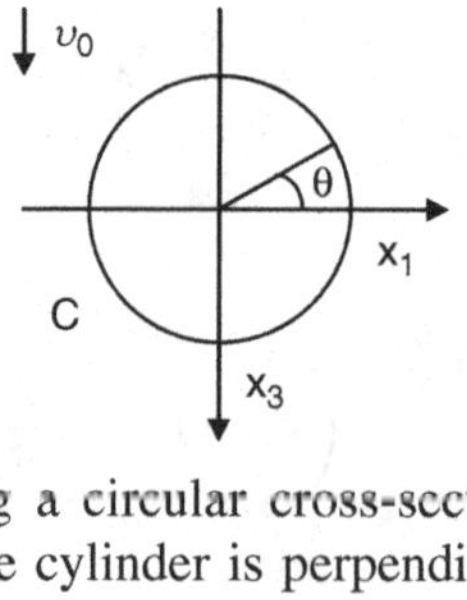

Figure 5.C.1 A cylinder having a circular cross-section is placed in a velocity field such that the velocity far from the cylinder is perpendicular to the cylinder.

The squared-velocity-pondered length of the circle C in Figure 5.C.1 is

$$\int_0^{2\pi} \upsilon^2(\theta)R \, d\theta = 4\pi \upsilon_o^2 R \tag{5.C.5}$$

If the porosity is close to 1, Λ is given by

$$\Lambda = \frac{2\upsilon_o^2}{4\pi \upsilon_o^2 RL} = \frac{1}{2\pi LR} \tag{5.C.6}$$

where L is the total length of the cylinder per unit volume of fibrous material.

The perturbation in the velocity field created by one fibre decreases as R^2/D^2, where D is the distance from the axis of the fibre. The order of magnitude of the distance between the closest fibres is 10 times, or more, as large as R for the usual glass wools and rock wools, and the interaction between the different fibres can be neglected.

The radius R expressed in metres can be obtained from the empirical expression in Bies an Hansen (1980)

$$\sigma R^2 \rho_1^{-1.53} = 0.79 \times 10^{-9} \tag{5.C.7}$$

where σ is the flow resistivity expressed in N m^{-4} s and ρ_1 the density expressed in kg/m^3.

References

Atalla N., Sgard, F., Olny, X. and Panneton, R. (2001) Acoustic absorption of macro-perforated porous materials. J. Sound Vib. **243**(4), 659–678.

Attenborough, K. (1982) Acoustical characteristics of porous materials. *Phys. Rep.*, **82**, 179–227.

Auriault, J. L. (1980) Dynamic behaviour of a porous medium saturated by a newtonian fluid, *Int. J. Engn. Sci*. **18**, 775–785.

Auriault, J. L. (1983) Effective macroscopic decription for heat conduction in periodic composites. *J. Heat Mass Transfer*. **26**, 861–869.

Auriault, J. L., Borne, L., and Chambon R. (1985) Dynamics of porous saturated media, checking of the generalized law of Darcy. *J. Acoust. Soc. Amer*. **77**, 1641–1650.

Auriault, J. L. (1986) Mecanique des milieux poreux saturés déformables. Université P. Fourier, Grenoble, (France).

Auriault, J. L. (1991). Heterogeneous medium. Is an equivalent macroscopic description possible?. *Int. J. Eng. Sci*. **29**, 785–795.

Auriault, J. L. & Boutin, C. (1994) Deformable media with double porosity – III: Acoustics. *Transp. Porous Media* **14**, 143–162.

Auriault, J. L. (2005) Transport in Porous media: Upscaling by Multiscale Asymptotic Expansions. In *CISM Lecture 480 Applied micromechanics of porous materials* Udine 19–23 July 2004, Dormieux L. and Ulm F. J. eds, pp 3–56, Springer.

Ayrault, C., Moussatov, A., Castagnede, B. and Lafarge, D. (1999) Ultrasonic characterization of plastic foams via measurements with static pressure variations. *Appl. Phys. Letters* **74**, 2009–2012.

Bensoussan, A., Lions, J. L. and Papanicolaou G. (1978) *Asymptotic Analysis for Periodic Structures*. North Holland, Amsterdam.

Bies, D. A. and Hansen, C. H. (1980) Flow resistance information for acoustical design. *Applied Acoustics*, **13**, 357–91.

Boutin, C., Royer, P. and Auriault, J. L. (1998) Acoustic absorption of porous surfacing with dual porosity. *Int. J. Solids Struct*. **35**, 4709–4733.

Carcione J. M. and Quiroga-Goode, G. (1996) Full frequency-range transient solution for compressional waves in a fluid-saturated viscoelastic porous medium. *Geophysical Prospecting*, **44**, 99–129.

Champoux, Y. and Allard, J. F. (1991) Dynamic tortuosity and bulk modulus in air-saturated porous media. *J. Appl. Physics*, **70**, 1975–9.

Cortis, A., Smeulders, D. M. J., Guermond, J. L. and Lafarge, D. (2003) Influence of pore roughness on high-frequency permeability. *Phys. Fluids* **15**, 1766–1775.

Debray, A., Allard, J. F., Lauriks, W. and Kelders, L. (1997) Acoustical measurement of the trapping constant of porous materials. *Rev. Scient. Inst*. **68**, 4462–4465. There is a typographical error in the paper, Equationn 14 should be replaced by $Z = \frac{i}{\phi\omega l C_0 C_1}\left[1 - \frac{1}{3}\frac{i\phi}{\sigma\omega}C_0 C_1 \omega^2 l^2\right]$.

Fellah, Z. E. A., Depollier, C., Berger, Lauriks, W., Trompette, P. and Chapelon, J. Y. (2003) Determination of transport parameters in air saturated porous media via ultrasonic reflected waves. *J. Acoust. Soc. Amer*., **114** (2003) 2561–2569.

Iwase, T., Yzumi, Y. and Kawabata, R. (1998) A new measuring method for sound propagation constant by using sound tube without any air space back of a test material. *Internoise 98*, Christchurch, New Zealand.

Johnson, D. L., Koplik, J. and Schwartz, L. M. (1986) New pore size parameter characterizing transport in porous media. *Phys. Rev. Lett*., **57**, 2564–2567.

Johnson, D. L., Koplik, J. and Dashen, R. (1987) Theory of dynamic permeability and tortuosity in fluid-saturated porous media. *J. Fluid Mechanics*, **176**, 379–402.

Joos, G. (1950) *Theoretical Physics*. Hafner Publishing Company, New York.

Keith-Wilson, D., Ostashev, V. D. and Collier S. L. (2004) Time-domain equations for sound propagation in rigid-frame porous media. *J. Acoust. Soc. Amer*., **116**, 1889–1892.

Keller, J. B. (1977) Effective behaviour of heterogeneous media. In Landman U., ed., *Statistical Mechanics and Statistical Methods in Theory and Application*, Plenum, New York, pp 631–644.

Lafarge, D. (1993) Propagation du son dans les matériaux poreux à structure rigide saturés par un fluide viscothermique: Définition de paramètres géométriques, analogie electromagnétique, temps de relaxation. Ph. D. Thesis, Université du Maine, Le Mans, France.

Lafarge, D., Lemarinier, P., Allard, J. F. and Tarnow, V. (1997) Dynamic compressibility of air in porous structures at audible frequencies. *J. Acoust. Soc. Amer*., **102**, 1995–2006.

Lafarge, D. (2002), in *Poromechanics II*, Proceedings of the Second Biot Conference on Poromechanics, J. L. Auriault ed. (Swets & Zeilinger, Grenoble), pp 703–708.

Lafarge D. (2006) in *Matériaux et Acoustique, I Propagation des Ondes Acoustiques*, Bruneau M., Potel C. (ed.), Lavoisier, Paris.

Leclaire, Ph., Kelders, L., Lauriks, W., Melon, M., Brown, N. and Castagnède, B. (1996) Determination of the viscous and thermal characteristic lengths of plastic foams by ultrasonic measurements in helium and air. *J. Appl. Phys*. **80**, 2009–2012.

Levy, T. (1979) Propagation of waves in a fluid saturated porous elastic solid. *Int. J. Engn. Sci*. **17**, 1005–1014.

Norris, A. N. (1986) On the viscodynamic operator in Biot's equations of poroelasticity. *J. Wave-Material Interaction* **1**, 365–380.

Olny, X. (1999) Acoustic absorption of porous media with single and double porosity – modelling and experimental solution. Ph.D. Thesis, ENTPE-INSA Lyon, 281 pp.

Olny, X., and Boutin, C. (2003) Acoustic wave propagation in double porosity media. *J. Acoust. Soc. Amer*. **114**, 73–89.

Olny, X. and Panneton, R. (2008) Acoustical determination of the parameters governing thermal dissipation in porous media. *J. Acoust. Soc. Amer.* **123**, 814–824.

Panneton, R. and Olny, X. (2006) Acoustical determination of the parameters governing viscous dissipation in porous media. *J. Acoust. Soc. Amer.* **119**, 2027–2040.

Perrot, C. (2006) Microstructure et Macro-Comportement Acoustique: Approche par Reconstruction d'une Cellule Elementaire Representative. Ph. D. Thesis, Université de Sherbrooke (Canada) and INSA de Lyon (France).

Perrot, C., Panneton, R. and Olny X. (2007) Computation of the dynamic thermal dissipation properties of porous media by Brownian motion simulation: Application to an open-cell aluminum foam. *J. of Applied Physics* **102**, 074917 1–13.

Pride, S. R., Morgan, F. D.and Gangi, F. A. (1993) Drag forces of porous media acoustics. *Physical Review B* **47**, 4964–4975.

Sanchez-Palencia, E. (1974). Comportement local et macroscopique d'un type de milieux physiques hétérogènes. *Int. J. Eng. Sci*., **12**, 331–351.

Sanchez-Palencia, E. (1980). Nonhomogeneous Media and Vibration Theory. *Lecture Notes in Physics* **127**, Springer, Berlin.

Sgard, F., Olny, X., Atalla, N. and Castel F. (2005) On the use of perforations to improve the sound absorption of porous materials. *Applied acoustics* **66**, 625–651.

Straley, C., Matteson, A., Feng S., Schwartz, L. M., Kenyon, W. E. and Banavar J. R. (1987) Magnetic resonance, digital image analysis, and permeability of porous media. *Appl. Phys. Lett.* **51**, 1146–1148.

Tarnow, V. (1996) Airflow resistivity of models of fibrous acoustic materials. *J. Acoust. Soc. Amer.* **100**(6), 3706–3713.

Tizianel, J., Allard, J. F., Castagnède, B., Ayrault, C., Henry, M. and Gedeon, A. (1999) Transport parameters and sound propagation in an air-saturated sand. *J. Appl. Phys.* **86**, 5829–5833.

Torquato, S. (1990) Relationship between permeability and diffusion-controlled trapping constant of porous media. *Phys. Rev. Lett.* **64**, 2644–2646.

Utsuno, H., Tanaka T. and Fujikawa, T. (1989) Transfer function method for mesuring characteristic impedance and propagation constat of porous materials. *J. Acoust. Soc. Amer.* **86**, 637–643.

Wilson, D. K. (1993) Relaxation-matched modeling of propagation through porous media, including fractal pore structure. *J. Acoust. Soc. Amer.* **94**, 1136–1145.

6

Biot theory of sound propagation in porous materials having an elastic frame

6.1 Introduction

For many materials having an elastic frame and set on a rigid floor, as shown in Figure 6.1(a), the frame can be almost motionless for large ranges of acoustical frequencies, thus allowing the use of models worked out for rigid framed materials. Nevertheless, this is not generally true for the entire range of acoustical frequencies. Moreover, for the material set between two elastic plates represented in Figure 6.1(b), and for many other similar situations, frame vibration is induced by the vibrations of the plates.

The transmission of sound through such a sandwich can be predicted only in the context of a model where the air and the frame move simultaneously. Such a model is provided by the Biot theory (Biot 1956) of sound propagation in elastic porous media. Only the case of isotropic porous structures is considered in this chapter. In the context of the Biot theory, the deformations of the structure related to wave propagation are supposed to be similar to those in an elastic solid, i.e. in a representative elementary volume there is no dispersion of the velocity in the solid part, in contrast to the velocity in air. This leads to a description of the air–frame interaction very similar to that used for rigid structures in Chapter 5.

6.2 Stress and strain in porous materials

6.2.1 Stress

In an elastic solid, or in a fluid, stresses are defined as being tangential and normal forces per unit area of material. The same definition will be used for porous materials,

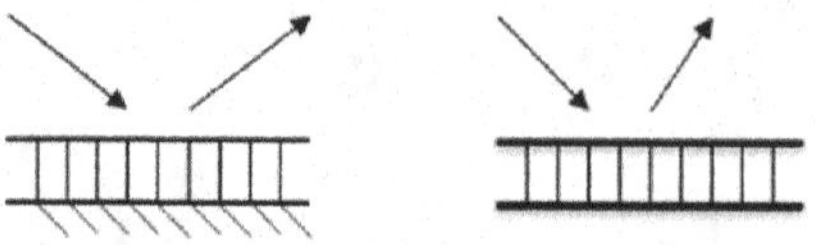

Figure 6.1 (a) Material set on a rigid floor; (b) material set between two elastic plates.

and stresses are defined as forces acting on the frame or the air per unit area of porous material. As a consequence, the stress tensor components for air are

$$\sigma_{ij}^{f} = -\phi p \delta_{ij} \tag{6.1}$$

p being the pressure and ϕ the porosity.

The stress tensor σ^s at a point M for the frame is an average of the different local tensors in the frame in the neighbourhood of M.

6.2.2 Stress–strain relations in the Biot theory: The potential coupling term

The displacement vectors for the frame will be denoted by $\boldsymbol{u}^{\mathrm{s}}$. The macroscopic average dispacement of air will be denoted by $\boldsymbol{u}^{\mathrm{f}}$, while the corresponding strain tensors have elements represented by e_{ij}^{s} and e_{ij}^{f}.

Biot developed an elegant Lagrangian model where the stress–strain relations are derived from a potential energy of deformation. A detailed description of the model is given by Johnson (1986). It has been shown by Pride and Berryman (1998) that, as for the description of the fluid-rigid frame interaction in Chapter 5, the validity of the Biot stress–strain relations is restricted to the case where the wavelengths are much larger than the dimensions of the volume of homogenization. The stress–strain relations in the Biot theory are

$$\sigma_{ij}^{s} = [(P - 2N)\theta^{s} + Q\theta^{f}]\delta_{ij} + 2Ne_{ij}^{s} \tag{6.2}$$

$$\sigma_{ij}^{f} = (-\phi p)\delta_{ij} = (Q\theta^{s} + R\theta^{f})\delta_{ij} \tag{6.3}$$

In these equations, θ^{s} and θ^{f} are the dilatations of the frame and of the air, respectively. If $Q = 0$, Equation (6.2) is identical to Equation (1.78) and becomes the stress–strain relation in elastic solids. Equation (6.3) becomes the stress–strain relation in elastic fluids. The coefficient Q is a potential coupling coefficient. The two terms $Q\theta^{f}$ and $Q\theta^{s}$ give the contributions of the air dilatation to the stress in the frame, and of the frame dilatation to the pressure variation in the air in the porous material. The same coefficient Q appears in both Equations (6.2) and (6.3), because $Q\theta^{s}$ and $Q\theta^{f}$ are obtained by the derivation of a potential energy of interaction per unit volume of material E_{PI}, which must be given in the context of a linear model by

$$E_{PI} = Q(\nabla \cdot \boldsymbol{u}^{\mathrm{s}})(\nabla \cdot \boldsymbol{u}^{\mathrm{f}}) \tag{6.4}$$

where $\nabla \cdot \boldsymbol{u}$ denotes the divergence of $\boldsymbol{u}$

$$\nabla \cdot \boldsymbol{u} = \frac{\partial u_1}{\partial x_1} + \frac{\partial u_2}{\partial x_2} + \frac{\partial u_3}{\partial x_3} \tag{6.5}$$

The 'gedanken experiments' suggested by Biot provide an evaluation of the elasticity coefficients P, N, Q and R. These experiments are static, but they give a description which remains valid for wavelengths large compared with the characteristic dimension of the representative elementary volume. There are three gedanken experiments which are described in Biot and Willis (1957).

First, the material is subjected to a pure shear ($\theta^s = \theta^f = 0$). We then have

$$\sigma_{ij}^s = 2Ne_{ij}^s \qquad \text{and} \qquad \sigma_{ij}^f = 0 \tag{6.6}$$

It is clear that N is the shear modulus of the material, and consequently the shear modulus of the frame, since the air does not contribute to the shear restoring force.

In the second experiment, the material is surrounded by a flexible jacket that is subjected to a hydrostatic pressure p_1. As shown in Figure 6.2, the pressure of the air inside the jacket remains constant and equal to p_o.

This experiment provides a definition for the bulk modulus K_b of the frame at constant pressure in air

$$K_b = -p_1/\theta_1^s \tag{6.7}$$

θ_1^s being the dilatation of the frame, and σ_{11}^s, σ_{22}^s, σ_{33}^s being equal to $-p_1$. For the case of the materials studied in this book, K_b is the bulk modulus of the frame in vacuum. Equations (6.2) and (6.3) can be rewritten

$$-p_1 = (P - \tfrac{4}{3}N)\theta_1^s + Q\theta_1^f \tag{6.8}$$

$$0 = Q\theta_1^s + R\theta_1^f \tag{6.9}$$

In these equations, θ_1^f is the dilatation of the air in the material, and is generally unknown *a priori*. This dilatation is due to the variation in the porosity of the frame, which is not directly predictable, the microscopic field of stresses in the frame for this experiment being very complicated.

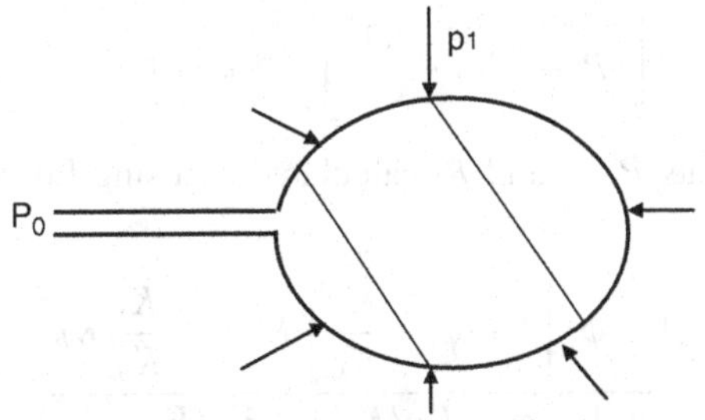

Figure 6.2 The frame of the jacketed material is subjected to a hydrostatic pressure p_1 while the pressure in the air in the jacket is equal to p_o.

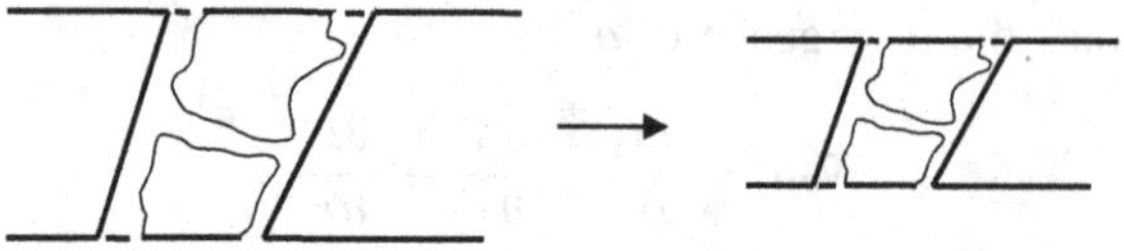

Figure 6.3 A nonjacketed material subjected to an increase of pressure.

In the third experiment, represented in Figure 6.3, the material is nonjacketed, and is subjected to an increase of pressure p_1 in air. This variation in pressure is transmitted to the frame, and the components of the stress tensor for the frame become

$$\tau_{ij}^s = -p_f(1-\phi)\delta_{ij} \tag{6.10}$$

Equations (6.2) and (6.3) can be rewritten

$$-p_f(1-\phi) = (P - \tfrac{4}{3}N)\theta_2^s + Q\theta_2^f \tag{6.11}$$

$$-\phi p_f = Q\theta_2^s + R\theta_2^f \tag{6.12}$$

In these equations, θ_2^s and θ_2^f are the dilatations of the frame and the air, respectively. The quantity $-p_f/\theta_2^s$, which will be denoted by K_s, is the bulk modulus of the elastic solid from which the frame is made

$$K_s = -p_f/\theta_2^s \tag{6.13}$$

In this last experiment there is no variation of the porosity, the deformation of the frame is the same as if the material were not porous, and can be associated with a simple change of scale.

The quantity $-p_f/\theta_2^s$ is the bulk modulus K_f of the air

$$K_f = -p_f/\theta_2^f \tag{6.14}$$

From Equations (6.7)–(6.14), a system of three equations containing the three unknown parameters P, Q and R can be written

$$Q/K_s + R/K_f = \phi \tag{6.15}$$

$$(P - \tfrac{4}{3}N)/K_s + Q/K_f = 1 - \phi \tag{6.16}$$

$$\left[(P - \tfrac{4}{3}N) - \frac{Q^2}{R}\right]\Big/ K_b = 1 \tag{6.17}$$

The elastic coefficients P, Q and R, calculated by using Equations (6.15)–(6.17) are given by

$$P = \frac{(1-\phi)\left[1-\phi-\dfrac{K_b}{K_s}\right]K_s + \phi\dfrac{K_s}{K_f}K_b}{1-\phi-K_b/K_s+\phi K_s/K_f} + \frac{4}{3}N \tag{6.18}$$

$$Q = \frac{[1-\phi-K_b/K_s]\phi K_s}{1-\phi-K_b/K_s+\phi K_s/K_f} \tag{6.19}$$

$$R = \frac{\phi^2 K_s}{1 - \phi - K_b/K_s + \phi K_s/K_f} \tag{6.20}$$

Equations (6.2)–(6.3) were replaced by Biot in a latter work (Biot 1962) by equivalent stress–strain relations involving new definitions of stresses and strains. This second representation of the Biot theory is given in Appendix 6.A and is used in Chapter 10 for transversally isotropic media.

6.2.3 A simple example

Let us consider a glass wool. The glass is very stiff, compared to the glass wool itself. In a first approximation, the volume of glass can be assumed to be constant in the second gedanken experiment, represented in Figure 6.4. The material the frame is made of is not compressible. This property remains valid for most of the frames of sound absorbing porous media. A unit volume of material contains a volume $(1 - \phi)$ of frame at $p_1 = 0$. The same volume of frame, at p_1 different from zero, is in a volume of porous material equal to $1 + \theta_1^s$. The porosity ϕ' is given by

$$1 - \phi = (1 - \phi')(1 + \theta_1^s) \tag{6.21}$$

The dilatation of the air in the material is due to the variation of porosity, and θ_1^f is given by

$$\phi'(1 + \theta_1^f) = \phi \tag{6.22}$$

For this material, the description of the second 'gedanken experiment' is simpler than in the general case. By using Equations (6.21) and (6.22), the following relation relating θ_1^f to θ_1^s can be obtained in the case of small dilatations

$$\theta_1^f = \frac{(1 - \phi)}{\phi}\theta_1^s \tag{6.23}$$

Equations (6.8) and (6.9) can be rewritten

$$\left[(P - \tfrac{4}{3}N) - Q\frac{(1 - \phi)}{\phi}\right] \Big/ K_b = 1 \tag{6.24}$$

$$Q = R(1 - \phi)/\phi \tag{6.25}$$

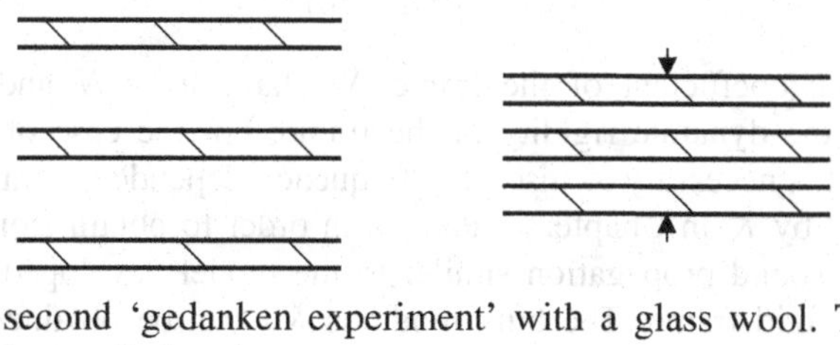

Figure 6.4 The second 'gedanken experiment' with a glass wool. The fibres are displaced, but the volume of glass is constant.

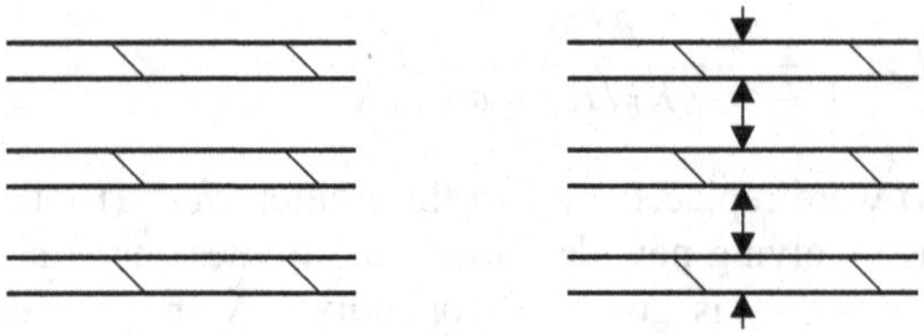

Figure 6.5 The third 'gedanken experiment' with a glass wool. Due to the stiffness of the glass, the frame is not affected by the increase of pressure.

The third 'gedanken experiment' is represented in Figure 6.5. Due to the stiffness of the glass, at a first approximation, one assumes that the frame is not affected by the increase of pressure.

Equation (6.12) can be rewritten

$$R = \phi K_f \tag{6.26}$$

By using Equations (6.24)–(6.26), Q and P can be written

$$Q = K_f(1 - \phi) \tag{6.27}$$

$$P = \tfrac{4}{3}N + K_b + \frac{(1-\phi)^2}{\phi}K_f \tag{6.28}$$

These expressions for P, Q and R can be obtained directly from Equations (6.18)–(6.20) for K_s infinite (the material the frame is made of is not compressible), and can be used for most of the sound-absorbing porous materials. It may be pointed out that the description of the third 'gedanken experiment' is valid only under the hypothesis that the frame is homogeneous. A more complicated formulation has been developed by Brown and Korringa (1975) and Korringa (1981) for the case of a frame made from different elastic materials.

6.2.4 Determination of P, Q and R

The complex dynamic coefficients of elasticity must be used at acoustical frequencies. The rigidity of an isotropic elastic solid is commonly characterized by a shear modulus and a Poisson coefficient. The bulk modulus K_b in Equation (6.28) can be evaluated by the following equation:

$$K_b = \frac{2N(\nu + 1)}{3(1 - 2\nu)} \tag{6.29}$$

ν being the Poisson coefficient of the frame. We have used N and ν instead of N and K_b to specify the dynamic rigidity of the frame. For the case of porous materials saturated with air, it is necessary to insert the frequency-dependent parameter K_f defined previously (denoted by K in Chapters 4 and 5), in order to obtain from the Biot theory a correct model of sound propagation similar to the model developed in Chapter 5 for materials having a rigid frame. The bulk modulus K_f can be obtained from Equations (5.34), and (5.35) or (5.38).

6.2.5 Comparison with previous models of sound propagation in porous sound-absorbing materials

It has been shown by Depollier *et al.* (1988) that, in the simple monodirectional case, the stress–strain equations (6.2) and (6.3), used with the simplified evaluation at P, Q and R by Equations (6.26)–(6.28), are very similar to the stress–strain equations from Beranek (1947) and Lambert (1983), but are not compatible with the equation proposed by Zwikker and Kosten (1949), (Equation 3.05 in Zwikker and Kosten 1949 is incorrect, due to a term P_o which must be removed.)

6.3 Inertial forces in the Biot theory

Biot introduced an inertial interaction between the frame and the fluid, which is not related to the viscosity of the fluid, but to the inertial forces. The porous frame is saturated by an nonviscous fluid. Denoting the velocity in the medium by $\dot{\mathbf{u}}$ and the density by ρ_0, the components of the inertial force per unit volume can be written as

$$F_i = \frac{\partial}{\partial t}\frac{\partial E_c}{\partial \dot{\mathrm{u}}_i} \qquad i = 1, 2, 3 \tag{6.30}$$

For the case of a porous material, similar expressions can be used to calculate the inertial forces, but the kinetic energy is not obtained with the summation of the two terms

$$\tfrac{1}{2}\rho_1 \left|\dot{\boldsymbol{u}}^{\mathbf{s}}\right|^2 + \tfrac{1}{2}\phi\rho_o \left|\dot{\boldsymbol{u}}^{\mathbf{f}}\right|^2 \tag{6.31}$$

where ρ_0 is the density of air. This is because the velocity $\dot{\boldsymbol{u}}^{\mathbf{f}}$ is not the true velocity of the air in the material, but is a macroscopic velocity. In the context of a linear model, the kinetic energy has been given by Biot:

$$E_c = \tfrac{1}{2}\rho_{11} \left|\dot{\boldsymbol{u}}^{\mathbf{s}}\right|^2 + \rho_{12}\dot{\boldsymbol{u}}^{\mathbf{s}} \cdot \dot{\boldsymbol{u}}^{\mathbf{f}} + \tfrac{1}{2}\rho_{22} \left|\dot{\boldsymbol{u}}^{\mathbf{f}}\right|^2 \tag{6.32}$$

ρ_{11}, ρ_{12} and ρ_{22} being parameters depending on the nature and the geometry of the porous medium and the density of the fluid. This expression presents the invariance of the kinetic energy, and, by using a Lagrangian formulation, leads to inertial forces that do not contain derivatives higher than second order in t. The components of the inertial forces acting on the frame and on the air are, respectively

$$q_i^s = \frac{\partial}{\partial t}\frac{\partial E_c}{\partial \dot{u}_i^s} = \rho_{11}\ddot{u}_i^s + \rho_{12}\ddot{u}_i^f \qquad i = 1, 2, 3 \tag{6.33}$$

$$q_i^f = \frac{\partial}{\partial t}\frac{\partial E_c}{\partial \dot{u}_i^f} = \rho_{12}\ddot{u}_i^s + \rho_{22}\ddot{u}_i^f \qquad i = 1, 2, 3 \tag{6.34}$$

An inertial interaction exists between the frame and the air that creates an inertial force on one element due to the acceleration of the other element. This interaction can appear in the absence of viscosity. It has been described by Landau and Lifschitz (1959) for the case of a sphere moving in a fluid. The interaction creates an apparent increase in the mass of the sphere. The coefficients ρ_{11}, ρ_{12} and ρ_{22} are related to the geometry of the frame, and do not depend on the frequency. The following 'gedanken experiment' in Biot (1956) supplies relations between ρ_o, ρ_1, and these coefficients. If the frame and the air move together at the same velocity,

$$\dot{\boldsymbol{u}}^{\mathrm{s}} = \dot{\boldsymbol{u}}^{\mathrm{f}} \tag{6.35}$$

there is no interaction between the frame and the air, and the macroscopic velocity $\dot{\mathbf{u}}^{\mathrm{f}}$ is identical to the microscopic velocity. The porous material moves as a whole, and the kinetic energy is given by

$$E_c = \frac{1}{2}(\rho_1 + \phi\rho_o)|\dot{\boldsymbol{u}}^{\mathrm{s}}|^2 \tag{6.36}$$

A comparison of Equations (6.36) and (6.32) yields

$$\rho_{11} + 2\rho_{12} + \rho_{22} = \rho_1 + \phi\rho_o \tag{6.37}$$

The components q_i^f of the inertial force per unit volume of material are given by

$$q_i^f = \phi\rho_o \frac{\partial^2 u_i^f}{\partial t^2} \tag{6.38}$$

the microscopic velocity of the air being $\dot{\boldsymbol{u}}^{\mathrm{f}}$. A comparison of Equations (6.38) and (6.34) yields

$$\phi\rho_o = \rho_{22} + \rho_{12} \tag{6.39}$$

and thus ρ_1 is given by

$$\rho_1 = \rho_{11} + \rho_{12} \tag{6.40}$$

This description can be related to the case of materials having a rigid frame. If the frame does not move, Equation (6.34) becomes

$$\boldsymbol{q}^{\mathrm{f}} = \rho_{22}\ddot{\boldsymbol{u}}^{\mathrm{f}} \tag{6.41}$$

with

$$\rho_{22} = \phi\rho_o - \rho_{12} \tag{6.42}$$

The quantity $\boldsymbol{q}^{\mathrm{f}}$ is the inertial force acting on a mass $\phi\rho_0$ of nonviscous fluid, the comparison of Equations (6.41) and (5.19), or of Equations (6.32) and (5.22), yields

$$\rho_{22}\ddot{\boldsymbol{u}}^{\mathrm{f}} = \alpha_\infty\phi\rho_o\ddot{\boldsymbol{u}}^{\mathrm{f}} \tag{6.43}$$

From Equations (6.42) and (6.43), the quantity ρ_{12} can be rewritten

$$\rho_{12} = -\phi\rho_o(\alpha_\infty - 1) \tag{6.44}$$

This quantity is the opposite of the inertial coupling term ρ_a previously defined by Equation (4.146)

$$-\rho_{12} = \rho_a \tag{6.45}$$

6.4 Wave equations

The equations of motion (1.44) in an elastic solid without external forces are

$$\rho\frac{\partial^2 u_i^s}{\partial t^2} = (\lambda+\mu)\frac{\partial\theta^s}{\partial x_i} + \mu\nabla^2 u_i^s \qquad i = 1, 2, 3 \tag{6.46}$$

Using Equation (1.76), these equations can be rewritten

$$\rho\frac{\partial^2 u_i^s}{\partial t^2} = (K_c-\mu)\frac{\partial\theta^s}{\partial x_i} + \mu\nabla^2 u_i^s \tag{6.47}$$

The equations of motion of the frame can be obtained by modifying Equation (6.47) in the following way: by comparing Equations (6.2) and (1.21), where $\lambda = K_c - 2\mu$, it appears that an extra term $Q(\partial\theta^f/\partial x_i)$ must be placed on the right-hand side of Equation (6.47), and $K_c - 2\mu$ and μ must be replaced by P and N, respectively. For an nonviscous fluid the inertial force at the left-hand side of Equation (6.47) is given by Equation (6.33), where ρ_{11} is equal to $\rho_1 + \rho_a$

$$\rho\frac{\partial^2 u_i^s}{\partial t^2} \rightarrow (\rho_1+\rho_a)\frac{\partial^2 u_i^s}{\partial t^2} - \rho_a\frac{\partial^2 u_i^f}{\partial t^2} \qquad i = 1, 2, 3 \tag{6.48}$$

with $\rho_a = \phi\rho_0(\alpha_\infty - 1)$. For a viscous fluid, in the frequency domain, α_∞ in ρ_a is replaced by $\alpha_\infty + \frac{\nu\phi}{j\omega q_0}G(\omega)$ (see Eqautions (5.50)–(5.57)). The inertial coupling term is replaced by

$$-\omega^2\rho_a(u_i^s - u_i^f) \rightarrow -\omega^2\rho_a(u_i^s - u_i^f) + \sigma\phi^2 G(\omega)j\omega(u_i^s - u_i^f) \tag{6.49}$$

Equation (6.47) becomes

$$\begin{aligned} &-\omega^2 u_i^s(\rho_1+\rho_a) + \omega^2\rho_a u_i^f \\ &= (P-N)\frac{\partial\theta^s}{\partial x_i} + N\nabla^2 u_i^s + Q\frac{\partial\theta^f}{\partial x_i} - \sigma\phi^2 G(\omega)j\omega(u_i^s - u_i^f) \\ &\qquad i = 1, 2, 3 \end{aligned} \tag{6.50}$$

In the same way, the following equations can be obtained for the air in the porous material:

$$\begin{aligned} -\omega^2 u_i^f(\phi\rho_o+\rho_a) + \omega^2\rho_a u_i^s = R\frac{\partial\theta^f}{\partial x_i} + Q\frac{\partial\theta^s}{\partial x_i} + \sigma\phi^2 G(\omega)j\omega(u_i^s - u_i^f) \\ i = 1, 2, 3 \end{aligned} \tag{6.51}$$

In vector form, Equations (6.50) and (6.52) can be rewritten

$$\begin{aligned} &-\omega^2\boldsymbol{u}^s(\rho_1+\rho_a) + \omega^2\rho_a\boldsymbol{u}^f \\ &= (P-N)\nabla\nabla\cdot\boldsymbol{u}^s + Q\nabla\nabla\cdot\boldsymbol{u}^f + N\nabla^2\boldsymbol{u}^s - j\omega\sigma\phi^2 G(\omega)(\boldsymbol{u}^s - \boldsymbol{u}^f) \end{aligned} \tag{6.52}$$

$$\begin{aligned} &-\omega^2(\phi\rho_o+\rho_a)\boldsymbol{u}^f + \omega^2\rho_a\boldsymbol{u}^s \\ &= R\nabla\nabla\cdot\boldsymbol{u}^f + Q\nabla\nabla\cdot\boldsymbol{u}^s + j\omega\sigma\phi^2 G(\omega)(\boldsymbol{u}^s - \boldsymbol{u}^f) \end{aligned} \tag{6.53}$$

Equations (6.52) and (6.53) become

$$-\omega^2(\tilde{\rho}_{11}\boldsymbol{u}^{\mathrm{s}} + \tilde{\rho}_{12}\boldsymbol{u}^{\mathrm{f}}) = (P - N)\nabla\nabla\cdot\boldsymbol{u}^{\mathrm{s}} + N\nabla^2\boldsymbol{u}^{\mathrm{s}} + Q\nabla\nabla\cdot\boldsymbol{u}^{\mathrm{f}} \tag{6.54}$$

$$-\omega^2(\tilde{\rho}_{22}\boldsymbol{u}^{\mathrm{f}} + \tilde{\rho}_{12}\boldsymbol{u}^{\mathrm{s}}) = R\nabla\nabla\cdot\boldsymbol{u}^{\mathrm{f}} + Q\nabla\nabla\cdot\boldsymbol{u}^{\mathrm{s}} \tag{6.55}$$

where

$$\begin{aligned}\tilde{\rho}_{11} &= \rho_1 + \rho_a - j\sigma\phi^2\frac{G(\omega)}{\omega}\\ \tilde{\rho}_{12} &= -\rho_a + j\sigma\phi^2\frac{G(\omega)}{\omega}\\ \tilde{\rho}_{22} &= \phi\rho_o + \rho_a - j\sigma\phi^2\frac{G(\omega)}{\omega}\end{aligned} \tag{6.56}$$

Three other formalisms of the Biot theory are presented in Appendix 6.A: (i) Biot's second formulation (Biot 1962), (ii) the Dazel representation (Dazel *et al.* 2007) and (iii) the mixed displacement pressure formulation (Atalla *et al.* 1998). The latter will be used in Chapter 13 to illustrate the finite element implementation of the Biot theory.

6.5 The two compressional waves and the shear wave

6.5.1 The two compressional waves

As in the case for an elastic solid, the wave equations of the dilatational and the rotational waves can be obtained by using scalar and vector displacement potentials, respectively. Velocity potentials are used in Chapter 8. Two scalar potentials for the frame and the air, φ^s and φ^f, are defined for the compressional waves, giving

$$\boldsymbol{u}^s = \nabla\varphi^s \tag{6.57}$$

$$\boldsymbol{u}^f = \nabla\varphi^f \tag{6.58}$$

By using the relation

$$\nabla\nabla^2\varphi = \nabla^2\nabla\varphi \tag{6.59}$$

in Equations (6.54) and (6.55), it can be shown that φ^s and φ^s are related as follows:

$$-\omega^2(\tilde{\rho}_{11}\varphi^s + \tilde{\rho}_{12}\varphi^f) = P\nabla^2\varphi^s + Q\nabla^2\varphi^f \tag{6.60}$$

$$-\omega^2(\tilde{\rho}_{22}\varphi^f + \tilde{\rho}_{12}\varphi^s) = R\nabla^2\varphi^f + Q\nabla^2\varphi^s \tag{6.61}$$

Let us denote by $[\varphi]$ the vector

$$[\varphi] = [\varphi^s, \varphi^f]^T \tag{6.62}$$

Equations (6.60) and (6.61) can then be reformulated as

$$-\omega^2[\rho][\varphi] = [M]\nabla^2[\varphi] \tag{6.63}$$

where $[\rho]$ and $[M]$ are respectively

$$[\rho] = \begin{bmatrix} \tilde{\rho}_{11} & \tilde{\rho}_{12} \\ \tilde{\rho}_{12} & \tilde{\rho}_{22} \end{bmatrix}, \qquad [M] = \begin{bmatrix} P & Q \\ Q & R \end{bmatrix} \tag{6.64}$$

Equation (6.63) can be rewritten

$$-\omega^2[M]^{-1}[\rho][\varphi] = \nabla^2[\varphi] \tag{6.65}$$

Let δ_1^2 and δ_2^2 be the eigenvalues, and $[\varphi_1]$ and $[\varphi_2]$ the eigenvectors, of the left-hand side of Equation (6.65). These quantities are related by

$$\begin{aligned} -\delta_1^2[\varphi_1] &= \nabla^2[\varphi_1] \\ -\delta_2^2[\varphi_2] &= \nabla^2[\varphi_2] \end{aligned} \tag{6.66}$$

The eigenvalues δ_1^2 and δ_2^2 are the squared complex wave numbers of the two compressional waves, and are given by

$$\delta_1^2 = \frac{\omega^2}{2(PR - Q^2)}[P\tilde{\rho}_{22} + R\tilde{\rho}_{11} - 2Q\tilde{\rho}_{12} - \sqrt{\Delta}] \tag{6.67}$$

$$\delta_2^2 = \frac{\omega^2}{2(PR - Q^2)}[P\tilde{\rho}_{22} + R\tilde{\rho}_{11} - 2Q\tilde{\rho}_{12} + \sqrt{\Delta}] \tag{6.68}$$

where Δ is given by

$$\Delta = [P\tilde{\rho}_{22} + R\tilde{\rho}_{11} - 2Q\tilde{\rho}_{12}]^2 - 4(PR - Q^2)(\tilde{\rho}_{11}\tilde{\rho}_{22} - \tilde{\rho}_{12}^2) \tag{6.69}$$

The two eigenvectors can be written

$$[\varphi_1] = \begin{bmatrix} \varphi_1^s \\ \varphi_1^f \end{bmatrix}, \qquad [\varphi_2] = \begin{bmatrix} \varphi_2^s \\ \varphi_1^f \end{bmatrix} \tag{6.70}$$

Using Equation (6.60), one obtains

$$\varphi_i^f/\varphi_i^s = \mu_i = \frac{P\delta_i^2 - \omega^2\tilde{\rho}_{11}}{\omega^2\tilde{\rho}_{12} - Q\delta_i^2} \qquad i = 1, 2 \tag{6.71}$$

or

$$\varphi_i^f/\varphi_i^s = \mu_i = \frac{Q\delta_i^2 - \omega^2\tilde{\rho}_{12}}{\omega^2\tilde{\rho}_{22} - R\delta_i^2} \qquad i = 1, 2 \tag{6.72}$$

These equations give the ratio of the velocity of the air over the velocity of the frame for the two compressional waves and indicate in what medium the waves propagate preferentially. Four characteristic impedances can be defined, because both waves simultaneously propagate in the air and the frame of the porous material. In the case of waves propagating in the x_3 direction, the characteristic impedance related to the propagation in the air is

$$Z^f = p/(j\omega u_3^f) \tag{6.73}$$

The macroscopic displacements of the frame and the air are parallel to the x_3 direction, and by the use of Equation (6.3), Equation (6.73) can be rewritten for the two compressional waves

$$Z_1^f = (R + Q/\mu_1)\frac{\delta_1}{\phi\omega} \tag{6.74}$$

$$Z_2^f = (R + Q/\mu_2)\frac{\delta_2}{\phi\omega} \tag{6.75}$$

The characteristic impedance related to the propagation in the frame is

$$Z^s = -\sigma_{33}^s/(j\omega u_3^s) \tag{6.76}$$

By the use of Equation (6.2), Equation (6.76) can be rewritten for the two compressional waves

$$Z_i^s = (P + Q\mu_i)\frac{\delta_i}{\omega} \quad i = 1, 2 \tag{6.77}$$

6.5.2 The shear wave

As in the case for an elastic solid, the wave equation for the rotational wave can be obtained by using vector potentials. Two vector potentials, $\boldsymbol{\psi}^s$ and $\boldsymbol{\psi}^f$, for the frame and for the air, are defined as follows:

$$\boldsymbol{u}^s = \nabla \wedge \boldsymbol{\psi}^s \tag{6.78}$$

$$\boldsymbol{u}^f = \nabla \wedge \boldsymbol{\psi}^f \tag{6.79}$$

Substitution of the displacement representation, Equations (6.78) and (6.79), into Equations (6.54) and (6.55) yields

$$-\omega^2\tilde{\rho}_{11}\boldsymbol{\psi}^s - \omega^2\tilde{\rho}_{12}\boldsymbol{\psi}^f = N\nabla^2\boldsymbol{\psi}^s \tag{6.80}$$

$$-\omega^2\tilde{\rho}_{12}\boldsymbol{\psi}^s - \omega^2\tilde{\rho}_{22}\boldsymbol{\psi}^f = 0 \tag{6.81}$$

The wave equation for the shear wave propagating in the frame is

$$\nabla^2\boldsymbol{\psi}^s + \frac{\omega^2}{N}\left(\frac{\tilde{\rho}_{11}\tilde{\rho}_{22} - \tilde{\rho}_{12}^2}{\tilde{\rho}_{22}}\right)\boldsymbol{\psi}^s = 0 \tag{6.82}$$

The squared wave number for the shear wave is given by

$$\delta_3^2 = \frac{\omega^2}{N}\left(\frac{\tilde{\rho}_{11}\tilde{\rho}_{22} - \tilde{\rho}_{12}^2}{\tilde{\rho}_{22}}\right) \tag{6.83}$$

and the ratio μ_3 of the amplitudes of displacement of the air and of the frame is given by Equation (6.81)

$$\mu_3 = -\tilde{\rho}_{12}/\tilde{\rho}_{22} \tag{6.84}$$

or by

$$\mu_3 = \frac{N\delta_3^2 - \omega^2\tilde{\rho}_{11}}{\omega^2\tilde{\rho}_{22}} \tag{6.85}$$

6.5.3 The three Biot waves in ordinary air-saturated porous materials

For the case where a strong coupling exists between the fluid and the frame, the two compressional waves exhibit very different properties, and are identified as the slow wave and the fast wave (Biot 1956, Johnson 1986). The ratio μ of the velocities of the fluid and the frame is close to 1 for the fast wave, while these velocities are nearly opposite for the slow wave. The damping due to viscosity is much stronger for the slow wave which, in addition, propagates more slowly than the fast wave. With ordinary porous materials saturated with air, it is more convenient to refer to the compressional waves as a frame-borne wave and an airborne wave.

This new nomenclature is obviously fully justified if there is no coupling between the frame and air. For such a case, one wave propagates in the air and the other in the frame. For the case where a weak coupling exists, the partial decoupling previously predicted by Zwikker and Kosten (1949) occurs. With the frame being heavier than air, the frame vibrations will induce vibrations of the air in the porous material, yet the frame can be almost motionless when the air circulates around it. More precisely, one of the two waves, the airborne wave, propagates mostly in the air, whilst the frame-borne wave propagates in both media. The wave number of the frame-borne wave and its characteristic impedance corresponding to the propagation in the frame can be close to the wave number and the characteristic impedance of the compressional wave in the frame when in vacuum. The shear wave is also a frame-borne wave, and is very similar to the shear wave propagating in the frame when in vacuum.

6.5.4 Example

The two compressional waves propagating in a fibrous material at normal incidence are described in the context of the Biot theory. The material is a layer of glass wool 'Domisol Coffrage' manufactured by St Gobain-Isover (BP19 60290 Rantigny France). The material is anisotropic, but the compressional waves in the normal direction are identical in this material and in an equivalent isotropic material which presents the same stiffness in the case of normal displacements, and whose other acoustical parameters are the same as for the fibrous material in the normal direction. The parameters α_∞, ρ_1, σ, ϕ, N and ν are indicated in Table 6.1. The shear modulus N is evaluated from acoustic measurements, as indicated in Section 6.6, and the Poisson coefficient is equal to zero (Sides *et al.* 1971).

The diameter of the fibres, calculated by use of Equation (5.C.7) is

$$d = 12 \times 10^{-6}\ \text{m}$$

The characteristic dimensions Λ and Λ' are obtained by the use of Equations (5.29) and (5.30)

$$\Lambda = 0{\cdot}56 \times 10^{-4}\ \text{m}, \qquad \Lambda' = 2\Lambda = 1{\cdot}1 \times 10^{-4}\ \text{m}$$

Table 6.1 Values of parameters α_∞, ρ_1, σ, ϕ, N and ν for the glass wool 'Domisol Coffrage'.

Tortuosity α_∞	Density of frame ρ_1 (kg m^{-3})	Flow resistivity σ (N m^{-4} s)	Porosity ϕ	Shear modulus N (N cm^{-2})	Poisson coefficient ν
1·06	130·0	40 000	0·94	220(1 + j0·1)	0

The bulk modulus K_f of the air in the fibrous material, $G(\omega)$, and the parameters P, Q and R are evaluated using Eqs (5.38), (5.36), (6.28), (6.27) and (6.26), respectively. The subscripts a and b will be used to specify the quantities related to the airborne wave and the frame-borne wave. At high frequencies, for the airborne wave, the ratio μ_a of the velocities of the frame and the air is μ_2 given by Equation (6.71), and the wave number is δ_2 given by Equation (6.68). At frequencies lower than 495 Hz, the airborne wave is related to μ_1 and δ_1.

The quantity $|\mu_a|$ is larger than 40 for frequencies higher than 50 Hz. The velocity of the frame is negligible, compared with the velocity of the air, and this wave is very similar to the one that would propagate if the material were rigidly framed. The wave number k_a is represented in Figure 6.6. The wave number k'_a for the same material with a rigid frame is given by

$$k'_a = \omega \left(\frac{\tilde{\rho}_{22}}{R} \right)^{1/2} \tag{6.86}$$

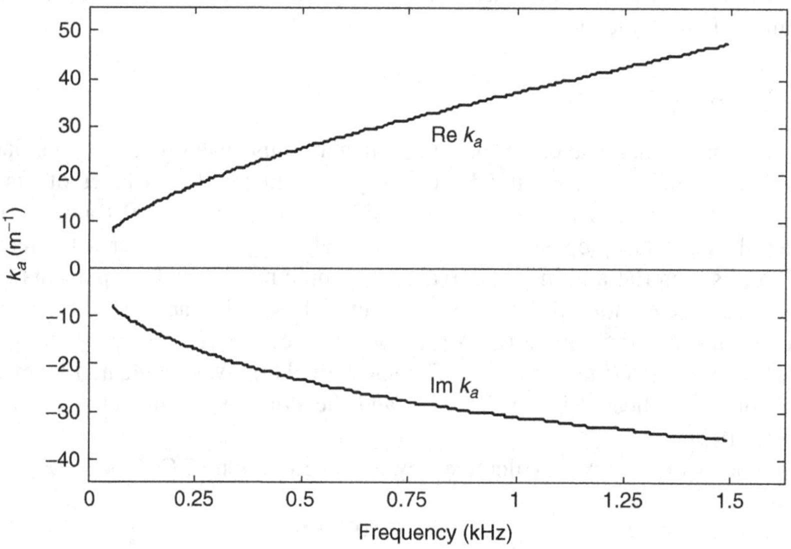

Figure 6.6 The wave number k_a of the airborne wave in the fibrous material.

R being given by Equation (6.26). The wave numbers k_a and k'_a are represented by the same curve in Figure 6.6. It should be noticed that at low frequencies this wave is strongly damped, the imaginary part and the real part of k_a being nearly equal. The characteristic impedance Z_a^f related to the propagation of the airborne wave in the air in the material is represented in Figure 6.7.

The related characteristic impedance $Z_2'^f$ for the same material with a rigid frame is given by

$$Z_2'^f = \frac{(\tilde{\rho}_{22}R)^{1/2}}{\phi} \tag{6.87}$$

This evaluation is represented by the same curve in Figure 6.7.

The ratio modulus $|\mu_b|$ of the velocities of the frame and the air for the frame-borne wave decreases from 1·0 at 50 Hz to 0·82 at 1500 Hz. The frame-borne wave, as indicated in Section 6.5.3, induces a noticeable velocity of the air in the material. On the other hand, the wave number δ_b and the characteristic impedance Z_b^s, evaluated by Equation (6.67) at high frequencies and Equation (6.77), are very close to the wave number and the characteristic impedance for longitudinal waves propagating in the frame in vacuum

$$\delta'_1 = \omega\sqrt{\frac{\rho_1}{K_c}} \tag{6.88}$$

$$Z_1'^s = \sqrt{\rho_1 K_c} \tag{6.89}$$

K_c being the elasticity coefficient of the frame in the vacuum given by Equation (1.76).

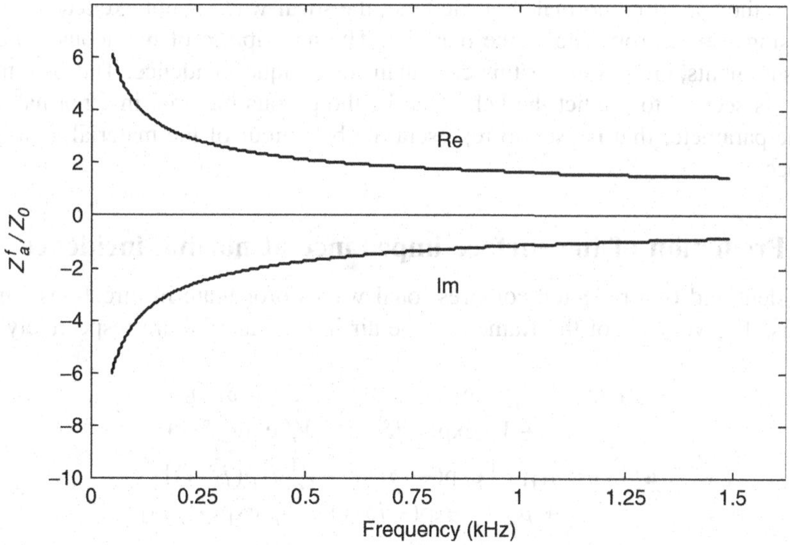

Figure 6.7 The normalized characteristic impedance Z_a^f/Z_0 related to the propagation of the airborne wave in the air in the fibrous material.

The chosen material provides a good illustration of partial decoupling, because the frame is much heavier and stiffer than the air. Some porous materials used for sound absorption have a frame whose bulk modulus K_s has the same order of magnitude as the bulk modulus K_f of the air, and a density ρ_1 which is about 10 times larger than the density ρ_o of the air. For these materials, the partial decoupling does not exist at low frequencies, up to an upper bound depending on the flow resistivity and the density of the material. A simple expression is given by Zwikker and Kosten (1949) for this frequency

$$f_o = \frac{1}{2\pi}\frac{\phi^2\sigma}{\rho_1} \tag{6.90}$$

It may be pointed out that at frequencies higher than f_o, the frame-borne wave can be noticeably different from the compressional wave propagating in the frame in vacuum, and the airborne wave can be noticeably different from the compressional wave in the same material with a rigid frame.

6.6 Prediction of surface impedance at normal incidence for a layer of porous material backed by an impervious rigid wall

6.6.1 Introduction

A layer of porous material in a normal plane acoustic field is represented in Figure 6.8. In order to obtain simple boundary conditions at the wall–material interface, the material is glued to the wall. In a normal acoustic field, the shear wave is not excited and only the compression waves propagate in the material. The description of the acoustic field, and the measurements, are easier for this case than for oblique incidence. The Biot theory is used in this section to predict the behaviour of the porous material in a normal acoustic field. The parameter that is used to represent the behaviour of the material is the surface impedance.

6.6.2 Prediction of the surface impedance at normal incidence

Two incident and two reflected compressional waves propagate in directions parallel to the x axis. The velocity of the frame and the air in the material are respectively

$$\begin{aligned}\dot{u}^s(x) = {} & V_i^1 \exp(-j\delta_1 x) + V_r^1 \exp(j\delta_1 x) \\ & + V_i^2 \exp(-j\delta_2 x) + V_r^2 \exp(j\delta_2 x)\end{aligned} \tag{6.91}$$

$$\begin{aligned}\dot{u}^f(x) = {} & \mu_1[V_i^1 \exp(-j\delta_1 x) + V_r^1 \exp(j\delta_1 x)] \\ & + \mu_2[V_i^2 \exp(-j\delta_2 x) + V_r^2 \exp(j\delta_2 x)]\end{aligned} \tag{6.92}$$

In these equations, the time dependence exp(jωt) has been removed, δ_1 and δ_2 are given by Equations (6.67) and (6.68), and μ_1 and μ_2 by Equation (6.71). The quantities V_i^1, V_r^1, V_i^2 and V_r^2 are the velocities of the frame at $x = 0$ associated with the incident (subscript i) and the reflected (subscript r) first (index 1) and second (index 2) Biot

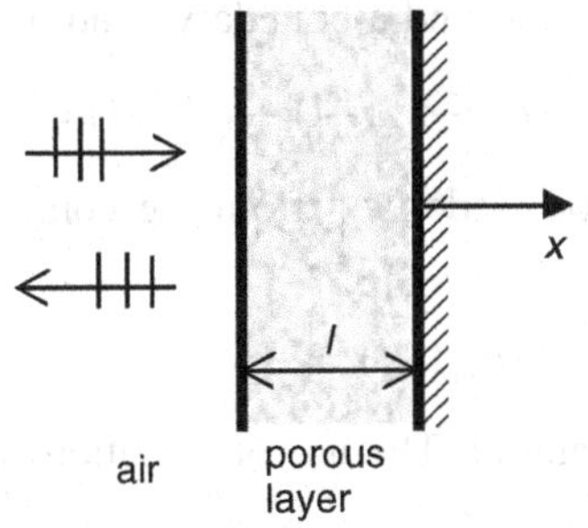

Figure 6.8 A layer of porous material bonded on to an impervious rigid wall, in a normal acoustic field.

compressional waves. The stresses in the material are given by

$$\begin{aligned}\sigma_{xx}^s(x) = &-Z_1^s[V_i^1\exp(-j\delta_1 x) - V_r^1\exp(j\delta_1 x)]\\ &- Z_2^s[V_i^2\exp(-j\delta_2 x) - V_r^2\exp(j\delta_2 x)]\end{aligned} \tag{6.93}$$

$$\begin{aligned}\sigma_{xx}^f(x) = &-\phi Z_1^f\mu_1[V_i^1\exp(-j\delta_1 x) - V_r^1\exp(j\delta_1 x)]\\ &- \phi Z_2^f\mu_2[V_i^2\exp(-j\delta_2 x) - V_r^2\exp(j\delta_2 x)]\end{aligned} \tag{6.94}$$

At $x = 0$, where the wall and the material are in contact, the velocities are equal to zero

$$\dot{u}^s(0) = \dot{u}^f(0) = 0 \tag{6.95}$$

At $x = -l$, the porous material is in contact with the free air. Let us consider a thin layer of air and porous material, including this boundary. This layer is represented in Figure 6.9.

Let us denote by $p(-l - \varepsilon)$ the pressure in the air on the left-hand side of the thin layer, while $\sigma_{xx}^s(-l + \varepsilon)$ and $\sigma_{xx}^f(-l + \varepsilon)$ are the stresses acting on the air and on the frame on the right-hand side. The resulting force ΔF acting on the thin layer is

$$\Delta F = p(-l - \varepsilon) + \sigma_{xx}^s(-l + \varepsilon) + \sigma_{xx}^f(-l + \varepsilon) \tag{6.96}$$

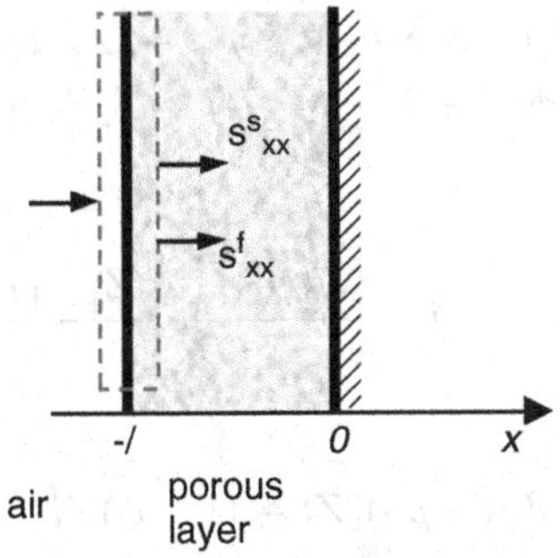

Figure 6.9 A thin layer of air and porous material including the boundary.

This force tends to zero with ε, and a boundary condition for the stress at $x = -l$ is

$$p(-l) + \sigma_{xx}^s(-l) + \sigma_{xx}^f(-l) = 0 \tag{6.97}$$

Another boundary condition is derived from the continuity of pressure and can be expressed

$$\sigma_{xx}^f(-l) = -\phi p(-l) \tag{6.98}$$

ϕ being the porosity of the material. The use of Equations (6.97) and (6.98) yields

$$\sigma_{xx}^s(-l) = -(1-\phi)p(-l) \tag{6.99}$$

The conservation of the volume of air and frame through the plane $x = -l$ yields

$$\phi\dot{u}^f(-l) + (1-\phi)\dot{u}^s(-l) = \dot{u}^a(-l) \tag{6.100}$$

$\dot{u}^a(-l)$ being the velocity of the free air at the boundary. The surface impedance Z of the material is given by

$$Z = p(-l)/\dot{u}^a(-l) \tag{6.101}$$

This surface impedance can be evaluated in the following way. At first, it can easily be shown that Equations (6.91), (6.92) and (6.95) together yield

$$V_i^1 = -V_r^1, \qquad V_i^2 = -V_r^2 \tag{6.102}$$

Equations (6.98)–(6.102) yield

$$\begin{aligned} -(1-\phi)\dot{u}^a(-l)Z = -Z_1^s V_i^1[\exp(j\delta_1 l) + \exp(-j\delta_1 l)] \\ - Z_2^s V_i^2[\exp(j\delta_2 l) + \exp(-j\delta_2 l)] \end{aligned} \tag{6.103}$$

$$\begin{aligned} -\phi\dot{u}^a(-l)Z = -Z_1^f \phi\mu_1 V_i^1[\exp(j\delta_1 l) + \exp(-j\delta_1 l)] \\ - Z_2^f \phi\mu_2 V_i^2[\exp(j\delta_2 l) + \exp(-j\delta_2 l)] \end{aligned} \tag{6.104}$$

$$\begin{aligned} &[\phi\mu_1 + (1-\phi)]V_i^1[\exp(j\delta_1 l) - \exp(-j\delta_1 l)] \\ &\quad + [\phi\mu_2 + (1-\phi)]V_i^2[\exp(j\delta_2 l) - \exp(-j\delta_2 l)] = \dot{u}^a(-l) \end{aligned} \tag{6.105}$$

This system of three equations (6.103)–(6.105) has a solution (V_i^1, V_i^2) if

$$\begin{vmatrix} -(1-\phi)Z & -2Z_1^s\cos\delta_1 l & -2Z_2^s\cos\delta_2 l \\ -Z & -2Z_1^f\mu_1\cos\delta_1 l & -2Z_2^f\mu_2\cos\delta_2 l \\ 1 & 2j\sin\delta_1 l(\phi\mu_1 + 1 - \phi) & 2j\sin\delta_2 l(\phi\mu_2 + 1 - \phi) \end{vmatrix} = 0 \tag{6.106}$$

and Z is given by

$$Z = -j\frac{(Z_1^s Z_2^f \mu_2 - Z_2^s Z_1^f \mu_1)}{D} \tag{6.107}$$

where D is given by

$$\begin{aligned} D = (1 - \phi + \phi\mu_2)[Z_1^s - (1-\phi)Z_1^f\mu_1]\mathrm{tg}\delta_2 l \\ + (1 - \phi + \phi\mu_1)[Z_2^f\mu_2(1-\phi) - Z_2^s]\mathrm{tg}\delta_1 l \end{aligned} \tag{6.108}$$

A systematic method of calculating the surface impedance at oblique incidence is based on transfer matrices. It is presented in Chapter 11.

6.6.3 Example: Fibrous material

The surface impedances at normal incidence calculated by Equation (6.107), of two samples of different thicknesses made up of the material described in Section 6.5.4, are represented for $l = 10$ cm and $l = 5{\cdot}4$ cm in Figures 6.10 and 6.11, and compared with measured values (Allard *et al.* 1991). Measurements were performed in a free field on samples of large lateral dimensions.

The agreement between measurement and prediction by Equation (6.107) is good in the entire range of frequencies where the measurement was performed. A peak appears in the real and the imaginary parts of the impedance around 470 Hz for $l = 10$ cm and 860 Hz for $l = 5{\cdot}6$ cm. The surface impedances calculated by Equation (6.107), and by Equations (4.137), (6.86) and (6.87) for the same material with a rigid frame, are close to each other, except around the peaks which are not predicted by the one-wave model. The peaks appear around the $\lambda/4$ resonance of the frame-borne wave which is located at the frequency f_r such that

$$l\mathrm{Re}(\delta_b) = \frac{\pi}{2} \tag{6.109}$$

The quantity δ_b is very close to δ_1' given by Equation (6.88), and f_r can be written as

$$f_r = \frac{1}{4l}\sqrt{\frac{\mathrm{Re}(K_c)}{\rho_1}} \tag{6.110}$$

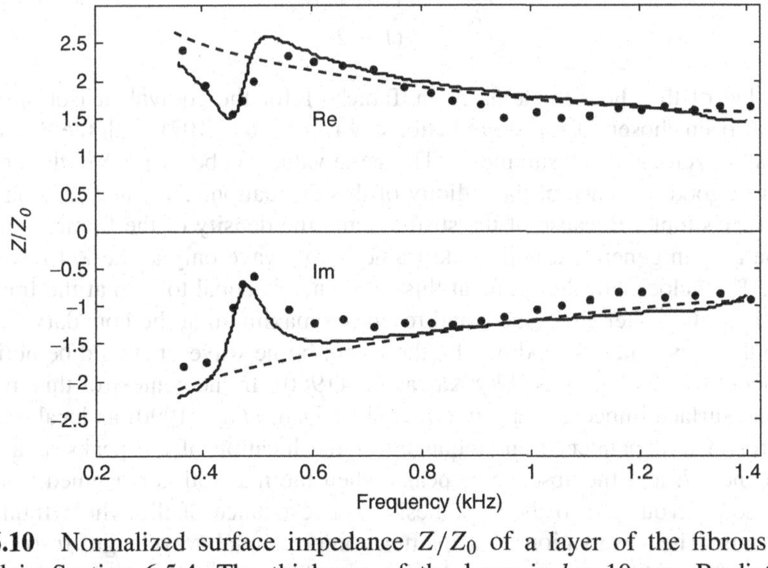

Figure 6.10 Normalized surface impedance Z/Z_0 of a layer of the fibrous material described in Section 6.5.4. The thickness of the layer is $l = 10$ cm. Prediction with Equation (6.107). ———. Prediction for the same material with a rigid frame: ------. Measurements: • ••. (Measurement taken from Allard *et al.* 1991).

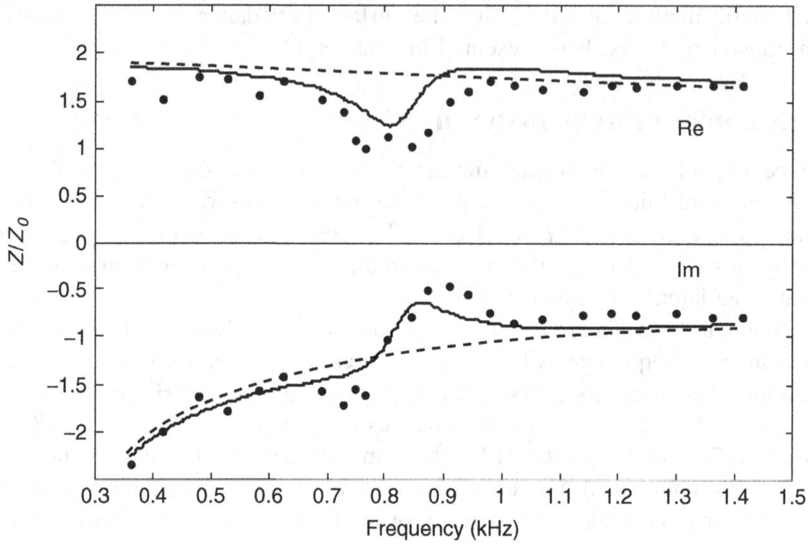

Figure 6.11 Normalized surface impedance Z/Z_0 of a layer of the fibrous material described in Section 6.5.4. The thickness of the layer is $l = 5 \cdot 6$ cm. Prediction with Equation (6.107). ______. Prediction for the same material with a rigid frame: ------. Measurements: • • •. (Measurement taken from Allard *et al.*1991).

K_c is given by Eq. (1.76). In terms of the shear modulus and the Poisson ratio, it is given by

$$K_c = \frac{2(1-\nu)N}{(1-2\nu)} \tag{6.111}$$

The value of the shear modulus N in Table 6.1 for the equivalent isotropic porous material has been chosen to adjust predictions by Equation (6.107), with the Poisson coefficient equal to zero, and measurements. The same value can be used for both thicknesses. This fact is a good evidence of the validity of this evaluation. The interpretation of these results is very simple. Because of the stiffness and the density of the frame, the acoustic field in the air can generate a noticeable frame-borne wave only at the $\lambda/4$ resonance of the frame. The velocity of the frame at this resonance is equal to zero at the frame–wall contact, where the material is glued, and reaches a maximum at the boundary frame-free air, where the impedance is modified by the frame-borne wave. It should be noticed that another model has been proposed by Kawasima (1960). In the context of this model, the peaks in the surface impedance are interpreted by Dahl *et al.* (1990) as local resonances of the fibres. The dependence on frequency of the location of the peaks as a function of the thickness l, and the absence of peaks when the material is not glued to the wall, could be more favourable to the hypothesis of a resonance of the whole frame in our measurements. The observation of a quarter compressional wavelength resonance was possible in a free field on a large sample. It seems difficult to observe the compressional resonance in a Kundt tube, due to the contact with the lateral surface of the tube, and the geometry of the porous samples in the tube. A numerical study on the effect of the lateral boundary conditions of the tube on the impedance curve has been conducted by Pilon *et al.* (2003) using the finite element method (see Chapter 13).

Appendix 6.A: Other representations of the Biot theory

The Biot second representation (Biot 1962)

The total stress components $\sigma^t_{ij} = \sigma^s_{ij} + \sigma^f_{ij} = \sigma^s_{ij} - \phi p \delta_{ij}$ and the pressure p are used instead of σ^s_{ij} and σ^f_{ij}. The displacements $\boldsymbol{u}^{\mathrm{s}}$ and $\boldsymbol{w} = \phi(\boldsymbol{u}^f - \boldsymbol{u}^s)$ are used instead of the couple $\boldsymbol{u}^{\mathrm{s}}$, $\boldsymbol{u}^{\mathrm{f}}$. The medium the frame is made of is not compressible. The stress–strain Equation (6.3) can be replaced by

$$-\phi p = K_f(\mathrm{div}\, \boldsymbol{u}^s - \zeta) \tag{6.A.1}$$

where $\zeta = -\mathrm{div}\boldsymbol{w}$. The stress elements of the frame in vacuum are given by

$$\hat{\sigma}_{ij} = \delta_{ij}\left(K_b - \frac{2}{3}N\right)\mathrm{div}\, \boldsymbol{u}^{\mathrm{s}} + 2Ne^s_{ij} \tag{6.A.2}$$

The stress elements of the saturated frame are given by

$$\sigma^s_{ij} = \delta_{ij}\left[\left(K_b - \frac{2}{3}N\right)\mathrm{div}\, \boldsymbol{u}^{\mathrm{s}} - (1-\phi)p\right] + 2Ne^s_{ij} \tag{6.A.3}$$

and the total stress components are given by

$$\sigma^t_{ij} = \delta_{ij}\left[\left(K_b - \frac{2}{3}N\right)\mathrm{div}\, \boldsymbol{u}^{\mathrm{s}} - p\right] + 2Ne^s_{ij} \tag{6.A.4}$$

Equations (6.A.1) and (6.A.4) provide a simple description of the stress in a porous medium when the bulk modulus K_s of the elastic solid from which the frame is made is much larger than the other coefficients of rigidity. For instance, the third 'gedanken experiment' can be described as follows. A variation $d\theta^f$ creates a variation $\Delta\xi = -\phi \mathrm{d}\theta^f$ and a variation $\mathrm{d}p = -K_f \mathrm{d}\theta^f$. This variation is related to a variation of the diagonal elements of the total stress which is the sum of the variation in air and in the frame of these elements $-\phi \mathrm{d}p - (1-\phi)\mathrm{d}p = -\mathrm{d}p$. A general description of the second representation, when K_{s} is not very large compared with the other rigidity coefficients, and when the porous structure is anisotropic, was performed by Cheng (1997). A description of the different waves with the second representation is performed in Chapter 10 for transversally isotropic porous media.

The Dazel representation (Dazel *et al.* 2007)

Only the simple case where the medium the frame is made of is not compressible is considered. The total displacement $\boldsymbol{u}^t = (1-\phi)\boldsymbol{u}^s + \phi \boldsymbol{u}^f$ and $\boldsymbol{u}^s$ are used instead of the couple $\boldsymbol{u}^{\mathrm{s}}$, $\boldsymbol{u}^{\mathrm{f}}$. The normal velocity in air at an air porous layer interface is equal to the normal component of the total displacement. With $\varsigma^t = \nabla.\boldsymbol{u}^t$ and $K_{eq} = K_f/\phi$ the pressure is given by

$$p = -K_{eq}\varsigma^t \tag{6.A.5}$$

and the stress components are given by Equations (6.A.2)–(6.A.4). The wave equations can be written

$$\left(K_b + \frac{1}{3}N\right)\nabla\nabla.\boldsymbol{u}^s + N\nabla^2\boldsymbol{u}^s = -\omega^2\tilde{\rho}_s\boldsymbol{u}^s - \omega^2\tilde{\rho}_{eq}\tilde{\gamma}\boldsymbol{u}^t \tag{6.A.6}$$

$$K_{eq}\nabla\nabla.\boldsymbol{u}^t = -\omega^2\tilde{\rho}_{eq}\tilde{\gamma}\boldsymbol{u}_s - \omega^2\tilde{\rho}_{eq}\boldsymbol{u}^t \tag{6.A.7}$$

In the previous equations, $\tilde{\gamma}$, $\tilde{\rho}_{eq}$, and $\tilde{\rho}_s$ are given respectively by

$$\tilde{\gamma} = \phi\left(\frac{\tilde{\rho}_{12}}{\tilde{\rho}_{22}} - \frac{1-\phi}{\phi}\right) \tag{6.A.8}$$

$$\tilde{\rho}_{eq} = \frac{\tilde{\rho}_{22}}{\phi^2} \tag{6.A.9}$$

$$\tilde{\rho}_s = \tilde{\rho}_{11} - \frac{\tilde{\rho}_{12}^2}{\tilde{\rho}_{22}} + \tilde{\gamma}^2\tilde{\rho}_{eq} \tag{6.A.10}$$

Using two scalar potentials ϕ^s and ϕ^t for the compressional waves gives the following equation of motion

$$-\omega^2[\boldsymbol{\rho}]\begin{Bmatrix}\varphi^s\\ \varphi^t\end{Bmatrix} = [\boldsymbol{K}]\nabla^2\begin{Bmatrix}\varphi^s\\ \varphi^t\end{Bmatrix} \tag{6.A.11}$$

where $[\boldsymbol{\rho}]$ and $[\boldsymbol{K}]$ are given, respectively, by

$$[\rho] = \begin{bmatrix}\tilde{\rho}_s & \tilde{\rho}_{eq}\tilde{\gamma}\\ \tilde{\rho}_{eq}\tilde{\gamma} & \tilde{\rho}_{eq}\end{bmatrix}, [\boldsymbol{K}] = \begin{bmatrix}\hat{P} & 0\\ 0 & K_{eq}\end{bmatrix} \tag{6.A.12}$$

where

$$\hat{P} = K_b + \frac{4}{3}N$$

The matrix $[\boldsymbol{K}]$ is diagonal. The wave numbers of the Biot compressional waves and the ratios μ_i, i = 1, 2 are obtained from Equation (6.A.12). The wave number of the shear wave and the ratio μ_3 are obtained using a potential vector. It is shown that with the Dazel representation, the prediction of the surface impedance of a porous media can be performed with a mathematical formalism which is simpler than the one associated with the first formalism.

The mixed pressure–displacement representation (Atalla *et al.* 1998)

In this representation the displacement $\boldsymbol{u}^s$ and the pressure p are used instead of the couple $\boldsymbol{u}^s$, $\boldsymbol{u}^f$. The developments assume that the porous material properties are homogeneous. The derivation follows the presentation of Atalla *et al.* (1998). Note that that a more general time domain formulation valid for anisotropic materials is given by Gorog *et al.* (1997).

The system (Equations 6.54, 6.55) is first rewritten:

$$\begin{cases}\omega^2\tilde{\rho}_{11}\boldsymbol{u}^s + \omega^2\tilde{\rho}_{12}\boldsymbol{u}^f + div\,\sigma^s = 0\\ \omega^2\tilde{\rho}_{22}\boldsymbol{u}^f + \omega^2\tilde{\rho}_{12}\boldsymbol{u}^s - \phi\,\boldsymbol{grad}\;p = 0\end{cases} \tag{6.A.13}$$

Using the second equation in (6.A.13), the displacement vector of the fluid phase $\boldsymbol{u}^{\mathrm{f}}$ is expressed in terms of the pressure p in the pores and in terms of the displacement vector of the solid phase particle $\boldsymbol{u}^{\mathbf{s}}$:

$$\boldsymbol{u}^{\mathrm{f}} = \frac{\phi}{\tilde{\rho}_{22}\,\omega^2}\,\boldsymbol{grad}\,p - \frac{\tilde{\rho}_{12}}{\tilde{\rho}_{22}}\boldsymbol{u}^s \tag{6.A.14}$$

Using Equation (6.A.14), the first equation in (6.A.13) transforms into:

$$\omega^2\tilde{\rho}\boldsymbol{u}^s + \phi\frac{\tilde{\rho}_{12}}{\tilde{\rho}_{22}}\,\boldsymbol{grad}\mathrm{p} + div\,\sigma^s = 0 \tag{6.A.15}$$

where the following effective density is introduced:

$$\tilde{\rho} = \tilde{\rho}_{11} - \frac{(\tilde{\rho}_{12})^2}{\tilde{\rho}_{22}} \tag{6.A.16}$$

Equation (6.A.15) is still dependent on the fluid phase displacement $\boldsymbol{u}^{\mathrm{f}}$ because of the dependency $\sigma^s = \sigma^s(\boldsymbol{u}^s, \boldsymbol{u}^f)$. To eliminate this dependency, Equations (6.2) and (6.3), are combined to obtain:

$$\sigma_{ij}^s(\boldsymbol{u}^s) = \hat{\sigma}_{ij}^s(\boldsymbol{u}^s)\ \text{-}\phi\frac{\tilde{Q}}{\tilde{R}}p\delta_{ij} \tag{6.A.17}$$

with $\hat{\sigma}^s$ the stress of the frame in vacuum defined in Equation (6.A.2). Note that tilde is used here to account for damping and possible frequency dependence of the elastic coefficients Q and R (e.g. polymeric frame).

Equation (6.A.17) is next used to eliminate the dependency $\sigma^s = \sigma^s(\boldsymbol{u}^s, \boldsymbol{u}^f)$ in Equation (6.A.15). This leads to the solid phase equation in terms of the ($\boldsymbol{u}^s$, p) variables:

$$div\ \hat{\sigma}^s(\boldsymbol{u}^s) + \tilde{\rho}\omega^2\boldsymbol{u}^s + \tilde{\gamma}\ \boldsymbol{grad}\ p = 0 \tag{6.A.18}$$

with:

$$\tilde{\gamma} = \phi\left(\frac{\tilde{\rho}_{12}}{\tilde{\rho}_{22}} - \frac{\tilde{Q}}{\tilde{R}}\right) \tag{6.A.19}$$

In the case where $K_b/K_s \ll 1$, Equation (6.A.19) reduces to (6.A.8).

Next, to derive the fluid phase equation in terms of ($\boldsymbol{u}^s$, p) variables, the divergence of Equation (6.A.14) is taken:

$$div\ \boldsymbol{u}^f = \frac{\phi}{\omega^2\tilde{\rho}_{22}}\Delta p - \frac{\tilde{\rho}_{12}}{\tilde{\rho}_{22}}\mathrm{div}\ \boldsymbol{u}^s \tag{6.A.20}$$

Combining this equation with the second equation in (6.A.13), the fluid phase equation is obtained in terms of the ($\boldsymbol{u}^s$, p) variables:

$$\Delta p + \frac{\tilde{\rho}_{22}}{\tilde{R}}\omega^2 p + \frac{\tilde{\rho}_{22}}{\phi^2}\tilde{\gamma}\omega^2\mathrm{div}\ \boldsymbol{u}^s = 0 \tag{6.A.21}$$

This equation is the classical equivalent fluid equation for absorbing media with a source term. The first two terms of this equation may be obtained directly from Biot's equations in the limit of a rigid skeleton.

Grouping Equations (6.A.18) and (6.A.21), the Biot poroelasticity equations in terms of $(\boldsymbol{u}^s, p)$ variables are given by:

$$\begin{cases} div\ \hat{\sigma}^s(\boldsymbol{u}^s) + \tilde{\rho}\omega^2\boldsymbol{u}^s + \tilde{\gamma}\ \boldsymbol{grad}\, p = 0 \\ \Delta p + \dfrac{\tilde{\rho}_{22}}{\tilde{R}}\omega^2 p + \dfrac{\tilde{\rho}_{22}}{\phi^2}\tilde{\gamma}\omega^2 \text{div}\ \boldsymbol{u}^s = 0 \end{cases} \tag{6.A.22}$$

This system exhibits the classical form of a fluid-structure coupled equation. However, the coupling is of a volume nature since the poroelastic material is a superposition in space and time of the elastic and fluid phases. The first two terms of the structure equation represent the dynamic behaviour of the material in vacuum, while the first two terms of the fluid equation represent the dynamic behaviour of the fluid when the frame is supposed motionless. The third terms in both equations couple the dynamics of the two phases. It is shown in Chapter 13 that this formalism leads to a simple weak formulation for finite element based numerical implementations. An example of the application of this formalism to the optimization of the surface impedance of a porous material is given in Kanfoud and Hamdi (2009).

References

Allard, J.F., Depollier, C., Guignouard, P. and Rebillard, P. (1991) Effect of a resonance of the frame on the surface impedance of glass wool of high density and stiffness. *J. Acoust. Soc. Amer.*, **89**, 999–1001.

Atalla N., Panneton, R. and Debergue, P. (1998) A mixed displacement pressure formulation for poroelastic materials. *J. Acoust. Soc. Amer.*, **104**, 1444–1452.

Beranek, L. (1947) Acoustical properties of homogeneous isotropic rigid tiles and flexible blankets. *J. Acoust. Soc. Amer.*, **19**, 556–68.

Biot, M.A. (1956) The theory of propagation of elastic waves in a fluid-saturated porous solid. I. Low frequency range. II. Higher frequency range. *J. Acoust. Soc. Amer.*, **28**, 168–91.

Biot, M.A. and Willis, D.G. (1957) The elastic coefficients of the theory of consolidation. *J. Appl. Mechanics*, **24**, 594–601.

Biot, M.A. (1962) Generalized theory of acoustic propagation in porous dissipative media. *J. Acoust. Soc. Amer.*, **34**, 1254–1264.

Brown, R.J.S. and Korringa, J. (1975), On the dependence of the elastic properties of a porous rock on the compressibility of the pore fluid. *Geophysics*, **40**, 608–16.

Cheng, A.H.D. (1997) Material coefficients of anisotropic poroelasticity. *Int. J. Rock Mech. Min. Sci*. **34**, 199–205.

Dahl, M.D., Rice, E.J. and Groesbeck, D.E. (1990) Effects of fiber motion on the acoustical behaviour of an anisotropic, flexible fibrous material. *J. Acoust. Soc. Amer.*, **87**, 54–66.

Dazel, O., Brouard, B., Depollier., C. and Griffiths. S. (2007) An alternative Biot's displacement formulation for porous materials. *J. Acoust. Soc. Amer.*, **121**, 3509–3516.

Depollier, C., Allard, J.F. and Lauriks, W. (1988) Biot theory and stress–strain equations in porous sound absorbing materials. *J. Acoust. Soc. Amer.*, **84**, 2277–9.

Gorog S., Panneton, R. and Atalla, N. (1997) Mixed displacement–pressure formulation for acoustic anisotropic open porous media. *J. Applied Physics*, **82**(9), 4192–4196.

Johnson, D.L. (1986) Recent developments in the acoustic properties of porous media. In *Proc. Int. School of Physics Enrico Fermi, Course XCIII*, ed. D. Sette. North Holland Publishing Co., Amsterdam, pp. 255–90.

Kanfoud, J., Hamdi, M.A., Becot F.-X. and Jaouen L. (2009) Development of an analytical solution of modified Biot's equations for the optimization of lightweight acoustic protection. *J. Acoust. Soc. Amer.*, **125**, 863–872

Kawasima, Y. (1960) Sound propagation in a fibre block as a composite medium. *Acustica*, **10**, 208–17.

Korringa, J. (1981), On the Biot Gassmann equations for the elastic moduli of porous rocks. *J. Acoust. Soc. Amer.*, **70**, 1752–3.

Lambert, R.F. (1983) Propagation of sound in highly porous open-cell elastic foams. *J. Acoust. Soc. Amer.*, **73**, 1131–8.

Landau, L.D. and Lifshitz, E.M. (1959) *Fluid Mechanics*. Pergamon, New York.

Pilon, D., Panneton, R. and Sgard, F. (2003) Behavioral criterion quantifying the edge-constrained effects on foams in the standing wave tube. *J. Acoust. Soc. Amer.*, **114**(4), 1980–1987.

Pride, S.R., Berryman J.G. (1998) Connecting theory to experiment in poroelasticity. *J. Mech. Phys. Solids*, **46**, 19–747.

Sides, D.J., Attenborough, K. and Mulholland, K.A. (1971) Application of a generalized acoustic propagation theory to fibrous absorbants. *J. Sound Vib.*, **19**, 49–64.

Zwikker, C. and Kosten, C.W. (1949) *Sound Absorbing Materials*. Elsevier, New York.

7

Point source above rigid framed porous layers

7.1 Introduction

Sound propagation in a plane layered structure is described in the simplest way with plane waves, and the Kundt tube can be used to measure the reflection coefficient of a plane surface having limited lateral dimensions. In the medium audible frequency range, it is easy to build, with a pipe and a compression driver, a sound source which is a good approximation of a monopole source. The monopole field reflected by a porous layer can provide useful information about the porous structure if an adequate modelling of reflection is performed. An exact model for the reflected field and several approximations are presented.

7.2 Sommerfeld representation of the monopole field over a plane reflecting surface

The Sommerfeld representation provides an exact integral representation of the reflected field. The monopole pressure field p is created by an ideal unit point source S. The pressure field p only depends on the distance R from the source S (Figure 7.1)

$$p(R) = \frac{\exp(-jk_0 R)}{R} \tag{7.1}$$

where k_0 is the wave number in the free air. The spherical wave can be expended into plane waves using a two-dimensional Fourier transform (Brekhovskikh and Godin 1992),

Propagation of Sound in Porous Media: Modelling Sound Absorbing Materials, Second Edition J. F. Allard and N. Atalla

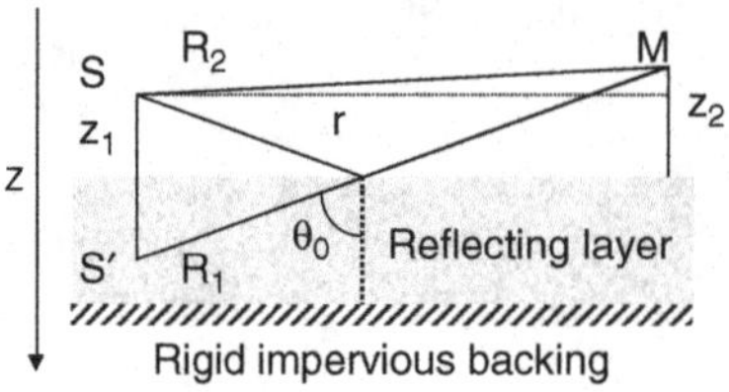

Figure 7.1 The source–receiver geometry, the monopole source at S, its image at S′, and the receiver at M above the layer. The angle θ_0 is the angle of specular reflection, R_1 is the distance from the image of the source to the receiver, and R_2 is the distance from the source to the receiver.

and p at M can be rewritten

$$p(R) = \frac{-j}{2\pi}\int_{-\infty}^{\infty}\int_{-\infty}^{\infty}\frac{\exp[-j(\xi_1 x + \xi_2 y + \mu|z_2 - z_1|)]}{\mu}\mathrm{d}\xi_1\mathrm{d}\xi_2,$$
$$\mu = \sqrt{k_0^2 - \xi_1^2 - \xi_2^2},\ \mathrm{Im}\,\mu \le 0, \mathrm{Re}\,\mu \ge 0 \tag{7.2}$$

Let $V(\xi_1, \xi_2)$ be the plane wave reflection coefficient of the layer. If the layer is isotropic or transversely isotropic with the axis of symmetry Z, V only depends on $\xi = (\xi_1^2 + \xi_2^2)^{1/2}$, and the reflected pressure p_r at M can be written

$$p_r = \frac{-j}{2\pi}\int_{-\infty}^{\infty}\int_{-\infty}^{\infty} V(\xi/k_0)\frac{\exp[-j(\xi_1 x + \xi_2 y - \mu(z_1 + z_2))]}{\mu}\mathrm{d}\xi_1\mathrm{d}\xi_2 \tag{7.3}$$

The variables ξ and μ are related to an angle of incidence θ defined by

$$\cos\theta = \mu/k_0$$
$$\sin\theta = \xi/k_0 \tag{7.4}$$

and $V(\xi/k_0)$ is the reflection coefficient for the angle of incidence θ. For $\xi \le k_0$, θ is a real angle, and for

$$\xi > k_0, \quad \theta = \frac{\pi}{2} + j\beta$$

where $\sinh\beta = j\mu/k_0$, $\cosh\beta = \xi/k_0$.

Using the polar coordinates (ψ, ξ) and (r, φ) defined by $\xi_1 = \xi\cos\psi$, $\xi_2 = \xi\sin\psi$, $x = r\cos\varphi$, $y = r\sin\varphi$, Equation (7.3) can be rewritten

$$p_r = \frac{-j}{2\pi}\int_0^{2\pi}\int_0^{\infty}\exp[j\mu(z_1 + z_2)]\frac{V(\xi/k_0)}{\mu}\exp[-jr\xi\cos(\psi - \varphi)]\xi\mathrm{d}\xi\mathrm{d}\psi \tag{7.5}$$

Using $\int_0^{2\pi}\exp[-jr\xi\cos(\psi - \varphi)]\mathrm{d}\psi = 2\pi \mathrm{J}_0(r\xi)$ (Abramovitz and Stegun 1972), Equation (7.5) becomes

$$p_r = -j\int_0^{\infty}\frac{V(\xi/k_0)}{\mu}\mathrm{J}_0(r\xi)\exp[j\mu(z_1 + z_2)]\xi\mathrm{d}\xi \tag{7.6}$$

The reflected pressure depends on the sum $z_1 + z_2$, not on each height separately. The right-hand side of this equation is referred to as the Sommerfeld integral. The integral can be evaluated up to a limit for ξ which depends on $z_1 + z_2$ because μ in the exponential is imaginary with a positive imaginary part for $\xi > k_0$ which increases with μ. The singularity for $\mu = 0$ is removed by using μ instead of ξ as a variable of integration. A simple test for the accuracy of the evaluation for a given geometry consists in the comparison of the evaluation performed for $V = 1$ and $\exp(-jk_0R_1)/R_1$. The Bessel function is related to the Hankel function of first order H_0^1 by $J_0(u) = 0.5(H_0^1(u) - H_0^1(-u))$ with $\mu(-\xi) = \mu(\xi)$, and $V(-\xi/k_0) = V(\xi/k_0)$. Therefore p_r can be rewritten as

$$p_r = \frac{j}{2}\int_{-\infty}^{\infty} \frac{\xi}{\mu} H_0^1(-\xi r) V(\xi/k_0) \exp[j\mu(z_1 + z_2)]\, d\xi \tag{7.7}$$

From Equation (3.39), the surface impedance at oblique incidence is given by

$$Z_s(\sin\theta) = -j\frac{Z}{\phi\cos\theta_1}\ \text{cotg}\ kl\cos\theta_1 \tag{7.8}$$

where l is the thickness of the layer, ϕ is the porosity, $Z = (\rho K)^{1/2}$ is the characteristic impedance in the air saturating the porous medium, k is the wave number in the porous medium, θ_1 is the refraction angle satisfying $k\sin\theta_1 = k_0\sin\theta$, and $\cos\theta_1$ is given by

$$\cos^2\theta_1 = 1 - \frac{1}{n^2} - \frac{1}{n^2}\cos^2\theta \tag{7.9}$$

where $n = k/k_0$. From Equation (3.44) the reflection coefficient V is given by

$$V(\xi/k_0) = \frac{Z_s(\sin\theta) - Z_0/\cos\theta}{Z_s(\sin\theta) + Z_0/\cos\theta} \tag{7.10}$$

7.3 The complex sinθ plane

Equation (7.6) is an integral over the real variable ξ. It can be considered as an integral on $\sin\theta = \xi/k_0$ in the complex $s = \sin\theta$ plane on the right-hand side of the real axis. Equation (7.7) is an integral over the whole real sinθ axis. It may be advantageous to use other paths of integration in this plane, to show the contribution of the poles, and/or to get approximate expressions of p_r more tractable for large r than Equation (7.7). A symbolic representation of the path of integration of Equation (7.7) is given in Figure 7.2.

Small displacements of the path and the cuts are performed to show their relative positions. The reflection coefficient V of a layer of finite thickness involves $\cos\theta$ and $\cos\theta_1$. For a layer of finite thickness, Z_s and V are even functions of $\cos\theta_1$, but V depends on the sign of $\cos\theta$. At each point in the $s = \sin\theta$ plane the reflection coefficient can take two values, depending on the choice of the sign of $\cos\theta$. Following Brekhovskikh and Godin (1992), we cut the s plane with the lines

$$1 - s^2 = u_1, \quad u_1 \text{ real} \geq 0 \tag{7.11}$$

These are shown as dotted lines in Figure 7.2. They lie on the whole imaginary axis and on the real axis for $0 \leq |s| \leq 1$. On these lines, the imaginary part of

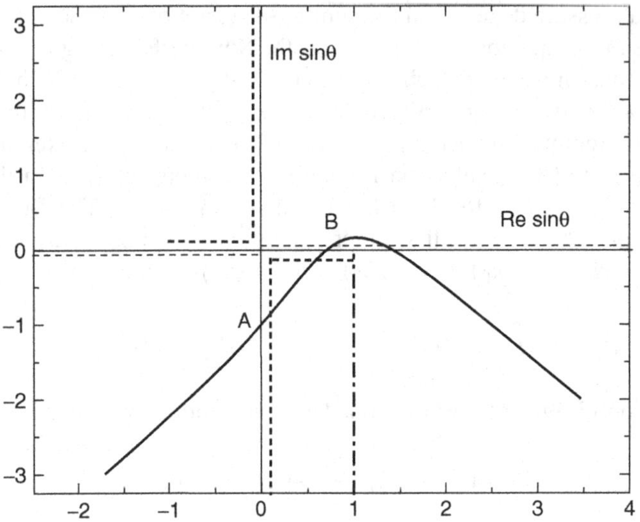

Figure 7.2 The complex $\sin\theta$ plane. The solid line is the path of steepest descent for $\theta_0 = \pi/4$, both cuts are represented by short dash lines along the imaginary axis and for $|Re \sin\theta| \leqslant 1$ along the real axis. The initial integration path is represented by dashed lines parallel to the real axis.

$\cos\theta = (1 - s^2)^{1/2}$ is equal to 0, and when one of the two lines is crossed, keeping constant the real part of $\cos\theta$ corresponds to a change of sign of the imaginary part. The dependence of the reflected wave in z_2 is $\exp(jz_2 \cos\theta)$. The amplitude of the wave tends to 0 when $z_2 \to -\infty$ if $\text{Im}\cos\theta < 0$. The s plane is a superposition of two planes, the physical Riemann sheet, where $\text{Im}\cos\theta < 0$, and the second Riemann sheet, where $\text{Im}\cos\theta > 0$. As indicated previously, the communication between both sheets can be performed without discontinuity by crossing the cuts. For a semi-infinite layer, the surface impedance becomes

$$Z_s = Z/\phi \cos\theta_1 \tag{7.12}$$

and Z_s and V depend on the sign of $\cos\theta_1$. New cuts must be added in the s plane, defined by

$$n^2 - s^2 = u_2, \quad u_2 \text{ real} \geq 0 \tag{7.13}$$

7.4 The method of steepest descent (passage path method)

The method has been used by Brekhovskikh and Godin (1992) for a similar problem, the prediction of the monopole field reflected by a semi-infinite fluid. The method provides predictions of p_r which are valid for $k_0 R_1 \gg 1$. The same calculations are performed in what follows, with the difference that, for layers of finite thickness, the surface impedance in Equation (7.10) is given by Equation (7.8) and the cut related to $\cos\theta_1$ is removed because both choices of $\cos\theta_1$ give the same impedance. The Hankel function in Equation (7.7) is replaced by its asymptotic expression

$$H_0^1(x) = \left(\frac{2}{\pi x}\right)^{1/2} \exp\left[j\left(x - \frac{\pi}{4}\right)\right] \tag{7.14}$$

and the new variable of integration $s = \xi/k_0$ is used instead of ξ. Using $z_2 + z_1 = -R_1 \cos\theta_0$ and $r = R_1 \sin\theta_0$, Equation (7.7) can be rewritten

$$p_r = \left(\frac{k_0}{2\pi r}\right)^{1/2} \exp\left(\frac{-j\pi}{4}\right) \int_{-\infty}^{\infty} F(s) \exp[k_0 R_1 f(s)]\, \mathrm{d}s \tag{7.15}$$

F and f are given by

$$f(s) = -j(s \sin\theta_0 + \sqrt{1 - s^2} \cos\theta_0) \tag{7.16}$$

$$F(s) = V(s)\sqrt{\frac{s}{1 - s^2}} \tag{7.17}$$

An asymptotic evaluation of the integral in Equation (7.15) is obtained in the following way. The initial contour of integration in the complex sinθ plane for Equation (7.15) can be modified within certain limits without modification of the result. The function f(s) is rewritten

$$f(s) = f_1(s) + jf_2(s) \tag{7.18}$$

where f_1 and f_2 are real. A new contour γ is used, where f_1 has a maximum at a point s_M, and decreases as rapidly as possible with $|s - s_M|$. If f is an analytic function, the line of steepest descent of f_1 is the line of constant value of f_2. The derivative $\mathrm{d}f(s)/\mathrm{d}s = 0$ for $s = s_M$, and s_M is the stationary point. The line of constant f_2 including the stationary point is the best choice for the path γ. This line is called the steepest descent path. For $k_0 R_1 \gg 1$, the contribution on γ to the integral is restricted to a small domain around s_M. The stationary point in the s plane is located at $s = \sin\theta_0$, where θ_0 is the angle of specular reflection represented in Figure 7.1. It has been shown by Brekhovskikh and Godin (1992) that the path of steepest descent γ is specified by

$$s \sin\theta_0 + (1 - s^2)^{1/2} \cos\theta_0 = 1 - ju_3^2, \quad -\infty < u_3 < \infty \tag{7.19}$$

In the previous equations and in what follows, $\sqrt{1 - s^2} = \cos\theta$, the choice of the determination depends on the location on the passage path. The passage path for $\theta_0 = \pi/4$ is represented in Figure 7.2. The choice for $\sqrt{1 - s^2}$ is $\mathrm{Im}\sqrt{1 - s^2} \le 0$, except between A and B, where the path has once crossed the cosθ cut, and $\mathrm{Im}\sqrt{1 - s^2} \ge 0$. The stationary point is B where $\sin\theta = \sin\theta_0$. For $\theta_0 = 0$ the passage path is the real s axis, i.e. the initial path of integration and for $\theta_0 = \pi/2$, the passage path is the half axis defined by $\mathrm{Re}\, s = 1$ located in the Ims ≤ 0 half plane. This axis is the dot–dash line in Figure 7.2. If a pole of the reflection coefficient crosses the path of integration when it is deformed, the pole residue must be added to the integral. The expression of residues is given in Section 7.5.3. There is no pole contribution for $\theta_0 = 0$ because the path of integration is not modified. As indicated in Section 7.5.3, the crossing is impossible for semi-infinite layers. Moreover, the pole contributions decrease exponentially when k_0R_1 increases and become negligible. The asymptotic expression for p_r obtained to second-order approximation in $1/(k_0R_1)$ by Brekhovskikh and Godin (1992), when the pole contributions are neglected, can be written

$$p_r = \frac{\exp(-jk_0R_1)}{R_1}\left[V(\sin\theta_0) + \frac{j}{k_0}\frac{N}{R_1}\right] \tag{7.20}$$

$$N = \left[\frac{1-s^2}{2}\frac{\partial^2 V}{\partial s^2} + \frac{1-2s^2}{2s}\frac{\partial V}{\partial s}\right]_{s=\sin\theta_0} \tag{7.21}$$

This result is valid independently of the dependence of the reflection coefficient V on the angle of incidence. It has been shown in Brekhovskikh and Godin (1992) that this result is also valid at small angles of incidence for large k_0R_1. The coefficient N is given by Equations (7.A.3)–(7.A.4) for layers of finite thickness. For a semi-infinite layer, when $l \to \infty$, N is given by

$$\begin{aligned} N = {} & m(1-n^2)[2m(n^2-1) + 3m\cos^2\theta_0 - m\cos^4\theta_0 \\ & + \sqrt{n^2-\sin^2\theta_0}\cos\theta_0(2n^2+\sin^2\theta_0)](m\cos\theta_0 + \sqrt{n^2-\sin^2\theta_0})^{-3} \\ & \times (n^2-\sin^2\theta_0)^{-3/2} \end{aligned} \tag{7.22}$$

where $m = \rho/(\phi\rho_0)$, ρ being the effective density of the air in the medium and ρ_0 being the free air density. This expression can be obtained by replacing the ratio of the densities in Equation (1.2.10) in Brekhovskikh and Godin (1992), which gives the coefficient N at a fluid–fluid interface, by m. In Brekhovskikh and Godin (1992), for a semi-infinite layer, a supplementary term, the lateral wave, is added in Equation (7.20). This contribution is neglected in the present work. For a layer of finite thickness, and for $\theta_0 = 0$, N is given by

$$N = \left[1 + \frac{2nk_0l}{(1-n^2)\sin(2nk_0l)}\right]M(0) \tag{7.23}$$

$$\begin{aligned} M(0) &= \frac{2m'(0)(1-n^2)}{(m'(0)+n)^2 n} \\ m'(0) &= -jm\cot(lk_0n) \end{aligned} \tag{7.24}$$

The coefficient N, for a given layer, only depends on the angle of specular reflection. Predictions of the reflected pressure p_r obtained with the exact formulation, Equation (7.6) and the passage path method are compared in Figures 7.3 and 7.4 and for a layer of material 1 defined in Table 7.1. The source is the unit source of Equation (7.1).

Table 7.1 Acoustic parameters for different materials.

Materials	Tortuosity α_∞	Flow resistivity σ (Nm^{-4} s)	Porosity ϕ	Viscous dimension Λ (μm)	Thermal dimension Λ'(μm)
Material 1	1.1	20 000	0.96	100	300
Material 2	1.32	5500	0.98	120	500

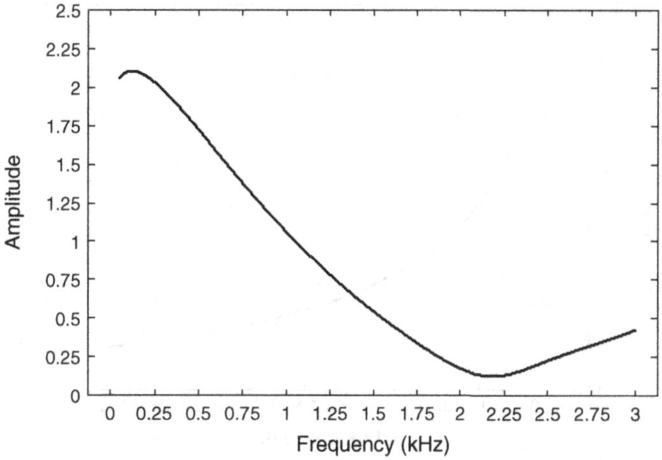

Figure 7.3 Amplitude of the reflected wave calculated with Equation (7.6), material 1, $l = 4$ cm, $z_1 + z_2 = -0.5$ m, $r = 0$.

The thickness of the sample is $l = 4$ cm. In this chapter, Equations (5.50), (5.52) are used for the effective density and Equations (5.51), (5.55) for the bulk modulus. Materials 1 and 2 could be ordinary porous sound absorbing materials. The modulus of the exact reflected pressure calculated with Equation (7.6) for a layer of thickness $l = 4$ cm at $\theta_0 = 0$ and $z_1 + z_2 = -0.5$ m is represented in Figure 7.3. The modulus of the difference between the exact evaluation and the evaluation by Equation (7.20) is represented in Figure 7.4, together with the modulus of the term $N/(k_0 R_1^2)$.

The difference is much smaller than $|N/(k_0 R_1^2)|$ for frequencies higher than 500 Hz. This shows the interest of the steepest descent method for large $k_0 R_1$. In Figures 7.5, 7.6

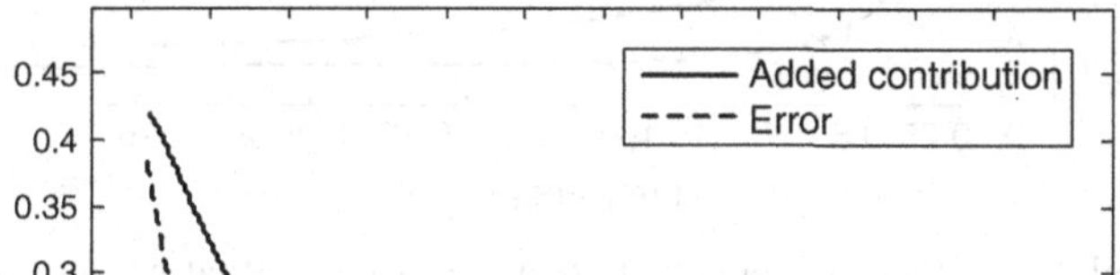

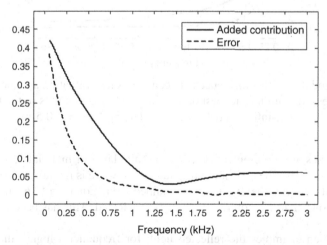

Figure 7.4 Modulus of the difference between the exact evaluation and the evaluation of the reflected pressure with the steepest descent method from Equation (7.20), and modulus of the term $N/k_0 R_1^2$ in Equation (7.20), material 1, $l = 4$ cm, $z_1 + z_2 = -0.5$ m, $r = 0$.

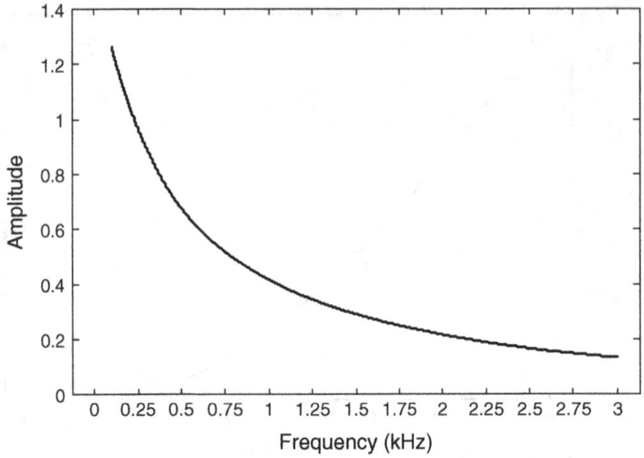

Figure 7.5 Amplitude of the reflected wave calculated with Equation (7.6), semi-infinite layer of material 1, $z_1 + z_2 = -0.5$ m, $r = 0.5$ m.

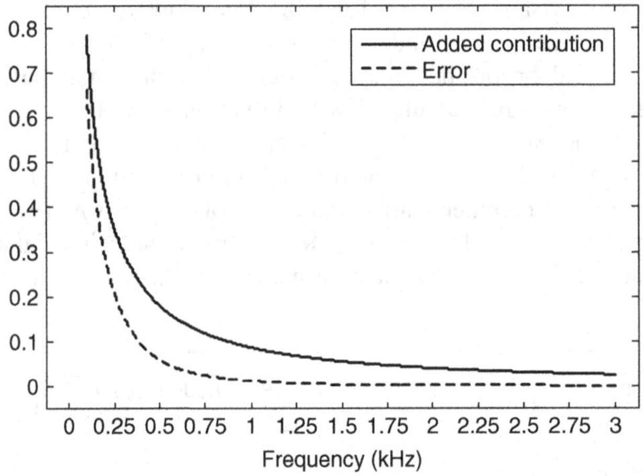

Figure 7.6 Modulus of the difference between the exact evaluation and the evaluation of the reflected pressure with the steepest descent method, and modulus of the term $N/k_0R_1^2$ in Equation (7.20), semi-infinite layer of material 1, $z_1 + z_2 = -0.5$ m, $r = 0.5$ m.

the same quantities are represented for a semi-infinite layer of material 1, with $r = 0.5$ m and $z_1 + z_2 = -0.5$ m. The contribution of the lateral wave is not taken into account. The good agreement between the exact result obtained from Equation (7.6) and the passage path method shows that this contribution can be neglected. This has been always verified in our previous studies.

For the chosen examples the reflected field, for frequencies higher than 500 Hz, is obtained with good precision with the steepest descent method. For frequencies higher than 1 kHz, the reflected pressure is close to the pressure created by the unit image source multiplied by the plane wave reflection coefficient. These frequencies decrease when the distance R_1 from the image of the source to the receiver increases. Similar results are obtained for semi-infinite layers. However predictions performed with the steepest descent method in the same range of frequencies and similar R_1 for θ_0 close to $\pi/2$ can give wrong results. A new formulation of the problem must be performed, which takes into account the location of the singularities of the reflection coefficient.

7.5 Poles of the reflection coefficient

7.5.1 Definitions

The singularities of the plane wave reflection coefficient are located at $\sin\theta_p = s_p$ satisfying

$$\cos\theta_p = -\frac{Z_0}{Z_s(s_p)} \tag{7.25}$$

leading to a denominator of V in Equation (7.10) equal to 0.

For a locally reacting medium, Z_s does not depend on the angle of incidence and a pole can only exists at s_p satisfying $\cos\theta_p = -Z_0/Z_s$. There are two poles in the complex s plane for both determinations of $\sqrt{1-\cos^2\theta_p}$.

For a semi-infinite layer, it may be shown that only one $\cos\theta_p$ is related to a zero of the denominator of V and $\cos\theta_p$ is given by

$$\cos\theta_p = -\left(\frac{n^2-1}{\rho^2/(\rho_0\phi)^2-1}\right)^{1/2} \tag{7.26}$$

There is a minus sign before the square root because $\mathrm{Re}\cos\theta_p$ is negative. Both related s_p are given by

$$\sin\theta_p = \pm\left(\frac{\rho^2/(\rho_0\phi)^2-n^2}{\rho^2/(\rho_0\phi)^2-1}\right)^{1/2} \tag{7.27}$$

The pole trajectory as a function of frequency is represented in the complex $\sin\theta$ plane in Figure 7.7(a) and in the complex $\cos\theta$ plane in Figure 7.7(b).

For a layer of finite thickness, there is at any frequency an infinite number of poles. It will be shown in Section 7.6 that if one pole is located at θ_p close to $\pi/2$, the one parameter which characterizes the porous layer for the prediction of the reflected field at an angle of specular reflection θ_0 close to $\pi/2$ is $\cos\theta_p$. When $|Z_s(s_p)| \gg Z_0$ in Equation (7.25), the pole related to s_p is located at θ_p close to $\pi/2$. This happens for instance for media with a large flow resistivity, and for porous layers having a small thickness. A thin layer is a layer where $|k_1 l| \ll 1$. Any layer having a finite thickness is a thin layer at a sufficiently low frequency. The first-order development of $1/Z_s(s_p)$ is $j\phi kl\cos^2\theta_1/Z$, and $\cos^2\theta_1$, neglecting the second-order term $\cos^2\theta/n^2$ in Equation (7.9), can be replaced by $1-1/n^2$. The related solution of Equation (7.25) is (Allard and Lauriks 1997; Lauriks *et al.* 1998)

$$\cos\theta_p = -jk_0 l(n^2-1)\frac{\phi Z_0}{nZ} \tag{7.28}$$

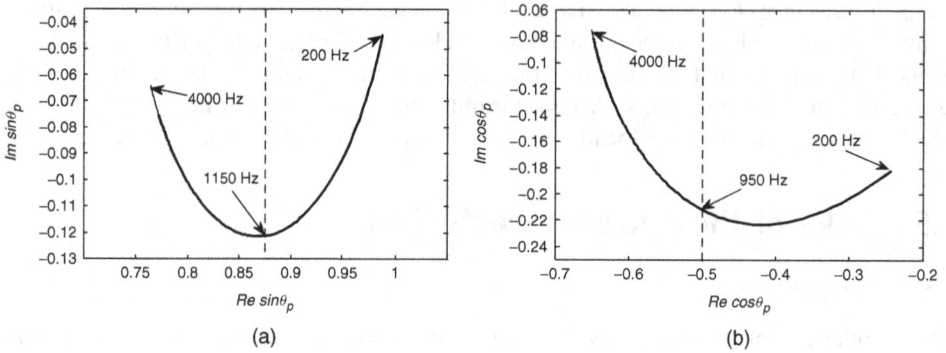

Figure 7.7 Trajectories of the pole of the reflection coefficient for a semi-infinite layer of material 1.

This pole is called the main pole. Simple iterative methods can be used to predict $\cos\theta_p$ without approximations. For thin layers, it was shown in Allard and Lauriks (1997) that the other poles are far from the real $\sin\theta$ axis and cannot contribute significantly to the reflected field. Equation (7.28) can be rewritten

$$\cos\theta_p = -j\phi k_0 l\left(\frac{\gamma P_0}{K} - \frac{\rho_0}{\rho}\right) \tag{7.29}$$

For a perfect fluid without viscosity and thermal conduction, $\cos\theta_p$ is given by

$$\cos\theta_p = -j\phi k_0 l(1 - \alpha_\infty^{-1}) \tag{7.30}$$

and is imaginary with a negative imaginary part.

7.5.2 Planes waves associated with the poles

Thin layers

For an angle of incidence $\theta = \theta_p$ the plane reflected wave satisfies the boundary conditions with no incident wave. The space dependence of this wave is $\exp[-jk_0(x\sin\theta_p - z\cos\theta_p)]$. For a thin layer without damping, the reflected wave is propagative in the x direction and evanescent in the direction opposite to z.

As indicated at the end of Section 3.2 the axes of evanescence and of propagation are perpendicular in a medium with a real wave number. In the present case the wave is evanescent in the direction perpendicular to the surface of the porous layer and opposite to the z axis. Using a complex wave number vector $\boldsymbol{k} = \boldsymbol{k}^{\mathbf{R}} + j\boldsymbol{k}^{\mathbf{I}}$ for the reflected wave, with $(k^R)^2 - (k^I)^2 = k_0^2$, the spatial dependence of the plane wave is given by

$$\exp[-j(\boldsymbol{k}^{\mathbf{R}} + j\boldsymbol{k}^{\mathbf{I}})\boldsymbol{OM}] = \exp\left[-j(k_x^R x + k_z^R z) + (k_x^I x + k_z^I z)\right] \tag{7.31}$$

The direction of $\boldsymbol{k}^{\mathbf{I}}$ is opposite to the direction of evanescence. The wave number is represented symbolically in Figure 7.8(a). For a thin layer saturated by a perfect fluid the

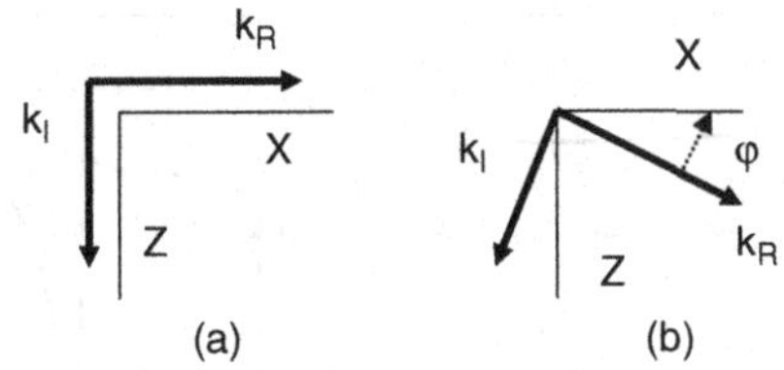

Figure 7.8 The wave number vector is $\boldsymbol{k} = \boldsymbol{k}_\mathrm{R} + j\boldsymbol{k}_\mathrm{I}$. The wave is evanescent in the direction opposite to $\boldsymbol{k}_\mathrm{I}$ and propagates in the direction $\boldsymbol{k}_\mathrm{R}$: (a) thin layer saturated by a perfect fluid, (b) air-saturated thin layer.

wave number components k_x and k_z are given by

$$k_x = k_0 \sin\theta_p = k_R \tag{7.32}$$

$$k_z = -k_0 \cos\theta_p = jk_I \tag{7.33}$$

The wave number vector of a thin layer saturated by air is represented in Figure 7.8(b). For a layer of material 1 of thickness $l = 4$ cm, the pole with $\sin\theta_p$ closest to 1 has been localized with a simple recursive algorithm; $\sin\theta_p$ is represented in the complex s plane in Figure 7.9(a) and $\cos\theta_p$ is represented in Figure 7.9(b) in the complex $\cos\theta_p$ plane in the 50–1000 Hz frequency range.

For small thicknesses, the imaginary and the real part of $\cos\theta_p$ are negative and the imaginary part of $\sin\theta_p$ is negative. As shown in Figure 7.8(b), the real part of the wave number vector makes an angle φ with the axis x. The wave number components are now given by

$$k_x = k_0 \sin\theta_p = k^R \cos\varphi - jk^I \sin\varphi \tag{7.34}$$

$$k_z = -k_0 \cos\theta_p = k^R \sin\varphi + jk^I \cos\varphi \tag{7.35}$$

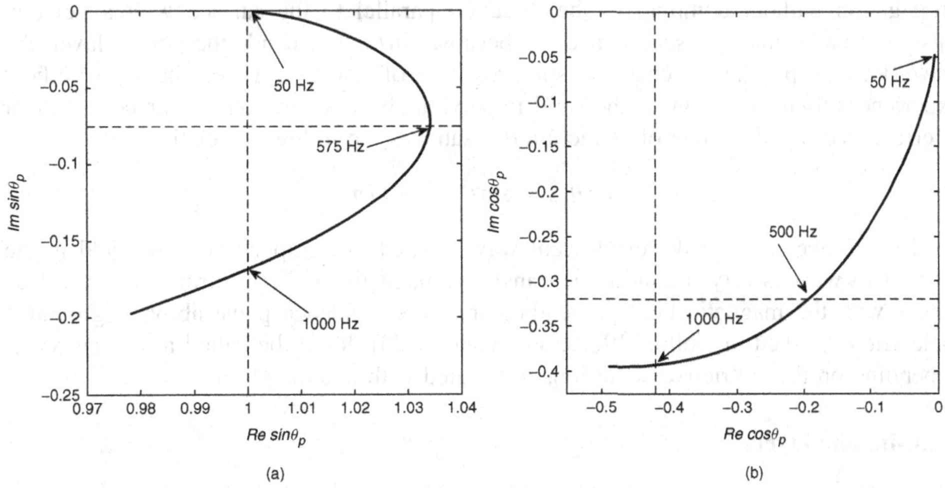

Figure 7.9 Trajectory of $\sin\theta_p$ and of $\cos\theta_p$. Material 1, $l = 4$ cm.

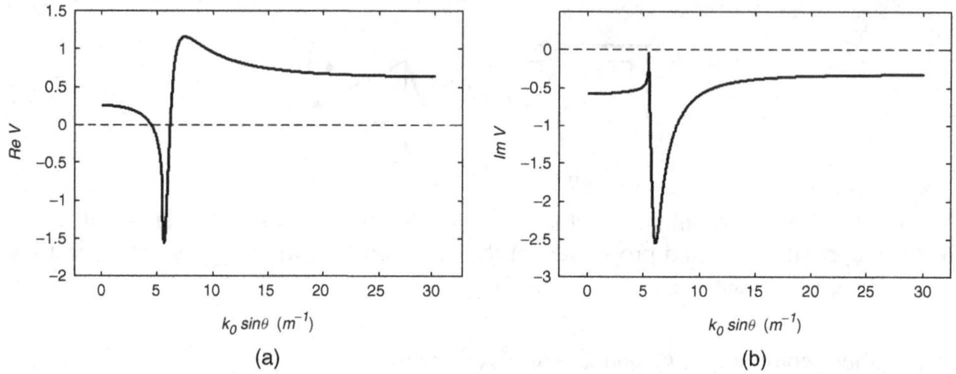

Figure 7.10 The predicted reflection coefficient for a layer of material 2 of thickness $l = 0.1$ m at 0.3 kHz as a function of $k_0 \sin\theta$. When $V = -1$, $\sin\theta = 1$.

Taking into account the signs of the real and the imaginary part of $\cos\theta_p$ and $\sin\theta_p$ leads to $0 \le \varphi \le \pi/2$, as in Figure 7.8(b). The plane wave reflection coefficient can be measured at $\sin\theta$ real and larger than 1 with the Tamura method (Tamura 1990, Brouard *et al.* 1996). In the $\sin\theta$ plane, $\sin\theta_p$ is close to the real axis at sufficiently low frequencies for a layer of finite thickness. The modulus of the reflection coefficient must present a peak when $\sin\theta$ is close to $|\sin\theta_p|$. The predicted reflection coefficient for a layer of thickness $l = 0.1$ m of material 2 is shown in Figure 7.10. The maximum of the modulus of the reflection coefficient is larger than 1 for $\sin\theta$ larger than 1. The reflected wave is purely evanescent for $\sin\theta$ real and larger than one, and does not carry energy, so there is no power created. Measurements performed by Brouard *et al.* (1996) on a similar material are in a good agreement with these predictions. Other experimental evidences of this peak can be found in Brouard (1994), Brouard *et al.* (1996), and Allard *et al.* (2002).

The wave number vector in the free air above a thin layer, in the absence of damping, is symbolically represented in Figure 7.8(a). This wave presents in air all the characteristic properties of a surface wave, i.e. damping in the direction normal to the surface and propagation without damping in the direction parallel to the surface with a velocity $c_0/\sin\theta_p$ smaller than the sound speed c_0 because $\sin\theta_p > 1$. Inside the porous layer, the wave does not present the characteristic properties of a surface wave. The acoustic field experiences total reflection on the rigid impervious backing and on the air porous layer interface because the angle of refraction θ_{1p} satisfies the following relations

$$\sin\theta_{1p} = \sin\theta_p/n > 1/n \tag{7.36}$$

The surface wave is the evanescent wave related to a trapped acoustic field in the layer. This wave is very similar to the transverse magnetic (T.M.) electromagnetic surface waves with the magnetic field perpendicular to the incidence plane above a grounded dielectric described in Collin (1960), and Wait (1970). It can be called a surface wave, depending on the restrictive conditions associated with this definition.

Semi-infinite layers

For a semi-infinite porous layer saturated by a perfect fluid, Equations (7.26)–(7.27) give $\sin\theta_p$ and $\cos\theta_p$ real, and $\sin\theta_p < 1$, $\cos\theta_p < 0$. The wave number of the associated wave

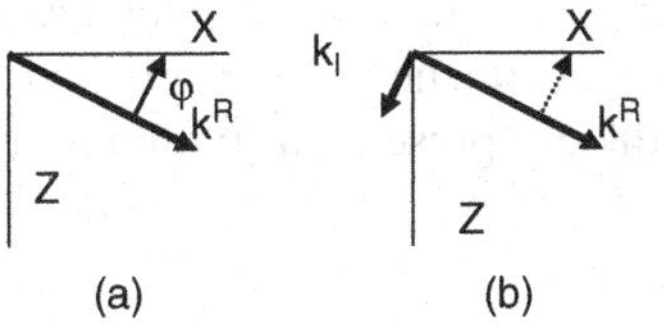

Figure 7.11 The real and imaginary wave number vector of the plane wave associated with the pole for a semi-infinite layer: (a) perfect fluid, (b) air.

is represented in Figure 7.11(a) for a semi-infinite layer saturated by a perfect fluid, and in Figure 7.11(b) for an air saturated semi-infinite layer. The modulus of the imaginary wave number vector is equal to 0 in Figure 7.11(a), $k^R = k_0$, $k_x = k_0 \cos\varphi$, $k_z = k_0 \sin\varphi$. Formally, changing the sign of cosθ changes V into $1/V$ (see Equation 7.10), and the reflected wave becomes the incident wave. The reflected wave related to the pole for the semi-infinite layer can be considered as a plane wave at an angle of incidence where the reflection coefficient is equal to 0. This angle is real in the absence of damping and equal to $\pi/2 - \varphi$. This angle is the Brewster angle θ_B of total refraction. A similar Brewster angle of total refraction exists for T. M. electromagnetic waves (Collin 1960). The angle is complex for air saturated porous media. From Equation (7.10) θ_B and θ_p are related by

$$\cos\theta_B = -\cos\theta_p \tag{7.37}$$

The Brewster angle is not real for air saturated porous media, but cosθ_B can lie close to the real cosθ plane in the complex cosθ plane. Then the modulus of the reflection coefficient can present a minimum around θ close to θ_B. The predicted modulus of the reflection coefficient is shown in Figure 7.12 at 4 kHz as a function of cosθ. The minimum appears close to cos$\theta_B = 0.65$.

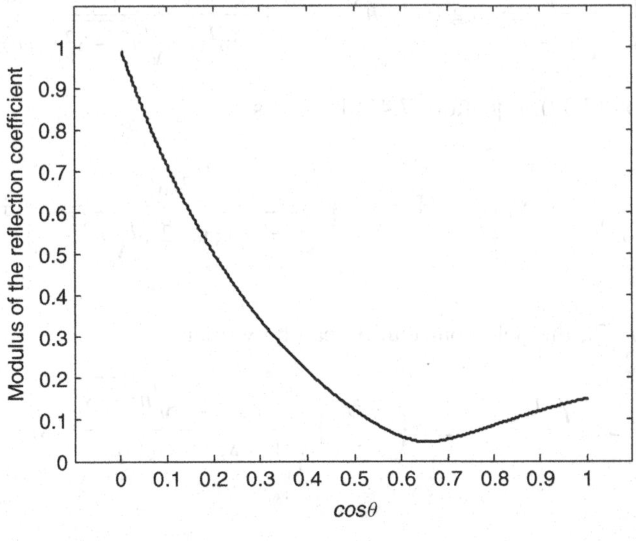

Figure 7.12 The modulus of the reflection coefficient of a semi-infinite layer of material 1 at 4 kHz. The predicted $\text{Re}\cos\theta_B = -\text{Re}\cos\theta_p = 0.65$.

Measurements of the reflection coefficient of a thick layer of sand at oblique incidence have been performed previously with the Tamura method by Allard *et al.* (2002). The modulus of the reflection coefficient presents a minimum at $\cos\theta$ close to the predicted $\mathrm{Re}\cos\theta_B = -\mathrm{Re}\cos\theta_p$.

7.5.3 Contribution of a pole to the reflected monopole pressure field

If a pole is crossed by the path of integration when the path is modified, the integral over the initial path is equal to the integral over the modified path plus a pole contribution. These poles are zeros of the first order of the denominator on the right-hand side of Eqution (7.10). There is also an apparent problem with the term $(s/(1-s^2))^{1/2}$ in $F(s)$ in Equation (7.17), but the zero at the denominator disappears if $\cos\theta$ is used instead of $\sin\theta$ as the variable of integration. The following expression is used for the reflection coefficient

$$V(s) = \frac{-\sqrt{n^2-s^2} - jn(Z/\phi Z_0)\sqrt{1-s^2}\,\mathrm{cotg}\,(k_0 l\sqrt{n^2-s^2})}{\sqrt{n^2-s^2} - jn(Z/\phi Z_0)\sqrt{1-s^2}\,\mathrm{cotg}\,(k_0 l\sqrt{n^2-s^2})} \tag{7.38}$$

At a pole location $s = s_p$, the following relation is fulfilled

$$\sqrt{n^2-s_p^2} = jn\frac{Z}{\phi Z_0}\sqrt{1-s_p^2}\,\mathrm{cotg}\,(k_0 l\sqrt{n^2-s_p^2}) \tag{7.39}$$

The derivative $G'_s(s_p)$ of the denominator can be written

$$G'_s(s_p) = -\frac{s_p}{\sqrt{n^2-s_p^2}} - \frac{jnZ}{\phi Z_0} \times \left[-\frac{s_p}{\sqrt{1-s_p^2}}\,\mathrm{cotg}\,(k_0 l\sqrt{n^2-s_p^2}) + \frac{\sqrt{1-s_p^2}}{\sin^2(k_0 l\sqrt{n^2-s_p^2})}\frac{k_0 l s_p}{\sqrt{n^2-s_p^2}} \right] \tag{7.40}$$

Using relation (7.39), Equation (7.40) becomes

$$G'_s(s_p) = \frac{s_p}{1-s_p^2}\sqrt{n^2-s_p^2}\left[1-(1-s_p^2)\left(\frac{2k_0 l}{\sqrt{n^2-s_p^2}\sin 2k_0 l\sqrt{n^2-s_p^2}} + \frac{1}{n^2-s_p^2}\right)\right] \tag{7.41}$$

Using Eq. (7.15), the pole contribution can be written

$$SW(s_p) = -2\pi j\left(\frac{k_0}{2\pi r}\right)^{1/2}\exp\left(\frac{-j\pi}{4}\right)\sqrt{\frac{s_p}{1-s_p^2}}\frac{-2\sqrt{n^2-s_p^2}}{G'_s(s_p)}\exp[k_0 R_1 f(s_p)] \tag{7.42}$$

The pole contribution $SW(s_p)$ can be written, if s_p is close to 1

$$SW(s_p) = 4\pi \left(\frac{k_0}{2\pi r}\right)^{1/2} \exp\left(\frac{j\pi}{4}\right) \sqrt{\frac{1-s_p^2}{s_p}} \exp(k_0 R_1 f(s_p)) \tag{7.43}$$

The same expression is obtained for a non-locally reacting medium. For the case of thin layers, Figure 7.9 shows that $\mathrm{Re}\sin\theta_p > 1$. However, $\mathrm{Re}\sin\theta_p$ remains close to 1 for ordinary reticulated foams and fibrous layers and the pole is crossed only for angles of incidence close to $\pi/2$. For semi-infinite layers, expression (7.41) is simplified because $l/\sin(2k_0 l\sqrt{n^2 - s_p^2})$ is replaced by 0. When the passage path method is used, there is no pole contribution. In Figure 7.7 Re $\sin\theta_p$ is always smaller than 1. When the initial path of integration becomes the passage path, the part of the path located at $\mathrm{Re}\sin\theta < 1$ can cross the pole, but the pole is in the physical sheet and the path is not in the physical sheet in the domain $0 < \mathrm{Re}\ \sin\theta < 1$, $0 > \mathrm{Im}\sin\theta$, because it has crossed the cut. The condition $\mathrm{Re}\sin\theta < 1$ has always been verified for semi-infinite layers when the causal effective densities presented in Chapter 5 have been used. However, we have not found any general proof of this condition.

7.6 The pole subtraction method

The passage path method is valid for $k_0 R_1 \gg 1$ only if the poles and the stationary point are sufficiently far from each other, allowing a slow variation of the reflection coefficient close to the stationary point on the path of integration. If a pole is close to the stationary point, the pole subtraction method can be used to predict the reflected pressure under the same condition, $k_0 R_1 \gg 1$. The passage path method remains valid for $|u| \gg 1$, where u is the numerical distance defined by

$$u = \sqrt{2k_0 R_1} \exp\left(-j\frac{3\pi}{4}\right) \sin\frac{\theta_p - \theta_0}{2} \tag{7.44}$$

This condition can be much restrictive than $k_0 R_1 \gg 1$ if θ_p and θ_0 are close to each other. The expressions for p_r obtained with the pole subtraction method are given for one pole of first order close to the stationary point in Appendix 7.B. Two cases are considered, a locally reacting surface with a constant impedance Z_L, and a porous layer of finite thickness. The reference integral method of Brekhovskikh and Godin (1992) is used. It is shown in Appendix 7.B that if one pole exists at θ_p close to $\pi/2$, for θ_0 close to $\pi/2$ the monopole reflected field is the same as the monopole reflected field above the locally reacting surface with a surface impedance Z_L given by

$$Z_L = -Z_0/\cos\theta_p \tag{7.45}$$

The pole of the reflection coefficient of the related locally reacting surface is located at the same angle θ_p. The following approximation for p_r, obtained by setting $\sqrt{s_0 s_p} = 1$

in Equation (7.B.19), is used for the case of the porous surface

$$p_r = \frac{\exp(-jk_0R_1)}{R_1} \times \left\{ V_L(s_0) - \frac{\sqrt{1-s_p^2}\sqrt{2k_0R_1}\exp(-3\pi j/4)(1+\sqrt{\pi}u\exp(u^2)\mathrm{erfc}(-u))}{u} \right\}$$

$$u = \exp\left(\frac{-j3\pi}{4}\right)\sqrt{2k_0R_1}\sin\frac{\theta_p-\theta_0}{2} \tag{7.46}$$

and erfc is the complement of the error function. The expression $W(u) = 1 + \sqrt{\pi}u\exp(u^2)\mathrm{erfc}(-u)$ is obtained for small $|u|$ with the series development

$$W(u) = 1 + \sqrt{\pi}u\exp(u^2) + 2u^2\exp(u^2)\left[1 - \frac{u^2}{3} + \frac{u^4}{2!5} - \frac{u^6}{3!7} + \cdots\right] \tag{7.47}$$

and for large $|u|$ with the development

$$W(u) = [1 + \mathrm{sgn}(\mathrm{Re}(u))]\sqrt{\pi}u\exp(u^2) + \frac{1}{2u^2} - \frac{1\times 3}{(2u^2)^2} + \frac{1\times 3\times 5}{(2u^2)^3} + \cdots \tag{7.48}$$

The following approximations for V and u are used

$$V(\sin\theta) = \frac{\cos\theta + \cos\theta_p}{\cos\theta - \cos\theta_p} \tag{7.49}$$

$$u = \exp\left(\frac{-j3\pi}{4}\right)\sqrt{k_0R_1/2}(\cos\theta_0 - \cos\theta_p) \tag{7.50}$$

the term $\sqrt{1-s_p^2}\sqrt{2k_0R_1}\exp(-3\pi j/4)/u$ in Equation (7.46) can be replaced by $1 - V(\sin\theta) = -2\cos\theta_p/(\cos\theta_0 - \cos\theta_p)$, and Equation (7.46) can be rewritten

$$p_r = \frac{\exp(-jk_0R_1)}{R_1}[V_L(\sin\theta_0) + (1 - V_L(\sin\theta_0))(1 + \sqrt{\pi}u\exp(u^2)\mathrm{erfc}(-u))] \tag{7.51}$$

where V_L is the reflection coefficient for an impedance plane $Z_s = -Z_0/\cos\theta_p$. A similar equation has been given by Chien and Soroka (1975) for the case of a locally reacting surface. The one difference is that for the porous layer $\cos\theta_p$ is obtained from Equation (7.25) where Z_s depends on the angle of incidence. Equation (7.49) is valid only for θ_0 close to $\pi/2$, and if θ_p is close to θ_0. It is difficult to define the limits of validity of Equations (7.49) – (7.51), due to the numerous parameters involved. Limits to the accuracy when θ_p or θ_0 are too far from $\pi/2$ will appear in the following section. For a semi-infinite layer, and more generally when a pole is close to the stationary point without being crossed, there is, at small numerical distances, a contribution to the reflected pressure in W given by $\sqrt{\pi}u\exp(u^2)$, half the contribution added for large $|u|$ when $\mathrm{Re}(u) > 0$. A similar contribution, the Zenneck wave, exists in the electric dipole field reflected by a conducting surface. Descriptions of the Zenneck wave have been carried out by Baños (1966), and by Brekhovskikh (1960). The close similarities between the acoustic and the electromagnetic case are shown in Allard and Henry (2006).

Some points are summarized. At sufficiently low frequencies, any layer having a finite thickness can be replaced for θ_0 close to $\pi/2$ by an impedance plane having the same pole as the main pole of the layer. The contribution of the term $[1 + \mathrm{sgn}(\mathrm{Re}(u))]\sqrt{\pi}u\exp(u^2)$ in Equation (7.48) corresponds to the contribution of a crossed pole in the passage path method (see Appendix 7.C). This contribution does not exist if the porous layer is semi-infinite, because in this case $Re\sin\theta_0 < 1$. However, in this case, for small $|u|$, there is a contribution equal to half the contribution of the pole if it were crossed.

7.7 Pole localization

7.7.1 Localization from the *r* dependence of the reflected field

Using Equations (7.49)–(7.50), Equation (7.51) can be rewritten

$$p_r = \frac{\exp(-jk_0R_1)}{R_1}\{1 - \cos\theta_p\sqrt{2k_0R_1}\exp\left(\frac{-3j\pi}{4}\right)\sqrt{\pi}\exp(u^2)\mathrm{erfc}(-u)\} \tag{7.52}$$

and $\cos\theta_p$ is related to the reflected pressure by

$$\cos\theta_p = [1 - R_1p_r\exp(jk_0R_1)]/\left[(2\pi k_0R_1)^{1/2}\exp\left(\frac{-3\pi j}{4}\right)\exp(u^2)\mathrm{erfc}(-u)\right] \tag{7.53}$$

Two simulated measurements of $\cos\theta_p$ with Equation (7.53) are presented in Figures 7.13 and 7.14. The reflected pressure is calculated with Equation (7.6) and θ_p related to the main pole is evaluated with an iterative method from Equation (7.53). A comparison between the exact $\cos\theta_p$ and $\cos\theta_p$ evaluated with Equation (7.53) is performed, for a layer of thickness $l = 3$ cm of material 1. In Figure 7.13, $r = 1$ m and $z_1 = z_2 = -1$ cm. In Figure 7.14, $z_1 = z_2 = -5$ cm. In both figures, when frequency increases, $|\cos\theta_p|$ also increases and the systematic error increases. With thinner layers, the range of frequencies where the systematic error can be neglected is larger. The angle of specular reflection is larger in Figure 7.13 than in Figure 7.14, and the systematic error is larger. Measurements

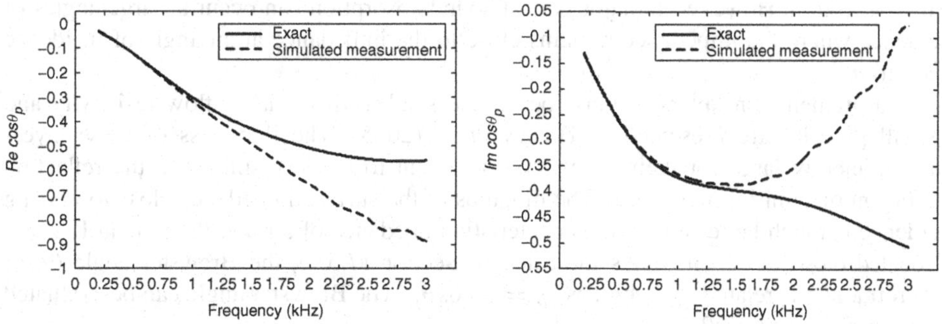

Figure 7.13 Comparison between the exact $\cos\theta_p$ and a simulated measurement obtained with Equation (7.53). Material 1, $l = 3$ cm, $r = 1$ m, $z_1 = z_2 = -2$ cm.

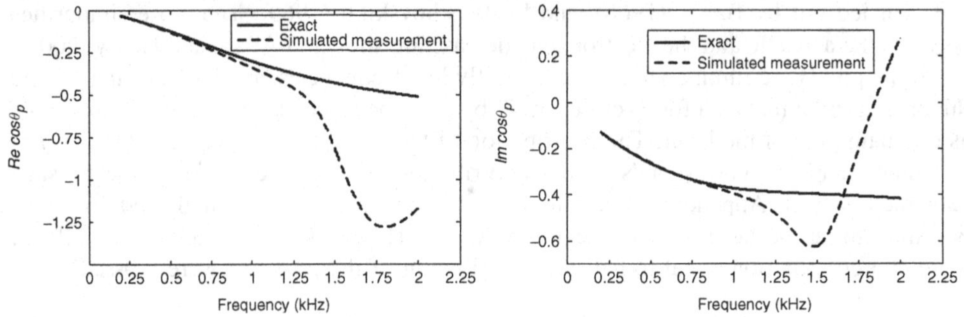

Figure 7.14 Comparison between the exact $\cos\theta_p$ and a simulated measurement obtained with Equation (7.53). Material 1, $l = 3$ cm, $r = 1$ m, $z_1 = z_2 = -5$ cm.

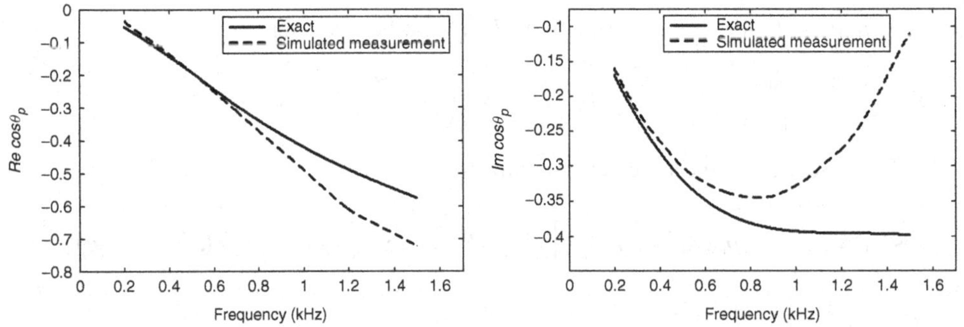

Figure 7.15 Comparison between the exact $\cos\theta_p$ and a simulated measurement obtained with Equation (7.56) from the variation of the total pressure on an axis perpendicular to the surface of the layer. Material 1, $l = 3$ cm, $z_1 = z_2 = -0.5$ cm, $z_2' = -2.5$ cm, $r = 1$ m.

are presented in Allard *et al.* (2003a, b) for thin layers of porous foam. In the domain of validity of Equation (7.53), θ_p is close to $\pi/2$, and from Equation (7.25) $Z_s(\sin\theta_p) = -Z_0/\cos\theta_p$, the surface impedance at an angle of incidence θ_p close to $\pi/2$, can be evaluated from $\cos\theta_p$. This surface impedance can be an important parameter in room acoustics. The reflection coefficient at an angle of incidence equal to $\pi/2$ is − 1, and there is no absorption. However, a major part of sound absorption can occur at large angles of incidence where the impedance remains close to the impedance at an angle of incidence equal to $\pi/2$.

Measurements on layers of glass beads and sand having a large flow resistivity and a small porosity are presented in Hickey *et al.* (2005). The thickness of these layers was sufficiently large for their reflection coefficient to be very similar to the reflection coefficient of semi-infinite layers. The modulus of the surface impedance close to grazing incidence is much larger than the characteristic impedance of air and the pole at $\mathrm{Re}\theta > 0$ is located close to $\theta = \pi/2$. As indicated in Section (7.5.2), the Brewster angle θ_B of total refraction is related to θ_p by $\cos\theta_B = -\cos\theta_p$. The Brewster angle can be evaluated from the measured $\cos\theta_p$.

7.7.2 Localization from the vertical dependence of the total pressure

If the pressure field above the porous layer is the same as that above a locally reacting medium of surface impedance $Z_s(\sin\theta_p)$, the surface impedance of the layer is $Z_s(\sin\theta_p)$. This is verified in what follows when $\theta_p \approx \pi/2$ and $\theta_0 \approx \pi/2$, if $|z_1|$ and $|z_2|$ are much smaller than r. The total pressure p_t is the sum of p_r and of the direct field $\exp(-jk_0R_2)/R_2$, R_2 being the distance from the source to the receiver

$$p_t = \frac{\exp(-jk_0R_2)}{R_2} + \frac{\exp(-jk_0R_1)}{R_1}\{1 - \cos\theta_p\sqrt{2k_0R_1}\exp\left(\frac{-3j\pi}{4}\right)\sqrt{\pi}\exp(u^2)\mathrm{erfc}(-u)\} \quad (7.54)$$

where u is given by Equation (7.50). At $\theta_0 = \pi/2$, $\partial R_2/\partial z_2 = \partial R_1/\partial z_2 = 0$, and $\partial\theta_0/\partial z_2 = 1/r$ (the z axis is directed toward the porous layer). The derivative of $w(u) = \exp(u^2)\mathrm{erfc}(-u)$ is $w'(u) = 2uw(u) + 2/\sqrt{\pi}$ (Abramovitz and Stegun 1972, Chapter 7), where u can be replaced by $-\exp(-j3\pi/4)\sqrt{k_0r/2}\cos\theta_p$.

The derivative $\partial u/\partial z_2$ at $\theta_0 = \pi/2$ is given by

$$\frac{\partial u}{\partial z_2} = -\sqrt{2k_0r}\exp\left(-\frac{j3\pi}{4}\right)\frac{1}{2r}\sin\theta_0 \quad (7.55)$$

where $\sin\theta_0$ is close to 1, and from Equation (7.55)

$$\partial p_t/\partial z_2 = p_t jk_0\cos\theta_p \quad (7.56)$$

The surface impedance is the ratio $p_t/v_z = -p_t j\omega\rho_0/(\partial p_t/\partial z_2) = -Z_0/\cos\theta_p$, which is equal to $Z_s(\sin\theta_p)$. This impedance can be evaluated in a free field from pressure measurements close to the surface on a normal to the surface at z_2 and z_2'. The velocity v_z is given by $(j/\omega\rho_0)\partial p_t/\partial z_2$ where the pressure derivative $\partial p_t/\partial z_2$ can be approximated by $[p(z_2') - p(z_2)]/(z_2' - z_2)$. The measured surface impedance Z_s can be obtained from

$$Z_s = \frac{j(z_2' - z_2)}{\omega\rho_0}\frac{p(z_2)}{(p(z_2') - p(z_2))} \quad (7.57)$$

or by the equivalent expression equivalent for small $z_1 - z_2$

$$Z_s = \frac{j(z_2' - z_2)}{\omega\rho_0}\frac{1}{\ln(p(z_2')/p(z_2))} \quad (7.58)$$

In Figure 7.15, the evaluated quantity is not Z_s, but $\cos\theta_p = -Z_0/Z_s$. Simulated measurements are shown for a layer of thickness $l = 3$ cm of material 1 the exact total pressure being calculated with Equation (7.6). Equation (7.58) is used to evaluate Z_s. The systematic error increases faster with frequency than with the previous method. Measurements are presented in Allard *et al.* (2004).

The z_2 dependence of a pole contribution p_{pole} is $\exp(jk_0\cos\theta_p z_2)$. This gives for the pole contribution derivative the following expression

$$\partial p_{pole}/\partial z_2 = p_{pole} jk_0\cos\theta_p \quad (7.59)$$

The ratio $[\partial p_{pole}/\partial z_2]/p_{pole}$ has the same expression as $[\partial p_t/\partial z_2]/p_t$ in Equation (7.56). The difference is that the impedance condition concerns the total field and must be satisfied close to the reflecting surface only.

7.8 The modified version of the Chien and Soroka model

In Section 7.1 an exact integral expression for the reflected field is given. An approximate expression obtained with the passage path method and valid for large k_0R_1 is given in Section 7.4. Several equivalent expressions valid for θ_0 close to $\pi/2$ when one pole is located at an angle of incidence θ_p close to grazing incidence are given in Section 7.6. The validity of a modified version of the Chien and Soroka model suggested by Nicolas *et al.* (1985) and Li *et al.* (1998), currently used to predict the reflected monopole field above porous layers in the context of long-range sound propagation and also for the evaluation of acoustic surface impedance, is studied in this section. The initial work by Chien and Soroka (1975) concerns the monopole field reflected by an impedance plane Z_s. In the initial formulation, the reflected field is given by

$$p_r = \frac{\exp(-jk_0R_1)}{R_1}[V_L(\sin\theta_0) + (1 - V_L(\sin\theta_0))(1 + \sqrt{\pi}u\exp(u^2)\mathrm{erfc}(-u))] \quad (7.60)$$

where V_L is the reflection coefficient

$$V_L(\sin\theta) = \frac{\cos\theta - Z_0/Z_s}{\cos\theta + Z_0/Z_s} \quad (7.61)$$

and u is given by

$$u = \exp\left(\frac{-j3\pi}{4}\right)\sqrt{k_0R_1/2}\left(\cos\theta_0 + \frac{Z_0}{Z_s}\right) \quad (7.62)$$

There is one pole of the reflection coefficient and $\cos\theta_p = -Z_0/Z_s$. This expression can be obtained from Equation (1.4.10) in Brekhovskikh and Godin (1992) for θ_p and θ_0 close to $\pi/2$. For nonlocally reacting media it has been suggested to use $Z_s(\sin\theta_0)$ instead of the constant impedance Z_s in Equations (7.60)–(7.62). In the modified Chien and Soroka formulation $Z_s(\sin\theta_0)$ is substituted for Z_s which does not depend on the angle of incidence. The reflection coefficient V_L of the impedance plane becomes the reflection coefficient V of the porous layer at an angle of incidence θ_0. In the modified formulation the reflected pressure is given by

$$p_r = \frac{\exp(-jk_0R_1)}{R_1}[V(\sin\theta_0) + (1 - V(\sin\theta_0))(1 + \sqrt{\pi}u\exp(u^2)\mathrm{erfc}(-u))] \quad (7.63)$$

where V is the reflection coefficient

$$V(\sin\theta) = \frac{\cos\theta - Z_0/Z_s(\sin\theta)}{\cos\theta + Z_0/Z_s(\sin\theta)} \quad (7.64)$$

and u is now given by

$$u = \exp\left(\frac{-j3\pi}{4}\right)\sqrt{k_0R_1/2}\left(\cos\theta_0 + \frac{Z_0}{Z_s(\sin\theta_0)}\right) \quad (7.65)$$

For materials having an impedance which does not depend significantly on θ the modified simulation is similar to the initial formulation of Chien and Soroka. A detailed description of the domain of validity of the modified formulation is beyond the scope of the book, due to the large number of parameters that characterize a porous layer. Some trends can be shown with the following examples. A precise correction due to the sphericity is necessary when the direct field $\exp(-jk_0R_2)/R_2$ and the reflected field $V(\sin\theta_0)\exp(-jk_0R_1)/R_1$ almost cancel each other. This happens for $\theta_0 \approx \pi/2$ because at $\theta_0 = \pi/2$, $R_1 = R_2$ and $V = -1$. In the first examples θ_0 is close to $\pi/2$. Let p_t be the exact total field above the layer given by

$$p_t = p_r + \frac{\exp(-jk_0R_2)}{R_2} \tag{7.66}$$

where the reflected field p_r is calculated with Equation (7.6). Let p_t' be the total field obtained with the modified formulation. In Figure 7.16, $|(p_t' - p_t)/p_t|$ is represented as a function of frequency for a layer of material 1 and a layer of material 2 of thickness 2 cm. The geometry is defined by $z_1 = z_2 = -2.5$ cm, $r = 1$ m.

The error is small for both materials in the low frequency range. At low frequencies there is one pole at θ_p close to $\pi/2$ and Equations (7.49)–(7.51) can be used. The modified formulation corresponds to the same set of equations where $\cos\theta_p = -Z/Z_s(\sin\theta_p)$ is replaced by $-Z/Z_s(\sin\theta_0)$. The modified formulation can also be used close to grazing incidence for thin layers because θ_0 and θ_p are close to each other and $Z_s(\theta_p)$ is close to $Z_s(\theta_0)$. The error is also small for the modified formulation at high frequencies for both materials. This can be explained by using the passage path method to predict the reflected pressure. The use of the passage path method can be justified at high frequencies for media with a low flow resistivity when θ_0 is close to $\pi/2$ because, in Equation (7.25) $|Z_0/Z_s(s\approx 1)| \approx 1$, there is no pole close to $\pi/2$. It is shown in Figure 7.17 that in the high-frequency range there is a good agreement between the exact pressure p_t obtained with Equation (7.6) and the prediction p_t'' obtained with Equation (7.20) from

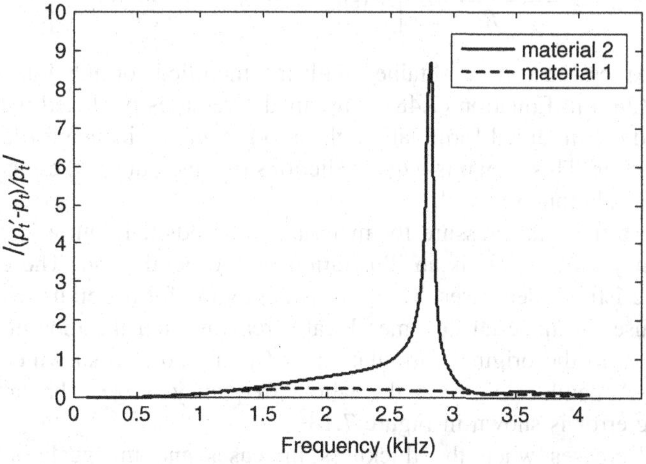

Figure 7.16 Modulus of the normalized error in the pressure evaluation with the modified formulation. Thickness $l = 2$ cm, $z_1 = z_2 = -2.5$ cm, $r = 1$ m.

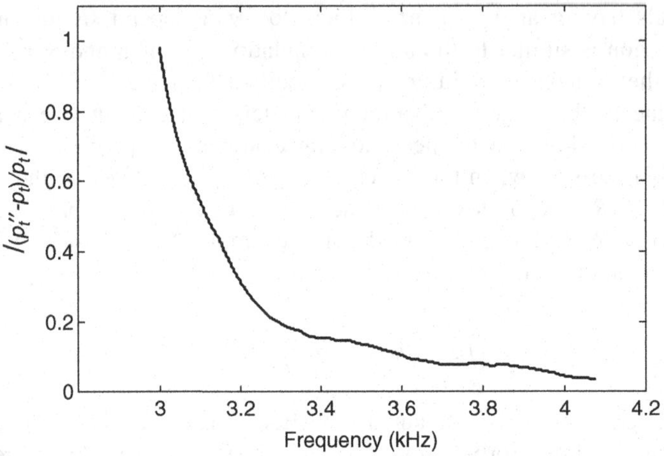

Figure 7.17 Normalized difference between the exact total pressure p_t and the pressure p_t'' predicted with the passage path method. Material 2, $l = 2$ cm, $z_1 = z_2 = -2.5$ cm, $r = 1$ m.

the passage path method though the contributions of the numerous poles crossing the path of integration are neglected.

This good agreement also shows that the pole contributions are negligible for the chosen geometry at high frequencies. Using Equation (7.A.3) for $\theta_0 = \pi/2$ gives at zeroth order in $\cos\theta_0$

$$N = -2\left(\frac{Z_s(\sin\theta_0)}{Z_0}\right)^2 \tag{7.67}$$

and from Equation (7.20) p_r is given by

$$p_r = \frac{\exp(-jk_0R_1)}{R_1}\left[V(\sin\theta_0) - \frac{-2j(Z_s(s_0)/Z_0)^2}{k_0R_1}\right] \tag{7.68}$$

The same expression can be obtained with the modified formulation when only the leading term $1/(2u^2)$ in Equation (7.48) is retained. Pressures predicted with the passage path method and the modified formulation close to grazing incidence for large numerical distances are similar. This explains why predictions obtained under these conditions with the modified formulation are valid.

In Figure 7.16 the total pressure for material 2 is predicted with a large error in the medium-frequency range. This is an illustration of a general trend. The error increases when the flow resistivity decreases. The error is negligible for materials with a large flow resistivity because the material becomes locally reacting and the modified formulation becomes identical to the original formulation of Chien and Soroka which can be used, as indicated in Appendix 7.C, over the whole frequency range. The influence of the thickness on the error is shown in Figure 7.18.

The error decreases when the thickness increases and the peak in the medium-frequency range disappears for semi-infinite layers. There is only one pole for a semi-infinite layer and the surface impedance is less dependent on the angle of incidence

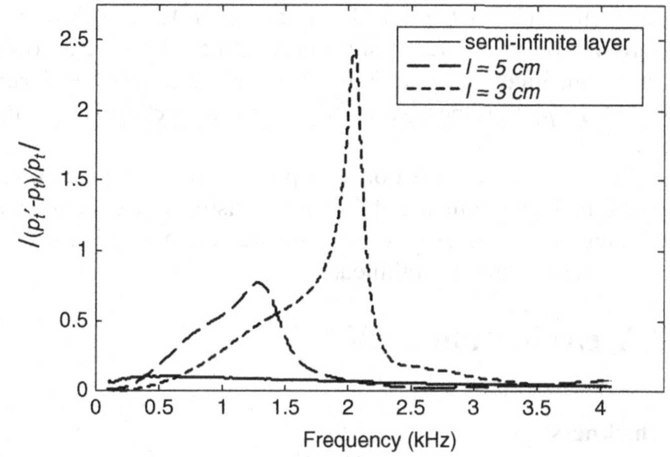

Figure 7.18 Modulus of the normalized error $|(p_t' - p_t)/p_t|$ in the pressure evaluation with the modified formulation for different thicknesses. Material 2, $z_1 = z_2 = -2.5$ cm, $r = 1$ m.

than for a thin layer. This can explain why the error decreases when the thickness increases.

The normalized difference between the exact total pressure and the pressure predicted with the modified formulation is shown in Figure 7.19 for a small angle of incidence. A comparison is performed with the normalized difference between the exact total pressure and the pressure p_L given by

$$p_L = \frac{\exp(-jk_0R_2)}{R_2} + \frac{\exp(-jk_0R_1)}{R_1}V(\sin\theta_0) \tag{7.69}$$

where the correction related to the sphericity of the source is omitted.

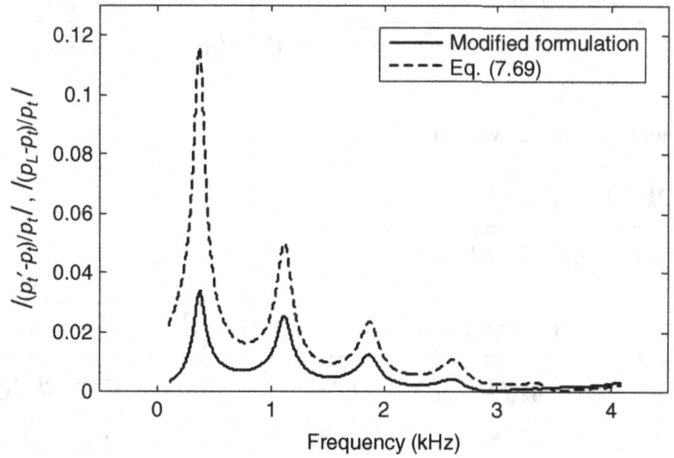

Figure 7.19 Normalized error in the estimation of the total pressure with the modified formulation and with Equation (7.69). Material 2, $l = 2$ cm, $z_1 = z_2 = -0.5$ m, $r = 1$ m.

Adding the correction $(1 - V(\sin\theta_0))(1 + \sqrt{\pi}u\exp(u^2)\text{erfc}(-u))\exp(-jk_0R_1)/R_1$ of the modified formulation to p_L noticeably decreases the difference at low frequencies. The large peak in the medium frequency range at θ_p close to $\pi/2$ has disappeared. The reflected field is close to p_L and the contribution of the correction is less important than for θ_0 close to $\pi/2$.

In summary, the modified formulation can provide reliable prediction, except for layers having a small thickness of media of low flow resistivity close to grazing incidence. In this case it is better to use the exact expression (Equation 7.6) if the conditions of a precise evaluation of the integral are fulfilled.

Appendix 7.A Evaluation of *N*

Layer of finite thickness

$$\frac{\partial V}{\partial s} = \frac{2m'(s_0)s(1-n^2)}{\sqrt{n^2-s^2}\sqrt{1-s^2}(m'(s_0)\sqrt{1-s^2}+\sqrt{n^2-s^2})^2}\left[1+\frac{2\sqrt{n^2-s^2}(1-s^2)k_0l}{\sin(2l\sqrt{n^2-s^2}(1-n^2)}\right]$$

$$m'(s) = -j\frac{m}{\phi}\cot(k_0l\sqrt{n^2-s^2}) \tag{7.A.1}$$

$$\begin{aligned}\frac{\partial^2 V}{\partial s^2} = {} & \frac{2(1-n^2)m'(s_0)s}{\sqrt{n^2-s^2}\sqrt{1-s^2}(m'(s_0)\sqrt{1-s^2}+\sqrt{n^2-s^2})^2}\frac{2k_0nl}{(1-n^2)} \\ & \times\left[\frac{2k_0nl(1-s^2)s\cos(2k_0l\sqrt{n^2-s^2})}{n^2\sin^2(2k_0l\sqrt{n^2-s^2})}-\frac{s((1-s^2)/\sqrt{n^2-s^2}+2\sqrt{n^2-s^2}))}{n\sin 2k_0l\sqrt{n^2-s^2}}\right] \\ & +\left[1+\frac{2(1-s^2)k_0l\sqrt{n^2-s^2}}{(1-n^2)\sin^2(2k_0l\sqrt{n^2-s^2})}\right]\left[\frac{2((1-n^2)s}{\sqrt{n^2-s^2}\sqrt{1-s^2}(m'(s_0)\sqrt{1-s^2}+\sqrt{n^2-s^2})^2}\right. \\ & \left.-\frac{4(1-n^2)m'^2(s_0)s}{\sqrt{n^2-s^2}\sqrt{1-s^2}(m'(s_0)\sqrt{1-s^2}+\sqrt{n^2-s^2})^3}\right]\frac{2m'k_0ls}{\sqrt{n^2-s^2}\sin(2k_0l\sqrt{n^2-s^2})}\end{aligned} \tag{7.A.2}$$

The coefficient N can be written

$$\begin{aligned}N = {} & \left[1+\frac{2k_0l(1-s_0^2)\sqrt{n^2-s_0^2}}{(1-n^2)\sin(2k_0l\sqrt{n^2-s_0^2})}\right] \\ & \times\left[M(s_0)+\frac{2(1-s_0^2)(1-n^2)s_0^2(-m'(s_0)\sqrt{1-s_0^2}+\sqrt{n^2-s_0^2})m'(s_0)k_0l}{(n^2-s_0^2)\sqrt{1-s_0^2}(m'(s_0)\sqrt{1-s_0^2}+\sqrt{n^2-s_0^2})^3\sin(2k_0l\sqrt{n^2-s_0^2})}\right] \\ & +\frac{1-s_0^2}{2}\frac{2m'(s_0)s_0(1-n^2)}{\sqrt{n^2-s_0^2}\sqrt{1-s_0^2}(m'(s_0)\sqrt{1-s_0^2}+\sqrt{n^2-s_0^2})^2}\frac{2k_0l}{(1-n^2)}n\end{aligned}$$

$$\times\left[\frac{2k_0ls_0(1-s_0^2)\cos(2k_0l\sqrt{n^2-s_0^2})}{n\sin^2[2k_0l\sqrt{n^2-s_0^2}]}-\frac{s_0(1-s_0^2+2(n^2-s_0^2))}{n\sin[2k_0l(n^2-s_0^2)]}\right] \tag{7.A.3}$$

where $M(s_0)$ is given by

$$\begin{aligned}M(s_0) =&m'(s_0)(1-n^2)\{2m'(s_0)(n^2-1)+(1-s_0^2)^{1/2}\\&\times[3m'(s_0)(1-s_0^2)^{1/2}-m'(s_0)(1-s_0^2)^{3/2}+\sqrt{n^2-s_0^2}(2n^2+s_0^2)]\}\\&\times(m'(s_0)\sqrt{1-s_0^2}+\sqrt{n^2-s_0^2})^{-3}(n^2-s_0^2)^{-3/2}\end{aligned} \tag{7.A.4}$$

Locally reacting medium of constant impedance Z_L

The pole is located at θ_p satisfying

$$\cos\theta_p=\sqrt{1-s_p^2}=-Z/Z_L \tag{7.A.5}$$

The reflection coefficient V_L can be written

$$V_L(s)=\frac{\cos\theta+\cos\theta_p}{\cos\theta-\cos\theta_p} \tag{7.A.6}$$

Equation (7.21) gives

$$N=\frac{2\sqrt{1-s_p^2}(1-\sqrt{1-s_0^2}\sqrt{1-s_p^2})}{(\sqrt{1-s_0^2}-\sqrt{1-s_p^2})^3} \tag{7.A.7}$$

Appendix 7.B Evaluation of p_r by the pole subtraction method

We follow the reference integral method described in Appendix A in Brekhovskikh and Godin (1992). The expressions for p_r are simultaneously obtained for a porous layer and for a locally reacting surface with an impedance Z_L. A subscript L is used for the locally reacting surface. The following expressions are used for the reflection coefficients

$$V(q)=1-\frac{2\sqrt{n^2-s^2}}{\sqrt{n^2-s^2}-j\sqrt{1-s^2}\cot(\sqrt{n^2-s^2}k_0l)nZ/(\phi Z_0)} \tag{7.B.1}$$

$$V_L(s)=1-\frac{2Z_0}{Z_L\sqrt{1-s^2}+Z_0} \tag{7.B.2}$$

The unit term provides a contribution exp(-jk_0R_1)/R_1 to p_r and is discarded in what follows. In Equation (7.15), $f(s)$, $f_L(s)$, $F(s)$, and $F_L(s)$, are now given by

$$f(s) = f_L(s) = -j(s\sin\theta_0 + \sqrt{1-s^2}\cos\theta_0) \tag{7.B.3}$$

$$F(s) = -2\sqrt{\frac{s}{1-s^2}}\sqrt{n^2-s^2}\left[\sqrt{n^2-s^2} - j\frac{Zn}{\phi Z_0}\sqrt{1-s^2}\cot(k_0l\sqrt{n^2-s^2})\right]^{-1} \tag{7.B.4}$$

$$F_L(s) = -2Z_0\sqrt{\frac{s}{1-s^2}}[Z_L\sqrt{1-s^2}+Z_0]^{-1} \tag{7.B.5}$$

Using Equations (A.3.9)–(A.3.14) of Brekhovskikh and Godin (1992) with the same notations, the integral in Equation (7.15) can be written

$$\int_{-\infty}^{\infty} F(s)\exp[k_0R_1f(s)]\,\mathrm{d}s = \exp(k_0R_1f(s_0))\left[a\mathrm{F}_1(1,kR_1,q_p) + \left(\frac{\pi}{k_0R_1}\right)^{1/2}\Phi_1(0)\right] \tag{7.B.6}$$

$$a = \lim_{s\to s_0}[F(s)(s-s_P)] \tag{7.B.7}$$

$$q_p = \{j[\cos(\theta_p-\theta_0)-1]\}^{1/2} \qquad \mathrm{Im}(q_p) < 0 \tag{7.B.8}$$

This equation can be replaced, for the case of thin layers and semi-infinite layers, by

$$q_p = \exp\left(\frac{-j\pi}{4}\right)\sqrt{2}\sin\frac{\theta_p-\theta_0}{2} \tag{7.B.9}$$

$$\mathrm{F}_1(1,k_0R_1,q_p) = -j\pi\exp(-k_0R_1q_p^2)[\mathrm{erfc}(j\sqrt{k_0R_1}q_p)] \tag{7.B.10}$$

$$\Phi_1(0) = F(s_0)\sqrt{\frac{-2}{f''(s_0)}} + a/q_p \tag{7.B.11}$$

where $s_0 = \sin\theta_0$. For the porous layer, a is given by

$$a = -2\sqrt{\frac{s_p}{1-s_p^2}}\frac{\sqrt{n^2-s_p^2}}{G_s'(s_p)} \tag{7.B.12}$$

where G' is given by Equation (7.40), and Equation (7.B.12) can be rewritten

$$a = -2\sqrt{\frac{1-s_p^2}{s_p}}\left[1-(1-s_p^2)\left(\frac{2k_0l}{\sqrt{n^2-s_p^2}\sin 2k_0l\sqrt{n^2-s_p^2}} + \frac{1}{n^2-s_p^2}\right)\right]^{-1} \tag{7.B.13}$$

For the locally reacting medium, a_L is given by

$$a_L = -2\sqrt{\frac{1-s_p^2}{s_p}} \tag{7.B.14}$$

Using $f(s) = -j\cos(\theta - \theta_0)$ gives $f''(s_0) = j/(1-s_0^2)$, and

$$F(s_0)\sqrt{\frac{-2}{f''(s_0)}} = \sqrt{\frac{s_0}{1-s_0^2}}(-1+V(s_0))\sqrt{2j(1-s_0^2)} \tag{7.B.15}$$

For the porous layer, the reflected pressure is given by

$$\begin{aligned} p_r = & \frac{\exp(-jk_0R_1)}{R_1} + \left(\frac{k_0}{2\pi r}\right)^{1/2} \exp\left(\frac{-j\pi}{4}\right)\exp(-jk_0R_1) \\ & \times \left\{-j\pi a \exp(-k_0R_1q_p^2)\mathrm{erfc}(j\sqrt{k_0R_1}q_p)\right. \\ & \left. + \left(\frac{\pi}{k_0R_1}\right)^{1/2}\left[\sqrt{\frac{s_0}{1-s_0^2}}(-1+V(s_0))\sqrt{2j(1-s_0^2)} + \frac{a}{q_p}\right]\right\} \end{aligned} \tag{7.B.16}$$

After some rearrangement, p_r can be rewritten

$$\begin{aligned} p_r = & \frac{\exp(-jk_0R_1)}{R_1} \\ & \times \left\{V(s_0) - \frac{\sqrt{1-s_p^2}\sqrt{2k_0R_1}\exp(-3\pi j/4)(1+\sqrt{\pi}u\exp(u^2)erfc(-u))}{u\sqrt{s_0s_p}[1-(1-s_p^2)[2k_0l/(\sqrt{n^2-s_p^2}\sin 2k_0l\sqrt{n^2-s_p^2}) + 1/(n^2-s_p^2)]]}\right\} \end{aligned} \tag{7.B.17}$$

where u is the numerical distance, defined by

$$u = -j\sqrt{k_0R_1}q_p \tag{7.B.18}$$

For the locally reacting surface, the expression for p_r is very similar

$$\begin{aligned} p_r = & \frac{\exp(-jk_0R_1)}{R_1} \\ & \times \left\{V_L(s_0) - \frac{\sqrt{1-s_p^2}\sqrt{2k_0R_1}\exp(-3\pi j/4)(1+\sqrt{\pi}u\exp(u^2)erfc(-u))}{u\sqrt{s_0s_p}}\right\} \end{aligned} \tag{7.B.19}$$

The reflection coefficient $V(s_0)$ in Equation (7.B.17) is given by Equation (7.38). If θ_0 is close to $\pi/2$, at the first order approximation in $\cos\theta_0$, $\sin\theta_0 = 1$ and

$\cos\theta_1 = \sqrt{1 - 1/n^2}$, and $V(s_0)$ is given with the same approximation by

$$V(s_0) = \frac{-(n^2-1)^{1/2} - jn(Z/\phi Z_0)\cos\theta_0 \cot(k_0 l\sqrt{n^2-1})}{(n^2-1)^{1/2} - jn(Z/\phi Z_0)\cos\theta_0 \cot(k_0 l\sqrt{n^2-1})} \tag{7.B.20}$$

If θ_p is also close to $\pi/2$, Equation (7.B..20) can be used to calculate $\cos\theta_p$ which is given by

$$\cos\theta_p = -j\frac{\phi Z_0}{nZ}\sqrt{n^2-1}\tan k_0 l\sqrt{n^2-1} \tag{7.B.21}$$

(see Equation 7.28 for the case of thin layers). The approximate reflection coefficient V has the same expression as V_L

$$V(s_0) = \frac{\cos\theta_0 + \cos\theta_p}{\cos\theta_0 - \cos\theta_p} \tag{7.B.22}$$

Neglecting the term multiplied by $1 - s_p^2 = \cos^2\theta_p$ in the denominator at the right-hand side of Equation (7.B.17), Equation (7.B.17) and (7.B.19) become identical.

Appendix 7.C From the pole subtraction to the passage path: locally reacting surface

In the case of a locally reacting surface, for large $|u|$,the reflected pressure evaluated with the pole subtraction method has the same expression as that obtained with the passage path method. The pole contribution, for the pole subtraction method, corresponds to the term $2\sqrt{\pi}u\exp(u^2)$ of Equation (7.48), and can be written

$$SW_P(s_p) = -\frac{\exp(-jk_0R_1)}{R_1}\frac{\sqrt{1-s_p^2}\sqrt{2k_0R_1}\exp\left(\frac{-3j\pi}{4}\right)2\sqrt{\pi}\exp(u^2)}{\sqrt{s_0 s_p}} \tag{7.C.1}$$

where $u^2 = jk_0R_1[1 - \cos(\theta_p - \theta_0)]$. This contribution is the same as $SW(s_p)$ given by Equation (7.43) in the context of the passage path method. This contribution exists under the same conditions. It may be shown that the condition Re$u > 0$ in Equation (7.48) corresponds to the fact that the pole has been crossed when the initial path of integration is deformed into the passage path. Moreover, with θ_0 and θ_p close to $\pi/2$ and u given by Equation (7.50), the contribution of the term $1/2u^2$ in Equation (7.48) to p_r corresponds to the contribution of N in Equation (7.20), N for an impedance plane being given in Brekhovskikh and Godin (1992) by

$$N = \frac{2\cos\theta_p(1 - \cos\theta_0\cos\theta_p)}{(\cos\theta_0 - \cos\theta_p)^3} \tag{7.C.2}$$

and the product $\cos\theta_0\cos\theta_p$ being neglected at the numerator. The expression of the reflected pressure obtained with the pole subtraction method remains valid for large numerical distances, and can replace the expression obtained with the passage path method. A medium with a high flow resistivity can be replaced by an impedance plane with $|Z_s| \gg Z_0$ over the whole audible frequency range and the expressions (7.49)–(7.51) of Chien and Soroka can be used for small and large numerical distances if θ_0 is close to $\pi/2$.

References

Abramowitz, M. and Stegun, I.A. (1972) *Handbook of Mathematical Functions*. National Bureau of Standards, Washington D. C.

Allard, J.F. and Lauriks, W. (1997) Poles and zeros of the plane wave reflection coefficient for porous surfaces. *Acta Acustica* **83**, 1045–1052.

Allard, J.F., Henry, M., Tizianel, J., Nicolas, J. and Miki, Y. (2002) Pole contribution to the field reflected by sand layers. *J. Acoust. Soc. Amer.* **111**, 685–689.

Allard, J.F., Henry, M., Jansens, G. and Lauriks, W. (2003a) Impedance measurement around grazing incidence for nonlocally reacting thin porous layers, *J. Acoust. Soc. Amer.* **113**, 1210–1215.

Allard, J.F., Gareton, V., Henry, M., Jansens, G. and Lauriks, W. (2003b) Impedance evaluation from pressure measurements near grazing incidence for nonlocally reacting porous layers. *Acta Acustica* **89**, 595–603.

Allard, J.F., Henry, M. and Gareton V. (2004) Pseudo-surface waves above thin porous layers. *J. Acoust. Soc. Amer.* **116**, 1345–1347.

Allard, J.F. and Henry, M. (2006) Fluid-fluid interface and equivalent impedance plane. *Wave Motion* **43**, 232–240.

Baños, A. (1966) *Dipole Radiation in the Presence of a Conducting Haf-Space*. Pergamon Press, New York.

Brekhovskikh, L.M. (1960) *Waves in Layered Media*. Academic, New York.

Brekhovskikh, L.M. and Godin, O.A. (1992) *Acoustic of Layered Media II, Point Source and Bounded Beams*. Springer Series on Waves Phenomena, Springer, New York.

Brouard, B. (1994) Validation par holographie acoustique de nouveaux modèles pour la propagation des ondes dans les materiaux poreux stratifiés. Ph.D. Thesis, Université du Maine, Le Mans.

Brouard, B., Lafarge, D., Allard, J.F. and Tamura, M. (1996) Measurement and prediction of the reflection coefficient of porous layers at oblique incidence and for inhomogeneous waves. *J. Acoust. Soc. Amer.* **99**, 100–107.

Chien, C.F. and Soroka, W.W. (1975) Sound propagation along an impedance plane. *J. Sound Vib.* **43**, 9–20.

Collin, R.E. (1960) *Field Theory of Guided Waves*. McGraw-Hill, New York.

Hickey, C., Leary, D. and Allard, J.F. (2005) Impedance and Brewster angle measurement for thick porous layers, *J. Acoust. Soc. Amer.* **118**, 1503–509.

Lauriks, W., Kelders, L. and Allard, J.F. (1998) Poles and zeros of the reflection coefficient of a porous layer having a motionless frame in contact with air. *Wave Motion* **28**, 59–67.

Li, K.M., Waters-Fuller, T. and Attenborough, K. (1998) Sound propagation from a point source over extended reaction ground. *J. Acoust. Soc. Amer.* **104**, 679–685.

Nicolas, J., Berry, J.F. and Daigle, G. (1985) Propagation of sound above a layer of snow. *J. Acoust. Soc. Am.* **77**, 67–73.

Tamura, M. (1990) Spatial Fourier transform method of measuring reflection coefficient at oblique incidence. *J. Acoust. Soc. Amer.* **88**, 2259–2264.

Wait, J.R. (1970) *Electromagnetic Waves in Stratified Media*. Pergamon Press, New York.

8

Porous frame excitation by point sources in air and by stress circular and line sources – modes of air saturated porous frames

8.1 Introduction

Some of the modes of air-saturated rigid framed porous layers have been described in Chapter 7. These modes are related to poles of the plane wave reflection coefficient and can be observed in the spherical reflected field at large angles of incidence. A real porous structure is not motionless, but the results of Chapter 7 will not be noticeably modified, for most of the porous media, if the Biot theory is used instead of a rigid-framed model. In this chapter, the Biot theory is used to describe the frame displacement created by a point source pressure field in air, or by a normal stress applied in a limited domain at the surface of a layer. The porous media are supposed to be isotropic. Anisotropy effects are considered in Chapter 10. A description of the displacement field created by a normal stress field at the surface of the layers with a given wave number parallel to the surface is discussed. The same description is performed when the stress field is replaced by a plane pressure field in air. The Sommerfeld representation can be used to adapt the results to the case of a point source in air. The Fourier transform or the Hankel transform can be used for the case of a circular normal stress.

Another aim of this chapter is to present experiments which can be used to evaluate the rigidity parameters of a porous frame in the audible frequency range. Many methods of measuring the rigidity coefficients at very low frequencies have been carried out

Propagation of Sound in Porous Media: Modelling Sound Absorbing Materials, Second Edition J. F. Allard and N. Atalla

(Melon *et al.* 1998, Park 2005, Pritz 1986, 1994; Sfaoui 1995a, b, Langlois *et al.* 2001, Tarnow 2005), but the rigidity coefficients can present noticeable variations between the quasi-static regime and the audible frequency range.

8.2 Prediction of the frame displacement

8.2.1 Excitation with a given wave number component parallel to the faces

In Figure 8.1(a), a plane acoustic field in air impinges upon a porous layer of finite thickness l at an angle of incidence θ. The layer is bonded onto a rigid impervious backing. The geometry of the problem is two-dimensional in the incident plane. The total pressure field p (incident plus reflected) at the surface of the layer is given by $p = p^e \exp(-jk_x x)$, the time dependence $\exp(j\omega t)$ is discarded and $k_x = \xi = k_0 \sin\theta$. The first representation of the Biot theory is used in this chapter. Three Biot waves can propagate in the porous layer, toward the backing and toward the air–porous material boundary. In Figure 8.1(b), the incident plane wave is replaced by a normal stress field $\tau_{zz}^s = \exp(-j\xi x)$, with the same dependence on x, applied to the surface. In a first step, the Biot theory is used to predict the radial and the vertical displacement components u_x^s and u_z^s of the porous frame created by a plane wave in air with a horizontal (trace) wave number ξ. Let k_1 and k_2 be the wave numbers of the two Biot compressional waves, and k_3 the wave number of the shear wave. Each plane wave is related to a displacement $\boldsymbol{u}^s$ of the frame and $\boldsymbol{u}^f$ of air, with $\boldsymbol{u}^f = \mu \boldsymbol{u}^s$. Let N be the shear modulus and ν the Poisson ratio of the frame. Let P, Q, and R be the Biot rigidity coefficients. The quantities k_1, k_2, k_3, μ_1, μ_2, μ_3, P, Q, and R, are described in Chapter 6.

The excitation by the pressure field or the stress field is related to Biot waves propagating up and down the layer with the same horizontal wave number component. The superscript + will be used for waves propagating from the upper to the lower face and the superscript – for waves propagating in the opposite direction. Let $\varphi_i^{\pm}$, i = 1, 2 be the velocity potentials of the compressional waves, with the related displacement $\boldsymbol{u}^s = -j\nabla\varphi/\omega$, and let $\boldsymbol{\Psi}^{\pm} = \boldsymbol{n}\varphi_3^{\pm}$ be the vector potentials, $\boldsymbol{n}$ being the unit vector on the y axis and $\boldsymbol{u}^s = -j\nabla \wedge \boldsymbol{\Psi}/\omega$ the related frame displacement. The scalar functions

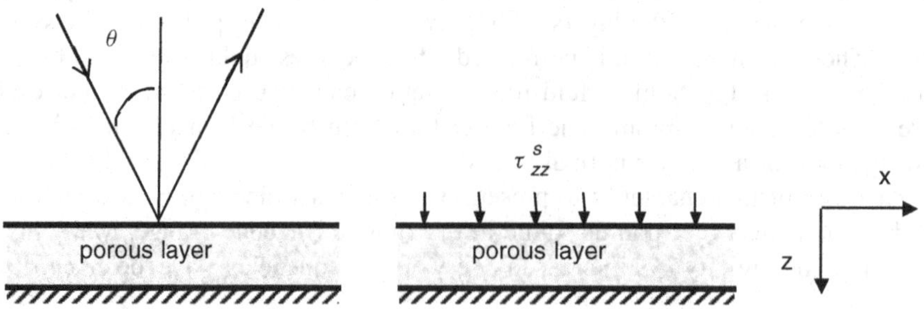

Figure 8.1 Two sources inducing Biot wave in a porous medium: (a) incident plane field, (b) normal stress field.

$\varphi_i^{\pm}$ can be written

$$\begin{aligned}\varphi_i^{\pm} &= a_i^{\pm}\exp(\mp j\alpha_i z - j\xi x)\\ \alpha_i &= (k_i^2 - \xi^2)^{1/2}\\ i &= 1, 2, 3\end{aligned} \tag{8.1}$$

Let $\sigma_{\alpha\beta}^s$ and $\sigma_{\alpha\beta}^f$ be the components of the Biot stress tensors for the frame, and the air, respectively. They are related to forces acting on the frame and on the air, for a unit surface of porous medium. The stress components in the Biot theory are given by

$$\sigma_{ij}^s = [(P - 2N)\theta^s + Q\theta^f]\delta_{ij} + 2Ne_{ij}^s \tag{8.2}$$

$$\sigma_{ij}^f = (Q\theta_s + R\theta_f)\delta_{ij} \tag{8.3}$$

where θ^f and θ^s are the dilatation of the air, and the dilatation of the frame, respectively, and the terms e_{ij}^s are the strain components of the frame.

In order to simplify the calculations for layers of finite thickness, three independent displacement fields which satisfy the boundary conditions at the lower face of the layer are defined. They are obtained by associating to each potential function $\exp(-j\alpha_i z)$, $i = 1,2,3$, three potential functions $r_{ik}\exp(j\alpha_k z)$, $k = 1,2,3$. The boundary conditions at the lower face when the layer is glued to a rigid impervious backing are

$$u_z^s = 0 \tag{8.4}$$

$$u_z^f = 0 \tag{8.5}$$

$$u_x^s = 0 \tag{8.6}$$

The coefficients r_{ji} are given in Appendix 8.A. Let $\boldsymbol{b}_1$, $\boldsymbol{b}_2$, and $\boldsymbol{b}_3$ be the frame displacement fields defined by (the factor $\exp(-j\xi x)$ being removed)

$$\begin{aligned}\boldsymbol{b}_1 = &-\frac{j}{\omega}[\nabla(\exp(-j\alpha_1 z) + r_{11}\exp(j\alpha_1 z)\\ &+ r_{12}\exp(j\alpha_2 z)) + \nabla\wedge\boldsymbol{n}r_{13}\exp(j\alpha_3 z)]\end{aligned} \tag{8.7}$$

$$\begin{aligned}\boldsymbol{b}_2 = &\frac{-j}{\omega}[\nabla(r_{21}\exp(j\alpha_1 z) + \exp(-j\alpha_2 z)\\ &+ r_{22}\exp(j\alpha_2 z)) + \nabla\wedge\boldsymbol{n}r_{23}\exp(j\alpha_3 z)]\end{aligned} \tag{8.8}$$

$$\begin{aligned}\boldsymbol{b}_3 = &\frac{-j}{\omega}[\nabla(r_{31}\exp(j\alpha_1 z) + r_{32}\exp(j\alpha_2 z))\\ &+ \nabla\wedge\boldsymbol{n}(\exp(-j\alpha_3 z) + r_{33}\exp(j\alpha_3 z))]\end{aligned} \tag{8.9}$$

Any linear combination $\lambda_1\boldsymbol{b}_1 + \lambda_2\boldsymbol{b}_2 + \lambda_3\boldsymbol{b}_3$ satisfies the boundary conditions at the lower face. Let p^e be the pressure and v_z^e be the normal component of the velocity of air in the free air close to the upper face of the layer. At the upper face, the boundary conditions for the case of an incident plane wave can be written (see Equations 6.98–6.100)

$$\phi\sum\dot{u}_z^f + (1-\phi)\sum\dot{u}_z^s = v_z^e \tag{8.10}$$

$$\sum \sigma_{zz}^f = -\phi p^e \tag{8.11}$$

$$\sum \sigma_{zz}^s = -(1-\phi)p^e \tag{8.12}$$

$$\sum \sigma_{xz}^s = 0 \tag{8.13}$$

where the summation is performed on the six Biot waves. If the excitation is created by a normal unit stress field τ_{zz}^s applied on the frame, Equation (8.12) must be replaced by

$$\Sigma\sigma_{zz}^s = -(1-\phi)p + \tau_{zz}^s \tag{8.14}$$

At the upper face, the velocity components and the stress components related to $\sum \lambda_j \boldsymbol{b}_j$ can be written, the x dependence being discarded

$$\Sigma\sigma_{xz}^s = M_{1,1}\lambda_1 + M_{1,2}\lambda_2 + M_{1,3}\lambda_3 \tag{8.15}$$

$$\Sigma\dot{u}_z^s = M_{2,1}\lambda_1 + M_{2,2}\lambda_2 + M_{2,3}\lambda_3 \tag{8.16}$$

$$\Sigma\dot{u}_z^f = M_{3,1}\lambda_1 + M_{3,2}\lambda_2 + M_{3,3}\lambda_3 \tag{8.17}$$

$$\Sigma\sigma_{zz}^f = M_{4,1}\lambda_1 + M_{4,2}\lambda_2 + M_{4,3}\lambda_3 \tag{8.18}$$

$$\Sigma\sigma_{zz}^s = M_{5,1}\lambda_1 + M_{5,2}\lambda_2 + M_{5,3}\lambda_3 \tag{8.19}$$

$$\Sigma\dot{u}_x^s = M_{6,1}\lambda_1 + M_{6,2}\lambda_2 + M_{6,3}\lambda_3 \tag{8.20}$$

The coefficients $M_{i,j}$ are given in Appendix 8.A. For the case of the impinging plane wave (Figure 8.1a), the coefficients λ_j related to the unit total pressure field p with $p^e = 1$ can be obtained from Equations (8.11)–(8.13). Inserting these values in Equation (8.10) gives the velocity v_z^e, the surface impedance $Z_s(\xi/k_0) = p^e/v_z^e$, and the reflection coefficient $V(\xi/k_0)$

$$V(\xi/k_0) = \frac{Z_s(\xi/k_0) - Z_0/\cos\theta}{Z_s(\xi/k_0) + Z_0/\cos\theta} \tag{8.21}$$

where Z_0 is the characteristic impedance of the free air and $\cos\theta = (1-(\xi/k_0)^2)^{1/2}$. The normal total velocity components $\dot{U}_z^s = \sum \dot{u}_z^s$, $\dot{U}_z^f = \sum \dot{u}_z^f$, and the reflection coefficient can be obtained with the parameters T_{ij} given by

$$T_{ij} = M_{i+1,j} - \frac{M_{1,j}}{M_{1,3}} M_{i+1,3} \qquad i = 1, \ldots, 5,\ j = 1, 2 \tag{8.22}$$

Using Equations (8.13) and (8.15), Equations (8.18)–(8.19) can be rewritten

$$\sum \sigma_{zz}^f = T_{31}\lambda_1 + T_{32}\lambda_2 \tag{8.23}$$

$$\sum \sigma_{zz}^s = T_{41}\lambda_1 + T_{42}\lambda_2 \tag{8.24}$$

From Equations (8.11)–(8.12) the ratio λ_2/λ_1 is given by

$$\frac{\lambda_2}{\lambda_1} = s_1 = -\frac{\phi T_{41} - (1-\phi)T_{31}}{\phi T_{42} - (1-\phi)T_{32}} \tag{8.25}$$

The coefficient λ_1 is obtained from Equation (8.23) where $\sum \sigma_{zz}^f = -\phi$. The velocity components, the surface impedance and the reflection coefficient are given by

$$\dot{U}_z^s = (T_{11} + T_{12}s_1)\frac{-\phi T_{42} + (1-\phi)T_{32}}{T_{31}T_{42} - T_{32}T_{41}} \tag{8.26}$$

$$\dot{U}_x^s = (T_{21} + T_{22}s_1)\frac{-\phi T_{42} + (1-\phi)T_{32}}{T_{31}T_{42} - T_{32}T_{41}} \tag{8.27}$$

$$v_z^e = \frac{D_1}{T_{31}T_{42} - T_{32}T_{41}} \tag{8.28}$$

$$Z_s(\xi/k_0) = (T_{31}T_{42} - T_{32}T_{41})/D_1 \tag{8.29}$$

$$V(\xi/k_0) = \frac{-D_1 Z_0 + (T_{31}T_{42} - T_{32}T_{41})\cos\theta}{D_1 Z_0 + (T_{31}T_{42} - T_{32}T_{41})\cos\theta} \tag{8.30}$$

where D_1 is given by

$$\begin{aligned} D_1 = &[-\phi T_{42} + (1-\phi)T_{32}][(1-\phi)T_{11} + \phi T_{21}] \\ &+ [\phi T_{22} + (1-\phi)T_{12}][\phi T_{41} - (1-\phi)T_{31}] \end{aligned} \tag{8.31}$$

When the external source is the unit stress $\tau_{zz}^s = \exp(-j\xi x)$, a plane wave similar to the reflected wave in the previous case exists, with a z dependence $\exp(jzk_0\cos\theta)$. At the surface of the layer in the free air, the pressure p^e and the normal velocity v_z^e are related by

$$p^e = -v_z^e Z_0/\cos\theta \tag{8.32}$$

Using Equation (8.10), Equation (8.32) can be rewritten

$$p^e = s_2\lambda_1 + s_3\lambda_2 \tag{8.33}$$

with s_2 and s_3 given by

$$s_2 = -\frac{Z_0}{\cos\theta}[(1-\phi)T_{11} + \phi T_{21}] \tag{8.34}$$

$$s_3 = -\frac{Z_0}{\cos\theta}[(1-\phi)T_{12} + \phi T_{22}] \tag{8.35}$$

Equation (8.23) becomes

$$(T_{31} + \phi s_2)\lambda_1 + (T_{32} + \phi s_3)\lambda_2 = 0 \tag{8.36}$$

Equation (8.24) with $\tau_s = 1$ becomes

$$[T_{41} + (1-\phi)s_2]\lambda_1 + [T_{42} + (1-\phi)s_3]\lambda_2 = 1 \tag{8.37}$$

The parameters λ_1 and λ_2 can be obtained from this set of equations. For an excitation of unit amplitude, the frame velocity components at the surface of the layer are given by

$$\dot{U}_z^s = [T_{11}(T_{32} + \phi s_3) - T_{12}(T_{31} + \phi s_2)]/D_2 \tag{8.38}$$

$$\dot{U}_x^s = [T_{51}(T_{32} + \phi s_3) - T_{52}(T_{31} + \phi s_2)]/D_2 \tag{8.39}$$

were D_2 is given by

$$D_2 = [T_{41} + (1-\phi)s_2][T_{32} + \phi s_3] - [T_{42} + (1-\phi)s_3][T_{31} + \phi s_2] \quad (8.40)$$

The denominators D_1 and D_2 are related by

$$D_2 = T_{41}T_{32} - T_{31}T_{42} - D_1 \frac{Z_0}{\cos\theta} \quad (8.41)$$

A comparison of Equations (8.30) and (8.41) shows that the singularities of the velocity components $\dot{U}_z^s$ and $\dot{U}_x^s$ are located at the same x wave number component ξ as the singularities of the reflection coefficient.

The frame velocity components created by a plane field in air are used at the end of the chapter to predict the displacements induced by a point source in air. The frame velocity components obtained for a mechanical excitation are used in what follows for different geometries of the excitation.

8.2.2 Circular and line sources

Let g(r) be the radial spatial dependence of the circular source τ_s. Using the direct and inverse Hankel transforms

$$\begin{aligned} G(\xi) &= \int_0^\infty r g(r) J_0(\xi r) dr \\ g(r) &= \int_0^\infty \xi G(\xi) J_0(\xi r)\, d\xi \end{aligned} \quad (8.42)$$

the axisymmetric source can be replaced by a superposition of excitations with a spatial dependence given by the Bessel function J_0 $(r\xi)$. The following equivalence (see Equations (7.3)–(7.6)) and Appendix 8.B

$$\begin{aligned} \int_0^\infty G(\xi) J_0(r\xi)\xi\, d\xi &= \frac{1}{2\pi}\int_{-\infty}^{\infty}\int_{-\infty}^{\infty} G(\xi)\exp(-j(\xi_1 x + \xi_2 y) d\xi_1\, d\xi_2 \\ \xi &= \sqrt{\xi_1^2 + \xi_2^2} \end{aligned} \quad (8.43)$$

shows that $J_0(\xi r)$ can be replaced by a superposition of unit fields with different orientations and the same radial wave number ξ. The total vertical velocity of the frame $\dot{\overline{U}}_z^s(r)$ can be written (see Appendix 8.B)

$$\dot{\overline{U}}_z^s(r) = \frac{1}{2\pi}\int_0^\infty \xi G(\xi) J_0(\xi) \dot{U}_z^s(\xi) d\xi \quad (8.44)$$

Using

$$J_1(u) = \frac{j}{2\pi}\int_0^{2\pi} \exp(-ju\cos\theta)\cos\theta\, d\theta$$

(see Appendix 8.B) gives the following expression for the radial component $\dot{\overline{U}}_r^s(r)$

$$\dot{\overline{U}}_r^s(r) = \frac{-j}{2\pi} \int_0^\infty \xi J_1(r\xi) G(\xi) \dot{U}_x^s(\xi)\, d\xi \tag{8.45}$$

For a line source in the direction Y with a spatial dependence h(x), the direct and inverse spatial Fourier transforms can be used

$$\begin{aligned} H(\xi) &= \int_{-\infty}^{\infty} h(x) \exp(j\xi x)\, dx \\ h(x) &= \frac{1}{2\pi} \int_{-\infty}^{\infty} H(\xi) \exp(-j\xi x)\, d\xi \end{aligned} \tag{8.46}$$

The z and x components of the velocity displacement at the surface are now given by

$$\dot{\overline{U}}_z^s(x) = \frac{1}{2\pi} \int_{-\infty}^{\infty} H(\xi) \dot{U}_z^s(\xi) \exp(-j\xi x)\, d\xi \tag{8.47}$$

$$\dot{\overline{U}}_x^s(x) = \frac{1}{2\pi} \int_{-\infty}^{\infty} H(\xi) \dot{U}_x^s(\xi) \exp(-j\xi x)\, d\xi \tag{8.48}$$

8.3 Semi-infinite layer – Rayleigh wave

Rayleigh pole

It was shown by Feng and Johnson (1983) that the Rayleigh wave is the one surface mode that could be detected experimentally at a semi-infinite porous frame–light fluid interface. In Feng and Johnson (1983) the different losses predicted by the Biot theory are not accounted for. A description of the Rayleigh wave for the lossy air-saturated porous sound absorbing media is performed by Allard *et al.* (2002). Illustrations are given with the porous medium denoted as material 1 in Table 8.1. Predictions for a semi-infinite layer are obtained by setting all r_{ij} in Equations (8.7)–(8.9) equal to 0. *A priori*, due to the partial decoupling, the Rayleigh wave must be similar to the one that could exist at the surface of the porous frame in vacuum. For the case of the frame in vacuum, the velocity of the Rayleigh wave must be slightly smaller than the velocity of the shear wave. For material 1, at 2 kHz, the wave number k_3 of the shear Biot wave is equal to $k_3 = 230.6 - j24.0\ m^{-1}$.

Using for the porous medium the empirical formula in Victorov (1967) that relates, for an elastic solid, the wave number k_R of the Rayleigh wave to the wave number of the shear wave

$$k_R = k_3 \frac{1+\nu}{0.87 + 1.12\nu} \tag{8.49}$$

Table 8.1 Parameters for different materials.

Material	Frame density ρ_1 (kg/m^3)	Poisson ratio ν	Shear modulus N (kPa)	Tortuosity α_∞	Porosity ϕ	Flow resistivity σ (N m^{-4} s)	Characteristic dimensions Λ, Λ' (μm)
1	25.	0.3	75 + j15	1.4	0.98	50 000	50, 150
2	24.5	0.44	80 + j12	2.2	0.97	22 100	39, 275

one obtains $k_R = 248 - \text{j}25.9\ \text{m}^{-1}$. There are no explicit expressions for the location of the singularities of $\dot{U}_z^s$ and $\dot{U}_x^s$ given by Equations (8.38)–(8.39), but the x wave number components close to the predicted k_R can be obtained by recursive methods. Two poles are obtained at $\xi_R = 249.1 - \text{j}26.5\ \text{m}^{-1}$ and $\xi_R = 247.7 + \text{j}25.7\ \text{m}^{-1}$ for a set of three Biot waves having a decreasing amplitude with increasing z in the porous medium. The wave in air above the porous medium has a dependence on z given by $\exp(jkz\cos\theta)$, with $\cos\theta = -0.73 - j6.73$ for the first pole, and $\cos\theta = 0.71 + j6.69$for the second pole. The first pole is on the physical Riemann sheet. Both $\cos\theta$ are almost opposite, both $\sin\theta$ are almost equal.

Rayleigh wave contribution for a circular source

The simplest source is a normal periodic point force $F\exp(j\omega t)\delta(x)\delta(y)$. It can be replaced by a superposition of axisymmetric fields

$$F\delta(x)\delta(y) = \frac{F}{2\pi}\int_0^\infty \text{J}_0(\xi r)\xi\, d\xi \tag{8.50}$$

Using Equation (8.44), the vertical velocity at a radial distance r from the source can be written

$$\dot{\overline{U}}_z^s(r) = \frac{F}{2\pi}\int_0^\infty \text{J}_0(\xi r)\dot{U}_z^s(\xi)\xi\, d\xi \tag{8.51}$$

where $\dot{U}_z^s$ is given by Equation (8.38) and presents a singularity at $\xi = \xi_R$. Using the relation $\text{J}_0(u) = 0.5(H_0^1(u) - H_0^1(-u))$ the integral on half the real ξ axis can be replaced, as in Chapter 7, by an integral on the whole ξ axis. This path of integration can be deformed in different ways to cut the Rayleigh poles. An example of integration on different paths is given in Allard *et al.* (2004) for water–poroelastic interfaces. Predictions obtained for a radial distance $r = 25$ cm with Equation (8.51) and a fast Fourier transform are shown in Figure 8.2 for a semi-infinite layer of material 1 in air. The vertical force is a burst centred at 2 kHz.

The part of the signal denoted as the 'Rayleigh wave' is almost identical to the contribution of the residues related to the Rayleigh pole in the physical Riemann sheet. It is shown in Appendix 8.C how to evaluate this contribution. Simulated measurements for different radial distances give a velocity of the Rayleigh wave close to $\omega/\text{Re}\xi_R \approx 52$ m/s at 2 kHz.

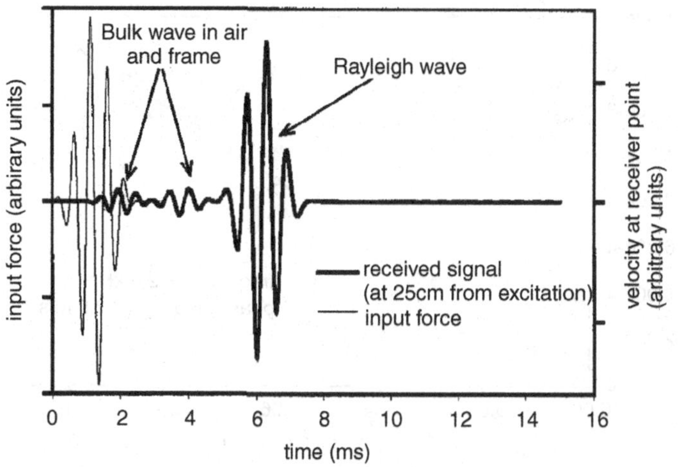

Figure 8.2 The point force $F(t)$ and the vertical velocity $\dot{U}_z^s(r)$ of the surface of the frame at $r = 25$ cm (Allard *et al.* 2002). Reprinted with permission from Allard, J. F., Jansens, G., Vermeir, G. & Lauriks, W. Frame-borne surface wave in air-saturated porous media. *J. Acoust. Soc. Amer.* **111**, 690–696. Copyright 2002. Acoustical Society of America.

Measurements have been performed with different sources, line sources and circular sources having a radius around 5 mm. The first measurement of the velocity and the damping of a Rayleigh wave over an open cell foam is described in Allard *et al.* (2002). The experimental set-up is represented in Figure 8.3.

The thickness of the layer is $l = 10$ cm. The amplitude and phase of the normal velocity is measured at several distances from the source with a laser vibrometer and a scanning mirror. A small patch of retro-reflecting tape is glued to the foam to obtain a good signal. A sine burst with 10–20 periods is used as excitation and the detected signal is cross-correlated with the excitation in order to obtain the time of flight. The results of the measurements are shown in Figure 8.4. These measurements give a phase velocity of 68 m/s and a damping of 24 m^{-1}.

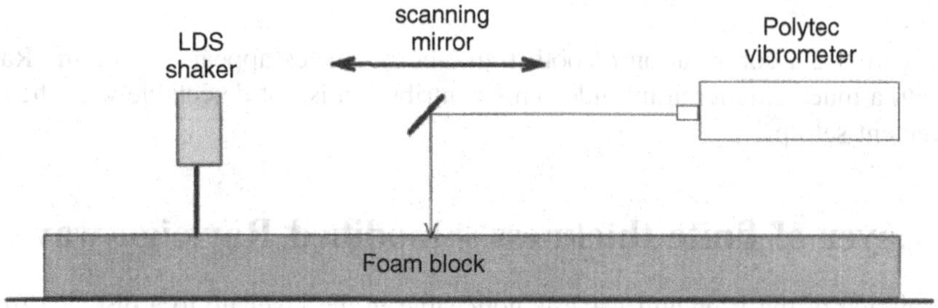

Figure 8.3 Experimental set-up for the observation of surface waves over open-cell foams (Allard *et al.* 2002). Reprinted with permission from Allard, J. F., Jansens, G., Vermeir, G. & Lauriks, W. Frame-borne surface wave in air-saturated porous media. *J. Acoust. Soc. Amer.* **111**, 690–696. Copyright 2002. Acoustical Society of America.

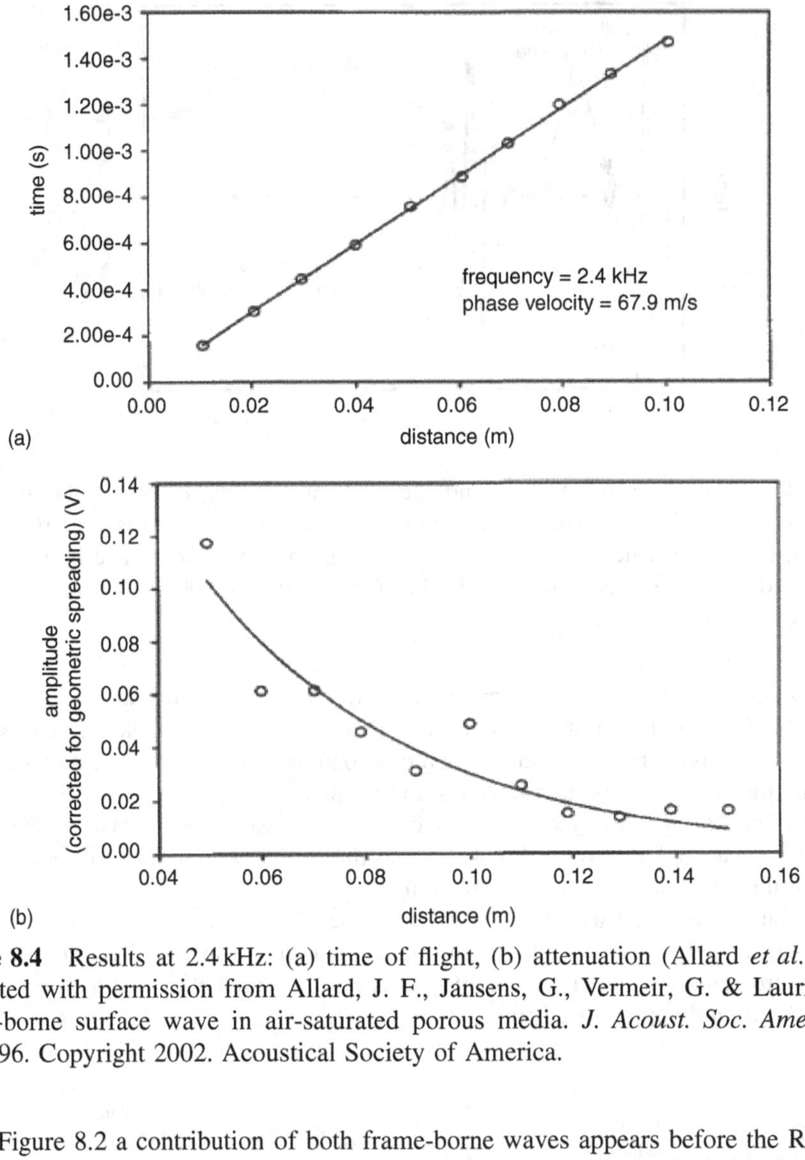

Figure 8.4 Results at 2.4 kHz: (a) time of flight, (b) attenuation (Allard *et al.* 2002). Reprinted with permission from Allard, J. F., Jansens, G., Vermeir, G. & Lauriks, W. Frame-borne surface wave in air-saturated porous media. *J. Acoust. Soc. Amer.* **111**, 690–696. Copyright 2002. Acoustical Society of America.

In Figure 8.2 a contribution of both frame-borne waves appears before the Rayleigh wave with a much smaller amplitude. This contribution is not detectable with the current measurement set-ups.

8.4 Layer of finite thickness – modified Rayleigh wave

The amplitude of the Rayleigh wave is noticeable in the layer up to a distance from the surface equal to twice the Biot shear wavelength. The coefficients r_{ij} for the material given in Table 8.1 are negligible for large thicknesses, but their moduli increase when the thickness decreases, and the location of the Rayleigh poles is modified because the coefficients $M_{i,j}$ depend on the thickness via the coefficients r_{ij}. The location of the

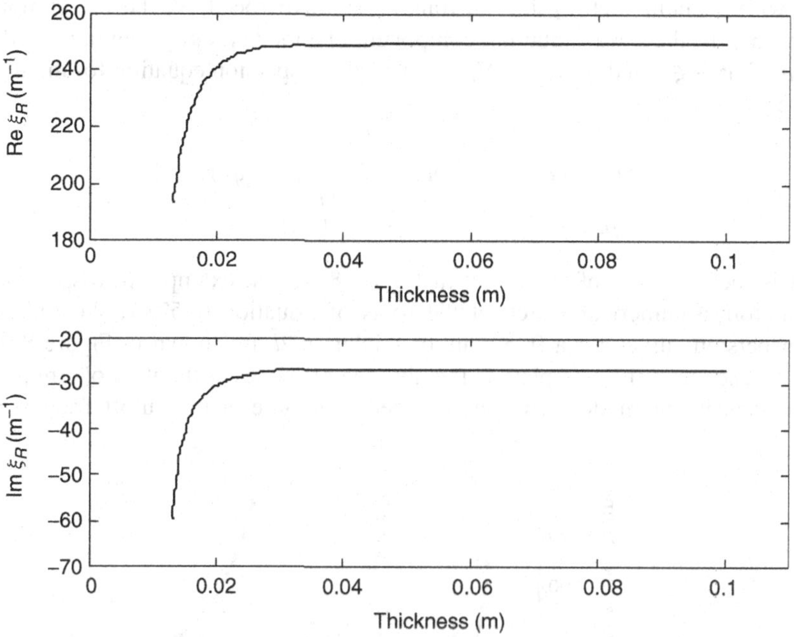

Figure 8.5 Wave number of the modified Rayleigh wave as a function of thickness for material 1, frequency fixed at 2 kHz.

modified pole in the physical Riemann ξ plane is shown in Figure 8.5 as a function of the thickness for a layer of material 1 (see Table 8.1) at 2 kHz. The wavelength of the shear Biot wave is equal to 2.73 cm. Noticeable modifications appear when the thickness has the same order of magnitude as the wavelength.

Sound absorbing porous media are generally available in porous layers of thickness ranging from 1 to 5 cm. This creates a limitation for the measurement of the wave speed of the Rayleigh waves at low frequencies. Another limitation is due to the large structural damping of the porous frames. As a consequence, the Rayleigh wave can only be detected at small distances from the source. It has been shown by Geebelen *et al.* (2008) that measurements performed at distances from the source as small as three Rayleigh wavelengths can provide reasonable orders of magnitude for the phase speeds.

8.5 Layer of finite thickness – modes and resonances

8.5.1 Modes and resonances for an elastic solid layer and a poroelastic layer

In a layer of finite thickness, free on both faces or bonded onto a rigid substrate, there is an infinite number of frame modes. Due to the partial decoupling, the frame modes in air or in vacuum must involve, at the same frequencies, similar deformation fields of the frame.

Modes have been studied at low frequencies by Boeckx *et al.* (2005), for a soft lossless elastic solid with a face bonded on a rigid substrate, the other face being in contact with vacuum. Let V_T be the transverse wave speed, V_L be the compressional wave speed, ξ be the x wave number component of a mode propagating in the x direction, $p^2 = (\omega/V_L)^2 - \xi^2$, and $q^2 = (\omega/V_T)^2 - \xi^2$. The dispersion equation for the modes can be written

$$\begin{aligned} &-4\xi^2(\xi^2 - q^2) - \sin pl \sin ql[\frac{\xi^2}{pq}(\xi^2 - q^2)^2 + 4\xi^2 pq \\ &+ \cos pl \cos ql[4\xi^4 + (\xi^2 - q^2)^2] = 0 \end{aligned} \tag{8.52}$$

where l is the thickness of the layer. In Figure 8.6(a) an example of dispersion curves obtained from a numerical search of the roots of Equation (8.52) is shown. The usual Lamb dispersion curves for a free plate of thickness $2l$ are shown in Figure 8.6(b). The main difference with the free plate is the absence of modes without cutoff frequency and a smaller density of modes. The phase speed is infinite at the cutoff frequencies, and

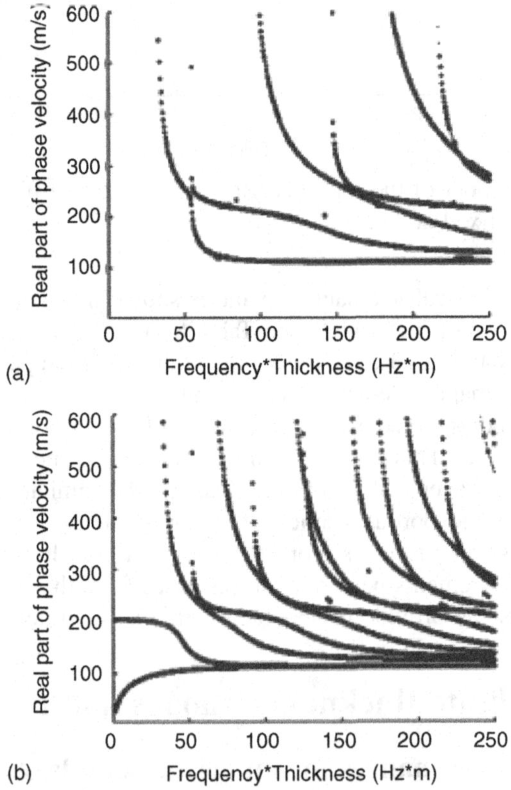

Figure 8.6 Phase speed as a function of frequency × thickness for: (a) a layer of thickness l of an elastic medium on a rigid substrate, (b) a free layer of thickness $2l$. The material density is 14 kg/m^3, $V_L = 222.$ m/s, $V_R = 122.$ m/s. (Boeckx *et al.* 2005). Reprinted with permission from Boeckx, L. , Leclaire, P. , Khurana, P., Glorieux, C., Lauriks, W. & Allard, J. F. Investigation of the phase velocity of guided acoustic waves in soft porous layers. *J. Acoust. Soc. Amer.* **117**, 545–554. Copyright 2005, Acoustical Society of America.

from Equation (8.52) the cutoff frequencies are given by

$$f_c = (2m+1)\frac{V_T}{4l} \text{or} f_c = (2m+1)\frac{V_L}{4l}, m = 0, 1, 2\ldots \tag{8.53}$$

At the cutoff frequencies, the phase speed being infinite, the displacements are identical on the whole free surface. These displacement correspond to the $(2m+1)\lambda_T/4$ and $(2m+1)\lambda_L/4$ resonances of the layer. For a medium with a small Poisson ratio the first resonance is the $\lambda_T/4$ resonance followed by the $\lambda_L/4$ resonance. If the Poisson coefficient is sufficiently close to 0.5, the second resonance is the $3\lambda_T/4$ resonance.

An experimental set-up for the case of bonded layers is described in Boeckx *et al.* (2005) with phase speed measurements of the two first modes.

The main difference with the modes of a porous frame saturated by air is due to the large loss angle. The phase speed does not tend to infinity when frequency tends to the cutoff, but presents a maximum.

8.5.2 Excitation of the resonances by a point source in air

A heavy porous frame is not strongly coupled to the acoustic field in the surrounding air. Moreover, the loss angle for usual sound absorbing porous layers is large; a current order of magnitude is 1/10. However, the frame velocity created at the free surface of a porous layer by a point source in air can be measured with a laser velocimeter, and large peaks can appear in the velocity distributions. In contrast to the usual acoustic impedance measurements, the chosen boundary conditions for the frame are important and if the layer is bonded on a rigid backing, a careful gluing of the frame is mandatory. The experimental set-up for the excitation and the measurement of the frame displacement is shown in Figure 8.7. The velocity component $\dot{\bar{U}}_z^s$ created by the monopole field over the porous frame is given by

$$\dot{\bar{U}}_z^s(r) = -j\int_0^\infty \frac{J_0(r\xi)}{\mu}[1+V(\xi/k_0)]\dot{U}_z^s(\xi)\exp[j\mu(z_1+z_2)]\xi\, d\xi \tag{8.54}$$

where V is the reflection coefficient for $\cos\theta = \mu/k_0$ and $\dot{U}_z^s$ is given by Equation (8.38). Using

$$J_1(z) = \frac{j}{2\pi}\int_0^{2\pi} \exp(-jz\cos\theta)\cos\theta\, d\theta$$

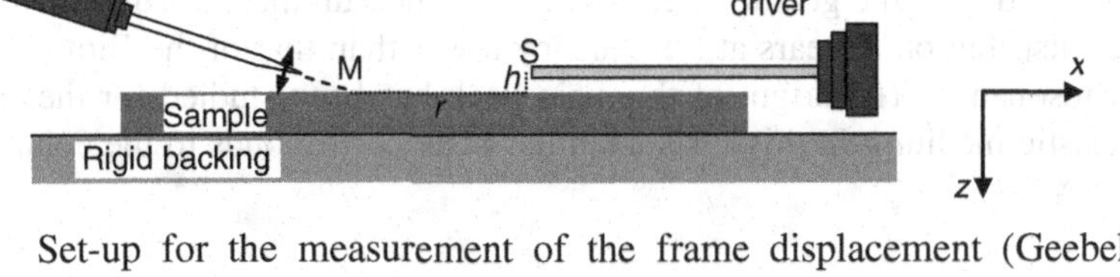

Figure 8.7 Set-up for the measurement of the frame displacement (Geebelen *et al.* 2007). Reproduced by Permission of MedPharm Scientific Publishers - S. Hirzel Verlag.

(see Equation 8.B.12 in Appendix 8.B) the following expression for the radial component is obtained

$$\dot{\bar{U}}_r^s(r) = -\int_0^\infty \frac{J_1(r\xi)}{\mu}[1 + V(\xi/k_0)]\dot{U}_x^s(\xi)\exp[j\mu(z_1 + z_2)]\xi \, d\xi \tag{8.55}$$

In Equations (8.54) and (8.55), $(1 + V)\dot{U}_z^s$ and $(1 + V)\dot{U}_x^s$ can be written, respectively

$$[1 + V(\xi/k_0)]\dot{U}_z^s(\xi) = 2\cos\theta \frac{T_{11}[(1 - \phi)T_{32} - \phi T_{42}] + T_{12}[\phi T_{41} - (1 - \phi)T_{31}]}{D_1 Z + (T_{31}T_{42} - T_{32}T_{41})\cos\theta} \tag{8.56}$$

$$[1 + V(\xi/k_0)]\dot{U}_x^s(\xi) = 2\cos\theta \frac{T_{21}[(1 - \phi)T_{32} - \phi T_{42}] + T_{22}[\phi T_{41} - (1 - \phi)T_{31}]}{D_1 Z + (T_{31}T_{42} - T_{32}T_{41})\cos\theta} \tag{8.57}$$

The singularities of $[1 + V(\xi/k_0)]\dot{U}_z^s(\xi)$ and $[1 + V(\xi/k_0)]\dot{U}_x^s(\xi)$ are the same as those of V given by Equation (8.30).

Measurements and predictions are reported in Allard *et al.* (2007) and Geebelen *et al.* (2007). Measurements of the normal velocity are performed with the laser beam normal to the surface using a height of the source around 5 cm, and a radial distance from the source to the point where the velocity is measured of 1 or 2 cm. Measurements of the radial velocity are performed with an angle of incidence of the laser beam larger than 60°, and a radial distance larger than 5 cm. Due to the Bessel function J_1 in Equation (8.45), the radial displacement is negligible at small r. A microphone is mounted in the tube of the source, at the side opposite to the compression driver, and in a first step the measured velocity is normalized with the pressure signal provided by the microphone. In a second step, this measured velocity is normalized with respect to the predicted velocity, to make the comparison easier.

In Figures 8.8 and Figures 8.9, previsions and measurements are presented. Predictions are performed with the acoustic parameters of material 2 in Table (8.1). The thickness of the layer is $l = 2$ cm. The acoustic parameters that characterize the rigidity are the Poisson ratio and the shear modulus. They have been adjusted to fit the measured distribution. The mode manifestations are the broad peaks in the velocity distributions. For the radial velocity in Figure 8.9, the maximum is close to the Biot shear wave $\lambda/4$ resonance. It may be noticed that the quarter shear wave resonance is generally not observable, contrary to the Biot frame-borne compressional wave $\lambda/4$ resonance. For the normal velocity, distributions for materials with a Poisson coefficient equal to 0.35 or less are peaked at this resonance (Allard *et al.* 2007, Geebelen *et al.* 2007). For material 2, the Poisson ratio is large and with the geometry chosen for the measurement set-up, the main peak in the normal distribution appears at a frequency lower than that of the Biot compressional wave $\lambda/4$ resonance. The origin of the main peak has been studied for the simpler case of a soft elastic medium. In this case, a similar peak corresponds to the contribution of a

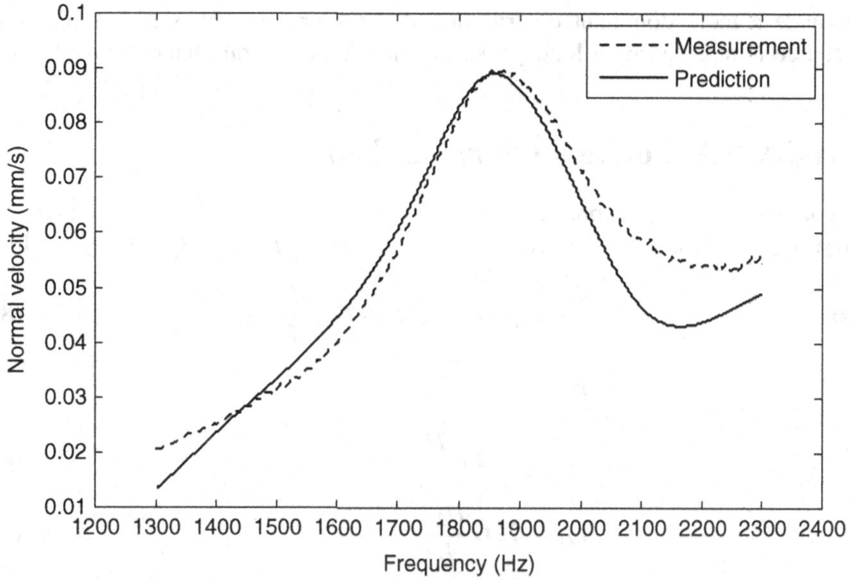

Figure 8.8 Normal velocity, $r = h = 3.5$ cm, prediction vs measurement (taken from Allard *et al.* 2007).

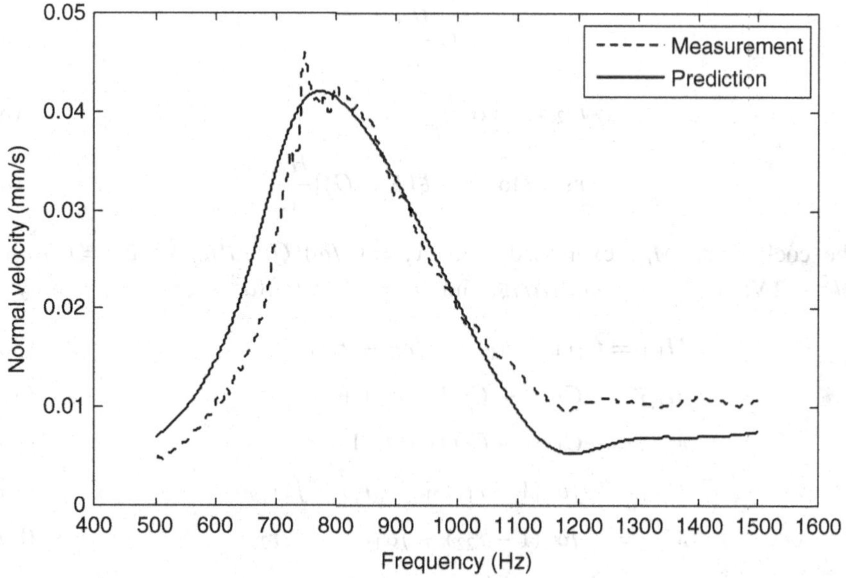

Figure 8.9 Radial velocity, $r = 16.5$ cm, $h = 3.5$ cm, prediction vs measurement (taken from Allard *et al.* 2007).

mode related to the shear wave $3\lambda/4$ resonance. The compressional wave $\lambda/4$ resonance will provide a more dominant contribution if the source height is increased, or if the point source is replaced by a loudspeaker set at one or several meters above the porous layer.

Appendix 8.A Coefficients r_{ij} and $M_{i,j}$

The coefficients r_{ij} expressed with $E = \alpha_1\alpha_2(\mu_1 - \mu_2)$, $F = \alpha_1\xi(\mu_1 - \mu_3)$, $G = \alpha_2\xi(\mu_2 - \mu_3)$, $L = \alpha_3 E + \xi(F - G)$, $H_{ij} = \exp(-jl(\alpha_i + \alpha_j))$ are given by:

$$r_{11} = [\alpha_3 E + \xi(F + G)]\frac{H_{11}}{L} \tag{8.A.1}$$

$$r_{12} = -2\xi F\frac{H_{12}}{L} \tag{8.A.2}$$

$$r_{13} = -2\xi E\frac{H_{13}}{L} \tag{8.A.3}$$

$$r_{21} = 2\xi G\frac{H_{21}}{L} \tag{8.A.4}$$

$$r_{22} = [\alpha_3 E - \xi(F + G)]\frac{H_{22}}{L} \tag{8.A.5}$$

$$r_{23} = -2\xi E\frac{H_{23}}{L} \tag{8.A.6}$$

$$r_{31} = -2\alpha_3 G\frac{H_{31}}{L} \tag{8.A.7}$$

$$r_{32} = 2\alpha_3 F\frac{H_{32}}{L} \tag{8.A.8}$$

$$r_{33} = [\alpha_3 E - \xi(F - G)]\frac{H_{33}}{L} \tag{8.A.9}$$

The coefficients $M_{i,j}$ expressed with $A_j = (j/\omega)(Q + R\mu_j)k_j^2$, $B_j = (j/\omega)[(P + Q\mu_j)k_j^2 - 2N\xi^2]$, $C_j = (j/\omega)2N\alpha_j\xi$, and $H = -(jN/\omega)(\alpha_3^2 - \xi^2)$ are given by:

$$M_{1,1} = C_1(1 - r_{11}) - C_2 r_{12} + H r_{13} \tag{8.A.10}$$

$$M_{1,2} = -C_1 r_{21} + C_2(1 - r_{22}) + H r_{23} \tag{8.A.11}$$

$$M_{1,3} = -C_1 r_{31} - C_2 r_{32} + H(1 + r_{33}) \tag{8.A.12}$$

$$M_{2,1} = -j\alpha_1(1 - r_{11}) + j\alpha_2 r_{12} - j\xi r_{13} \tag{8.A.13}$$

$$M_{2,2} = -j\alpha_2(1 - r_{22}) + j\alpha_1 r_{21} - j\xi r_{23} \tag{8.A.14}$$

$$M_{2,3} = j\alpha_1 r_{31} + j\alpha_2 r_{32} - j\xi(1 + r_{33}) \tag{8.A.15}$$

$$M_{3,1} = -j\mu_1\alpha_1(1 - r_{11}) + j\mu_2\alpha_2 r_{12} - j\mu_3\xi r_{13} \tag{8.A.16}$$

$$M_{3,2} = -j\mu_2\alpha_2(1 - r_{22}) + j\mu_1\alpha_1 r_{21} - j\mu_3\xi r_{23} \tag{8.A.17}$$

$$M_{3,3} = j\mu_1\alpha_1 r_{31} + j\mu_2\alpha_2 r_{32} - j\mu_3\xi(1 + r_{33}) \tag{8.A.18}$$

$$M_{4,1} = A_1(1 + r_{11}) + A_2 r_{12} \tag{8.A.19}$$

$$M_{4,2} = A_2(1 + r_{22}) + A_1 r_{21} \tag{8.A.20}$$

$$M_{4,3} = A_1 r_{31} + A_2 r_{32} \tag{8.A.21}$$

$$M_{5,1} = B_1(1 + r_{11}) + B_2 r_{12} - C_3 r_{13} \tag{8.A.22}$$

$$M_{5,2} = B_2(1 + r_{22}) + B_1 r_{21} - C_3 r_{23} \tag{8.A.23}$$

$$M_{5,3} = C_3(1 - r_{33}) + B_1 r_{31} + B_2 r_{32} \tag{8.A.24}$$

$$M_{6,1} = -j\xi(1 + r_{11} + r_{12}) - j\alpha_3 r_{13} \tag{8.A.25}$$

$$M_{6,2} = -j\xi(1 + r_{21} + r_{22}) - j\alpha_3 r_{23} \tag{8.A.26}$$

$$M_{6,3} = j\alpha_3(1 - r_{33}) - j\xi(r_{31} + r_{32}) \tag{8.A.27}$$

Appendix 8.B Double Fourier transform and Hankel transform

Double Fourier transform

$$\mathrm{f}(x, y) = \frac{1}{4\pi^2} \int_{-\infty}^{\infty} \int_{-\infty}^{\infty} \mathrm{F}(u, v) \exp[-j(ux + vy)]\, \mathrm{d}u\, \mathrm{d}v \tag{8.B.1}$$

$$\mathrm{F}(u, v) = \int_{-\infty}^{\infty} \int_{-\infty}^{\infty} \mathrm{f}(x, y) \exp[j(ux + vy)]\, \mathrm{d}x\, \mathrm{d}y \tag{8.B.2}$$

If f depends only on $r = \sqrt{x^2 + y^2}$, F depends only on $w = \sqrt{u^2 + v^2}$.

Hankel transform

$$\mathrm{f}(r) = \int_{0}^{\infty} w\overline{\mathrm{F}}(w)\mathrm{J}_0(rw)\, \mathrm{d}w \tag{8.B.3}$$

$$\overline{\mathrm{F}}(w) = \int_{0}^{\infty} r\mathrm{f}(r)\mathrm{J}_0(rw)\, \mathrm{d}r \tag{8.B.4}$$

Relationship between the two transforms

Equations (8A.1) and (8A.2) can be rewritten with the polar coordinates r, θ instead of x, y, and w, ψ instead of u, w

$$\mathrm{f}(r) = \frac{1}{4\pi^2} \int_{0}^{\infty} \mathrm{F}(w) w\, \mathrm{d}w \int_{0}^{2\pi} \exp[-j(rw\cos(\psi - \varphi))]\, \mathrm{d}\psi \tag{8.B.5}$$

$$F(w) = \int_0^{\infty} f(r) r dr \int_0^{2\pi} \exp[j(rw\cos(\psi - \varphi))]\, d\varphi \tag{8.B.6}$$

Using (Abramovitz and Stegun 1972)

$$\int_0^{2\pi} \exp[ju\cos(\psi - \varphi)]\, d\psi = 2\pi J_0(u) \tag{8.B.7}$$

Equations (8B.5) and (8B.6) can be rewritten

$$f(r) = \frac{1}{2\pi} \int_0^{\infty} F(w) w J_0(wr)\, dw \tag{8.B.8}$$

$$F(w) = 2\pi \int_0^{\infty} f(r) r J_0(wr)\, dr \tag{8.B.9}$$

The functions F and $\overline{F}$ are related by $F = 2\pi \overline{F}$.

Response of a linear system

If the response to an excitation $\exp[-j(ux + vy)]$ is a Z vector component like u_z^s or a scalar which only depends on r, the total response U_z of the system for the excitation $f(r)$ is given by

$$U_z(r) = \frac{1}{4\pi^2} \int_0^{\infty} F(w) u_z(w) w\, dw \int_0^{2\pi} \exp[j(rw\cos(\psi - \varphi))]\, d\psi \tag{8.B.10}$$

which can be rewritten

$$U_z(r) = \frac{1}{2\pi} \int_0^{\infty} F(w) u_z(w) J_0(wr) w\, dw \tag{8.B.11}$$

where $F/2\pi$ can be replaced by $\overline{F}$.

If the response is a radial component which only depends on r and not on θ, the total response in the direction x is obtained by projecting in the direction x the responses for the different orientations of the excitation in the XY plane. Using

$$J_1(u) = \frac{j}{2\pi} \int_0^{2\pi} \exp(-ju\cos\theta)\cos\theta\, d\theta$$

(Abramovitz and Stegun 1972) leads to

$$U_x(r) = \frac{-j}{2\pi} \int_0^{\infty} F(w) u_r(w) J_1(wr) w\, dw \tag{8.B.12}$$

Appendix 8.C Rayleigh pole contribution

Equation (8.51) can be rewritten (see Equations 7.6, 7.7)

$$\dot{\overline{U}}_z^s(r) = -\frac{F}{4\pi}\int_{-\infty}^{\infty} H_0^1(-\xi r)\dot{U}_z^s(\xi)\xi\, d\xi \tag{8.C.1}$$

where $\dot{U}_z^s$ is given by Equation (8.38). There is a simple zero of D_2 given by Equation (8.41) at $\xi = \xi_R$. Let a be the parameter defined by

$$a = \frac{\partial D_2}{\partial \xi}/_{\xi=\xi_R} \tag{8.C.2}$$

The Rayleigh wave contribution is given in the frequency domain by

$$\dot{\overline{U}}_{Rz}^s(r) = -\frac{F}{4\pi}(-2\pi j)H_0^1(-\xi_R r)\frac{T_{11}(T_{32}+\phi s_3) - T_{12}(T_{31}+\phi s_2)/_{\xi=\xi_R}}{a} \tag{8.C.3}$$

The variables in the derivation of D_2 are $\alpha_1, \alpha_2, \alpha_3$and $\cos\theta$. The derivatives of these parameters are $\partial\alpha_i/\partial\xi = -\xi/\alpha_i$ and $\partial\cos\theta/\partial\xi = -\xi/k_0^2\cos\theta$.

References

Abramowitz, M. and Stegun, I.A. (1972) *Handbook of Mathematical Functions*. National Bureau of Standards, Washington D. C.

Allard, J.F., Jansens, G., Vermeir, G. and Lauriks, W. (2002) Frame-borne surface wave in air-saturated porous media. *J. Acoust. Soc. Amer*. **111**, 690–696.

Allard, J.F., Henry, M., Glorieux, C., Lauriks, W. and Petillon, S. (2004) Laser induced surface modes at water-elastic and poroelastic solid interfaces, *J. Appl. Phys*. **95**, 528–535.

Allard, J.F., Brouard, B., Atalla, N. and Ginet, S. (2007) Excitation of soft porous frame resonances and evaluation of rigidity coefficients. *J. Acoust. Soc. Amer*. **121**, 78–84.

Boeckx, L., Leclaire, P., Khurana, P., Glorieux, C., Lauriks, W. and Allard, J.F. (2005) Investigation of the phase velocity of guided acoustic waves in soft porous layers. *J. Acoust. Soc. Amer*. **117**, 545–554.

Feng, S. and Johnson, D.L. (1983) High-frequency acoustic properties of a fluid/porous solid interface. I. New surface mode. II. The 2D Green's function. *J. Acoust. Soc. Amer*. **74**, 906–924.

Geebelen., N., Boeckx, L., Vermeir, G., Lauriks, W., Allard, J.F. and Dazel, O. (2007) Measurement of the rigidity coefficients of a melamine foam. *Acta Acustica* **93**, 783–788.

Geebelen N., Boeckx, L., Vermeir, G., Lauriks, W., Allard, J.F. and Dazel, O. (2008), Near field Rayleigh wave on soft porous layers. *J. Acoust. Soc. Amer*. **123**, 1241–1247.

Langlois, C., Panneton, R. and Atalla, N. (2001) Polynomial relations for quasi-static mechanical characterization of poroelastic materials. *J. Acoust. Soc. Amer*., **109**(6), 3032–3040.

Melon, M., Mariez, M., Ayrault, C. and Sahraoui, S. (1998) Acoustical and mechanical characterization of anisotropic open-cell foams. *J. Acoust. Soc. Amer*. **104**, 2622–2627.

Park, J. (2005) Measurement of the frame acoustic properties of porous and granular materials. *J. Acoust. Soc. Amer*. **118**, 3483–3490.

Pritz, T. (1986) Frequency dependence of frame dynamic characteristics of mineral and glass wool materials. *J. Sound Vib*. **106**, 161–169.

Pritz, T. (1994) Dynamic Young's modulus and loss factor of plastic foams for impact sound isolation. *J. Sound Vib.* **178**, 315–322.

Sfaoui, A. (1995a) On the viscosity of the polyurethane foam. *J. Acoust. Soc. Amer.* **97**, 1046–1052.

Sfaoui, A. (1995b), Erratum: On the viscosity of the polyurethane foam. *J. Acoust. Soc. Amer.* **98**, 665.

Tarnow, V. (2005) Dynamic measurements of the elastic constants of glass wool. *J. Acoust. Soc. Amer.* **118**, 3672–3678.

Victorov, I.A. (1967) *Rayleigh and Lamb Waves*. Plenum, New York.

9

Porous materials with perforated facings

9.1 Introduction

Sound absorbing porous materials with perforated facings have been used for many years, because they can present a high absorption coefficient at low frequencies. Models of predicting the surface impedance and the absorption coefficient have been carried out and checked, at normal and oblique incidence, for different configurations (Bolt 1947, Zwikker and Kosten 1949, Ingard and Bolt 1951, Brillouin 1949, Callaway and Ramer 1952, Ingard 1954, Velizhanina 1968, Davern 1977, Byrne 1980, Guignouard *et al.* 1991). The effect of a perforated facing was described in an elegant and physical way by Ingard (1954) at normal incidence. In this chapter, the results in Ingard (1954) are generalized at oblique incidence, and for anisotropic stratified porous media. As in all the previous modelling, the facing and the frames of the porous layers are supposed to be motionless. Nonlinear effects are not studied in this chapter. Information about this subject can be found in Ingard and Labate (1950), Ingard and Ising (1967) and Ingard (1968, 1970).

9.2 Inertial effect and flow resistance

9.2.1 Inertial effect

Let R be the radius of the perforations, and d the thickness of the facing represented in Figure 9.1.

Let s be the fraction of perforated open area, also called the open area ratio, equal to $n\pi R^2$ if n is the number of perforations per unit area. The radius R of the perforations and the thickness d of the facing are supposed to be much smaller than the wavelength λ in the air:

$$d \ll \lambda, \quad R \ll \lambda$$

Propagation of Sound in Porous Media: Modelling Sound Absorbing Materials, Second Edition J. F. Allard and N. Atalla

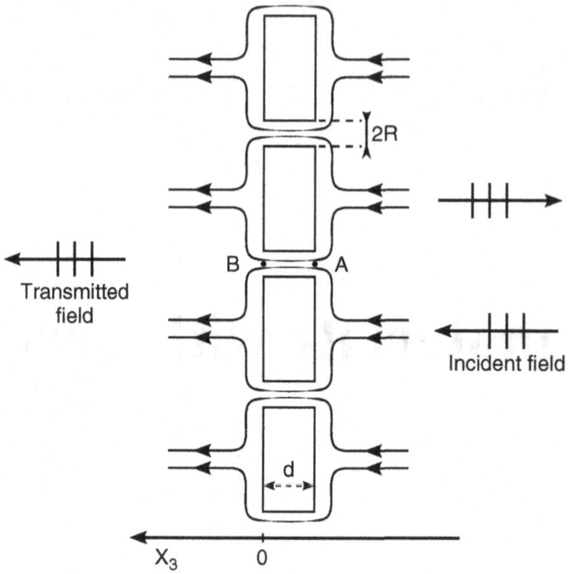

Figure 9.1 A perforated facing in a normal acoustic field.

The facing distorts the flow of air in a region having a thickness which is generally smaller than the diameter of a perforation. This modification of the flow creates an increase of kinetic energy. This effect is similar to the effect of tortuosity in porous media. If a plane wave propagates at the left-hand side of the facing, parallel to the x_3 axis (see Figure 9.1), the impedance p/v_3 in the free air is equal to the characteristic impedance Z_c. At B, at the boundary of the facing in front of a perforation, the impedance would be equal to sZ_c if the inertial effects were not present. An evaluation of Z_B, carried out in Section 9.2.2, indicates that Z_B is given by

$$Z_B = Z_c s + j\omega\varepsilon_e\rho_0 \tag{9.1}$$

where $\varepsilon_e\rho_0$ is an added mass per unit area of perforation, ρ_0 is the density of air, and ε_e is a length that must be added to the cylindrical perforation at the left of B in order to create the same effect as the distortion of the flow.

9.2.2 Calculation of the added mass and the added length

For the case of normal incidence, if the arrangement of the perforations is the square grid represented in Figure 9.2 (the perforations are periodically distributed in the two directions with a spatial period D equal to the distance between two holes), it is possible to divide the space around the facing in separate cylinders having a square shaped cross-section without modifying the acoustic field.

At the boundary between two cylinders, the velocity components perpendicular to the boundary are equal to zero, due to the symmetry of the grid, and the different cylinders can be separated by rigid sheets (it may be noticed that these sheets do not exist, and the viscous and the thermal interaction with the sheet cannot be taken into account). An elementary cell for the partition is represented in Figure 9.3. In the plane $x_3 = 0$,

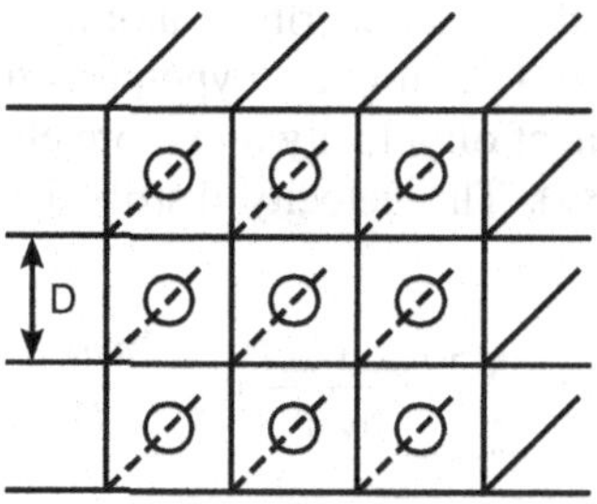

Figure 9.2 A square grid of holes at normal incidence and the related partition.

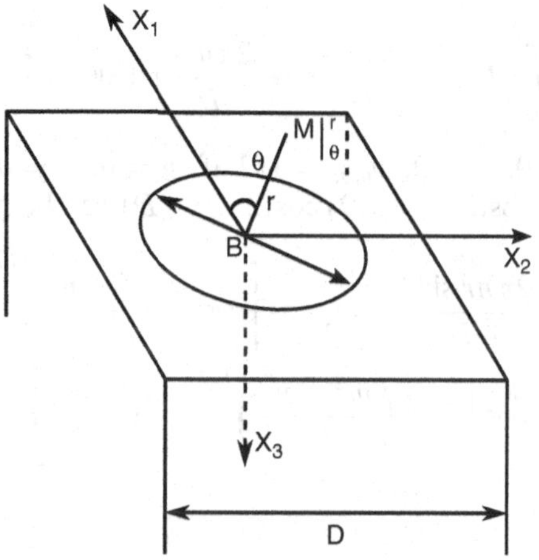

Figure 9.3 The elementary cell of the partition. The circumference at $x_3 = 0$ is the boundary of the aperture in contact with the free air at the left hand side of Figure 9.1.

the polar coordinates r, θ and the cartesian coordinates x_1,x_2 can be used. On the faces located at $x_1 = \pm D/2$, $x_2 = \pm D/2$, the velocity components in the directions x_1 and x_2, respectively, are equal to zero. A simple pressure field in the cell is

$$p_{m,n}(x_1,x_2,x_3) = A_{m,n} \exp(-jk_{m,n}x_3) \cos \frac{2\pi m}{D} x_1 \cos \frac{2\pi n}{D} x_2 \tag{9.2}$$

In this equation, $k_{m,n}$ is given by

$$k_{m,n} = \left(k^2 - \frac{4\pi^2 m^2}{D^2} - \frac{4\pi^2 n^2}{D^2} \right)^{1/2} \tag{9.3}$$

where k is the wave number in the free air. The set of these functions for m and n taking the values 0, 1, 2, ..., etc., is an orthogonal basis for the pressure field in the cylinder.

A pressure field in the cylinder can be written

$$p(x_1,x_2,x_3) = \sum_{m,n} C_{m,n} \exp(-jk_{m,n}x_3) \cos \left(\frac{2\pi m}{D} x_1 \right) \cos \left(\frac{2\pi n}{D} x_2 \right) \tag{9.4}$$

As indicated by Ingard (1953), the exact distribution of the velocity in the aperture located at $x_3 = 0$ is not known *a priori*. The simplest hypothesis involves considering a uniform amplitude U for the x_3 component of velocity and more elaborate models provide similar results (Norris and Sheng 1989). The velocity distribution on the plane $x_3 = 0$ can be written

$$\mathrm{Y}(R-r)U = \sum_{m,n} \frac{k_{m,n}C_{m,n}}{\omega\rho_0} \cos\frac{2\pi m x_1}{D} \cos\frac{2\pi n x_2}{D} \tag{9.5}$$

In this equation, $\mathrm{Y}(R-r)$ is the unit step equal to 1 if $r < R$ and 0 if $r > R$.

By multiplying both sides of this equation by $\cos\frac{2\pi m x_1}{D} \cos\frac{2\pi n x_2}{D}$ and integrating over the aperture, one obtains

$$\nu_{m,n} \int_0^{2\pi} \int_0^R U \cos\frac{2\pi m x_1}{D} \cos\frac{2\pi n x_2}{D} r\mathrm{d}r\mathrm{d}\theta = \frac{k_{m,n}C_{m,n}}{\omega\rho_0}\frac{D^2}{4} \tag{9.6}$$

where $\nu_{m,n} = 1$ if $m \neq 0$, $n \neq 0$; $\nu_{m,n} = 1/2$ if $m = 0$, $n \neq 0$ or $m \neq 0$, $n = 0$; and $\nu_{0,0} = 1/4$. The product $\cos(2\pi m x_1/D)\cos(2\pi n x_2/D)$ can be rewritten (Ingard, 1953)

$$\begin{aligned}\cos\frac{2\pi m r\cos\theta}{D} \cos\frac{2\pi n r\sin\theta}{D} &= \frac{1}{2}\cos\left[2\pi r\left(\frac{m^2+n^2}{D^2}\right)^{1/2}\sin(\theta+\gamma)\right] \\ &\quad + \frac{1}{2}\cos\left[2\pi r\left(\frac{m^2+n^2}{D^2}\right)^{1/2}\sin(\theta-\gamma)\right]\end{aligned} \tag{9.7}$$

where $\gamma = \mathrm{arctg}\; m/n$.

Using the relations:

$$\int_0^{2\pi} \cos\left[2\pi\frac{r}{D}(m^2+n^2)^{1/2}\sin(\theta\pm\gamma)\right]d\theta = 2\pi \mathrm{J}_0\left[2\pi\frac{r}{D}(m^2+n^2)^{1/2}\right] \tag{9.8}$$

$$\int_0^R \mathrm{J}_0\left[2\pi\frac{r}{D}(m^2+n^2)^{1/2}\right] r dr = \mathrm{J}_1\left[2\pi\frac{R}{D}(m^2+n^2)^{1/2}\right]\frac{RD}{2\pi(m^2+n^2)^{1/2}} \tag{9.9}$$

Equation (9.6) can be rewritten

$$\nu_{m,n}\frac{URD}{(m^2+n^2)^{1/2}}\mathrm{J}_1\left[2\pi\frac{R}{D}(m^2+n^2)^{1/2}\right] = \frac{k_{m,n}C_{m,n}}{\omega\rho_0}\frac{D^2}{4} \tag{9.10}$$

and $C_{m,n}$ is given by

$$C_{m,n} = \frac{4\nu_{m,n} U R \omega\rho_0 \mathrm{J}_1\left[2\pi\frac{R}{D}(m^2+n^2)^{1/2}\right]}{D(m^2+n^2)^{1/2}k_{m,n}} \tag{9.11}$$

For the case $m = n = 0$, Equation (9.6) yields

$$C_{0,0} = \omega\rho_0 U s/k = Z_c s U \tag{9.12}$$

where $s = \pi R^2/D^2$.

The average pressure over the aperture is

$$\bar{p} = \frac{1}{\pi R^2}\int_0^R\int_0^{2\pi}\sum_{m,n} C_{m,n}\cos\frac{2\pi m x_1}{D}\cos\frac{2\pi n x_2}{D} r\,dr\,d\theta \tag{9.13}$$

and the impedance $\bar{p}/U$ is given by

$$\frac{\bar{p}}{U} = \frac{s\omega\rho_0}{k} + 4\sum_{m,n}{}' v_{m,n}\frac{\mathrm{J}_1^2\left[2\pi\frac{R}{D}(m^2+n^2)^{1/2}\right]}{\pi(m^2+n^2)k_{m,n}}\omega\rho_0 \tag{9.14}$$

The prime over the symbol Σ means that the term $(m = 0, n = 0)$ is excluded from the summation.

For frequencies well below c_0/D, $k_{m,n}$ given by Equation (9.3) is nearly equal to

$$k_{m,n} = -\mathrm{j}\frac{2\pi}{D}(m^2+n^2)^{1/2} \tag{9.15}$$

and Equation (9.14) becomes

$$\frac{\bar{p}}{U} = Z_c s + j\omega\rho_0\sum_{m,n}{}' v_{m,n}\frac{2D\mathrm{J}_1^2[2\pi(R/D)(m^2+n^2)^{1/2}]}{\pi^2(m^2+n^2)^{3/2}} \tag{9.16}$$

A comparison of Equations (9.1) and (9.16) yields the following expression for the added length:

$$\varepsilon_e = \sum_{m,n}{}' v_{m,n}\frac{2D\mathrm{J}_1^2[2\pi(R/D)(m^2+n^2)^{1/2}]}{\pi^2(m^2+n^2)^{3/2}} \tag{9.17}$$

The added length ε_e can be approximated for $s^{1/2} < 0.4$ by

$$\varepsilon_e = 0.48S^{1/2}(1 - 1.14s^{1/2}) \tag{9.18}$$

where S is the area of the aperture. The limit $0.48S^{1/2}$ for $s = 0$ can be obtained by calculating the radiation impedance of a single circular aperture in a plane.

9.2.3 Flow resistance

Due to the viscous dissipation in the aperture and the surface of the plate, which was neglected in Section 9.2.2, the impedance Z_B presents also a resistive component R_B that must be added to $Z_c s$ in Equation (9.1). A calculation by Nielsen (1949) gives the following expression for R_B:

$$R_B = R_s \tag{9.19}$$

where R_s is a surface resistance defined in Lord Rayleigh (1940) Vol II, p. 318:

$$R_s = \frac{1}{2}(2\eta\rho_0\omega)^{1/2} \tag{9.20}$$

It was pointed out by Ingard (1953) that R_B given by Equation (9.19) is too small, and that a better prediction of the measured values is obtained with $R_B = 2R_s$. At A, the two added masses at the two sides of the facing, the two added viscous corrections, and the mass and the flow resistivity in the aperture must be taken into account, and Z_A is given by

$$Z_A = \left(\frac{2d}{R} + 4\right) R_s + (2\varepsilon_e + d) j\omega\rho_0 + Z_c s \tag{9.21}$$

In this equation, $(2d/R)R_s$ is the flow resistance of the circular hole of length d. This term can be obtained using Equation (5.26) with $\alpha_\infty = 1$ and $\Lambda = R$. The effective density can be written

$$\rho_0\bar{\alpha}(\omega) = \rho_0 + \rho_0\frac{\delta}{\Lambda} - j\rho_0\frac{\delta}{\Lambda} \tag{9.22}$$

The term $\rho_0\delta/\Lambda$ can be neglected, because it is very small compared with ρ_0 at acoustical frequencies for usual perforations. The density ρ_0 corresponds to the term $j\omega d\rho_0$ in Equation (9.21). In the same way, $-j\rho_0\delta/\Lambda$ will correspond to $\rho_0\delta\omega d/R$ in Equation (9.21). By substituting $(2\eta/\rho_0\omega)^{1/2}$ for δ in this term, one obtains $2dR_s/R$. It may be noticed that when the material is in contact with a porous layer, the viscous term $(2d/R + 4)R_s$ is generally negligible, compared with $Z_c s$.

9.2.4 Apertures having a square cross-section

An elementary cell having a square aperture is represented in Figure 9.4. As in the case of a circular aperture, the amplitude U of the velocity in the hole is considered uniform. Equation (9.6) can be rewritten

$$v_{m,n}\int_{-a/2}^{a/2}\int_{-a/2}^{a/2} U\cos\frac{2\pi m x_1}{D}\cos\frac{2\pi n x_2}{D}\mathrm{d}x_1\mathrm{d}x_2 = \frac{k_{m,n}C_{m,n}}{\omega\rho_0}\frac{D^2}{4} \tag{9.23}$$

and $C_{m,n}$ is given by

$$C_{m,n} = 4v_{m,n}U\frac{\omega\rho_0}{k_{m,n}}\frac{1}{\pi^2 mn}\sin\left(\frac{\pi m a}{D}\right)\sin\left(\frac{\pi n a}{D}\right) \quad m \neq 0, n \neq 0 \tag{9.24}$$

$$C_{m,n} = 4\frac{v_{m,n}U\omega\rho_0}{Dk_{m,n}}\frac{a}{\pi n}\sin\left(\frac{\pi n a}{D}\right) \quad m = 0, n \neq 0 \tag{9.25}$$

$$C_{m,n} = \frac{U\omega\rho_0 a^2}{D^2 k} \quad m = n = 0 \tag{9.26}$$

The average pressure over the aperture is

$$\bar{p} = \frac{1}{a^2}\int_{-a/2}^{a/2}\int_{-a/2}^{a/2}\sum_{m,n} C_{m,n}\cos\frac{2\pi m x_1}{D}\cos\frac{2\pi n x_2}{D}dx_1 dx_2 \tag{9.27}$$

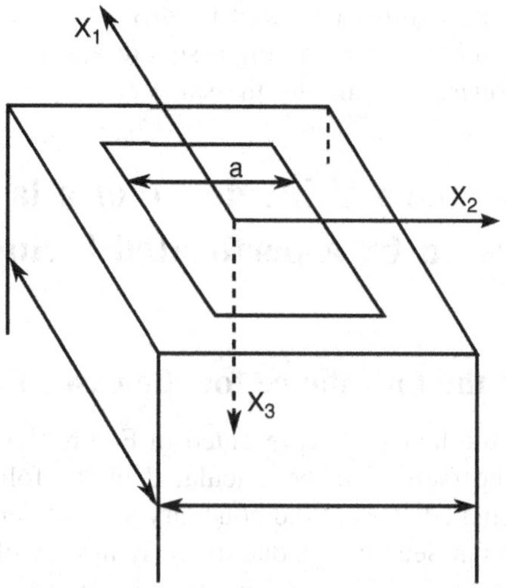

Figure 9.4 The elementary cell of the partition. The square at $x_3 = 0$ is the boundary of the aperture in contact with the free air.

This equation can be rewritten

$$\bar{p} = U\left[\frac{\omega\rho_0 s}{k} + 8\sum_{n=1}^{\infty}\frac{\nu_{0,n}\omega\rho_0}{k_{0,n}}\frac{1}{\pi^2 n^2}\sin^2\left(\frac{\pi na}{D}\right)\right.$$
$$\left. + \sum_{m=1}^{\infty}\sum_{n=1}^{\infty} 4\frac{D^2}{a^2}\nu_{m,n}\frac{\omega\rho_0}{k_{m,n}}\frac{1}{\pi^4 m^2 n^2}\sin^2\left(\frac{\pi ma}{D}\right)\sin^2\left(\frac{\pi na}{D}\right)\right] \quad (9.28)$$

where $k_{m,n}$ at sufficiently low frequencies is given by

$$k_{m,n} = -j\frac{2\pi}{D}(m^2+n^2)^{1/2} \quad (9.29)$$

A comparison of Equations (9.1) and (9.28) yields

$$\varepsilon_e = \frac{4D}{\pi^3}\sum_{n=1}^{\infty}\frac{\nu_{0,n}}{n^3}\sin^2\left(\frac{\pi na}{D}\right)$$
$$+ \frac{2D^3}{a^2\pi^5}\sum_{m=1}^{\infty}\sum_{n=1}^{\infty}\frac{\nu_{m,n}}{m^2 n^2 (m^2+n^2)^{1/2}}\sin^2\left(\frac{\pi ma}{D}\right)\sin^2\left(\frac{\pi na}{D}\right) \quad (9.30)$$

Ingard (1953) indicated that ε_e can be approximated for $s^{1/2} < 0.4$ by

$$\varepsilon_e = 0.48(S)^{1/2}(1 - 1.25s^{1/2}) \quad (9.31)$$

where S is the area of the aperture. The limit $0.48(S)^{1/2}$ for $s = 0$ can be obtained by calculating the radiation impedance of a single square aperture in a plane. Both limits for the circular and the square aperture are the same.

9.3 Impedance at normal incidence of a layered porous material covered by a perforated facing – Helmoltz resonator

9.3.1 Evaluation of the impedance for the case of circular holes

The material covered by the facing is represented in Figure 9.5. The surface impedance in the free air close to the facing can be calculated in the following way. At first the surface impedance is calculated at B, at the boundary surface between the porous layered material and the facing. As in Section 9.2, due to the symmetry of the problem, a partition of the material can be performed. The elementary cell, represented in Figure 9.6, is a cylinder having a square cross-section and a circular aperture.

Two different cases were considered by Ingard (1954). For the simplest case (Figure 9.6a), the facing is in contact with a layer of air. In this case, the modes of higher order responsible for the added length are present only in the layer of air if the frequency is much lower than the cutoff frequency $f_c = c_0/D$. The effect of these modes is the same as in the free air. Moreover, the acoustic field in the porous layer is plane, homogeneous and propagates perpendicular to the layer. The value of the

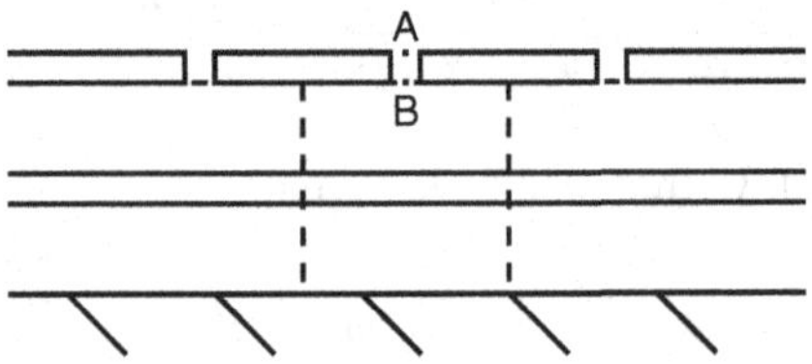

Figure 9.5 A layered porous material covered by a perforated facing.

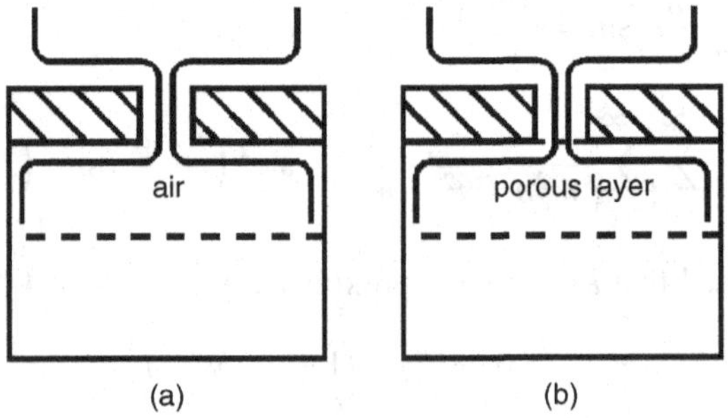

Figure 9.6 The elementary cell of the partition compatible with the symmetry of the problem. (a) The facing is in contact with air, (b) the facing is in contact with a porous layer.

impedance at A can be obtained by Equation (9.21) modified in the following way. The impedance Z_c in Equation (9.21) must be replaced by the impedance at B at normal incidence of the layered material, including the layer of air, with the facing removed. The second case is represented in Figure 9.6(b). A porous layer is in contact with the facing. The distortion of the flux under the facing is now located in a porous layer, and this distortion creates not only an inertial effect, but also a resistive effect. The same procedure as in Section 9.2 can be used to evaluate these different effects. The difference is the presence of porous layers, the finite length of the elementary cell, and the presence of two waves in every medium with opposite values of the x_3 component of the wave number vector.

The different layers of porous material are represented in Figure 9.7.

Let $k(1)$be the wave number in layer 1. The pressure field in this layer can be written

$$p(x_1, x_2, x_3) = \sum_{m,n} A_{m,n}(1)[\exp(-jx_3k_{m,n}(1)) + \exp(-j(2X_3 - x_3)k_{m,n}(1))] \times \cos\frac{2m\pi x_1}{D}\cos\frac{2n\pi x_2}{D} \tag{9.32}$$

In this equation, $k_{m,n}(1)$ is given by Equation (9.3) which can be rewritten

$$k_{m,n}(1) = \left[k^2(1) - \frac{4m^2\pi^2}{D^2} - \frac{4n^2\pi^2}{D^2}\right]^{1/2} \tag{9.33}$$

and X_3 is the total thickness of the layered material. The modes related to different sets (m, n) propagate independently in the cell.

The field $v_{3,m,n}(x_1, x_2, x_3)$ of the x_3 velocity component related to the (m, n) mode is equal to zero in layer 1 at the contact surface with the backing. Let M_1 and M'_1 be two points located close to the contact surface between the layers 1 and 2, M_1 in medium 1 and M'_1 in medium 2. Let $p_{m,n}(M)$ and $v_{3,m,n}(M)$ be the pressure and the x_3 velocity component at M related to the (m, n) mode. The impedance at M_1 related to the (m, n) mode is given by

$$Z_{m,n}(M_1) = \frac{p_{m,n}(M_1)}{v_{3,m,n}(M_1)} = -jZ_c(1)\frac{k(1)}{k_{m,n}(1)}\cot k_{m,n}(1)l_1 \tag{9.34}$$

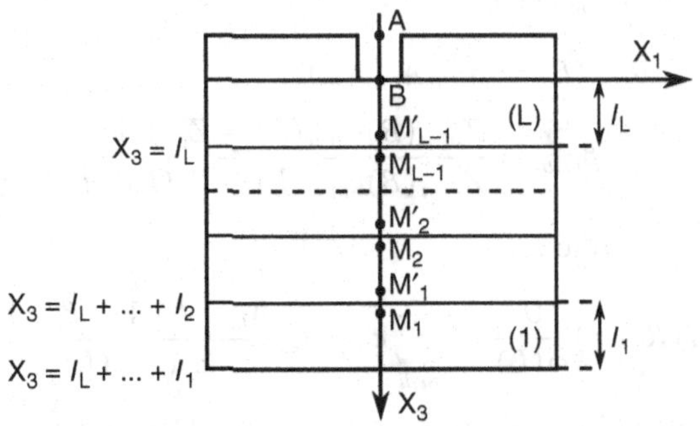

Figure 9.7 The different layers of the stratified material in the elementary cell.

where $Z_c(1)$ is the characteristic impedance and l_1 the thickness of layer 1 which is fixed on a rigid impervious wall. Let $\phi(1)$ and $\phi(2)$ be the porosities of layer 1 and 2, respectively. The pressures and the x_3 components of the velocities at M_1 and M_1' for the (m, n) mode are related by

$$v_{3,m,n}(M_1') = \frac{\phi(1)}{\phi(2)} v_{3,m,n}(M_1) \tag{9.35}$$

$$p_{m,n}(M_1') = p_{m,n}(M_1) \tag{9.36}$$

and $Z_{m,n}(M_1')$ is related to $Z_{m,n}(M_1)$ by

$$Z_{m,n}(M_1') = \frac{\phi(2)}{\phi(1)} Z_{m,n}(M_1) \tag{9.37}$$

Let M_2 and M_2' be two points located close to the boundary surface between layers 3 and 2, M_2 in medium 2, and M_2' in medium 3. Using the impedance translation formula Equation (3.36), the impedance at M_2 related to the mode (m, n) can be written

$$\begin{aligned} Z_{m,n}(M_2) = {} & \frac{Z_c(2)(k(2)/k_{m,n}(2))}{Z_{m,n}(M') - jZ_c(2)(k(2)/k_{m,n}(2)) \text{ cotg } k_{m,n}(2)l_2} \\ & \times [-jZ_{m,n}(M_1') \text{ cotg } k_{m,n}(2)l_2 + Z_c(2)(k(2)/k_{m,n}(2))] \end{aligned} \tag{9.38}$$

In this equation l_2 is the thickness and $Z_c(2)$ the characteristic impedance of the second layer, and $k_{m,n}(2)$ and $k(2)$ are related by Equation (3.4). The impedances at the boundaries of the layers related to each mode can be obtained by this way up to B in the Lth layer in contact with the facing. The pressure field in the Lth layer can be written

$$\begin{aligned} p(x_1, x_2, x_3) = {} & \sum_{m,n} [A_{m,n}(L) \exp(-jk_{m,n}(L)x_3) \\ & + B_{m,n}(L) \exp(jk_{m,n}(L)x_3)] \cos \frac{2\pi m x_1}{D} \cos \frac{2\pi n x_2}{D} \end{aligned} \tag{9.39}$$

The coefficients $A_{m,n}(L)$ and $B_{m,n}(L)$ are related to the impedance $Z_{m,n}(B)$ by

$$Z_{m,n}(B) = \left(\frac{A_{m,n}(L) + B_{m,n}(L)}{A_{m,n}(L) - B_{m,n}(L)} \right) \frac{Z_c(L)k(L)}{k_{m,n}(L)} \tag{9.40}$$

and $B_{m,n}(L)$ is related to $A_{m,n}(L)$ by

$$B_{m,n}(L) = \beta_{m,n} A_{m,n}(L) \tag{9.41}$$

$$\beta_{m,n} = \frac{Z_{m,n}(B)k_{m,n}(L) - Z_c(L)k(L)}{Z_{m,n}(B)k_{m,n}(L) + Z_c(L)k(L)} \tag{9.42}$$

Equation (9.5) can be rewritten

$$\begin{aligned} \Upsilon(R - r)\frac{U}{\phi(L)} = {} & \sum_{m,n} k_{m,n}(L) \frac{A_{m,n}(L)(1 - \beta_{m,n})}{Z_c(L)k(L)} \\ & \times \cos \frac{2\pi m x_1}{D} \cos \frac{2\pi n x_2}{D} \end{aligned} \tag{9.43}$$

where $\phi(L)$ is the porosity of the Lth medium, and Equation (9.11) becomes for the case n or $m \neq 0$

$$A_{m,n}(L) = \frac{4\nu_{m,n} U R Z_c(L) k(L) \mathrm{J}_1[(2\pi R/D)(m^2+n^2)^{1/2}]}{\phi(L) D (m^2+n^2)^{1/2} k_{m,n}(L)(1-\beta_{m,n})} \tag{9.44}$$

$$\nu_{1,0} = \nu_{0,1} = 1/2,\ \nu_{\mathrm{m,n}} = 1 \text{ if } n \text{ and } m \neq 0$$

For the case $n = m = 0$, $A_{0,0}(L)$ is given by

$$A_{0,0}(L) = \frac{\pi R^2 U}{\phi(L)(1-\beta_{0,0})D^2} Z_c(L) = \frac{Z_c(L) s U}{\phi(L)(1-\beta_{0,0})} \tag{9.45}$$

Let B' be a point located close to B, in a perforation above the Lth layer. The impedance $Z(B')$ is $\bar{p}/U$, where $\bar{p}$ is the average pressure over the aperture

$$\bar{p} = \frac{1}{\pi R^2} \int_0^R \int_0^{2\pi} \sum_{m,n} A_{m,n}(L)(1+\beta_{m,n}) \cos\frac{2\pi m x_1}{D} \cos\frac{2\pi n x_2}{D} r\,dr\,d\theta \tag{9.46}$$

By the use of Equations (9.8) and (9.9), $Z(B')$ can be written

$$Z(B') = \frac{1}{\phi(L)}\left[s Z_c(L)\frac{1+\beta_{0,0}}{1-\beta_{0,0}}\right] + \frac{4}{\pi}\sum_{m,n}{}' \frac{\nu_{m,n}}{\phi(L)} \frac{\mathrm{J}_1^2\left[2\pi \frac{R}{D}(m^2+n^2)\right]}{(m^2+n^2)} \times \left[\frac{Z_c(L)k(L)}{k_{m,n}(L)}\frac{(1+\beta_{m,n})}{(1-\beta_{m,n})}\right] \tag{9.47}$$

This equation can be simplified by using Equation (9.40):

$$Z(B') = \frac{s Z_{0,0}(B)}{\phi(L)} + \frac{4}{\pi}\sum_{m,n}{}' \frac{\nu_{m,n}}{\phi(L)} \frac{\mathrm{J}_1^2\left[2\pi \frac{R}{D}(m^2+n^2)\right] Z_{m,n}(B)}{(m^2+n^2)} \tag{9.48}$$

The inertial term $\mathrm{j}(\varepsilon_e + d)\rho_0\omega$, where ε_e is given by Equation (9.17), must be taken into account, but the effect of the viscous forces in the aperture and on the faces of the facing around the aperture can be neglected when $Z(A)$ is evaluated

$$Z(A) = Z(B') + j(\varepsilon_e + d)\rho_0\omega \tag{9.49}$$

Finally, the impedance Z in the free air close to the facing is

$$Z = Z(A)/s \tag{9.50}$$

If the porous materials are transversally isotropic with a symmetry axis Ox_3, like the glass wools described in Chapter 3, two wave numbers $k(i)$ and $k_p(i)$ and two characteristic impedances $Z_c(i)$ and $Z_p(i)$ in the x_3 direction and the directions parallel to the plane x_1Ox_2 exist. The previous description is always valid but $k_{m,n}(i)$ is now given by

$$k_{m,n}(i) = k(i)\left[1 - \left(\frac{2\pi m}{D}\right)^2 \frac{1}{k_p^2(i)} - \left(\frac{2\pi n}{D}\right)^2 \frac{1}{k_p^2(i)}\right]^{1/2} \tag{9.51}$$

It may be noticed that Equations (9.48) and (9.14) can be related in a simple way. Equation (9.14) can be rewritten

$$\frac{\bar{p}}{U} = sZ_c + 4\sum_{m,n}{}' v_{m,n}\frac{J_1^2\left[2\pi\frac{R}{D}(m^2+n^2)^{1/2}\right]}{\pi(m^2+n^2)}\frac{kZ_c}{k_{m,n}} \tag{9.52}$$

and Equation (9.48) can be obtained from Equation (9.52) by substituting $Z_{m,n}(B)/\phi(L)$ for $Z_c k/k_{m,n}$ (the ratios $k/k_{0,0}$ and $k(L)/k_{0,0}(L)$ do not appear explicitly in the contributions of the (0, 0) mode because they are equal to 1). Both quantities $Z_{m,n}(B)/\phi(L)$ and $Z_c k/k_{m,n}$ have the same physical meaning since $Z_{m,n}(B)/\phi(L)$ for the case of a semi-infinite layer of air is equal to $Z_c k/k_{m,n}$.

9.3.2 Evaluation at normal incidence of the impedance for the case of square holes

From Equation (9.28), the impedance $Z(B)$ for the case of square holes and a semi-infinite layer of air can be written

$$\begin{aligned} Z(B) = sZ_c &+ 8\sum_{n=1}^{\infty} v_{0,n}\frac{\sin^2\left(\frac{\pi na}{D}\right)}{\pi^2 n^2}\frac{kZ_c}{k_{0,n}} \\ &+ \frac{4D^2}{a^2}\sum_{m=1}^{\infty}\sum_{n=1}^{\infty} v_{m,n}\frac{\sin^2\left(\frac{\pi ma}{D}\right)\sin^2\left(\frac{\pi na}{D}\right)}{\pi^4 m^2 n^2}\frac{kZ_c}{k_{m,n}} \end{aligned} \tag{9.53}$$

The previously defined substitution can be used to calculate $Z(B)$ when a porous layered material is substituted for the semi-infinite layer of air. Equation (9.53) becomes

$$\begin{aligned} Z(B') = \frac{s}{\phi(L)}Z_{0,0}(B) &+ \frac{8}{\phi(L)}\sum_{n=1}^{\infty} v_{0,n}\frac{\sin^2\left(\frac{\pi na}{D}\right)}{\pi^2 n^2}Z_{0,n}(B) \\ &+ \frac{4}{\phi(L)}\frac{D^2}{a^2}\sum_{m=1}^{\infty}\sum_{n=1}^{\infty} v_{m,n}\frac{\sin^2\left(\frac{\pi ma}{D}\right)\sin^2\left(\frac{\pi na}{D}\right)}{\pi^4 m^2 n^2}Z_{m,n}(B) \end{aligned} \tag{9.54}$$

As in the case of circular apertures, the inertial term $j(\varepsilon_e + d)\rho_0\omega$, where ε_e is now given by Equation (9.30), must be taken into account, but the effect of the viscous forces in the aperture and on the facing around the aperture can be neglected when $Z(A)$ is evaluated

$$Z(A) = Z(B') + j(\varepsilon_e + d)\rho_0\omega \tag{9.55}$$

The impedance Z in the free air close to the facing is

$$Z = Z(A)/s \tag{9.56}$$

9.3.3 Examples

The validity of similar models was verified by comparing prediction and measurement of the impedance at normal incidence for different configurations (Bolt 1947, Zwikker and Kosten 1949, Ingard and Bolt 1951, Brillouin 1949, Callaway and Ramer 1952, Ingard 1954, Velizhanina 1968, Davern 1977, Byrne 1980). The trends predicted by the model appear clearly in the measured impedances. Simple configurations of layered media covered by perforated facings having circular apertures are modelled in what follows, showing the effect of the open area ratio, the diameter of the holes, and the flow resistivity of the material in contact with the facing. For the first example, the normalized impedance and the absorption coefficient calculated from Equation (9.50) are represented as a function of frequency in Figures 9.8 and 9.9 for different open area ratios. An isotropic porous material, set on a rigid impervious wall, is in contact with the facing at its upper face. The porous material is characterized by the following parameters: thickness $e = 2$ cm, flow resistivity $\sigma = 50\,000$ N m^{-4} s, characteristic dimensions $\Lambda = 0.034$ mm and $\Lambda' = 0.13$ mm, porosity $\phi = 0.98$, tortuosity $\alpha_\infty = 1.5$. The Johnson *et al.* model is used for the effective density and the Champoux and Allard model is used for the bulk modulus (see Chapter 5). The facing has a thickness $d = 1$ mm, and circular apertures of radius $R = 0.5$ mm. The open area ratio takes the following values: $s = 0.4$, 0.1, 0.025 and 0.005.

The real part of the impedance strongly increases when the open area ratio decreases, due to the resistive effects in the porous material close to the facing. The imaginary part also increases when the open area ratio decreases, due to the added masses at both sides of the facing that are multiplied by $1/s$. The absorption coefficient A_0 increases at low frequencies and decreases at high frequencies; the maximum value of A_0 simultaneously decreases because the real part of the impedance becomes much larger than Z_c.

In the second example, an air gap and a screen are successively inserted between the facing and the porous material. The three configurations are represented in Figure 9.10.

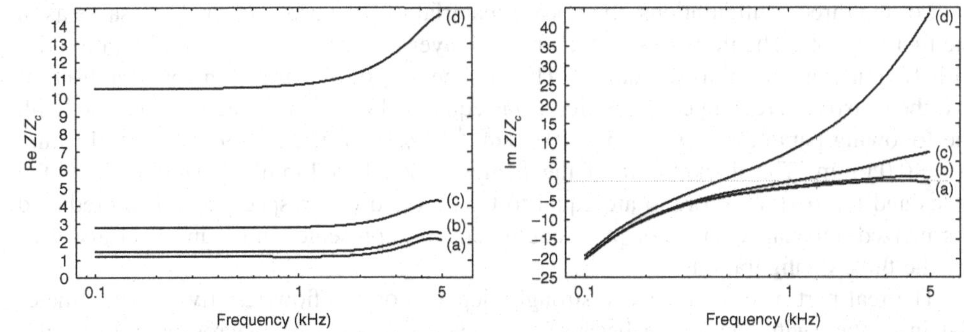

Figure 9.8 Influence of the open area ratio on the normalized impedance Z/Z_c of an isotropic porous material covered by a facing. Parameters for the porous material: thickness $e = 2$ cm, flow resistivity $\sigma = 50\,000$ N m^{-4} s, tortuosity $\alpha_\infty = 1.5$, characteristic dimensions $\Lambda = 0.034$ mm and $\Lambda' = 0.13$ mm, porosity $\phi = 0.98$. Parameters for the facing: thickness $d = 1$ mm, circular apertures of radius $R = 0.5$ mm, open area ratio $s =$ (a) 0.4, (b) 0.1, (c) 0.025, (d) 0.005.

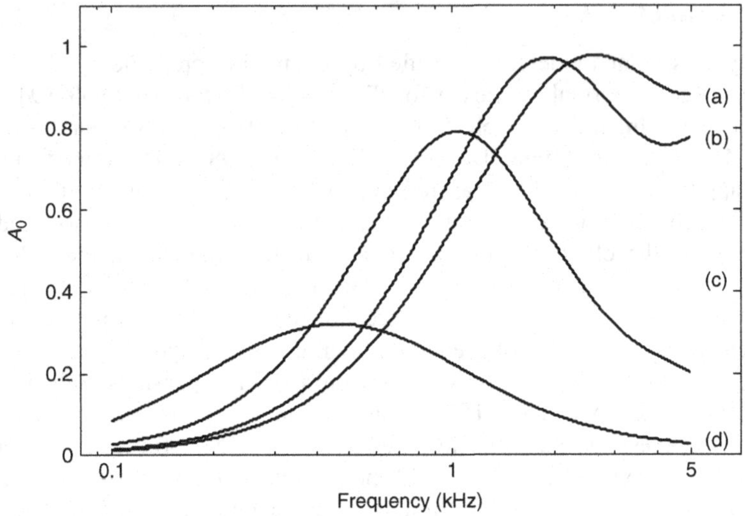

Figure 9.9 Influence of the open area ratio on the absorption coefficient A_0 of an isotropic porous material covered by a facing. Same material and same facing as in Figure 9.8. Open area ratio $s =$ (a) 0.4, (b) 0.1, (c) 0.025, (d) 0.005.

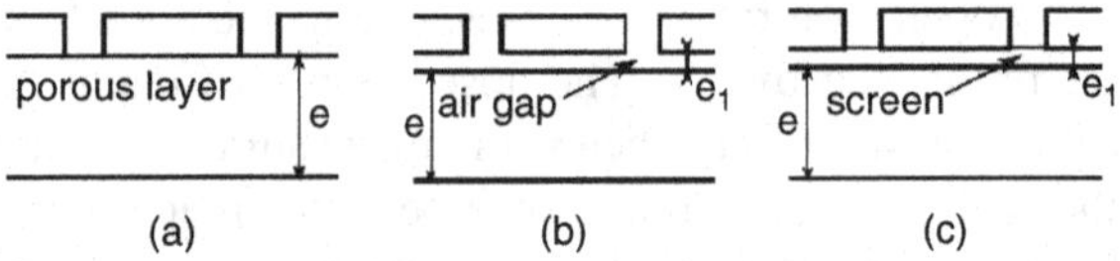

Figure 9.10 Three different configurations: (a) facing + porous layer, (b) facing + air gap + porous layer, (c) facing + resistive screen + porous layer.

For the three configurations, the porous material of the porous layer is the same as in the first example. The thickness e of the porous layer is equal to 2 cm in configuration (a), to 1.9 cm in configurations (b) and (c). The thickness e_1 of the air gap in configuration (b) and the resistive screen in configuration (c) is equal to 1 mm. The screen is isotropic, with the following parameters: $\sigma = 0.5 \times 10^6$ N m^{-4} s, $\alpha_\infty = 1.5$, $\phi = 0.95$, $\Lambda = 0.013$ mm, $\Lambda' = 0.05$ mm. The thickness d of the facing is equal to 1 mm. The radius R of the holes and the perforation ratio are equal to 0.5 mm and 0.1, respectively. The predicted normalized impedance and absorption coefficient are represented in Figures 9.11 and 9.12 for the three configurations.

The real part of the impedance strongly depends on the flow resistivity of the material in contact with the facing because the mean length of the macroscopic molecular trajectories in the screen in configuration (c), and in the air gap in configuration (b), is much larger than the thickness e_1. It may be noticed that the inertial effects under the facing do not depend on the flow resistivity of the material in contact with the facing. The imaginary parts of the impedances for the three configurations are nearly equal. The largest absorption coefficient A_0 at low frequencies is obtained with the resistive screen, but the real part of the impedance is too large for A_0 to be close to 1.

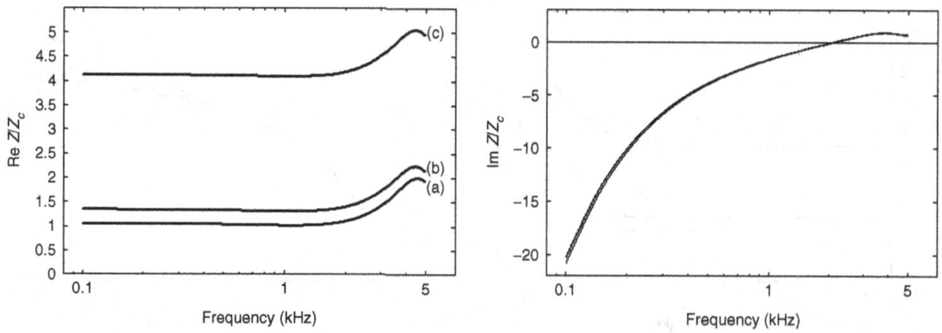

Figure 9.11 Normalized Impedance Z/Z_c for the three configurations represented in Figure 9.10: (a) porous layer, (b) porous layer + air gap, (c) porous layer + resistive screen. The effect on the reactance is negligible.

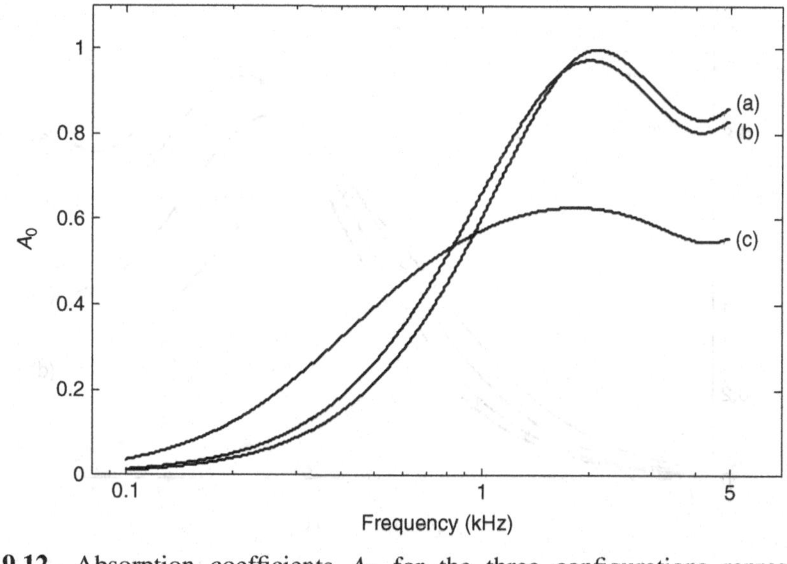

Figure 9.12 Absorption coefficients A_0 for the three configurations represented in Figure 9.10: (a) porous layer, (b) porous layer + air gap, (c) porous layer + resistive screen.

In the third example, the same porous material as in the first example is covered by a facing of thickness $d = 1$ mm, perforated by circular holes of radius R, with an open area ratio s = 0.1. The normalized impedance and the absorption coefficient are represented in Figures 9.13 and 9.14 for radii R equal to 0.5, 1, 2 and 4 mm. The distance D between the holes increases with R if s is constant, and the resistive effects in the porous material also increase. As a consequence, the real part of impedance strongly increases with R. The imaginary part of impedance, which is related to inertial effects, also increases with R, because the added length given by Equation (9.18) is proportional to R. The effect is small at low frequencies, the contribution of the added length to the imaginary part of impedance being equal to $\varepsilon_e \rho_0 \omega / s$.

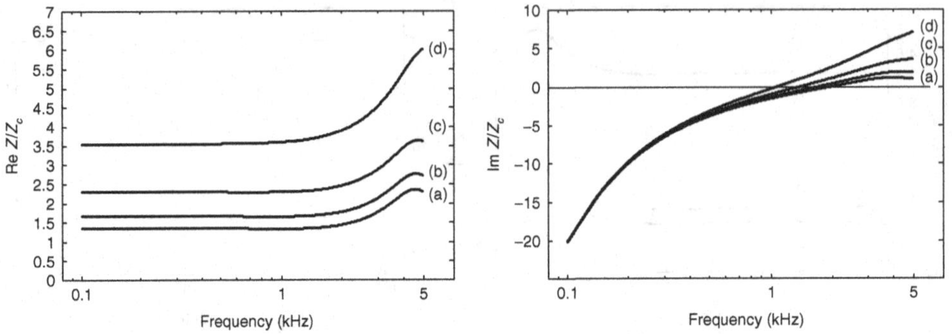

Figure 9.13 Influence of the radius of the aperture on the normalized impedance Z/Z_c. The configuration is the same as in Figure 9.10(a), and $s = 0.1$. The different curves correspond to $R =$ (a) 0.5 mm, (b) 1 mm, (c) 2 mm, (d) 4 mm.

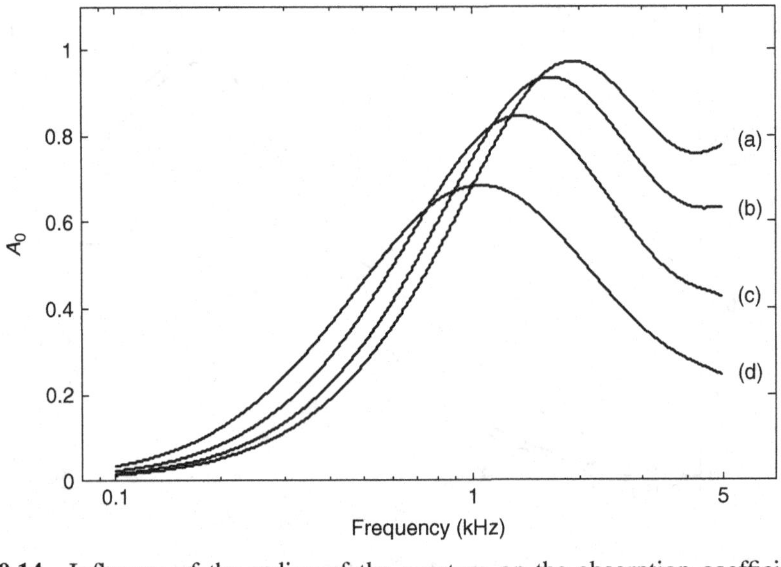

Figure 9.14 Influence of the radius of the aperture on the absorption coefficient. The configuration is the same as in Figure 9.10(a), and $s = 0.1$. The different curves correspond to $R =$ (a) 0.5 mm, (b) 1 mm, (c) 2 mm, (d) 4 mm.

9.3.4 Design of stratified porous materials covered by perforated facings

From the previous examples, some trends may be noticed. For values of the open area ratio larger than 0.2, the surface impedance of the porous layer is only slightly modified. For lower values of the open area ratio, an increase of the absorption coefficient at low frequencies can be obtained, but the absorption coefficient decreases at high frequencies. More precisely, the real part of the impedance increases with the radii of the apertures and the flow resistivity of the material close to the facing, where the velocity field is distorted. The real part of the impedance also increases when the open area ratio decreases. The imaginary part of the impedance does not depend at low frequencies on the radii of the apertures and the flow resistivity close to the facing. It increases when the open area ratio decreases. In order to increase the absorption coefficient at low frequencies, the imaginary part of the impedance, which is too large and negative, can be set close to zero by using a small value of the open area ratio and a porous layer of high flow resistivity. A value of the real part of the impedance close to the characteristic impedance of air can be obtained with a small value of R and by inserting between the porous layer of high flow resistivity and the facing a thin layer of material with a low flow resistivity. The thin layer of low flow resistivity and the small value of R are necessary for the real part of impedance not to be too large.

The predicted surface impedance and absorption coefficient of a layered material covered by a perforated facing is represented at normal and at oblique incidence in Figures 9.17 and 9.18. The material is made up of two layers, a thin layer M_1 of low flow resistivity in contact with the facing, and a thicker material M_2 having a large flow resistivity in contact with the impervious rigid backing. The parameters that characterize both layers are given in Table 9.1. The thickness of the facing is $d = 1$ mm, the radius of each hole is $R = 0.5$ mm, and the perforation ratio is $s = 0.005$.

The absorption coefficient A_0 reaches a maximum close to 1 around 500 Hz. The counterpart of this performance is a low value of A_0 at high frequencies. It may be noticed that Bolt (1947) designed a material having a large absorption at low frequencies which presented similar characteristics, i.e. small holes, a small open area ratio, and a flow resistivity lower than for the configuration designed for sound absorption at higher frequencies without facing.

9.3.5 Helmholtz resonators

The cell of Figure 9.6 is called a Helmholtz resonator if its lateral boundaries are impervious rigid surfaces. The Helmholtz resonator is a volume with an aperture. Different shapes for the volume and the aperture can be used; the acoustical properties of these resonators were studied by Ingard (1953). In practical cases, the viscous forces in the empty volume can be neglected since the lateral dimension of the volume is much larger than the viscous skin depth, and the effect of the viscous forces can be taken into account only at the neck of the resonator. It may be pointed out that the term $(4 + 2d/R)R_s$ in Equation (9.21) is generally much smaller than the real part of $Z(B)$ when porous media are present in the resonator. If there is no porous material in the resonator, this term cannot be neglected. Helmholtz resonators generally are designed to absorb sound at low frequencies. An array of resonators is represented in Figure 9.15.

Table 9.1 The parameters that characterize the two layers M_1 and M_2. The layer M_1 is in contact with the facing and the layer M_2 with the rigid impervious backing.

Material	Flow resistivity σ (N m^{-4} s)	Viscous dimension Λ (mm)	Thermal dimension Λ' (mm)	Tortuosity α_∞	Porosity ϕ	Thickness e (cm)
M_1	5000	0.12	0.27	1.1	0.99	0.1
M_2	50000	0.034	0.13	1.5	0.98	1.9

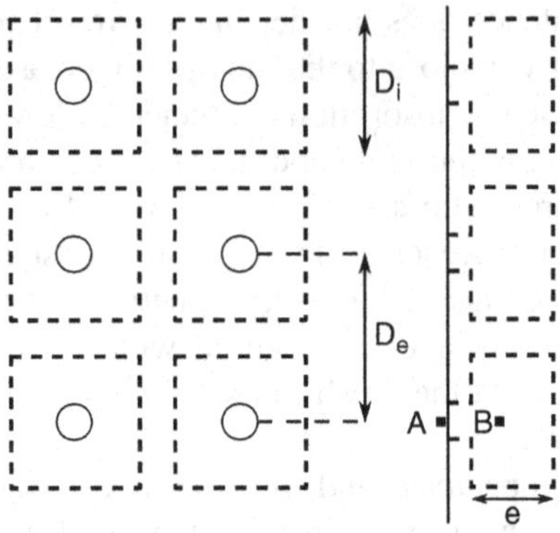

Figure 9.15 An array of resonators.

As in the case of the material described in Section 9.3.4, the absorption coefficient of an array of resonators presents a maximum when the imaginary part of the impedance of the array is equal to 0, at a frequency called the resonance frequency.

The frequency dependence of the impedance and the absorption coefficient of an array of resonators can be calculated by using the results of the previous sections if the resonator is a cylinder with a square cross-section. As an example, we consider the case where there is no porous material in the resonator. The impedance $Z(B')$ can be calculated by Equation (9.54) where $\phi(L) = 1$ and D_i substituted for D in Equations (9.3) and (9.54). The impedance $Z_{0,0}(B)$ is given by

$$Z_{0,0}(B) = -jZ_c \cot ke \tag{9.57}$$

where e is the length of the cylinder. For frequencies well below c_0/D_i and if the quantities $B_{m,n}$ can be neglected in Equation (9.40), the terms related to the modes of higher order can be replaced by an added length effect $j\omega\rho_0\varepsilon_i$, where ε_i can be calculated by Equation (9.18) if the apertures are circular.

It may be pointed out that the open area ratio s_i at B is the ratio of the area S of the aperture to the area of the cross-section of the 'internal' elementary cell, i.e. D_i^2, and the internal added length ε_i is given by

$$\varepsilon_i = 0.48S^{1/2}(1.0 - 1.14s_i) \tag{9.58}$$

$$\text{where } s_i = (S/D_i^2)^{1/2} \tag{9.59}$$

and $Z(B)$ is given by

$$Z(B) = -js_i Z_c \cot ke + j\varepsilon_i \rho_0 \omega \tag{9.60}$$

At A, the area of the 'external' elementary cell is D_e^2, and the open area ratio s_e and the added length at A are given by

$$s_e = (S/D_e^2)^{1/2} \tag{9.61}$$

$$\varepsilon_e = 0.48S^{1/2}(1.0 - 1.14s_e) \tag{9.62}$$

Equation (9.49) can be rewritten

$$Z = \frac{1}{s_e}\left[-js_i Z_c \operatorname{cotg} ke + j(\varepsilon_i + \varepsilon_e + d)\omega\rho_0 + \left(\frac{2d}{R} + 4\right) R_s\right] \tag{9.63}$$

The resonance frequency f_0 is implicitly defined by Im Z = 0:

$$\operatorname{cotg} \frac{2\pi f_0 e}{c_0} = \frac{1}{s_i c_0}[\varepsilon_i + \varepsilon_e + d]2\pi f_0 \tag{9.64}$$

If s is sufficiently small, this equation is verified for a value of the argument of the cotangent much smaller than 1, and the resonance frequency f_0 is given by

$$f_0 = \frac{c_0}{2\pi}\left(\frac{s_i}{e(\varepsilon_i + \varepsilon_e + d)}\right)^{1/2} \tag{9.65}$$

Multiplying the numerator and the denominator in the square root by D_i^2 gives

$$f_0 = \frac{c_0}{2\pi}\left(\frac{S}{V(d + \varepsilon_i + \varepsilon_e)}\right)^{1/2} \tag{9.66}$$

where V is the volume of the resonator. Equation (9.66) can be used for resonators of different shapes. The added lengths are calculated in Ingard (1953) for different positions and shapes of the aperture and different shapes of the resonator.

9.4 Impedance at oblique incidence of a layered porous material covered by a facing having circular perforations

9.4.1 Evaluation of the impedance in a hole at the boundary surface between the facing and the material

In the previous sections, at normal incidence, the pressure field in the porous medium is the sum of different waves with wave number components $k_1 = 2m\pi/D$ in the direction x_1 and $k_2 = 2n\pi/D$ in the direction x_2, n and m varying from $-\infty$ to ∞. These waves can be considered, by using the Floquet theorem (see for instance Collin (1960) pp 371 and 465), as the different modes transmitted at normal incidence by the doubly periodic perforated screen with the spatial period D. This theorem will be used here for evaluating the impedance at oblique incidence. The surface of the facing with a plane wave incident at an angle $\theta \neq 0$ is shown in Figure 9.16. The incidence plane is x_3Ox_2 and the circular perforations are periodically distributed in the two directions Ox_1 and Ox_2 with a spatial period D equal to the distance between two holes. The modes transmitted by the doubly periodic screen in the porous medium have now a space dependence $g(x_1, x_2)$ given by

$$g(x_1, x_2) = \exp\left\{j\left[\frac{2\pi m x_1}{D} + \left(\frac{2\pi n}{D} - k_0 \sin\theta\right) x_2\right]\right\} \tag{9.67}$$

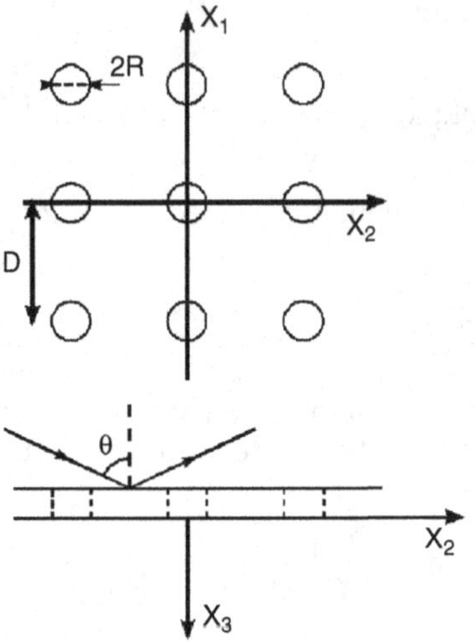

Figure 9.16 The surface of a perforated facing at oblique incidence.

where $m = 0, \pm 1, \pm 2, \ldots$ and $n = 0, \pm 1, \pm 2 \ldots$. As for the case of normal incidence the modes with opposite m can be associated with a cosine factorizing the common space dependence on x_2 and x_3. The pressure field in the layer L at the contact surface with the facing can be written

$$p(x_1, x_2, x_3) = \sum_{m=0}^{\infty} \sum_{n=-\infty}^{\infty} \cos\left(\frac{2\pi m x_1}{D}\right) \exp(j(2\pi n/D - k_0 \sin\theta)x_2) \times \{A_{m,n}(L)\exp(-j(k_{m,n}(L)x_3) + B_{m,n}(L)\exp(jk_{m,n}(L)x_3)\} \tag{9.68}$$

The x_1 component of velocity at $x_1 = \pm D/2$ is equal to zero, due to the symmetry of the problem. The wave number component $k_{m,n}(L)$ in the x_3 direction is given by

$$k_{m,n}(L) = \left(k^2(L) - \left(\frac{2\pi m}{D}\right)^2 - (2\pi n/D - k_0 \sin\theta)^2\right)^{1/2} \tag{9.69}$$

As in the case of normal incidence, the (m, n) mode of propagation has the same wave vector components in the different media of the stratified material, and Equations (9.41) and (9.42) relating $B_{m,n}(L)$ and $A_{m,n}(L)$ can be obtained in the same way as in the previous section, the one modification being the substitution of $2\pi n/D - k_0 \sin\theta$ for $2\pi n/D$.

The pressure field related to the (m, n) mode of propagation can be written at the contact surface with the facing

$$p(x_1, x_2, 0) = \sum_{m=0}^{\infty} \sum_{n=-\infty}^{\infty} \cos\frac{2\pi m x_1}{D} \exp\left(j\left(\frac{2\pi n}{D} - k_0 \sin\theta\right) x_2\right) \times A_{m,n}(L)(1 + \beta_{m,n}) \tag{9.70}$$

The x_3 component of the velocity field is given by

$$v_3(x_1, x_2, 0) = \sum_{m=0}^{\infty} \sum_{n=-\infty}^{\infty} \cos \frac{2\pi m x_1}{D} \exp\left(j\left(\frac{2\pi n}{D} - k_0 \sin\theta\right) x_2\right) \times A_{m,n}(L)(1 - \beta_{m,n}) \frac{k_{m,n}(L)}{Z_c(L)k(L)} \tag{9.71}$$

As in the case of normal incidence, the velocity amplitude is assumed to be uniform in each aperture. Let U_0 be the velocity on the aperture C_0. In the porous material of porosity $\phi(L)$ in contact with the facing, this velocity must be multiplied by $1/\phi(L)$. Multiplying the left-hand side of Equation (9.71) by $\cos(2\pi m x_1/D)\exp(-j[(2\pi n/D) - k_0 \sin\theta]x_2)$ and integrating over the aperture C_0 one obtains

$$I = \frac{U_0}{\phi(L)} \left[\int_0^R \int_0^{2\pi} \exp\left(-j\left(\frac{2\pi n}{D} - k_0 \sin\theta\right) x_2\right) \cos\left(\frac{2\pi m x_1}{D}\right) r dr d\theta \right] \tag{9.72}$$

In the integral the exponential function can be replaced by a cosine with the same argument, because the sine in the decomposition of the exponential is an odd function of x_2 and does not contribute to the integral. The use of Equations (9.7)–(9.9) yields

$$\int_0^R \int_0^{2\pi} \exp\left(-j\left(\frac{2\pi n}{D} - k_0 \sin\theta\right) x_2\right) \cos\left(\frac{2\pi m x_1}{D}\right) r dr d\theta = J_1\left[2\pi R\left(\frac{m^2}{D^2} + \left(\frac{n}{D} - \frac{k_0 \sin\theta}{2\pi}\right)^2\right)^{1/2}\right] \frac{R}{\left(\frac{m^2}{D^2} + \left(\frac{n}{D} - \frac{k_0 \sin\theta}{2\pi}\right)^2\right)} \tag{9.73}$$

The quantity I can also be evaluated by multiplying the right-hand side of Equation (9.71) by $\cos(2\pi m x_1/D)\exp(-j(2\pi n/D - k_0 \sin\theta)x_2)$ and integrating over the aperture. Equating these two evaluations of I yields

$$A_{m,n}(L) = \frac{2U_0}{\phi(L)D} \frac{Z_c(L)k(L)}{k_{m,n}(L)} \frac{1}{1 - \beta_{m,n}(L)} v'_m \times \frac{R}{\left(\frac{m^2}{D^2} + \left(\frac{n}{D} - \frac{k_0 \sin\theta}{2\pi}\right)^2\right)^{1/2}} J_1\left[2\pi R\left(\frac{m^2}{D^2} + \left(\frac{n}{D} - \frac{k_0 \sin\theta}{2\pi}\right)^2\right)^{1/2}\right] \tag{9.74}$$

where $v'_m = 1/2$ for $m = 0$ and $v'_m = 1$ for $m \neq 0$. As in the case of normal incidence, the impedance $Z(B')$ is estimated by

$$Z(B') = \frac{\int_0^R \int_0^{2\pi} p(x_1, x_2, 0) r dr d\theta}{\pi R^2 U_0} \tag{9.75}$$

Substituting Equation (9.70) for $p(x_1, x_2, 0)$ and using Equation (9.73) yields

$$Z(B') = \sum_{m=0}^{\infty} \sum_{n=0,\pm1,\pm2\cdots} \frac{A_{m,n}(L)}{\pi R^2 U_0}(1+\beta_{m,n}) \times \mathrm{J}_1\left[2\pi R\left(\frac{m^2}{D^2}+\left(\frac{n}{D}-\frac{k_0\sin\theta}{2\pi}\right)^2\right)^{1/2}\right] \frac{R}{\left(\frac{m^2}{D^2}+\left(\frac{n}{D}-\frac{k_0\sin\theta}{2\pi}\right)^2\right)^{1/2}} \tag{9.76}$$

By the use of Equations (9.74) and (9.40), this equation can be rewritten

$$Z(B') = \frac{2}{\pi\phi(L)} \sum_{m=0}^{\infty} \sum_{n=0,\pm1,\pm2\cdots} \nu'_m Z_{m,n}(B) \frac{\mathrm{J}_1^2\left(2\pi R\left(\frac{m^2}{D^2}+\left(\frac{n}{D}-\frac{k_0\sin\theta}{2\pi}\right)^2\right)^{1/2}\right)}{\left(m^2+\left(n-\frac{k_0 D\sin\theta}{2\pi}\right)^2\right)} \tag{9.77}$$

This equation is similar to Equation (9.48). The (0, 0) mode contribution to $Z(B')$ in Equation (9.77) is

$$Z^{0,0}(B') = Z_{0,0}(B) \frac{\mathrm{J}_1^2(Rk_0\sin\theta)}{\pi\phi(L)\left(\frac{k_0 D\sin\theta}{2\pi}\right)^2} \tag{9.78}$$

To a first-order approximation, this equation can be rewritten

$$Z^{0,0}(B') = \frac{s}{\phi(L)} Z_{0,0}(B) \tag{9.79}$$

The components k_1 and k_2 for this mode are $k_1 = 0$ and $k_2 = k\sin\theta$. The impedance $Z_{0,0}(B)/\phi(L)$ would be the impedance at B if there were no facing.

9.4.2 Evaluation of the external added length at oblique incidence

The added length ε_e at oblique incidence can be evaluated by the following equation

$$\mathrm{j}\omega\rho_0\varepsilon_e = \frac{2}{\pi} \sum_{m,n}{}'' \nu'_m \frac{Z_c k}{k_{m,n}} \frac{\mathrm{J}_1^2\left(2\pi R\left(\frac{m^2}{D^2}+\left(\frac{n}{D}-\frac{k_0\sin\theta}{2\pi}\right)^2\right)^{1/2}\right)}{\left(m^2+\left(n-\frac{k_0 D\sin\theta}{2\pi}\right)^2\right)} \tag{9.80}$$

The double prime over the symbol Σ means that the term $m = 0, n = 0$ is excluded

$$\sum_{m,n}{}'' = \sum_{m=1}^{\infty} \sum_{n=0,\pm1,\pm2\ldots} + \sum_{m=0,n=\pm1,\pm2\ldots} \tag{9.81}$$

The right-hand side of this equation is the right-hand side of Equation (9.77) where the contribution of the (0, 0) mode has been removed and the impedance $Z_{m,n}(B)$ is related to a semi-infinite layer of air instead of a porous layer. Then, Z_c is the characteristic impedance of air, and $k_{m,n}$ for frequencies well below c_0/D is approximately equal to

$$k_{m,n} = -j\left(\frac{4\pi^2 m^2}{D^2} + \left(\frac{2\pi n}{D} - k_0 \sin\theta\right)^2\right)^{1/2} \tag{9.82}$$

Equations (9.80) and (9.82) yield the following expression for the added length

$$\varepsilon_e = \frac{D}{\pi^2}\sum_{m,n}{}'' v'_m \frac{J_1^2\left(2\pi R\left(\frac{m^2}{D^2} + \left(\frac{n}{D} - \frac{k_0 \sin\theta}{2\pi}\right)^2\right)^{1/2}\right)}{\left(m^2 + \left(n - \frac{k_0 D \sin\theta}{2\pi}\right)^2\right)^{3/2}} \tag{9.83}$$

It can be shown that when θ tends to zero, Equation (9.83) becomes identical to Equation (9.17) giving ε_e at normal incidence. The factor 2 in Equation (9.17) does not exist in Equation (9.83) because n can be positive or negative in this equation while it is always positive in Equation (9.14). Calculations for θ varying up to 80° indicate that the dependence of ε_e on θ is weak, and that Equation (9.18) can be used also at oblique incidence.

9.4.3 Evaluation of the impedance of a faced porous layer at oblique incidence

As in the case of normal incidence, Equations (9.55) and (9.56) can be used to evaluate the impedance Z in the free air close to the facing

$$Z = Z(A)/s = \frac{Z(B')}{s} + j\frac{(\varepsilon_e + d)\rho_0\omega}{s} \tag{9.84}$$

The impedance $Z(B')$ is now given by Equation (9.77). As an example, the impedance and the absorption coefficient of the faced material of Section 9.3.4 are represented in Figures 9.17 and 9.18 for three angles of incidence $\theta = 0, \pi/6, \pi/3$. The dependence of the impedance and the absorption coefficient on the angle of incidence is small, due to the high flow resistance of the layer M_2 which is much thicker than the layer M_1.

The model was tested at oblique incidence for different configurations using measurements by Guignouard *et al.* (1991), and as in the case of normal incidence, the trends predicted by the model are present in the measurements. The centres of the holes are set on axes parallel to the Ox_2 direction, defined by the intersection of the incidence plane and the facing. Measurements at oblique incidence indicate that the surface impedances do not noticeably vary when the direction of the Ox_2 axis is modified.

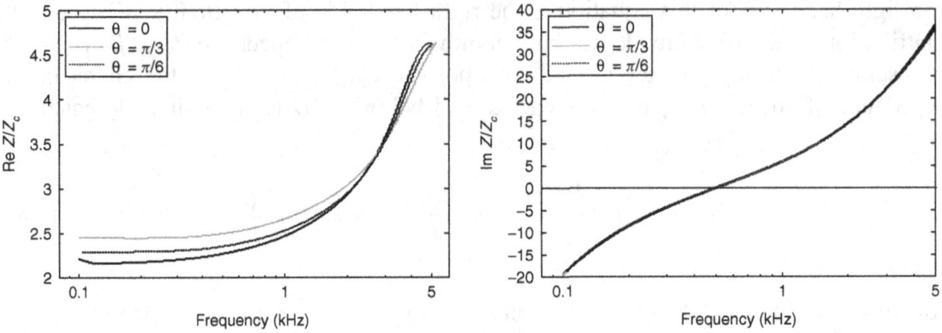

Figure 9.17 Normalized Impedance Z/Z_c at an angle of incidence $\theta = 0, \pi/6, \pi/3$ of a layered material made up of the two porous materials of Table 9.1 covered by a facing of thickness $d = 1$ mm, an open area ration $s = 0.005$ and circular apertures of radius $R = 0.5$ mm.

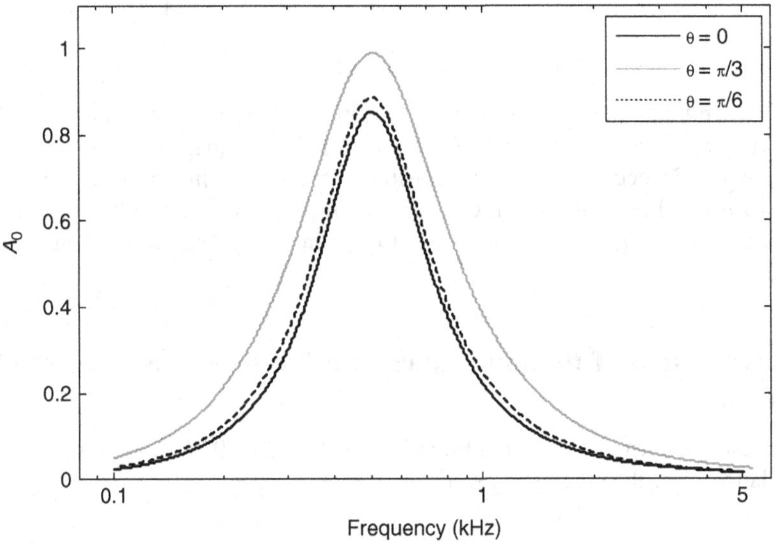

Figure 9.18 Absorption coefficient A_0 for the same configuration as in Figure 9.17.

9.4.4 Evaluation of the surface impedance at oblique incidence for the case of square perforations

At oblique incidence, Equation (9.54) becomes

$$Z(B') = \frac{1}{\phi(L)} \sum_{n=0,\pm1\pm2\ldots} \frac{\sin^2\left(\dfrac{\pi n a}{D} - \dfrac{k_0 a \sin\theta}{2}\right)}{\left(\pi n - \dfrac{k_0 D}{2}\sin\theta\right)^2} Z_{0,n}(B)$$

$$+\frac{2D^2}{\phi(L)a^2}\sum_{m=1}^{\infty}\sum_{n=0,\pm1,\pm2\ldots}\frac{\sin^2\left(\frac{\pi ma}{D}\right)\sin^2\left(\frac{\pi na}{D}-k_0\frac{a}{2}\sin\theta\right)}{\pi^4m^2\left(n-\frac{k_0D\sin\theta}{2\pi}\right)^2}Z_{m,n}(B) \tag{9.85}$$

The added external length ε_e can be evaluated in the same way as for circular apertures. The dependence on the incidence angle of ε_e can be neglected and Equation (9.31) can be used at oblique incidence. The impedance Z in the free air close to the facing is given by

$$Z=\frac{1}{s}[Z(B')+j\omega\rho_0(\varepsilon_e+d)] \tag{9.86}$$

References

Bolt, R.H. (1947) On the design of perforated facings for acoustic materials. *J. Acoust. Soc. Amer.*, **19**, 917–21.

Brillouin, J. (1949) Théorie de l'absorption du son par les structures à panneaux perforés. *Cahiers du Centre Scientifique et Technique du Bâtiment*, **31**, 1–15.

Byrne, K.P. (1980) Calculation of the specific normal impedance of perforated facing–porous backing constructions. *Applied Acoustics*, **13**, 43–55.

Callaway, D.B. and Ramer, L.G. (1952) The use of perforated facings in designing low frequency resonant absorbers. *J. Acoust. Soc. Amer.*, **24**, 309–12.

Collin, B. (1960) *Field Theory of Guided Waves*, McGraw-Hill, New York

Davern, W.A. (1977) Perforated facings backed with porous materials as sound absorbers-An experimental study. *Applied Acoustics*, **10**, 85–112.

Guignouard, P., Meisser, M., Allard, J.F., Rebillard, P. and Depollier, C. (1991) Prediction and measurement of the acoustical impedance and absorption coefficient at oblique incidence of porous layers with perforated facings. *Noise Control Eng. J.*, **36**, 129–35.

Ingard, U. (1953) On the theory and design of acoustic resonators. *J. Acoust. Soc. Amer.*, **25**, 1037–61.

Ingard, U. (1954) Perforated facing and sound absorption. *J. Acoust. Soc. Amer.*, **26**, 151–4.

Ingard, U. (1968) Absorption characteristics of non linear acoustic resonators. *J. Acoust. Soc. Amer.*, **44**, 1155–6.

Ingard, U. (1970) Non linear distorsion of sound transmitted through an orifice. *J. Acoust. Soc. Amer.*, **48**, 32–3.

Ingard, U. and Bolt, R.H. (1951) Absorption characteristics of acoustic material with perforated facing. *J. Acoust. Soc. Amer.*, **23**, 533–40.

Ingard, U. and Ising, H. (1967) Acoustic nonlinearity of an orifice. *J. Acoust. Soc. Amer.*, **42**, 6–17.

Ingard, U. and Labate, S. (1950) Acoustic circulation effects and the non linear impedance of orifices. *J. Acoust. Soc. Amer.*, **22**, 211–18.

Nielsen, A.K. (1949) *Trans Danish Acad. Tech. Sci. No. 10*.

Norris, A.N. and Sheng, I.C. (1989) Acoustic radiation from a circular pipe with infinite flange. *J. Sound Vib.*, **135**, 85–93.

Rayleigh, J.W.S. (1940) *Theory of Sound*. MacMillan, London, Vol. II, p. 318.

Velizhanina, K.A. (1968) Design calculation of sound absorbers comprising a porous material with a perforated facing. *Soviet Physics-Acoustics*, **14**, 37–41.

Zwikker, C. and Kosten, C.W. (1949) *Sound Absorbing Materials*. Elsevier, New York.

10

Transversally isotropic poroelastic media

10.1 Introduction

Fibrous media, glass wools and rockwools, are strongly anisotropic. Currently, due to the manufacturing process, they can be considered as transversally isotropic media (Tarnow 2005), with an axis of symmetry perpendicular to the faces of the layers. It has been shown by Melon *et al.* (1998) that porous foams can also present a noticeable anisotropy which is more complex, but in a first approach, the frame in vacuum can be considered as a transversally isotropic elastic medium. A more precise description is given by Liu and Liu (2004) with orthotropy, but it will not be easy to measure the numerous parameters used in the description. In an anisotropic medium the wave numbers of the different waves depend on the direction of propagation, or for instance on the wave number vector components at the surface of the medium. The source can often be defined by a spectrum of wave number components at the surface of the medium, and this last characterization of the wave number components or of the slowness components simplifies the prediction of the induced displacement field. In Sections (10.2)–(10.5) wave propagation in transversally isotropic media is described in a similar way as in the works by Sharma and Gogna (1991), and Vashishth and Khurana (2004), where the different waves are defined for given wave number components at the surface and where the symmetry axis is perpendicular to the surface. Sections (10.5)–(10.6) are concerned with the field created by a sound source in air and by a mechanical excitation, and with rigid-framed media. In Section (10.7), the case of a transversally isotropic porous medium with the axis of symmetry in an orientation different from the normal to the faces is considered. The Rayleigh surface waves in directions parallel and perpendicular to the axis of symmetry are described in Section (10.8). In Section (10.9) a matrix representation of transversally isotropic porous layers is given.

Propagation of Sound in Porous Media: Modelling Sound Absorbing Materials, Second Edition J. F. Allard and N. Atalla

10.2 Frame in vacuum

Stress–strain relations

The Z-axis is assumed to be the axis of rotational material symmetry. Let X and Y be two orthogonal axes in the symmetry plane. Let $\boldsymbol{u}^s$ be the solid displacement vector, the strain components are defined by $e_{ij} = 1/2(\partial u_i^s/\partial x_j + \partial u_j^s/\partial x_i)$. The stress-strain relations for the frame in vacuum can be written

$$\hat{\sigma}_{xx} = (2G + A)e_{xx} + Ae_{yy} + Fe_{zz} \tag{10.1}$$

$$\hat{\sigma}_{yy} = Ae_{xx} + (2G + A)e_{yy} + Fe_{zz} \tag{10.2}$$

$$\hat{\sigma}_{zz} = Fe_{xx} + Fe_{yy} + Ce_{zz} \tag{10.3}$$

$$\hat{\sigma}_{yz} = 2G'e_{yz} \tag{10.4}$$

$$\hat{\sigma}_{xz} = 2G'e_{xz} \tag{10.5}$$

$$\hat{\sigma}_{xy} = 2Ge_{xy} \tag{10.6}$$

where A, F, G, C and G', are the rigidity coefficients (see Equation 1.23 for a comparison with the isotropic case where $G = G'$, $F = A, C = A + 2G$). Using the engineering notation as in the work by Cheng (1997), the Young moduli are denoted as E_x, E_y, E_z and can be rewritten $E_x = E_y = E$, $E_z = E'$. The Poisson ratios are denoted as ν_{yx}, ν_{zy} and ν_{zx} and satisfy the relations $\nu_{yx} = \nu$, $\nu_{zy} = \nu_{zx} = \nu'$. The coefficients A, F, C and G are related to the new coefficients by

$$A = \frac{E(E'\nu + E\nu'^2)}{(1+\nu)(E' - E'\nu - 2E\nu'^2)} \tag{10.7}$$

$$F = \frac{EE'\nu'}{E' - E'\nu - 2E\nu'^2} \tag{10.8}$$

$$C = \frac{E'^2(1-\nu)}{E' - E'\nu - 2E\nu'^2} \tag{10.9}$$

$$G = \frac{E}{2(1+\nu)} \tag{10.10}$$

Waves in a transversally isotropic elastic medium

As for isotropic porous media, if the wavelength is much larger than the characteristic length of the representative elementary volume, the elastic frame in vacuum can be replaced by a homogeneous elastic medium with the same rigidity constants. The plane waves that propagate in a transversally isotropic elastic layer have been described by Royer and Dieulesaint (1996). The meridian plane contains the Z axis and the wave number vector. Let X be the direction perpendicular to Z in the meridian plane. With the notations of Vashishth and Khurana (2004), the space and time dependence is

$$\exp\left[j\omega\left(t - \frac{x}{c} - qz\right)\right]$$

The waves have a given slowness component $1/c$ in the X direction. The symbol q is used to denote the z slowness vector component. This should not lead to any confusion with the dynamic viscous permeability defined in Chapter 5 which does not appear in this chapter except in Table 10.1. Two kinds of waves exist with a polarization in the meridian plane, a pseudo-compressional wave (pseudo-P wave) and a pseudo-shear wave (pseudo-SV wave). The components of the slowness vector must satisfy the relation (see Equation 4.49 in Royer and Dieulesaint 1996)

$$2\rho_1 = G'\left(q^2 + \frac{1}{c^2}\right) + \frac{2G+A}{c^2} + Cq^2 \pm \left[\left(\frac{2G+A-G'}{c^2} + (G'-C)\,q^2\right)^2 + 4\,(F+G')^2\,\frac{q}{c}\right]^{1/2} \tag{10.11}$$

where ρ_1 is the density of the porous frame. The slowest wave is the pseudo-SV wave.

A transverse wave polarized perpendicular to the meridian plane (SH wave) can also propagate with components of the slowness vector satisfying the relation (see Equation 4.47 in Royer and Dieulesaint 1996)

$$\rho_1 = \frac{G}{c^2} + G'q^2 \tag{10.12}$$

The pseudo-transverse wave is purely transverse when the wave number vector is parallel to the axis of symmetry or perpendicular to this axis. The pseudo-compressional wave is purely compressional in the same conditions. It can be guessed that, if the frame is much heavier than air, three frame-borne waves similar to the P wave, the SV wave and the SH wave must propagate in the air saturated porous medium, plus an air borne wave similar to the wave which propagates in the air saturating the motionless frame.

10.3 Transversally isotropic poroelastic layer

10.3.1 Stress–strain equations

The bulk modulus and the acoustical parameters related to the bulk modulus, thermal characteristic length and viscous permeability, are scalar. As in Chapter 6, the material the frame is made of is supposed to be incompressible, and the stress–strain relations for the saturating air are given by Equation (6.3) with Q given by Equation (6.25) and R given by Equation (6.26)

$$\sigma_{ij}^f = -\phi p \delta_{ij} = K_f[\phi\theta_f + (1-\phi)\theta_s] \tag{10.13}$$

where θ_f is the air dilatation and θ_s is the frame dilatation. The second representation of the Biot (1962) theory (Appendix 6.A) is used. In this representation, the pressure and the total stress tensor acting on the porous medium are used. The total stress tensor is the sum of three tensors: the stress tensor of the frame in vacuum considered as an elastic solid, whose components $\hat{\sigma}_{ij}^s$ are given by Equations (10.1)–(10.6); the stress tensor components $\sigma_{ij}^f = -\phi p \delta_{ij}$; and the stress components $-p(1-\phi)\delta_{ij} = (1-\phi)\sigma_{ij}^f/\phi$ added to the frame stress tensor by the transmission of the pressure into the frame. The components $[1 + (1 - \phi)/\phi]\sigma_{ij}^f = -p\delta_{ij}$ are added to the stress tensor of the frame in vacuum to obtain the total stress tensor. The frame displacement vector $\boldsymbol{u}^s$ and the fluid-discharge displacement vector $\boldsymbol{w} = \phi(\boldsymbol{u}^f - \boldsymbol{u}^s)$ where $\boldsymbol{u}^f$ is the fluid displacement vector are used instead of $\boldsymbol{u}^s$ and $\boldsymbol{u}^f$. Using $\zeta = -\nabla.\mathbf{w}$, the total stress components and the pressure are given by

$$\sigma_{xx}^t = (2G + A)e_{xx} + Ae_{yy} + Fe_{zz} + \frac{\theta_s - \zeta}{\phi} K_f \tag{10.14}$$

$$\sigma_{yy}^t = Ae_{xx} + (2G + A)e_{yy} + Fe_{zz} + \frac{\theta_s - \zeta}{\phi} K_f \tag{10.15}$$

$$\sigma_{zz}^t = Fe_{xx} + Fe_{yy} + Ce_{zz} + \frac{\theta_s - \zeta}{\phi} K_f \tag{10.16}$$

$$\sigma_{yz}^t = 2G' e_{yz} \tag{10.17}$$

$$\sigma_{xz}^t = 2G' e_{xz} \tag{10.18}$$

$$\sigma_{xy}^t = 2Ge_{xy} \tag{10.19}$$

$$-p = \frac{K_f(\theta_s - \zeta)}{\phi} \tag{10.20}$$

where K_f is the bulk modulus of the saturating air. Equations (10.14)–(10.20) can be rewritten

$$\sigma_{xx}^t = (2B_1 + B_2)e_{xx} + B_2e_{yy} + B_3e_{zz} + B_6\zeta \tag{10.21}$$

$$\sigma_{yy}^t = B_2e_{xx} + (2B_1 + B_2)e_{yy} + B_3e_{zz} + B_6\zeta \tag{10.22}$$

$$\sigma_{zz}^t = B_3e_{xx} + B_3e_{yy} + B_4e_{zz} + B_7\zeta \tag{10.23}$$

$$\sigma_{xz}^t = 2B_5e_{xz} \tag{10.24}$$

$$\sigma_{yz}^t = 2B_5e_{yz} \tag{10.25}$$

$$\sigma_{xy}^t = 2B_1e_{xy} \tag{10.26}$$

$$p = B_6e_{xx} + B_6e_{yy} + B_7e_{zz} + B_8\zeta \tag{10.27}$$

where

$$B_1 = G,\ B_2 = A + K_f/\phi,\ B_3 = F + K_f/\phi,\ B_4 = C + K_f/\phi,$$
$$B_5 = G',\ B_6 = B_7 = -B_8 = -K_f/\phi.$$

10.3.2 Wave equations

It has been shown by Sanchez-Palencia (1980) that, for anisotropic porous media, the dynamic viscous permeability (or the dynamic tortuosity) are symmetrical tensors of the second order. This is also true for the different parameters that describe the visco-inertial interactions between both phases. These tensors are diagonal in three orthogonal directions.

For a transversally isotropic medium, one of these directions is the axis of symmetry Z. Any couple of orthogonal axes X and Y in the symmetry plane can complete the set. The superscript i, $i = x, y, z$ will be used to define the diagonal elements $\sigma^i, \Lambda^i, \alpha_\infty^i, \tilde{\rho}_{12}^i, \tilde{\rho}_{22}^i$. The viscous and inertial interactions between both phases when velocities are parallel to a direction where the parameters are diagonal are the same as for an isotropic medium with the parameters equal to the diagonal elements. The wave equations in the first representation of the Biot theory can be written

$$\frac{\partial \sigma_{xi}^s}{\partial x} + \frac{\partial \sigma_{yi}^s}{\partial y} + \frac{\partial \sigma_{zi}^s}{\partial z} = -\omega^2(\tilde{\rho}_{11}^i u_i^s + \tilde{\rho}_{12}^i u_i^f) \tag{10.28}$$

$$\frac{\partial \sigma_{xi}^f}{\partial x} + \frac{\partial \sigma_{yi}^f}{\partial y} + \frac{\partial \sigma_{zi}^f}{\partial z} = -\omega^2(\tilde{\rho}_{22}^i u_i^f + \tilde{\rho}_{12}^i u_i^s) \tag{10.29}$$

$$i = x, y, z$$

where

$$\tilde{\rho}_{22}^i = \phi\rho_0 - \tilde{\rho}_{12}^i \tag{10.30}$$

$$\tilde{\rho}_{11}^i = \rho_1 - \tilde{\rho}_{12}^i \tag{10.31}$$

In these equations ρ_0 is the density of air, ϕ is the porosity, and $\tilde{\rho}_{12}$ is defined similarly as in Chapter 6. Using Equations (10.28)–(10.29) with the stresses and the strains of the second representation gives the following wave equations (to alleviate the presentation we use the equivalent notations: $\boldsymbol{\sigma}_{ij,k} = \partial\boldsymbol{\sigma}_{ij}/\partial x_k, i, j, k = 1 \to x, 2 \to y, 3 \to z$)

$$\boldsymbol{\sigma}_{1i,1}^t + \boldsymbol{\sigma}_{2i,2}^t + \boldsymbol{\sigma}_{3i,3}^t = -\omega^2(\rho u_i^s + \rho_0 w_i) \tag{10.32}$$

where $\rho = \rho_1 + \phi\rho_0$, and

$$-p_{,i} = -\omega^2\left[\rho_0 u_i^s + \frac{\tilde{\rho}_{22}^i}{\phi^2} w_i\right] \tag{10.33}$$

Using Equation (5.50), this equation can be rewritten

$$-p_{,i} = -\omega^2(\rho_0 u_i^s + \mathrm{C}_i w_i) \tag{10.34}$$

where

$$C_i = \rho_0 \frac{\alpha_\infty^i}{\phi} + \sigma_i \frac{G_i(\omega)}{j\omega}$$

is related to the effective density in the direction i by $C_i = \rho_{ef}^i/\phi$.

10.4 Waves with a given slowness component in the symmetry plane

10.4.1 General equations

The meridian plane is defined in the same way as in Section (10.2). It contains the Z axis and the wave number vector. In this plane the direction perpendicular to Z is denoted by

X. The plane harmonic waves with a slowness $1/c$ in the direction X and a polarization in the meridian plane are described with the notations of Vashishth and Khurana (2004). The frame displacement components and the discharge displacement components are written

$$u_x^s = a_1 \exp\left[j\omega\left(t - \frac{x}{c} - qz\right)\right] \tag{10.35}$$

$$u_y^s = a_2 \exp\left[j\omega\left(t - \frac{x}{c} - qz\right)\right] \tag{10.36}$$

$$u_z^s = a_3 \exp\left[j\omega\left(t - \frac{x}{c} - qz\right)\right] \tag{10.37}$$

$$w_x = b_1 \exp\left[j\omega\left(t - \frac{x}{c} - qz\right)\right] \tag{10.38}$$

$$w_y = b_2 \exp\left[j\omega\left(t - \frac{x}{c} - qz\right)\right] \tag{10.39}$$

$$w_z = b_3 \exp\left[j\omega\left(t - \frac{x}{c} - qz\right)\right] \tag{10.40}$$

$$u_x^f = d_1 \exp\left[j\omega\left(t - \frac{x}{c} - qz\right)\right] \tag{10.41}$$

$$u_y^f = d_2 \exp\left[j\omega\left(t - \frac{x}{c} - qz\right)\right] \tag{10.42}$$

$$u_z^f = d_3 \exp\left[j\omega\left(t - \frac{x}{c} - qz\right)\right] \tag{10.43}$$

where $d_k = a_k + b_k/\phi$, $k = 1, 2, 3$. The symbol q is used to denote the z slowness vector component. The displacement components u_y and w_y, and the wave number component in the direction y, are equal to 0. The displacement components are solutions of a system of six equations which can be written

$$\begin{bmatrix} \rho - B_5 q^2 - \dfrac{2B_1 + B_2}{c^2} & -(B_3 + B_5)\dfrac{q}{c} & \dfrac{B_6}{c^2} + \rho_0 & B_6 \dfrac{q}{c} \\ -(B_3 + B_5)\dfrac{q}{c} & \rho - B_4 q^2 - \dfrac{B_5}{c^2} & B_7 \dfrac{q}{c} & \rho_0 + B_7 q^2 \\ \dfrac{B_6}{c^2} + \rho_0 & B_7 \dfrac{q}{c} & -\dfrac{B_8}{c^2} + C_1 & -B_8 \dfrac{q}{c} \\ B_6 \dfrac{q}{c} & \rho_0 + B_7 q^2 & -B_8 \dfrac{q}{c} & -B_8 q^2 + C_3 \end{bmatrix} \begin{bmatrix} a_1 \\ a_3 \\ b_1 \\ b_3 \end{bmatrix} = \begin{bmatrix} 0 \\ 0 \\ 0 \\ 0 \end{bmatrix} \tag{10.44}$$

$$\begin{bmatrix} \left(\dfrac{B_1}{c^2} + q^2 B_5 - \rho\right) & -\rho_0 \\ \rho_0 & C_1 \end{bmatrix} \begin{bmatrix} a_2 \\ b_2 \end{bmatrix} = 0 \tag{10.45}$$

The nondependence of a_2 and b_2 on a_1, a_3, b_1 and b_3 shows the decoupling of pseudo SH waves polarized in the direction y and the waves polarized in the meridian plane.

10.4.2 Waves polarized in a meridian plane

A nontrivial solution exists if the determinant of the square matrix of Equation (10.44) is equal to 0. Expansion of the determinant gives a cubic polynomial equation in q^2 which can be written with the notations of Vashishth and Khurana (2004)

$$T_0 q^6 + T_1 q^4 + T_2 q^2 + T_3 = 0 \tag{10.46}$$

The coefficients T_i are given in Appendix 10.A. For a given x slowness component $1/c$, Equation (10.46) gives three different q^2 and each q^2 is related to a wave propagating toward the increasing z with Re $q > 0$ and a wave propagating toward the decreasing z. The respective amplitudes a_1, a_3, b_1, b_3 can be obtained with the three first equations of the set (10.44). For instance, the waves can be normalized with a coefficient $N = a_3 + b_3$ which is the amplitude of the displacement component $(1-\phi)u_z^s + \phi u_z^f$. The displacement components are solutions of the following equations

$$\begin{aligned}&\left\{\rho - B_5 q^2 - \frac{2(B_1+B_2)}{c^2}\right\} a_1 - \left\{(B_3+B_5+B_6)\frac{q}{c}\right\} a_3 \\ &+ \left\{\frac{B_6}{c^2} + \rho_0\right\} b_1 = -B_6 \frac{q}{c} N\end{aligned} \tag{10.47}$$

$$\begin{aligned}&\left\{-(B_3+B_5)\frac{q}{c}\right\} a_1 + \left\{\rho - \rho_0 - (B_4+B_7)q^2 - \frac{B_5}{c^2}\right\} a_3 \\ &+ \left\{B_7 \frac{q}{c}\right\} b_1 = -N(\rho_0 + B_7 q^2)\end{aligned} \tag{10.48}$$

$$\begin{aligned}&\left\{\frac{B_6}{c^2} + \rho_0\right\} a_1 + \left\{(B_7+B_8)\frac{q}{c}\right\} a_3 \\ &+ \left\{-\frac{B_8}{c^2} + C_1\right\} b_1 = B_8 \frac{q}{c} N\end{aligned} \tag{10.49}$$

$$b_3 = N - a_3 \tag{10.50}$$

10.4.3 Waves with polarization perpendicular to the meridian plane

A nontrivial solution of Equation (10.45) exists if the determinant of the square matrix is equal to 0 and q^2 is given by

$$q^2 = \left(\rho - \frac{\rho_0^2}{C_1} - \frac{B_1}{c^2}\right) \Big/ B_5 \tag{10.51}$$

For instance, these waves can be normalized with $a_2 = N$. The displacement component b_2 is given by

$$b_2 = -N\rho_0 / C_1 \tag{10.52}$$

10.4.4 Nature of the different waves

For porous media with the frame much heavier than air, the partial decoupling has the same consequences for anisotropic and isotropic media. For isotropic media two kinds of Biot waves exist, the two frame-borne waves which are similar to the waves which propagate in the frame in vacuum, and the air-borne wave which is similar to the wave which propagates in the air saturating a rigid frame. Due to the different interactions between frame and air, both kinds of waves simultaneously propagate in the frame and in air, but the air-borne wave mainly propagates in air. For anisotropic porous media with the frame much heavier than the air, the Biot waves are of the same kind. More precisely, two waves polarized in the meridian plane are frame-borne waves and the third wave is an air-borne wave similar to that which propagates in a rigid framed porous medium which is described in Section (10.5.3). The wave polarized in a direction perpendicular to the meridian plane is the third frame-borne wave. Another consequence of the partial decoupling is shown in Section (10.8): the phase speeds of the Rayleigh waves for a usual glass wool are very close to the phase speeds for the frame in vacuum.

10.4.5 Illustration

A representation of the slowness as a function of the real angle θ between the direction of propagation and the axis of symmetry Z is given in Figures 10.1 and 10.2 for waves polarized in the meridian plane, and in Figure 10.3 for waves polarized perpendicular to the meridian plane. The porous medium is described by the parameters of Table 10.1. Similar parameters can describe a glass wool. The Lafarge model (Equation 5.35) is used for the bulk modulus and the Johnson *et al.* model (Equation 5.36) for the effective densities. The frequency is 0.5 kHz. Both slowness components q and $1/c$ are related to the slowness s and the real angle θ by

$$s \sin\theta = 1/c \tag{10.53}$$

$$s \cos\theta = q \tag{10.54}$$

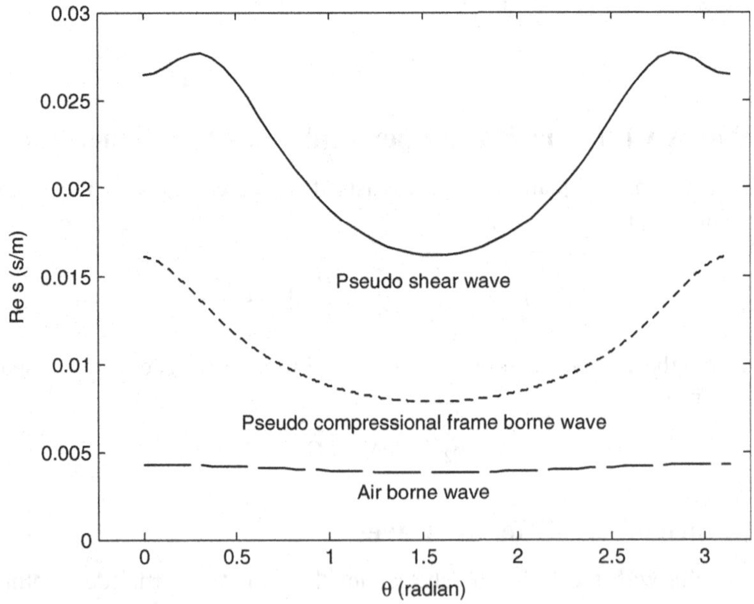

Figure 10.1 Re s as a function of θ for the waves polarized in the meridian plane.

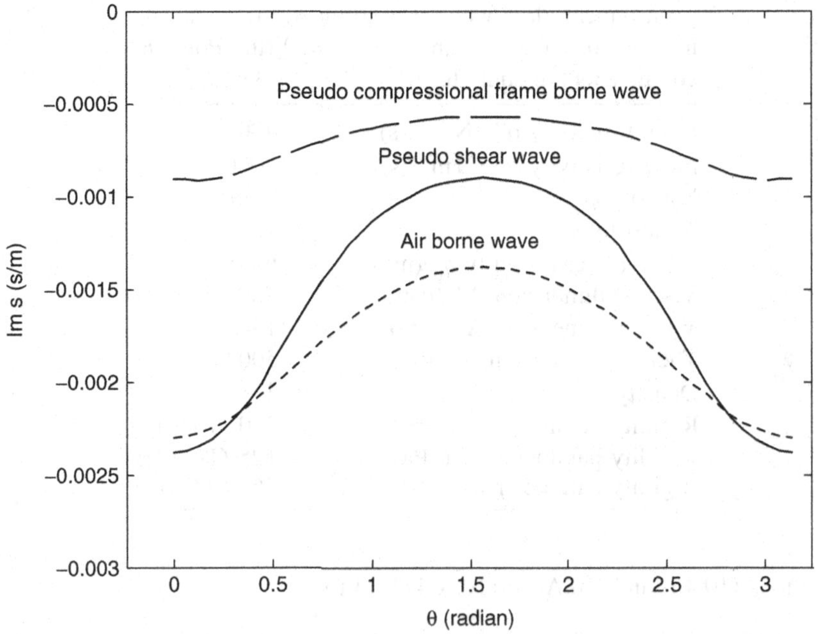

Figure 10.2 Im s as a function of θ for the waves polarized in the meridian plane.

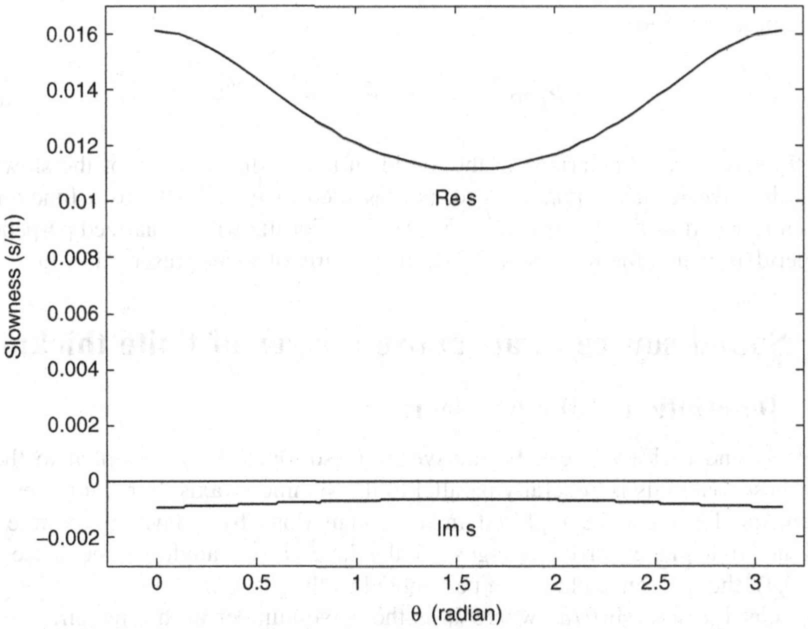

Figure 10.3 Slowness s as a function of θ for the wave polarized perpendicular to the meridian plane.

Table 10.1 Acoustical and mechanical parameters for a typical glass wool. The tortuosity is identical in directions x and z, and $F = A = 0$ (the Poisson coefficients are equal to 0).

Flow resistivity σ^x (N m^{-4} s)	4000
Flow resistivity σ^z (N m^{-4} s)	8000
Porosity ϕ	0.98
Tortuosity α_∞	1.1
Thermal permeability q'_0 (m^2)	6.10^{-9}
Viscous dimension Λ^x (μm)	200
Viscous dimension Λ^z (μm)	140
Thermal dimension Λ' (μm)	500
Density ρ_1 (kg/m^3)	32
Rigidity parameter G (kPa)	260 (1+0.1j)
Rigidity parameter G' (kPa)	125 (1+0.1j)
Rigidity parameter C (kPa)	46 (1+0.1j)

Equations (10.46) and (10.A.1)–(10.A.4) lead to

$$\begin{aligned} s^6[T_0\cos^6\theta + T_{12}\sin^2\theta\cos^4\theta + T_{23}\sin^4\theta\cos^2\theta + T_{34}\sin^6\theta] \\ +s^4[T_{11}\cos^4\theta + T_{22}\cos^2\theta\sin^2\theta + T_{33}\sin^4\theta] \\ +s^2[T_{21}\cos^2\theta + T_{32}\sin^2\theta] + T_{31} = 0 \end{aligned} \tag{10.55}$$

Equation (10.51) leads to

$$s^2[B_5\cos^2\theta + B_1\sin^2\theta] = \rho - \frac{\rho_0^2}{C_1} \tag{10.56}$$

For the three waves polarized in the meridian plane, the real part of the slowness s, which is the inverse of the phase speed, is presented in Figure 10.1 as a function of θ. The imaginary part of s is presented in Figure 10.2. For the wave polarized perpendicular to the meridian plane, the real and the imaginary parts of s are presented in Figure 10.3.

10.5 Sound source in air above a layer of finite thickness

10.5.1 Description of the problems

Glass wools and rockwools can be transversally isotropic and the normal to the large faces of these materials is generally parallel to the symmetry axis. For a sufficient lateral extension, the layer can be replaced in the calculations by a layer of infinite lateral dimensions. If a plane wave impinges on the layer at an angle of incidence θ (see Figure 10.4), the incidence plane can be denoted as the plane XZ and the slowness in the direction x is $1/c = k_0\sin\theta/\omega$, where k_0 is the wave number in the free air.

Due to the symmetry of the problem, only the waves polarized in the meridian plane exist in the layer. In a first step, the frame displacement created by the plane wave and the reflection coefficient of the layer are predicted. If the source in air is a point source,

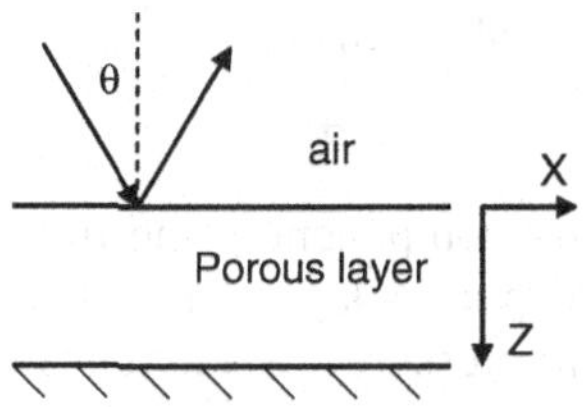

Figure 10.4 A plane wave in air impinges on a porous layer bonded on a rigid impervious backing with an angle of incidence θ. The axis of symmetry is parallel to the normal Z of the surface of the layer.

the results obtained for the plane wave can be used with the Sommerfeld representation as in Chapter 8.

10.5.2 Plane field in air

The layer has a finite thickness and is bonded onto a rigid impervious backing (see Figure 10.4). The boundary conditions are

$$\Sigma u_x^s = \Sigma u_z^s = \Sigma w_z = 0 \tag{10.57}$$

The summation is performed on all the waves at the bonded face. In what follows a superscript + is used for the waves propagating toward the backing and a superscript – for the waves propagating toward the free surface. At the free surface of the layer, $z = 0$ (see Figure 10.4) and the displacement components for a normalization factor $N = 1$ of the different waves are denoted as $a_1^{\pm}(i), a_3^{\pm}(i), b_1^{\pm}(i), b_3^{\pm}(i), i = 1, 2, 3$. The related z components of the slownesses are $q^{\pm}(i) = \pm q(i)$. As in Chapter 8, to avoid large systems of equations, each wave propagating from the free surface to the backing is associated with three waves propagating upward to satisfy the boundary conditions at the contact surface with the backing. In a similar way as for isotropic media, the normalization factors of the upward waves, related to a normalization factor $N = 1$ of each downward wave, are denoted as r_{ik} where i defines a downward wave and k one of the three associated upward waves. The coefficients r_{ik} satisfy the following equations

$$a_3^+(i) + \sum_{k=1,3} r_{ik} a_3^-(k) \exp[j\omega l(q(i) + q(k))] = 0 \tag{10.58}$$

$$a_1^+(i) + \sum_{k=1,3} r_{ik} a_1^-(k) \exp[j\omega l(q(i) + q(k))] = 0 \tag{10.59}$$

$$b_3^+(i) + \sum_{k=1,3} r_{ik} b_3^-(k) \exp[j\omega l(q(i) + q(k))] = 0 \tag{10.60}$$

where l is the thickness of the layer. Any superposition of the three downward waves, each one being associated with its three upward waves, satisfies the boundary conditions at the bonded face. At the surface in contact with air the boundary conditions can be written

$$v_z^e = j\omega[\Sigma u_z^s + \Sigma w_z] \tag{10.61}$$

$$\Sigma \sigma_{xz}^t = 0 \tag{10.62}$$

$$p^e = \Sigma p \tag{10.63}$$

$$-p^e = \Sigma \sigma_{zz}^t \tag{10.64}$$

where v_z^e and p^e are the z velocity component in the free air and the pressure in the free air at the contact surface with the porous layer. The coefficients N_i which satisfy the last three boundary conditions for unit pressure are solutions of the following set of equations

$$\sum_{i=1,2,3} N_i \left\{ q(i)a_1^+(i) + \frac{1}{c}a_3^+(i) - \sum_{k=1,2,3} \left(q(k)a_1^-(k) - \frac{1}{c}a_3^-(k) \right) r_{ik} \right\} = 0, \tag{10.65}$$

$$\begin{aligned} \sum_{i=1,2,3} N_i \Bigg\{ & \frac{1}{c}B_6a_1^+(i) + B_7q(i)a_3^+ - B_8\left(\frac{1}{c}b_1^+(i) + q(i)b_3^+(i)\right) \\ & + \sum_{k=1,2,3} r_{ik}\Bigg[\frac{1}{c}B_6a_1^-(k) - B_7q(k)a_3^-(k) \\ & - B_8\left(\frac{1}{c}b_1^-(k) - q(k)b_3^-(k)\right)\Bigg]\Bigg\} = -1/j\omega, \end{aligned} \tag{10.66}$$

$$\begin{aligned} \sum_{i=1,2,3} N_i \Bigg\{ & \frac{1}{c}B_3a_1^+(i) + B_4q(i)a_3^+ - B_7\left(\frac{1}{c}b_1^+(i) + q(i)b_3^+(i)\right) \\ & + \sum_{k=1,2,3} r_{ik}\Bigg[\frac{1}{c}B_3a_1^-(k) - B_4q(k)a_3^-(k) \\ & - B_7\left(\frac{1}{c}b_1^-(k) - q(k)b_3^-(k)\right)\Bigg]\Bigg\} = 1/j\omega. \end{aligned} \tag{10.67}$$

The coefficients N_1, N_2, N_3 are obtained from Equations (10.65)–(10.67). Using the relation $\boldsymbol{u}^s + \boldsymbol{w} = (1-\phi)\boldsymbol{u}^s + \phi\boldsymbol{u}^f$, the z component of velocity in the free air is given by

$$v_z^f = j\omega \sum_i N_i \left\{ (a_3^+(i) + b_3^+(i)) + \sum_k (a_3^-(k) + b_3^-(k))r_{ik} \right\} \tag{10.68}$$

and the surface impedance is given by

$$Z_s = \frac{p^e}{v_z^e} = 1 \Bigg/ \left[j\omega \sum_i N_i \left\{ (a_3^+(i) + b_3^+(i)) + \sum_k (a_3^-(k) + b_3^-(k))r_{ik} \right\} \right] \tag{10.69}$$

For unit pressure, the z component v_z^s of the frame velocity at the surface of the layer and the x component v_x^s are given by

$$v_z^s = j\omega \sum_{i=1,2,3} N_i \left\{ a_3^+(i) + \sum_{k=1,2,3} a_3^-(k) r_{ik} \right\} \tag{10.70}$$

$$v_x^s = j\omega \sum_{i=1,2,3} N_i \left\{ a_1^+(i) + \sum_{k=1,2,3} a_1^-(k) r_{ik} \right\} \tag{10.71}$$

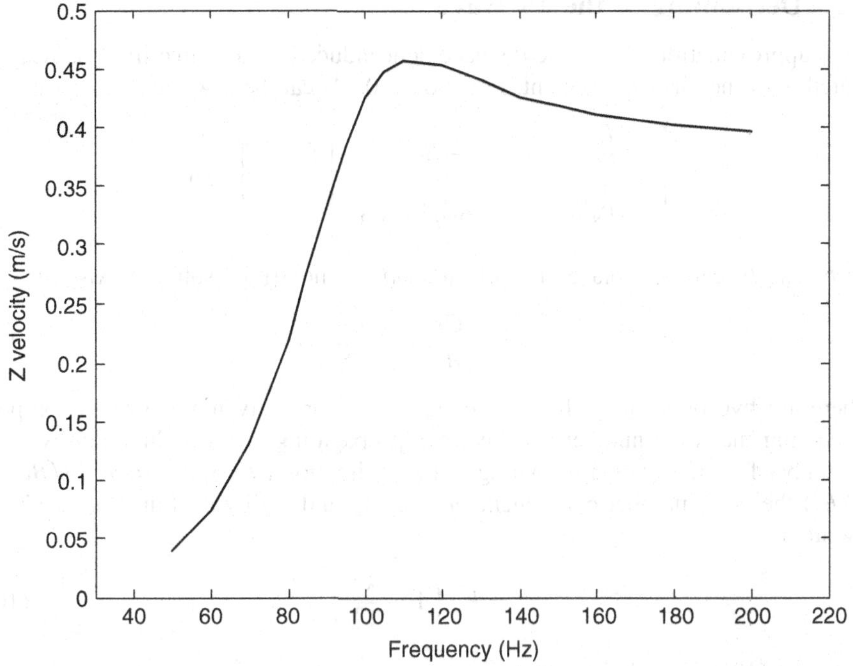

Figure 10.5 Normal surface frame velocity created by an incident unit pressure field.

For a unit incident wave, the acoustic pressure field at the contact surface with air is multiplied by $(1 + V)$, V being the reflection coefficient given by

$$V = \left(\frac{Z_s \cos\theta}{Z_0} - 1\right) \Big/ \left(\frac{Z_s \cos\theta}{Z_0} + 1\right) \tag{10.72}$$

and the velocity components given by Equations (10.70) and (10.71) must also be multiplied by $(1 + V)$. The modulus of the vertical velocity of the frame at the free surface of a layer of the material of Table 10.1 is given in Figure 10.5 as a function of frequency. The Lafarge model (Equation 5.35) is used for the bulk modulus and the Johnson *et al.* model (Equation 5.36) for the effective densities in the evaluation of the coefficients C_i. The thickness of the layer is $l = 10$ cm. The incident field is a unit pressure field at normal incidence.

There is a peak in the velocity distribution close to 110 Hz. This peak is due to the compressional quarter wavelength resonance of the frame. For the frame in vacuum, the speed of the compressional wave which propagates in the direction of the axis of symmetry is equal to $1/\mathrm{Re}(\rho_1/C)^{1/2}$, close to 120 m/s. The speed of the frame compressional Biot wave in the z direction is very close to this value. The peak is shifted toward high frequencies because, independently of the resonance, the vertical velocity induced by the unit field increases strongly with frequency.

For the case of a point source, the Sommerfeld representation can be used as in Chapter 8 to predict the frame displacement components.

10.5.3 Decoupling of the air wave

In a first approximation, the frame displacement induced by a source in air is negligible compared with the air displacement. Equation (10.44) can be rewritten

$$\begin{bmatrix} -\frac{B_8}{c^2} + C_1 & -B_8\frac{q}{c} \\ -B_8\frac{q}{c} & -B_8 q^2 + C_3 \end{bmatrix} \begin{bmatrix} b_1 \\ b_3 \end{bmatrix} = 0 \tag{10.73}$$

where b_1 and b_3 can be replaced by ϕd_1 and ϕd_3. A nontrivial solution exists if

$$q^2 = \frac{C_3}{B_8} - \frac{1}{c^2}\frac{C_3}{C_1} \tag{10.74}$$

There are two opposite z slowness components for a given x slowness component $1/c$. Denoting the wave number for the wave propagating in the x direction by k_P, the wave number for the waves propagating in the z direction by k_N, ($k_N^2 = \omega^2 C_3/B_8$, $k_P^2 = \omega^2 C_1/B_8$) the wave number components ω/c by k_x and ωq by k_z, Equation (10.74) can be rewritten

$$k_z^2 = k_N^2\left(1 - \frac{k_x^2}{k_P^2}\right) \tag{10.75}$$

Equation (10.27) can be rewritten

$$p = jB_8\phi\left(u_1^f k_x + u_3^f k_z\right) \tag{10.76}$$

The system of Equation (10.58)–(10.60) is replaced by

$$r_{11} = \exp(-2jk_z l) \tag{10.77}$$

The surface impedance Z_s can be written

$$\frac{p^e}{v_z^e} = \frac{j\phi B_8(d_1 k_x + d_3 k_z)(1 + \exp(-2jk_z l))}{j\omega\phi d_3(1 - \exp(-2jk_z l))} \tag{10.78}$$

From Equation (10.73), the ratio $d_1 k_x/(d_3 k_z)$ is given by

$$\frac{d_1 k_x}{d_3 k_z} = \frac{B_8/c^2}{C_1 - B_8/c^2} \tag{10.79}$$

which leads to

$$k_z\left(1 + \frac{d_1 k_x}{d_3 k_z}\right) = \frac{k_N}{\left(1 - \frac{k_x^2}{k_P^2}\right)^{1/2}} \tag{10.80}$$

The surface impedance can be rewritten

$$Z_s = \frac{(\rho_{ef}^z K_f)^{1/2}}{j\phi[1 - (k_x/k_P)^{1/2}]}\cot jlk_z \tag{10.81}$$

Equation (10.75) can be rewritten (see Equation (3.54))

$$\frac{k_z^2}{k_N^2} + \frac{k_x^2}{k_P^2} = 1 \tag{10.82}$$

Let θ_1 be the angle of the wave number vector $\boldsymbol{k}_1$ of components k_x and k_z with the Z axis. The angle θ_1 and the modulus of the wave number vector satisfy the relation

$$\frac{k_1^2 \cos^2\theta_1}{k_N^2} + \frac{k_1^2 \sin^2\theta_1}{k_P^2} = 1 \tag{10.83}$$

Using the asymptotic limit of the effective density at high frequency gives the following angular dependence for the apparent tortuosity $\alpha_\infty(\theta_1)$

$$\frac{\cos^2\theta_1}{\alpha_\infty^P} + \frac{\sin^2\theta_1}{\alpha_\infty^N} = \frac{1}{\alpha_\infty(\theta_1)} \tag{10.84}$$

where α_∞^P is the tortuosity in the plane (X, Y) and α_∞^N is the tortuosity in the direction Z. Tortuosity measurements at ultrasonic frequencies by Castagnede *et al.* (1998) show a good agreement with this angular dependence. Using the asymptotic limit of the effective density at low frequency gives the same relation for the apparent flow resistivity $\sigma(\theta_1)$

$$\frac{\cos^2\theta_1}{\sigma_P} + \frac{\sin^2\theta_1}{\sigma_N} = \frac{1}{\sigma(\theta_1)} \tag{10.85}$$

10.6 Mechanical excitation at the surface of the porous layer

The excitation is a unit stress field τ_{zz} applied at the free surface of the layer with a space dependence $\exp(-j\omega x/c)$. The boundary conditions at the free surface are given by Equations (10.61)–(10.63) and Equation (10.64) is replaced by

$$-p^e + \tau_{zz} = \sum \sigma_{zz}^t \tag{10.86}$$

The unit stress creates a plane wave in air with a spatial dependence $\exp[-j(\omega x/c - k_0 z\cos\theta)]$ where k_0 is the wave number in the free air and $\cos\theta$ is given by

$$\cos\theta = \pm\frac{\sqrt{(k_0/\omega)^2 - (1/c)^2}}{k_0/\omega} \tag{10.87}$$

The choice in the physical Riemann sheet corresponds to $\mathrm{Im}\cos\theta \le 0$. The pressure p^e and the velocity component v_z^e are related by

$$p^e = -v_z^e Z_0/\cos\theta \tag{10.88}$$

where Z_0 is the characteristic impedance in the free air. The new set of equations for the determination of the coefficients N_i is

$$\sum_{i=1,2,3} N_i \left\{ q(i)a_1^+(i) + \frac{1}{c}a_3^+(i) - \sum_{k=1,2,3} \left(q(k)a_1^-(k) - \frac{1}{c}a_3^-(k) \right) r_{ik} \right\} = 0 \quad (10.89)$$

$$\begin{aligned} &\sum_{i=1,2,3} N_i \left\{ \frac{1}{c}B_6 a_1^+(i) + B_7 q(i) a_3^+ - B_8 \left(\frac{1}{c} b_1^+(i) + q(i) b_3^+(i) \right) \right. \\ &\quad + \sum_{k=1,2,3} r_{ik} \left[\frac{1}{c} B_6 a_1^-(k) - B_7 q(k) a_3^-(k) \right. \\ &\quad \left. \left. - B_8 \left(\frac{1}{c} b_1^-(k) - q(k) b_3^-(k) \right) \right] \right\} = \frac{Z_0}{\cos\theta} \\ &\quad \times \sum_i N_i \left\{ (a_3^+(i) + b_3^+(i)) + \sum_k (a_3^-(k) + b_3^-(k)) r_{ik} \right\}, \end{aligned} \quad (10.90)$$

$$\begin{aligned} &\sum_{i=1,2,3} N_i \left\{ \frac{1}{c}B_3 a_1^+(i) + B_4 q(i) a_3^+ - B_7 \left(\frac{1}{c} b_1^+(i) + q(i) b_3^+(i) \right) \right. \\ &\quad + \sum_{k=1,2,3} r_{ik} \left[\frac{1}{c} B_3 a_1^-(k) - B_4 q(k) a_3^-(k) \right. \\ &\left. \left. - B_7 \left(\frac{1}{c} b_1^-(k) - q(k) b_3^-(k) \right) \right] \right\} = -\frac{1}{j\omega} - \frac{Z_0}{\cos\theta} \sum_i N_i \left\{ (a_3^+(i) + b_3^+(i)) \right. \\ &\quad \left. + \sum_k (a_3^-(k) + b_3^-(k)) r_{ik} \right\}. \end{aligned} \quad (10.91)$$

The displacement components a_i and b_i are given by Equations (10.47)–(10.50) and the frame displacement components at the free boundary are given by Equations (10.70) and (10.71). The velocity components induced by a circular field or by a line field can be obtained with the same expressions as in Chapter 8.

10.7 Symmetry axis different from the normal to the surface

10.7.1 Prediction of the slowness vector components of the different waves

The waves are defined by the slowness vector components q_x and q_y at the free surface of the porous layer. Let Z be the axis normal to the surface and Z' be the symmetry axis. The axis Y is chosen perpendicular to Z and Z'. There is no loss of generality because the y slowness component is not equal to 0 by definition as in the previous Sections. The two sets of axes are XYZ and $X'YZ'$ (see Figure 10.6). The stress–strain Equations (10.13)–(10.27) and the wave Equations (10.28)–(10.34) are always valid, but

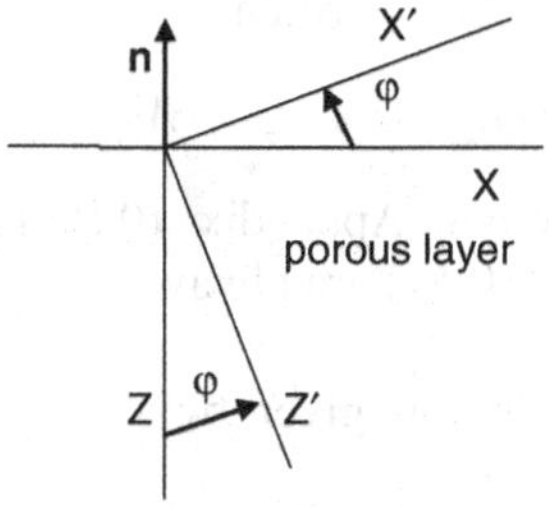

Figure 10.6 The axis of symmetry Z' and the new axis X' in a plane perpendicular to the surface of the layer. The axes Y and Y' are perpendicular to the figure.

the coordinates (x, y, z) must be replaced by (x', y, z') and the formalism of Section 10.5 must be modified.

The angle between Z and Z' is φ. As in the previous sections, the waves are characterized by a given slowness in the plane XY of the surface of the layer. The slowness components in this plane are $q_x = 1/c$ and q_y. The slowness vector components in both sets of axes are related by

$$q_{x'} = q_x \cos\varphi - q_z \sin\varphi \tag{10.92}$$

$$q_{y'} = q_y \tag{10.93}$$

$$q_{z'} = q_z \cos\varphi + q_x \sin\varphi \tag{10.94}$$

$$q_{x'} = \frac{q_x}{\cos\varphi} - q_{z'} \tan\varphi \tag{10.95}$$

The components q_x and q_y are defined in the description of the source in air or on the surface.

The slowness vector in the system $X'Y'Z'$ with Y' identical to Y is defined by the three components $q_{x'}$, $q_{y'}$ and $q_{z'}$. The new meridian plane is defined by the axis Z', and in the plane $x'y'$ by the vector $\boldsymbol{q}_{p'}$ of components $q_{x'}$ and $q_{y'}$ (see Figure 10.7).

The square modulus of $\boldsymbol{q}_{p'}$ is given by

$$q_{p'}^2 = q_y^2 + \frac{q_x^2}{\cos^2\varphi} - 2q_x \frac{\sin\varphi}{\cos^2\varphi} q_{z'} + q_{z'}^2 \tan^2\varphi \tag{10.96}$$

Let X'' be the axis parallel to $\boldsymbol{q}_{p'}$ and Y'' be the axis perpendicular to X'' and Z'. For the waves polarized in the new meridian plane Z', X'', Equation (10.42) is always valid, $1/c = q_x$ being replaced by $q_{x''} = q_{p'}$, and $q = q_z$ being replaced by $q_{z'}$. With this

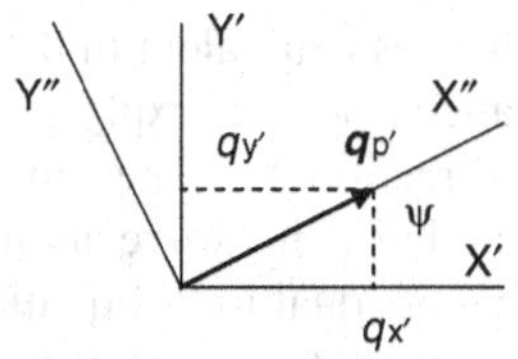

Figure 10.7 The vector $\boldsymbol{q}_{p'}$ in the plane $x'y'$.

substitution, Equation (10.46) can be rewritten

$$A_6 q_{z'}^6 + A_5 q_{z'}^5 + A_4 q_{z'}^4 + A_3 q_{z'}^3 + A_2 q_{z'}^2 + A_1 q_{z'} + A_0 = 0 \qquad (10.97)$$

The coefficients A_i are given in Appendix 10.B. For the waves polarized in the direction Y'', Equations (10.45), (10.51) are always valid with the same substitutions, 1/c by $q_{p'}$ and q by $q_{z'}$.

The slowness component $q_{z'}$ is now given by

$$q_{z'}^2(B_5 + B_1 \tan^2\varphi) - 2q_{z'} B_1 q_x \frac{\sin\varphi}{\cos^2\varphi} + \left[B_1\left(q_y^2 + \frac{q_x^2}{\cos^2\varphi}\right) - \left(\rho - \frac{\rho_0^2}{C_1}\right)\right] = 0 \qquad (10.98)$$

10.7.2 Slowness vectors when the symmetry axis is parallel to the surface

The previous model cannot be used. The axis X in the surface is chosen parallel to the axis of symmetry Z'. The slowness components $q_{z'} = q_x$ and q_y in the surface are known, and the components of the new vector $\boldsymbol{q}_{p'}$ perpendicular to the axis of symmetry Z' are q_y and q_z. Equation (10.46) can be rewritten

$$S_0 q_{P'}^6 + S_1 q_{P'}^4 + S_2 q_{P'}^2 + S_3 = 0 \qquad (10.99)$$

$$\begin{aligned} S_0 &= T_{34} \\ S_1 &= T_{23} q_x^2 + T_{33} \\ S_2 &= T_{12} q_x^4 + T_{22} q_x^2 + T_{32} \\ S_3 &= T_0 q_x^6 + T_{11} q_x^4 + T_{21} q_x^2 + T_{31} \end{aligned} \qquad (10.100)$$

The component q_z is given by

$$q_z = \pm(q_{P'}^2 - q_y^2)^{1/2} \qquad (10.101)$$

For the waves polarized perpendicular to the meridian plane, Equation (10.51) can be rewritten

$$q_{P'}^2 B_1 + q_{z'}^2 B_5 = \rho - \frac{\rho_0^2}{C_x} \qquad (10.102)$$

10.7.3 Description of the different waves

Let a_i'' and b_i'', $i = 1, 2, 3$, the quantities equivalent in the system X'', Y'', Z' to the a_i and b_i in the system X, Y, Z. For instance $u_{x''} = a_1'' \exp[j\omega(t - q_{x''}x'' - q_{z'}z')]$, where $u_{x''}$ is a displacement component in the direction X''. The same normalization as in the system X, Y, Z can be used for the a_i'' and b_i'', the wave number component in the direction perpendicular to the plane $Z'X''$ being equal to 0. Equations (10.47)–(10.50) and (10.52) can be used with the slowness components in the plane $X''Z'$. Each wave is associated with a specific system of axes X'', Y'', Z' and the boundary conditions are expressed in the system X, Y, Z. The coefficients N_j must be chosen for the boundary conditions to be satisfied. In a first step, the displacement components a_i' and b_i' are predicted in the system X', Y', Z' for each wave. Let ψ be the angle between the axes X'' and X' (see Figure 10.7). The angle ψ is defined by

$$\begin{aligned}\cos\psi &= q_{x'}/q_{P'}\\ \sin\psi &= q_{y'}/q_{P'}\end{aligned} \tag{10.103}$$

The displacement components in the system X', Y', Z' and the system X'', Y'', Z' are related by

$$\begin{aligned}a_1' &= a_1''\cos\psi - a_2''\sin\psi\\ a_2' &= a_1''\sin\psi + a_2''\cos\psi\\ a_3' &= a_3''\end{aligned} \tag{10.104}$$

The same relations are satisfied by the components b_i' and b_i''. The stress components can be evaluated in the system X', Y', Z' from the displacement components a_i', b_i' and the slowness vector components with Equations (10.21)–(10.27). In a last step, the contribution a_i and b_i of each wave to the total velocity are evaluated with the relations

$$\begin{aligned}a_1 &= a_1'\cos\varphi + a_3'\sin\varphi\\ a_2 &= a_2'\\ a_3 &= -a_1'\sin\varphi + a_3'\cos\varphi\end{aligned} \tag{10.105}$$

The same relations hold for the b_i and the b_i'. For each wave the total stress components in the system X, Y, Z and in the system X', Y', Z' are related by

$$\sigma_{zz}^t = \sigma_{x'x'}^t\sin^2\varphi - 2\sigma_{x'z'}^t\sin\varphi\cos\varphi + \sigma_{z'z'}^t\cos^2\varphi, \tag{10.106}$$

$$\begin{aligned}\sigma_{zx}^t = &-\sigma_{x'x'}^t\sin\varphi\cos\varphi + \sigma_{z'x'}^t(\cos^2\varphi - \sin^2\varphi)\\ &+\sigma_{z'z'}^t\sin\varphi\cos\varphi,\end{aligned} \tag{10.107}$$

$$\sigma_{zy}^t = -\sigma_{x'z'}^t\sin\varphi + \sigma_{x'y'}^t\cos\varphi \tag{10.108}$$

The unit stress field $\tau_{zz} = \exp(-j\omega(q_x x + q_y y))$ applied on the frame at the free surface of the layer creates a plane wave in air with a spatial dependence $\exp[-j(\omega q_x x + \omega q_y y - k_0 z\cos\theta)]$ where $\cos\theta$ is given by

$$\cos\theta = \pm\frac{\sqrt{(k_0/\omega)^2 - q_x^2 - q_y^2}}{k_0/\omega} \tag{10.109}$$

The choice in the physical Riemann sheet corresponds to $\mathrm{Im}\cos\theta \le 0$. The pressure p^e and the velocity component v_z^e are related by

$$p^e = -v_z^e Z_0/\cos\theta \tag{10.110}$$

where Z_0 is the characteristic impedance in the free air. If the predictions are restricted to the case of a semi-infinite layer, there are no upward waves and only the contributions in the porous medium of the four other waves must be taken into account. Let

$p(i), v_z(i), \sigma^t_{zz}(i), \sigma^t_{xz}(i), \sigma^t_{yz}(i), i = 1, 2, 3, 4$ be the contributions at the surface of the layer of the four waves with a normalization factor $N_i = 1$ to the pressure, the normal velocity and the related total stress components The four normalization coefficients are obtained from the set

$$-p^e + \tau_{zz} = \sum_{i=1,2,3,4} N_i \sigma^t_{zz}(i) \tag{10.111}$$

where

$$p^e = -j\omega \sum_{i=1,2,3,4} N_i (u_z(i) + w_z(i)) Z_0 / \cos\theta$$

$$\sum_{i=1,2,3,4} N_i \sigma^t_{xz}(i) = 0, \tag{10.112}$$

$$\sum_{i=1,2,3,4} N_i \sigma^t_{yz}(i) = 0, \tag{10.113}$$

$$-j\omega \sum_{i=1,2,3,4} N_i (u_z(i) + w_z(i)) Z_0 / \cos\theta = \sum_{i=1,2,3,4} N_i p(i) \tag{10.114}$$

The vertical displacement of the frame is $\sum N_i u_z(i)$.

10.8 Rayleigh poles and Rayleigh waves

When the symmetry axis Z' and the axis Z normal to the surface are parallel, the porous layer is invariant under rotations around Z. The Hankel transform can be used for circular excitations and the Sommerfeld representation for the point sources in air. When Z and Z' are not parallel, the invariance disappears, and Equation (7.5) cannot be replaced by the Sommerfeld integral. For similar reasons, the Hankel transform cannot be used for circular excitations. An excitation by a line source can be used to create Rayleigh waves which propagate in the direction perpendicular to the line source. The source can be a thin bar (blade) bonded on the porous surface on an axis $\mathbf{\Delta}$ which moves in the normal direction. A rough model for the source is a normal stress distribution applied to the frame, centered on $\mathbf{\Delta}$, and given by

$$\tau_{zz}(y, t) = g(r)h(t) \tag{10.115}$$

where r is the distance from the line source. The stress field can be rewritten

$$\tau_{zz}(r, t) = \frac{1}{4\pi^2} \int_{-\infty}^{\infty} \int_{-\infty}^{\infty} \tilde{h}(\omega)\tilde{g}(q_r) \exp(j\omega(t - q_r r))\omega \mathrm{d}q_r \mathrm{d}\omega \tag{10.116}$$

where $\tilde{h}$ and $\tilde{g}$ are Fourier tansforms given by

$$\tilde{h}(\omega) = \int_{-\infty}^{\infty} h(t) \exp(-j\omega t)\, \mathrm{d}t \tag{10.117}$$

$$\tilde{g}(q_r) = \int_{-\infty}^{\infty} g(r)\exp(i\omega q_r)\,\mathrm{d}r \tag{10.118}$$

The axes X and Y can be chosen with the origin on $\mathbf{\Delta}$. Let τ be the angle between Y and $\mathbf{\Delta}$. In Equation (10.116) $q_r r$ can be replaced by $q_x x + q_y y$, where $q_x = q_r \cos\tau$ and $q_y = q_r \sin\tau$. Equation (10.116) becomes

$$\tau_{zz}(r,t) = \frac{1}{4\pi^2}\int_{-\infty}^{\infty}\int_{-\infty}^{\infty} \tilde{h}(\omega)\tilde{g}(q_r)\exp(j\omega(t - q_x x - q_y y))\omega \mathrm{d}q_r \mathrm{d}\omega \tag{10.119}$$

The vertical displacement for a given direction $\mathbf{\Delta}$ of the source is a superposition of elementary displacements related to excitations with a space time dependence $\exp(j\omega(t - q_x x - q_y y))$. For a given ω let q_R be the the slowness at the Rayleigh pole. When $q_r \to q_R$ the elementary displacement tends to ∞. Iterative methods can be used to obtain the related slowness component in the plane XY. Many poles exist, and the iterations must start close to the true Rayleigh pole. For the usual porous frames of sound absorbing materials which are much heavier than air, the Rayleigh pole is close to the Rayleigh pole for the frame in vacuum. As an illustration the slowness q_R is predicted for the three geometries of Figure 10.8. For these geometries q_R for the frame in vacuum is a root of a polynomial in q_R^{-2}.

In Figure 10.8(a) the line source L is on the horizontal surface perpendicular to the axis of symmetry Z' and the Rayleigh wave propagates on this surface. In Figure 10.8(b) the Rayleigh wave propagates in a meridian plane in a direction perpendicular to Z', and in Figure 10.8(c) the Rayleigh wave propagates in a meridian plane in a direction parallel to Z'. The slowness component q_R on the free surface where the Rayleigh wave propagates is a root of a polynomial (see Equation 5.61 and Figure 5.1 for the definition of the free surface in Royer and Dieulesaint 1996) which can be rewritten with the present notations:

Case (a)

$$G'\left[(2G + A) - q_R^{-2}\rho_1\right]\left(q_R^{-2}\rho_1\right)^2 - C\left(G' - q_R^{-2}\rho_1\right)\left[(2G + A) - \frac{F^2}{C} - q_R^{-2}\rho_1\right]^2 = 0 \tag{10.120}$$

Case (b)

$$G\left[(2G + A) - q_R^{-2}\rho_1\right]\left(q_R^{-2}\rho_1\right)^2 - (2G + A)\left(G - q_R^{-2}\rho_1\right)\left[(2G + A) - \frac{A^2}{2G + A} - q_R^{-2}\rho_1\right]^2 = 0 \tag{10.121}$$

Case (c)

$$G'\left[C - q_R^{-2}\rho_1\right]\left(q_R^{-2}\rho_1\right)^2 - (2G + A)\left(G' - q_R^{-2}\rho_1\right)\left[C - \frac{F^2}{2G + A} - q_R^{-2}\rho_1\right]^2 = 0 \tag{10.122}$$

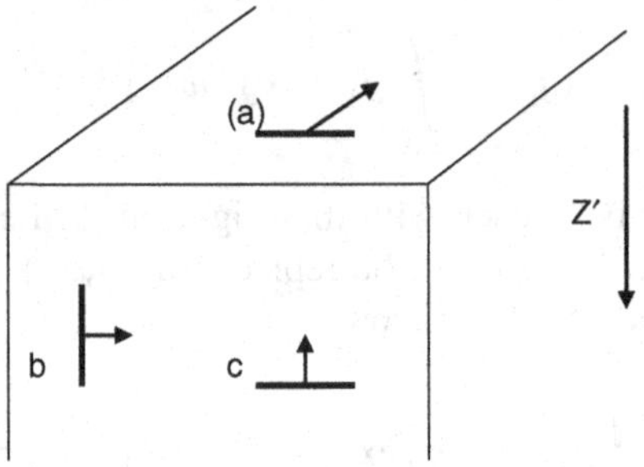

Figure 10.8 The line source (thick line) on a free surface perpendicular to Z' (a) and on a free surface parallel to Z' (b and c). The arrow indicates the direction of propagation on the free surface.

If the rigidity coefficients are real, the slowness component q_R is real and is located in the following intervals:

Case (a)

$$0 < q_R^{-2} < G', (2G + A) - \frac{F^2}{C} \tag{10.123}$$

Case (b)

$$0 < q_R^{-2} < G, (2G + A) - \frac{A^2}{2G + A} \tag{10.124}$$

Case (c)

$$0 < q_R^{-2} < G', C - \frac{F^2}{2G + A} \tag{10.125}$$

10.8.1 Example

The material is the glass wool described with the parameters of Table 10.1. A similar material was studied by Tarnow (2005). Predictions of the phase speed $1/\mathrm{Re}\ q_R$ and the damping $\mathrm{Im}\ q_R$ of the Rayleigh wave are given in Table 10.2 for the three cases defined in Figure 10.8 with Equations (10.120)–(10.122) for the frame in vacuum

The phase speed and the damping are represented in Figures 10.9 and 10.10 as a function of frequency for the porous medium and the frame in vacuum.

Table 10.2 Phase speed $1/\mathrm{Re}\ q_R$ and damping $\mathrm{Im}\ q_R$ for the Rayleigh wave for the frame in vacuum in the geometries a, b, c of Figure 10.8.

Geometry	Frame in vacuum Phase speed: $1/\mathrm{Re}\ q_R$ (m/s)	Frame in vacuum Damping: $\mathrm{Im}\ q_R$ (s/m)
a	51	-0.97×10^{-3}
b	79	-0.63×10^{-3}
c	36	-1.4×10^{-3}

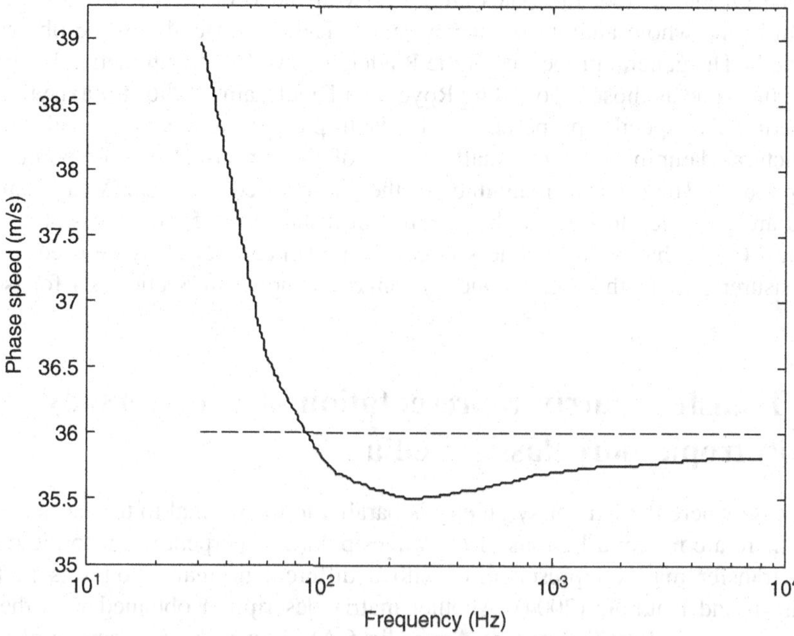

Figure 10.9 Phase speed $1/\text{Re}\ q_R$ of the Rayleigh wave in case c for the porous medium as a function of frequency (solid curve), and for the frame in vacuum (broken line).

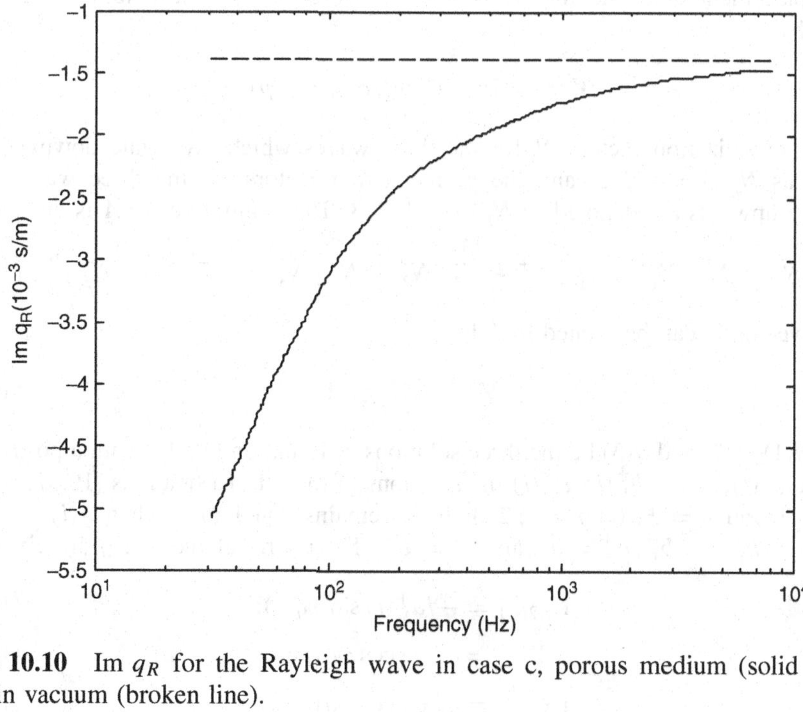

Figure 10.10 Im q_R for the Rayleigh wave in case c, porous medium (solid curve), frame in vacuum (broken line).

As for isotropic media, the poles are close with the Biot theory and for the frame in vacuum in the whole audible frequency range. The same trends can be observed for cases a and b. The general properties of the Rayleigh wave for axisymmetrical poroelastic media are the same as those reported by Royer and Dieulesaint (1996) for axisymmetrical elastic media. The specific properties, as for isotropic porous media, are related to the large structural damping and the small rigidity of the porous frames of porous sound absorbing media. The order of magnitude of the phase speed is 50 m/s. Measurements of the speed are possible close to the line source, at distances of 5 cm or less at acoustical frequencies. Layers having a thickness larger than 3 cm can generally be used for phase speed measurements in the high frequency range, as shown in Section 8.4 for isotropic media.

10.9 Transfer matrix representation of transversally isotropic poroelastic media

Only the case where the axis of symmetry is parallel to the normal to the faces is considered and there are no contributions of the waves polarized perpendicular to the meridian plane. A transfer matrix representation with a different normalization was performed by Vashishth and Khurana (2004). Another matrix description obtained with the Dazel representation of the Biot theory (see Appendix 6.A) is given by Khurana *et al.* (2009).

The transfer matrix

The displacement field and the stress field are described by the column vector $\boldsymbol{V}(z)$ defined by

$$\boldsymbol{V}(z) = [u_x^s, u_z^s, w_z, \sigma_{zz}^t, \sigma_{zx}^t, p]^T \tag{10.126}$$

The normalization factors N for the three waves which propagate downwards are denoted as $N_i^+, i = 1, 2, 3$, and the normalization factors for the three waves which propagate upwards are denoted as $N_i^-, i = 1, 2, 3$. The column vector $\boldsymbol{A}$ is defined by

$$\boldsymbol{A} = [N_1^+ + N_1^-, N_1^+ - N_1^-, N_2^+ + N_2^-, N_2^+ - N_2^-, N_3^+ + N_3^-, N_3^+ - N_3^-]^T \tag{10.127}$$

The vector $\boldsymbol{V}$ can be related to $\boldsymbol{A}$ by

$$\boldsymbol{V}(z) = [\Gamma(z)]\boldsymbol{A} \tag{10.128}$$

Let $q(1)$, $q(2)$ and $q(3)$ be the three solutions of Equation (10.46) with a positive real part. Let $a_1^{\pm}(i), a_3^{\pm}(i), b_1^{\pm}(i), b_3^{\pm}(i)$ the solutions of the set of Equations (10.47)–(10.50) for $N = 1$ and $q = \pm q(i), i = 1, 2, 3$. If N remains equal to 1 when $q(i) \to -q(i)$, $a_1^+ = -a_1^-, b_1^+ = -b_1^-, a_3^+ = a_3^-$ and $b_3^+ = b_3^-$. The matrix elements Γ_{kl} are given by

$$\Gamma_{1,2i-1} = -ja_1^+(i) \sin \omega q(i)z \tag{10.129}$$

$$\Gamma_{1,2i} = a_1^+(i) \cos \omega q(i)z \tag{10.130}$$

$$\Gamma_{2,2i-1} = a_3^+(i) \cos \omega q(i)z \tag{10.131}$$

$$\Gamma_{2,2i} = -ja_3^+(i)\sin\omega q(i)z \tag{10.132}$$

$$\Gamma_{3,2i-1} = b_3^+(i)\cos\omega q(i)z \tag{10.133}$$

$$\Gamma_{3,2i} = -jb_3^+(i)\sin\omega q(i)z \tag{10.134}$$

$$\begin{aligned}\Gamma_{4,2i-1} = {} & -B_7 j\frac{\omega}{c}jb_1^+(i)\sin\omega q(i)z + B_4 j\omega q(i)a_3^+(i)j\sin\omega q(i)z \\ & - B_7 j\omega q(i)b_3^+(i)j\sin\omega q(i)z + B_3 j\frac{\omega}{c}ja_1^+(i)\sin\omega q(i)z\end{aligned} \tag{10.135}$$

$$\begin{aligned}\Gamma_{4,2i} = {} & B_7 j\frac{\omega}{c}b_1^+(i)\cos\omega q(i)z - B_4 j\omega q(i)a_3^+(i)\cos\omega q(i)z \\ & + B_7 j\omega q(i)b_3^+(i)\cos\omega q(i)z - B_3 j\frac{\omega}{c}a_1^+(i)\cos\omega q(i)z\end{aligned} \tag{10.136}$$

$$\Gamma_{5,2i-1} = -B_5\left[\frac{j\omega}{c}a_3^+(i) + j\omega q a_1^+\right]\cos\omega q(i)z \tag{10.137}$$

$$\Gamma_{5,2i} = B_5\left[\frac{j\omega}{c}a_3^+(i) + j\omega q a_1^+\right]j\sin\omega q(i)z \tag{10.138}$$

$$\begin{aligned}\Gamma_{6,2i-1} = {} & -B_8 j\frac{\omega}{c}jb_1^+(i)\sin\omega q(i)z + B_7 j\omega q(i)a_3^+(i)j\sin\omega q(i)z \\ & - B_8 j\omega q(i)b_3^+(i)j\sin\omega q(i)z + B_6 j\frac{\omega}{c}ja_1^+(i)\sin\omega q(i)z\end{aligned} \tag{10.139}$$

$$\begin{aligned}\Gamma_{6,2i} = {} & B_8 j\frac{\omega}{c}b_1^+(i)\cos\omega q(i)z - B_7 j\omega q(i)a_3^+(i)\cos\omega q(i)z \\ & + B_8 j\omega q(i)b_3^+(i)\cos\omega q(i)z - B_6 j\frac{\omega}{c}a_1^+(i)\cos\omega q(i)z\end{aligned} \tag{10.140}$$

The vector $\boldsymbol{V}(l)$ at the upper surface is related to the vector $\boldsymbol{V}(0)$ at the lower surface of a layer of thickness l by

$$\boldsymbol{V}(l) = [\Gamma(l)][\Gamma(0)]^{-1}\boldsymbol{V}(0) \tag{10.141}$$

The transfer matrix for the porous layer in the second Biot representation is $[T] = [\Gamma(l)][\Gamma(0)]^{-1}$. All the components of V are equal at each side of the boundary between two porous layer with frames bonded together and the transfer matrix of a layered medium is the product of the transfer matrices of each layer.

Surface impedance $\mathbf{Z}_s$ of a layered porous medium

The porous medium is glued to a rigid impervious layer. Three components of $\boldsymbol{V}(l)$ are equal to 0, u_x, u_z, and w_z. In the free air close to the upper face, the z component of the air velocity v_z is related to the pressure by $p = Z_s v_z$ with $v_z = (w_z + u_z)j\omega$, the total stress component $\sigma_{zz}^t = -p$ and $\sigma_{zx}^t = 0$. The three components equal to 0 are related to the components of V at the upper surface by, respectively

$$0 = t_{11}u_x^s + (t_{12} - t_{13})u_z^s + \left(t_{13}\frac{1}{j\omega Z_s} - t_{14} + t_{16}\right)p \tag{10.142}$$

$$0 = t_{21}u_x^s + (t_{22} - t_{23})u_z^s + \left(t_{23}\frac{1}{j\omega Z_s} - t_{24} + t_{26}\right)p \tag{10.143}$$

$$0 = t_{31}u_x^s + (t_{32} - t_{33})u_z^s + \left(t_{33}\frac{1}{j\omega Z_s} - t_{34} + t_{36}\right)p \tag{10.144}$$

The components u_x^s, u_z^s, and p are different from 0 only if the determinant of the previous system of three equations is equal to 0. This leads to

$$Z_s = -\frac{\Delta_2}{j\omega\Delta_1} \tag{10.145}$$

with

$$\Delta_1 = \begin{vmatrix} t_{11} & t_{12} - t_{13} & t_{16} - t_{14} \\ t_{21} & t_{22} - t_{23} & t_{26} - t_{24} \\ t_{31} & t_{32} - t_{33} & t_{36} - t_{34} \end{vmatrix} \tag{10.146}$$

$$\Delta_2 = \begin{vmatrix} t_{11} & t_{12} - t_{13} & t_{13} \\ t_{21} & t_{22} - t_{23} & t_{23} \\ t_{31} & t_{32} - t_{33} & t_{33} \end{vmatrix} \tag{10.147}$$

Examples of applications can be found in Khurana *et al.* (2009) using a variant of the presented transfer matrix representation.

Appendix 10.A: Coefficients T_i in Equation (10.46)

Similar coefficients are given in the work by Vashishth and Khurana (2004).

$$T_0 = C_1B_5(B_7^2 - B_4B_8) \tag{10.A.1}$$

$$T_1 = T_{11} + T_{12}/c^2 \tag{10.A.2}$$

$$T_2 = T_{21} + T_{22}/c^2 + T_{23}/c^4 \tag{10.A.3}$$

$$T_3 = T_{31} + T_{32}/c^2 + T_{33}/c^4 + T_{34}/c^6 \tag{10.A.4}$$

$$\begin{aligned} T_{11} = {} & C_1(B_4B_8 - B_7^2)\rho + C_1\rho B_5B_8 \\ & + 2C_1\rho_0B_5B_7 + C_1C_3B_4B_5 - \rho_0^2(B_4B_8 - B_7^2) \end{aligned} \tag{10.A.5}$$

$$\begin{aligned} T_{12} = {} & C_1(2B_1 + B_2)(B_7^2 - B_4B_8) - C_3B_4B_5B_8 \\ & + C_3B_5B_7^2 + C_1(B_3^2B_8 + 2B_3B_5B_8) \\ & - C_1B_6(2B_3B_7 + 2B_5B_7 - B_4B_6) \end{aligned} \tag{10.A.6}$$

$$\begin{aligned} T_{21} = {} & -C_1\rho^2B_8 - 2C_1\rho\rho_0B_7 - C_3C_1\rho(B_4 + B_5) \\ & + C_1\rho_0^2B_5 + C_3\rho_0^2B_4 + \rho\rho_0^2B_8 + 2B_7\rho_0^3 \end{aligned} \tag{10.A.7}$$

$$\begin{aligned} T_{22} &= \rho(C_1 B_5 B_8 + C_3 B_4 B_8) + \rho C_1(2B_1 + B_2)B_8 \\ &\quad + 2C_1 B_7 \rho_0(2B_1 + B_2) - C_3 \rho B_7^2 + C_1 C_3 B_4(2B_1 + B_2) \\ &\quad + (C_3\rho - 4\rho_0^2)B_5 B_8 - 2\rho_0(B_3 + B_5)(C_1 B_6 + C_3 B_7) - 2\rho_0^2 B_3 B_8 \\ &\quad - C_1 C_3(B_3^2 + 2B_3 B_5) + 2C_3 \rho_0 B_4 B_6 + 2\rho_0^2 B_6 B_7 - C_1 \rho B_6^2 \end{aligned} \tag{10.A.8}$$

$$\begin{aligned} T_{23} &= C_3[-(B_3 + B_5)\{2B_6 B_7 - B_8(B_3 + B_5)\} + B_6^2 B_4 - B_8 B_5^2 \\ &\quad + (2B_1 + B_2)(B_7^2 - B_8 B_4)] - C_1[(2B_1 + B_2)B_5 B_8 - B_5 B_6^2] \end{aligned} \tag{10.A.9}$$

$$T_{31} = \rho^2 C_1 C_3 - \rho\rho_0^2(C_1 + C_3) + \rho_0^4 \tag{10.A.10}$$

$$\begin{aligned} T_{33} &= \rho C_3 B_8(B_5 + 2B_1 + B_2) + C_3 C_1 B_5(2B_1 + B_2) \\ &\quad + \rho_0^2[B_6^2 - B_8(2B_1 + B_2)] + C_3 B_6(2\rho_0 B_5 - \rho B_6) \end{aligned} \tag{10.A.11}$$

$$\begin{aligned} T_{32} &= (\rho_0^2 - C_3\rho)[\rho B_8 + (2B_1 + B_2)C_1 + 2\rho_0 B_6] \\ &\quad + B_5 C_3(\rho_0^2 - \rho C_1) \end{aligned} \tag{10.A.12}$$

$$T_{34} = C_3 B_5[B_6^2 - B_8(2B_1 + B_2)] \tag{10.A.13}$$

Appendix 10.B: Coefficients A_i in Equation (10.97)

The following symbols are used

$$\begin{aligned} G &= q_y^2 + \frac{q_x^2}{\cos^2\varphi} \\ H &= -2q_x \frac{\sin\varphi}{\cos^2\varphi} \\ I &= \tan^2\varphi \end{aligned} \tag{10.B.1}$$

Equation (10.96) can be rewritten

$$q_{P'}^2 = G + Hq_{z'} + Iq_{z'}^2 \tag{10.B.2}$$

Using the symbols

$$\begin{aligned} J &= G^2 \\ K &= 2GH \\ L &= H^2 + 2GI \\ M &= 2HI \\ N &= I^2 \end{aligned} \tag{10.B.3}$$

$q_{P'}^4$ is given by

$$q_{P'}^4 = J + Kq_{z'} + Lq_{z'}^2 + Mq_{z'}^3 + Nq_{z'}^4 \tag{10.B.4}$$

Using the symbols

$$\begin{aligned}
Pa &= G^3 \\
Pb &= 3HG^2 \\
Pc &= 3GH^2 + 3G^2 I \\
Pd &= 6GHI + H^3 \\
Pe &= 3GI^2 + 3H^2 I \\
Pf &= 3HI^2 \\
Pg &= I^3
\end{aligned} \tag{10.B.5}$$

$q_{P'}^6$ can be written

$$q_{P'}^6 = Pa + q_{z'}Pb + q_{z'}^2 Pc + q_{z'}^3 Pd + q_{z'}^4 Pe + q_{z'}^5 Pf + q_{z'}^6 Pg \tag{10.B.6}$$

The parameters T_1, T_2, T_3 are now given by

$$T_1 = T_{11} + T_{12} q_{P'}^2 \tag{10.B.7}$$

$$T_2 = T_{21} + T_{22} q_{P'}^2 + T_{23} q_{P'}^4 \tag{10.B.8}$$

$$T_3 = T_{31} + T_{32} q_{P'}^2 + T_{33} q_{P'}^4 + T_{34} q_{P'}^6 \tag{10.B.9}$$

The slowness component $q_{z'}$ satisfies the following equation

$$A_6 q_{z'}^6 + A_5 q_{z'}^5 + A_4 q_{z'}^4 + A_3 q_{z'}^3 + A_2 q_{z'}^2 + A_1 q_{z'} + A_0 = 0 \tag{10.B.10}$$

where

$$A_6 = T_0 + T_{12} I + T_{23} N + T_{34}\, Pg \tag{10.B.11}$$

$$A_5 = T_{12} H + T_{23} M + T_{34}\, Pf \tag{10.B.12}$$

$$A_4 = T_{11} + T_{22} I + T_{33} N + T_{12} G + T_{23} L + T_{34} Pe \tag{10.B.13}$$

$$A_3 = T_{22} H + T_{33} M + T_{23} K + T_{34} Pd \tag{10.B.14}$$

$$A_2 = T_{21} + T_{32} I + T_{22} G + T_{33} L + T_{23} J + T_{34} Pc \tag{10.B.15}$$

$$A_1 = T_{32} H + T_{33} K + T_{34} Pb \tag{10.B.16}$$

$$A_0 = T_1 + T_{32} G + T_{33} J + T_{34}\, Pa \tag{10.B.17}$$

References

Biot, M. A. (1962) Generalized theory of acoustic propagation in porous dissipative media. *J. Acoust. Soc. Amer.* **34**, 1254–1264.

Castagnede, B., Aknine, A., Melon, M. and Depollier, C. (1998) Ultrasonic characterization of the anisotropic behavior of air-saturated porous materials. *Ultrasonics* **36**, 323–343.

Cheng, A. H. D. (1997) Material coefficients of anisotropic poroelasticity. *Int. J. Rock Mech. Min. Sci.* **34**, 199–205.

Khurana, P., Boeckx L., Lauriks, W., Leclaire, P., Dazel, O. and Allard, J. F. (2009) A description of transversely isotropic sound absorbing porous materials by transfer matrices *J. Acoust. Soc. Amer*. **125**, 915–921.

Liu, K. and Liu, Y. (2004) Propagation characteristic of Rayleigh waves in orthotropic fluid-saturated porous media. *J. Sound Vib*. **271**, 1–13.

Melon, M., Mariez, E., Ayrault, C. and Sahraoui, S. (1998) Acoustical and mechanical characterization of anisotropic open-cell foams. *J. Acoust. Soc. Amer* **104**, 2622–2627.

Royer, D. and Dieulesaint, E. (1996) *Ondes Elastiques dans les Solides, Tome 1, Masson, Paris*.

Sanchez-Palencia, E. (1980) *Non-homogeneous Media and Vibration Theory*. Springer-Verlag (1980) p.151.

Sharma, M. D. and Gogna, M. L. (1991) Wave propagation in anisotropic liquid-saturated porous solids. *J. Acoust. Soc. Amer* **90**, 1068–1073.

Tarnow, V. (2005) Dynamic measurements of the elastic constants of glass wool. *J. Acoust. Soc. Amer* **118**, 3672–3678.

Vashishth, A., K. and Khurana, P. (2004) Waves in stratified anisotropic poroelastic media: a transfer matrix approach. *J. Sound Vib*. **277**, 239–275.

11

Modelling multilayered systems with porous materials using the transfer matrix method

11.1 Introduction

The description of the acoustic field in a porous layer is not simple, because the shear wave and the two compressional waves propagating forward and backward are present. In a layered medium with porous layers, elastic solid layers and fluid layers, a complete description can become very involved. A matrix representation of sound propagation, similar to that used by Brekhovskikh (1960), Folds and Loggins (1977), Scharnhorst (1983) and Brouard *et al.* (1995) is described in this chapter, and used to model plane acoustic fields in stratified media. The stratified media are assumed laterally infinite. They can be of different nature: elastic solid, thin plate, fluid, rigid porous, limp porous and poroelastic. The presented model is essentially based on the representation of plane wave propagation in different media, in terms of transfer matrices. The presentation of the transfer matrix for transversally isotropic porous media has been described in Chapter 10. However, the presentation in this chapter assumes the different media to be homogeneous and isotropic. The modelling is general in the sense that it can handle automatically arbitrary configurations of layered media. The theoretical background behind the transfer matrix method is first recalled. Next, transfer and coupling matrices of different types of layer will be given, followed by the algorithm used to solve for the acoustic indicators of the problem. Finally, basic application examples are presented. Extensions and examples of advanced applications will be discussed in Chapter 12.

Propagation of Sound in Porous Media: Modelling Sound Absorbing Materials, Second Edition J. F. Allard and N. Atalla

11.2 Transfer matrix method

11.2.1 Principle of the method

Figure. 11.1 illustrates a plane acoustic wave impinging upon a material of thickness h, at an incidence angle θ. The geometry of the problem is bidimensional, in the incident (x_1, x_3) plane. Various types of wave can propagate in the material, according to their nature. The x_1 component of the wave number for each wave propagating in the finite medium is equal to the x_1 component k_t of the incident wave in the free air:

$$k_t = k \sin\theta \tag{11.1}$$

where k is the wave number in free air. Sound propagation in the layer is represented by a transfer matrix $[T]$ such that

$$\mathbf{V}(M) = [T]\mathbf{V}(M') \tag{11.2}$$

where M and M' are set close to the forward and the backward face of the layer, respectively, and where the components of the vector $\mathbf{V}(M)$ are the variables which describe the acoustic field at a point M of the medium. The matrix $[T]$ depends on the thickness h and the physical properties of each medium.

11.3 Matrix representation of classical media

11.3.1 Fluid layer

The acoustic field in a fluid medium is completely defined in each point M by the vector:

$$V^f(M) = [p(M), \quad v_3^f(M)]^{\mathrm{T}} \tag{11.3}$$

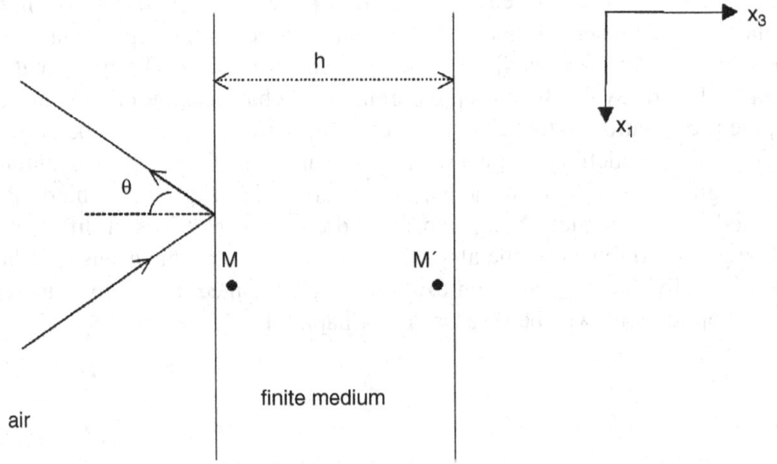

Figure 11.1 Plane wave impinging on a domain of thickness h.

where p and v_3^f are the pressure and the x_3 component of the fluid velocity, respectively. The superscript T indicates a transposition, $\mathbf{V}^f(M)$ being a column vector.

Let ρ and k be the density and the wave number of the fluid medium, respectively. Let k_3 be the x_3 component of the wave number vector in the fluid, equal to $\left(k^2 - k^2 \sin^2 \theta\right)^{1/2}$. The dependence in x_1 and time being removed, p and v_3 in the fluid can be written as

$$p(x_3) = A_1 \exp(-jk_3x_3) + A_2 \exp(jk_3x_3) \tag{11.4}$$

$$v_3^f(x_3) = \frac{k_3}{\omega\rho}[A_1 \exp(-jk_3x_3) - A_2 \exp(jk_3x_3)] \tag{11.5}$$

By arbitrarily setting the coordinate x_3 equal to zero at M', Equations (11.4) and (11.5) can be rewritten as

$$p(M') = A_1 + A_2 \tag{11.6}$$

$$v_3(M') = \frac{k_3}{\omega\rho}(A_1 - A_2) \tag{11.7}$$

Denoting by h the thickness of the layer, Equations (11.4) and (11.5) at $x_3 = -h$ can be rewritten as

$$V^f(M) = [T]V^f(M') \tag{11.8}$$

where the 2×2 transfer matrix $[T]$ is given by:

$$[T] = \begin{bmatrix} \cos(k_3h) & j\dfrac{\omega\rho}{k_3}\sin(k_3h) \\ j\dfrac{k_3}{\omega\rho}\sin(k_3h) & \cos(k_3h) \end{bmatrix} \tag{11.9}$$

11.3.2 Solid layer

In a layer consisting of an elastic solid, an incident and a reflected longitudinal wave, and an incident and a reflected shear wave can propagate. The acoustic field in the material can be completely described using the four amplitudes of these waves. The associated displacement potentials can be written as, respectively

$$\varphi = \exp(j\omega t - jk_1x_1)[A_1 \exp(-jk_{13}x_3) + A_2 \exp(jk_{13}x_3)] \tag{11.10}$$

$$\psi = \exp(j\omega t - jk_1x_1)[A_3 \exp(-jk_{33}x_3) + A_4 \exp(jk_{33}x_3)] \tag{11.11}$$

where the x_3 components k_{13} and k_{33} of the wave number vectors are

$$\begin{cases} k_{13} = (\delta_1^2 - k_t^2)^{1/2} \\ k_{33} = (\delta_3^2 - k_t^2)^{1/2} \end{cases} \tag{11.12}$$

In Equation (11.12) δ_1^2 and δ_3^3 are the squares of the wave numbers of the longitudinal and shear waves in the elastic solid layer, respectively. They are given by:

$$\begin{cases} \delta_1^2 = \dfrac{\omega^2 \rho}{\lambda + 2\mu} \\ \delta_3^2 = \dfrac{\omega^2 \rho}{\mu} \end{cases} \tag{11.13}$$

where ρ indicates the density for the elastic solid, and λ and μ are the first and second Lamé coefficients (these quantities are complex in general), respectively. The constants A_1, A_2, A_3 and A_4 represent the amplitudes of the four waves (incident and reflected) which can propagate in the layer. The acoustic field in the elastic solid layer can be predicted if these four amplitudes are known. Instead of these parameters, four mechanical variables may be chosen to express the sound propagation everywhere in the medium. However, different sets of four independent quantities may be chosen. Following Folds and Loggins (1977), the four chosen quantities are v_1, v_3, σ_{33} and σ_{13}. Let $\boldsymbol{V}^s$ be the vector

$$\boldsymbol{V}^s(M) = \begin{bmatrix} v_1^s(M) & v_3^s(\mathrm{M}) & \sigma_{33}^s(M) & \sigma_{13}^s(\mathrm{M}) \end{bmatrix}^{\mathrm{T}} \tag{11.14}$$

In these equation, v_1^s and v_3^s are the x_1 and x_3 components of the velocity at point M, respectively. σ_{33}^s and σ_{13}^s are the normal and tangential stresses at point M. These velocities and stresses are written as

$$\begin{cases} v_1^s = j\omega \left(\dfrac{\partial \varphi}{\partial x_1} - \dfrac{\partial \psi}{\partial x_3} \right) \\ v_3^s = j\omega \left(\dfrac{\partial \varphi}{\partial x_3} + \dfrac{\partial \psi}{\partial x_1} \right) \end{cases} \tag{11.15}$$

$$\begin{cases} \sigma_{33}^s = \lambda \left(\dfrac{\partial^2 \varphi}{\partial x_1^2} + \dfrac{\partial^2 \varphi}{\partial x_3^2} \right) + 2\mu \left(\dfrac{\partial^2 \varphi}{\partial x_3^2} + \dfrac{\partial^2 \psi}{\partial x_1 \partial x_3} \right) \\ \sigma_{13}^s = \mu \left(2\dfrac{\partial^2 \varphi}{\partial x_1 \partial x_3} + \dfrac{\partial^2 \psi}{\partial x_1^2} - \dfrac{\partial^2 \psi}{\partial x_3^2} \right) \end{cases} \tag{11.16}$$

To obtain the 4 × 4 transfer matrix $[T]$ of the elastic solid layer, the vector $\boldsymbol{V}^s(M)$ is first connected to vector $\boldsymbol{A} = [(A_1 + A_2), (A_1 - A_2), (A_3 + A_4), (A_3 - A_4)]$ by a matrix $[\Gamma(\mathrm{x}_3)]$ such that $\boldsymbol{V}^s(\mathrm{M}) = [\Gamma(\mathrm{x}_3)]\mathbf{A}$. Equations (11.15) and (11.16) can be used to evaluate $[\Gamma(\mathrm{x}_3)]$

$$[\Gamma(x_3] = \begin{bmatrix} \omega k_1 \cos(k_{13}x_3) & -j\omega k_1 \sin(k_{13}x_3) & j\omega k_{33} \sin(k_{33}x_3) & -\omega k_{33} \cos(k_{33}x_3) \\ -j\omega k_{13} \sin(k_{13}x_3) & \omega k_{13} \cos(k_{13}x_3) & \omega k_1 \cos(k_{33}x_3) & -j\omega k_1 \sin(k_{33}x_3) \\ -D_1 \cos(k_{13}x_3) & jD_1 \sin(k_{13}x_3) & jD_2 k_{33} \sin(k_{33}x_3) & -D_2 k_{33} \cos(k_{33}x_3) \\ jD_2 k_{13} \sin(k_{13}x_3) & -D_2 k_{13} \cos(k_{13}x_3) & D_1 \cos(k_{33}x_3) & -jD_1 \sin(k_{33}x_3) \end{bmatrix} \tag{11.17}$$

with $D_1 = \lambda(k_0{}^2 + k_{13}{}^2) + 2\mu k_{13}{}^2 = \mu(k_{13}{}^2 - k_0{}^2)$, and $D_2 = 2\mu k_0$. If the origin of the x_3 axis is fixed at point M, the vectors $\mathbf{V}^s(M)$ and $\mathbf{V}^s(M')$are expressed as

$$\begin{cases} \mathbf{V}^s(M) = [\Gamma(0)]\mathbf{A} \\ \mathbf{V}^s(M') = [\Gamma(h)]\mathbf{A} \end{cases} \tag{11.18}$$

Then, the transfer matrix $[T]$ which relates $\mathbf{V}^s(M)$ and $\mathbf{V}^s(M')$ is equal to $[T] = [\Gamma(0)][\Gamma(h)]^{-1}$. In order to alleviate the instability that may arise from the inversion of matrix $[\Gamma(h)]$, the origin of the x_3 axis can be rather fixed at point M', and the transfer matrix written:

$$[T^s] = [\Gamma(-h)][\Gamma(0)]^{-1} \tag{11.19}$$

The inversion of the matrix $[\Gamma(0)]$ is calculated analytically

$$[\Gamma(0)]^{-1} = \begin{bmatrix} \dfrac{2k_1}{\omega\delta_3^2} & 0 & -\dfrac{1}{\mu\delta_3^2} & 0 \\ 0 & \dfrac{k_{33}^2 - k_1^2}{\omega k_{13}\delta_3^2} & 0 & -\dfrac{k_1}{\mu k_{13}\delta_3^2} \\ 0 & \dfrac{k_1}{\omega\delta_3^2} & 0 & \dfrac{1}{\mu\delta_3^2} \\ \dfrac{k_{33}^2 - k_1^2}{\omega k_{33}\delta_3^2} & 0 & -\dfrac{k_1}{\mu k_{33}\delta_3^2} & 0 \end{bmatrix} \tag{11.20}$$

11.3.3 Poroelastic layer

The acoustic field in a layer of porous materials

In the context of Biot theory, three kinds of wave can propagate in a porous medium: two compressional waves and a shear wave (Chapter 6). Let us denote by $\mathbf{k}_1, \mathbf{k}_2, \mathbf{k}_3$, and $\mathbf{k}'_1, \mathbf{k}'_2, \mathbf{k}'_3$, the wave number vectors of the compressional waves (subscripts 1, 2) and the shear wave (subscript 3). The nonprimed vectors correspond to waves propagating forward while the primed vectors correspond to waves propagating backward. Let δ_1^2, δ_2^2 and δ_3^2, be the squared wave numbers of the two compressional waves and of the shear wave. These quantities are given by Equations (6.67), (6.68) and (6.83). The x_3 components of the wave number vectors are

$$\begin{cases} k_{i3} = (\delta_i^2 - k_t^2)^{1/2} & i = 1, 2, 3 \\ k'_{i3} = -k_{i3} & i = 1, 2, 3 \end{cases} \tag{11.21}$$

The square root symbol $(\ )^{1/2}$ represents the root which yields a positive real part. In the complex representation, the frame displacement potentials of the compressional waves can be written as

$$\varphi_i^s = A_i \exp(j(\omega t - k_{i3}x_3 - k_t x_1)) + A'_i \exp(j(\omega t + k_{i3}x_3 - k_t x_1)) \qquad i = 1, 2 \tag{11.22}$$

The displacements induced by the rotational waves are parallel to the $x_1 0 x_3$ plane, and only the x_2 component of the vector potential is different from zero. This component is

$$\psi_2^s = A_3 \exp(j(\omega t - k_{33}x_3 - k_t x_1)) + A_3' \exp(j(\omega t + k_{33}x_3 - k_t x_1)) \tag{11.23}$$

The air displacement potentials are related to the frame displacement potentials by

$$\varphi_i^f = \mu_i \varphi_i^s \qquad i = 1, 2 \tag{11.24}$$

and

$$\psi_2^f = \mu_3 \psi_2^s \tag{11.25}$$

The ratio μ_i of the velocity of the air over the velocity of the frame for the two compressional waves and μ_3 for the shear wave are given by Equations (6.71) and (6.84), respectively. The displacements field for the frame and the air are known completely if A_1, A_2, A_3, A_1', A_2' and A_3' are known, while the stresses can be calculated by Equation (6.2) and (6.3).

Matrix representation

The acoustic field in the porous layer consists of six waves and can be described by Equations (11.22) and (11.23). The acoustic field in the porous layer can be predicted everywhere if the six amplitudes A_1, A_2, A_3, A_1', A_2' and A_3' are known. However, instead of these parameters, one may choose six independent acoustic quantities. The six acoustic quantities that have been chosen are three velocity components and three elements of the stress tensors; the two velocity components v_1^s and v_3^s of the frame, the velocity component v_3^f of the fluid, the two components σ_{33}^s and σ_{13}^s of the stress tensor of the frame, and σ_{33}^f in the fluid. If these six quantities are known at a point M in the layer, the acoustic field can be predicted everywhere in the layer. Moreover, the values of these quantities anywhere in the layer depend linearly on the values of these quantities at M. Let $\mathbf{V^p}(M)$ be the vector

$$\mathbf{V^p}(M) = [\, v_1^s(\mathrm{M}) \quad v_3^s(\mathrm{M}) \quad v_3^f(\mathrm{M}) \quad \sigma_{33}^s(M) \quad \sigma_{13}^s(\mathrm{M}) \quad \sigma_{33}^f(M)\,]^{\mathrm{T}} \tag{11.26}$$

The superscript T indicates a transposition, $\mathbf{V}(M)$ being a column vector. These six quantities are written as:

$$\left\{ \begin{aligned} v_1^s &= j\omega\left(\frac{\partial \varphi_1^s}{\partial x_1} + \frac{\partial \varphi_2^s}{\partial x_1} - \frac{\partial \psi_2^s}{\partial x_3}\right) \\ v_3^s &= j\omega\left(\frac{\partial \varphi_1^s}{\partial x_3} + \frac{\partial \varphi_2^s}{\partial x_3} + \frac{\partial \psi_2^s}{\partial x_1}\right) \\ v_3^f &= j\omega\left(\frac{\partial \varphi_1^f}{\partial x_3} + \frac{\partial \varphi_2^f}{\partial x_3} + \frac{\partial \psi_2^f}{\partial x_1}\right) \end{aligned} \right. \tag{11.27}$$

and,

$$\left\{\begin{aligned}
\sigma_{33}^{s} &= (P-2N)\left(\frac{\partial^2(\varphi_1^s+\varphi_2^s)}{\partial x_1^2}+\frac{\partial^2(\varphi_1^s+\varphi_2^s)}{\partial x_3^2}\right)\\
&\quad + Q\left(\frac{\partial^2(\varphi_1^f+\varphi_2^f)}{\partial x_1^2}+\frac{\partial^2(\varphi_1^f+\varphi_2^f)}{\partial x_3^2}\right)+2N\left(\frac{\partial^2(\varphi_1^s+\varphi_2^s)}{\partial x_3^2}+\frac{\partial^2\psi_2^s}{\partial x_1\partial x_3}\right)\\
\sigma_{13}^{s} &= N\left(2\frac{\partial^2(\varphi_1^s+\varphi_2^s)}{\partial x_1\partial x_3}+\frac{\partial^2\psi_2^s}{\partial x_1^2}-\frac{\partial^2\psi_2^s}{\partial x_3^2}\right)\\
\sigma_{33}^{f} &= R\left(\frac{\partial^2(\varphi_1^f+\varphi_2^f)}{\partial x_1^2}+\frac{\partial^2(\varphi_1^f+\varphi_2^f)}{\partial x_3^2}\right)+Q\left(\frac{\partial^2(\varphi_1^s+\varphi_2^s)}{\partial x_1^2}+\frac{\partial^2(\varphi_1^s+\varphi_2^s)}{\partial x_3^2}\right)
\end{aligned}\right. \tag{11.28}$$

where N is the shear modulus of the material, and P, Q, and R are the elastic coefficients of Biot defined in Chapter 6.

If M and M' are set close to the forward and the backward face of the layer, respectively, a matrix $[T^p]$ which depends on the thickness h and the physical properties of the material relates $\mathbf{V^p}(M)$ to $\mathbf{V^p}(M')$:

$$\mathbf{V^p}(M) = [T^p]\mathbf{V^p}(M') \tag{11.29}$$

The matrix elements T_{ij} have been calculated by Depollier (1989) in the following way. Let $\mathbf{A}$ be the column vector

$$\mathbf{A} = [(A_1 + A_1'), (A_1 - A_1'), (A_2 + A_2'), (A_2 - A_2'), (A_3 + A_3'), (A_3 - A_3')]^{\mathrm{T}} \tag{11.30}$$

and let $[\Gamma(x_3)]$ be the matrix connecting $\mathbf{V^p}(M)$ at x_3 to $\mathbf{A}$:

$$\mathbf{V^p}(M) = [\Gamma(0)]\mathbf{A},\ \mathbf{V^p}(M') = [\Gamma(h)]\mathbf{A} \tag{11.31}$$

The vectors $\mathbf{V^p}(M)$ and $\mathbf{V^p}(M')$ are related by

$$\mathbf{V^p}(M) = [\Gamma(0)][\Gamma(h)]^{-1}\mathbf{V^p}(M') \tag{11.32}$$

and so $[T^p]$ is equal to

$$[T^p] = [\Gamma(0)][\Gamma(h)]^{-1} \tag{11.33}$$

In order to avoid a matrix inversion, the origin of the x_3 axis can be changed, and the following equation can be used:

$$[T^p] = [\Gamma(-h)][\Gamma(0)]^{-1} \tag{11.34}$$

Matrix $[\Gamma(0)]^{-1}$ can be evaluated analytically.

Evaluation of the matrices [Γ] and $[T^p]$

In order to calculate the elements of $[\Gamma(h)]^{-1}$, the quantities v_1^s, v_3^s, v_3^f, σ_{13}^s, σ_{33}^s and σ_{33}^f must be evaluated from the potentials φ_1^s, φ_2^s and ψ^s. The velocity component v_1^f will also be calculated for the sake of completeness. The velocity components v_1^s and v_3^s are obtained by using φ_1^s, φ_2^s and ψ_2^s as given by Eqs. (11.22) and (11.23) in Equation (11.27).

$$v_1^S = j\omega\left[\sum_{i=1,2}\{-jk_t(A_i + A_i')\cos k_{i3}x_3 - k_t(A_i - A_i')\sin k_{i3}x_3\}\right.$$
$$\left. + k_{33}(A_3 + A_3')\sin k_{33}x_3 + jk_{33}(A_3 - A_3')\cos k_{33}x_3\right] \tag{11.35}$$

$$v_3^S = j\omega\left[\sum_{i=1,2}\{-k_{i3}(A_i + A_i')\sin k_{i3}x_3 - jk_{i3}(A_i - A_i')\cos k_{i3}x_3\}\right.$$
$$\left. - jk_t(A_3 + A_3')\cos k_{33}x_3 - k_t(A_3 - A_3')\sin k_{33}x_3\right] \tag{11.36}$$

In these equations the dependence on time and x_1 has been removed. The velocities v_1^f and v_3^f can be evaluated from the displacement potentials φ_1^f, φ_2^f and ψ^f, which are related to φ_1^s, φ_2^s and ψ^s by Equation (11.27)

$$v_1^f = j\omega\left[\sum_{i=1,2}\{-jk_t\mu_i(A_i + A_i')\cos k_{i3}x_3 - k_t\mu_i(A_i - A_i')\sin k_{i3}x_3\}\right.$$
$$\left. + k_{33}\mu_3(A_3 + A_3')\sin k_{33}x_3 + jk_{33}\mu_3(A_3 - A_3')\cos k_{33}x_3\right] \tag{11.37}$$

$$v_3^f = j\omega\left[\sum_{i=1,2}\{-k_{i3}\mu_i(A_i + A_i')\sin k_{i3}x_3 - jk_{i3}\mu_i(A_i - A_i')\cos k_{i3}x_3\}\right.$$
$$\left. - jk_t\mu_3(A_3 + A_3')\cos k_{33}x_3 - k_t\mu_3(A_3 - A_3')\sin k_{33}x_3\right] \tag{11.38}$$

The two components σ_{33}^s and σ_{13}^s of the stress tensor of the frame, and the component σ_{33}^f in the fluid, can be calculated by Equations (6.2) and (6.3)

$$\sigma_{33}^s = (P - 2N)\nabla\cdot\mathbf{u}^s + Q\nabla\cdot\mathbf{u}^f + 2N\frac{\partial u_3^s}{\partial x_3} \tag{11.39}$$

$$\sigma_{13}^s = N\left(\frac{\partial u_1^s}{\partial x_3} + \frac{\partial u_3^s}{\partial x_1}\right) \tag{11.40}$$

$$\sigma_{33}^f = R\nabla\cdot\mathbf{u}^f + Q\nabla\cdot\mathbf{u}^s \tag{11.41}$$

The displacement potentials can be used to express the spatial derivatives of the x_3 and the x_1 components of $\mathbf{u^f}$ and $\mathbf{u^s}$ in Equations (11.39)–(11.41)

$$\nabla \cdot \mathbf{u^s} = \frac{\partial^2}{\partial x_1^2}(\varphi_1^s + \varphi_2^s) + \frac{\partial^2}{\partial x_3^2}(\varphi_1^s + \varphi_2^s) \tag{11.42}$$

$$\nabla \cdot \mathbf{u^f} = \frac{\partial^2}{\partial x_1^2}(\mu_1\varphi_1^s + \mu_2\varphi_2^s) + \frac{\partial^2}{\partial x_3^2}(\mu_1\varphi_1^s + \mu_2\varphi_2^s) \tag{11.43}$$

$$\frac{\partial u_3^s}{\partial x_3} = \frac{\partial^2}{\partial x_3^2}(\varphi_1^s + \varphi_2^s) + \frac{\partial^2 \psi_2}{\partial x_1 \partial x_3} \tag{11.44}$$

$$\frac{\partial u_1^s}{\partial x_3} + \frac{\partial u_3^s}{\partial x_1} = 2\frac{\partial^2(\varphi_1^s + \varphi_2^s)}{\partial x_1 \partial x_3} + \left(\frac{\partial^2 \psi_2}{\partial x_1^2} - \frac{\partial^2 \psi_2}{\partial x_3^2}\right) \tag{11.45}$$

and Equations (11.39)–(11.41) can be rewritten as

$$\begin{aligned}\sigma_{13}^s = N\Bigg[& 2k_t \sum_{i=1,2} k_{i3}\{j(A_i + A_i')\sin k_{i3} - (A_i - A_i')\cos k_{i3}x_3\} \\ & + (k_{33}^2 - k_t^2)[(A_3 + A_3')\cos k_{33}x_3 - j(A_3 - A_3')\sin k_{33}x_3]\Bigg]\end{aligned} \tag{11.46}$$

$$\begin{aligned}\sigma_{33}^s = \sum_{i=1,2} & \{[-(P + Q\mu_i)(k_t^2 + k_{i3}^2) + 2Nk_t^2](A_i + A_i')\cos k_{i3}x_3 \\ & + j[(P + Q\mu_i)(k_t^2 + k_{i3}^2) - 2Nk_t^2](A_i - A_i')\sin k_{i3}x_3\} \\ & + 2jNk_t k_{33}(A_3 + A_3')\sin k_{33}x_3 - 2Nk_t k_{33}(A_3 - A_3')\cos k_{33}x_3\end{aligned} \tag{11.47}$$

$$\begin{aligned}\sigma_{33}^f = \sum_{i=1,2} & (Q + R\mu_i)(k_t^2 + k_{i3}^2)\{-(A_i - A_i')\cos k_{i3}x_3 \\ & + j(A_i - A_i')\sin k_{33}x_3\}\end{aligned} \tag{11.48}$$

The coefficients of the terms $(A_1 \pm A_1')$, $(A_2 \pm A_2')$, and $(A_3 \pm A_3')$, in Equations (11.37), (11.38) and (11.46)–(11.48) are the matrix elements $\Gamma_{ij}(x_3)$ as given in Table 11.1. In this table, D_i and E_i are given by

$$D_i = (P + Q\mu_i)(k_t^2 + k_{i3}^2) - 2Nk_t^2 \qquad i = 1, 2 \tag{11.49}$$

$$E_i = (R\mu_i + Q)(k_t^2 + k_{i3}^2) \qquad i = 1, 2 \tag{11.50}$$

The matrix $[T^p] = [\Gamma(0)][\Gamma(h)]^{-1}$ is given in Appendix 11.A.

11.3.4 Rigid and limp frame limits

The rigid frame limit depicts the dynamic behaviour of the material when its frame is supposed motionless. This simplification, presented in Chapter 5, can be used for frequencies higher than the decoupling frequency, $F_d = \sigma \times \phi^2/(2\pi\rho_1)$. In this frequency domain, the visco-inertial coupling between the solid and the fluid phase is sufficiently

Table 11.1 The coefficients $\Gamma_{ij}(x_3)$.

(a) The three first columns of $[\Gamma]$

$\omega k_t \cos k_{i3}x_3$	$-j\omega k_t \sin k_{i3}x_3$	$\omega k_t \cos k_{23}x_3$
$-j\omega k_{i3} \sin k_{i3}x_3$	$\omega k_{i3} \cos k_{i3}x_3$	$-j\omega k_{23} \sin k_{23}x_3$
$-j\omega k_{i3}\mu_1 \sin k_{i3}x_3$	$\omega\mu_1 k_{i3} \cos k_{i3}x_3$	$-j\omega k_{23}\mu_2 \sin k_{23}x_3$
$-D_1 \cos k_{i3}x_3$	$jD_1 \sin k_{i3}x_3$	$-D_2 \cos k_{23}x_3$
$2jNk_t k_{i3} \sin k_{i3}x_3$	$-2Nk_t k_{i3} \cos k_{i3}x_3$	$2jNk_t k_{23} \sin k_{23}x_3$
$-E_1 \cos k_{i3}x_3$	$jE_1 \sin k_{i3}x_3$	$-E_2 \cos k_{23}x_3$

(b) The three last columns of $[\Gamma]$

$-j\omega k_t \sin k_{23}x_3$	$j\omega k_{33} \sin k_{33}x_3$	$-\omega k_{33} \cos k_{33}x_3$
$\omega k_{23} \cos k_{23}x_3$	$\omega k_t \cos k_{33}x_3$	$-j\omega k_t \sin k_{33}x_3$
$\omega\mu_2 k_{23} \cos k_{23}x_3$	$\omega k_t\mu_3 \cos k_{33}x_3$	$-j\omega k_t\mu_3 \sin k_{33}x_3$
$jD_2 \sin k_{23}x_3$	$2jNk_{33}k_t \sin k_{33}x_3$	$-2Nk_{33}k_t \cos k_{33}x_3$
$-2Nk_t k_{23} \cos k_{23}x_3$	$N(k_{33}^2 - k_t^2) \cos k_{33}x_3$	$-jN(k_{33}^2 - k_t^2) \sin k_{33}x_3$
$jE_2 \sin k_{23}x_3$	0	0

weak, so that an acoustical wave propagating in the fluid phase will not exert a force sufficient to generate vibration in the solid phase. In the rigid frame limit, the dynamic behaviour is represented by an equivalent fluid wave equation of the form (see Section 5.7):

$$\Delta p + \frac{\tilde{\rho}_{eq}}{\tilde{K}_{eq}}\omega^2 p = 0 \tag{11.51}$$

where $\tilde{\rho}_{eq} = \tilde{\rho}_f/\phi$ is the effective density and $\tilde{K}_{eq} = \tilde{K}_f/\phi$ is the effective bulk modulus of the rigid frame equivalent fluid medium. Using these effective properties, the matrix representation of the rigid frame limit is given by Equation (11.9) with the wave number given by $\omega\sqrt{\tilde{\rho}_{eq}/\tilde{K}_{eq}}$.

An equivalent fluid representation can also be used for limp materials, such as aeronautic grade fiberglass (density of the order of 0.3–0.5 pcf). It is however important to account for the inertia of the frame in the modelling of their dynamic behaviour. The limp model can be derived from the Biot theory, assuming that the stiffness of the frame is negligible. Various models have been proposed in the literature: Beranek (1947), Ingard (1994), Katragada *et al.* (1995), Panneton (2007). Here, a straightforward way to correct the equivalent fluid equation for this effect is presented. It starts from the mixed pressure–displacement formulation derived in Appendix 6.A (Equation 6A.22):

$$\begin{cases} div\ \hat{\sigma}^s(\boldsymbol{u}^s) + \omega^2\tilde{\rho}\boldsymbol{u} + \tilde{\gamma}\ \mathbf{grad}\ p = 0 \\ \Delta p + \omega^2\dfrac{\tilde{\rho}_{eq}}{\tilde{K}_{eq}}p - \omega^2\tilde{\rho}_{eq}\tilde{\gamma}\ div\ \boldsymbol{u}^s = 0 \end{cases} \tag{11.52}$$

where $\hat{\sigma}^s$ is the stress tensor of the solid phase in vacuum, $\tilde{\gamma} = \phi(\tilde{\rho}_{12}/\tilde{\rho}_{22} - \tilde{Q}/\tilde{R})$ and $\tilde{\rho} = \tilde{\rho}_{11} - (\tilde{\rho}_{12})^2/\tilde{\rho}_{22}$. Coefficients $\tilde{\rho}_{eq}$ and $\tilde{K}_{eq}$ are related to $\tilde{\rho}_{22}$ and $\tilde{R}$ in

Equation (6A.22) by $\tilde{\rho}_{22} = \phi^2 \tilde{\rho}_{eq}$ and $\tilde{\mathrm{R}} = \phi^2 \tilde{\mathrm{K}}_{eq}$. Assuming a limp frame, the stress tensor of the solid phase in vacuum is neglected ($\hat{\sigma}^s \approx 0$) and the first equation in (11.52) leads to $\omega^2 \tilde{\rho}$ div $\boldsymbol{u} = -\tilde{\gamma}\Delta p$. Substitution of this equation in the second equation of (11.52) yields the equivalent fluid equation for a limp material,

$$\Delta \mathrm{p} + \frac{\tilde{\rho}_{\mathrm{limp}}}{\tilde{K}_{\mathrm{eq}}} \omega^2 \mathrm{p} = 0 \tag{11.53}$$

where $\tilde{\rho}_{\mathrm{limp}}$ is an equivalent effective density accounting for the inertia of the frame,

$$\tilde{\rho}_{\mathrm{limp}} = \frac{\tilde{\rho}\tilde{\rho}_{eq}}{\tilde{\rho} + \tilde{\rho}_{eq}\tilde{\gamma}^2} \tag{11.54}$$

Assuming the bulk modulus of the elastic material from which the frame is made much larger than the bulk modulus of the frame (i.e. $K_b \ll K_s$), valid for the majority of porous materials, Panneton (2007) derives the following approximate expression for the limp effective density:

$$\tilde{\rho}_{\mathrm{limp}} \approx \frac{\rho_t \tilde{\rho}_{eq} - \rho_0^2}{\rho_t + \tilde{\rho}_{eq} - 2\rho_0} \tag{11.55}$$

where $\rho_t = \rho_1 + \phi\rho_0$ is the apparent total density of the equivalent fluid limp medium. Again, using these effective properties, the matrix representation of the limp frame limit is given by Equation (11.9). It is observed from Equation (11.55) that, when the frame is heavy, the rigid model is recovered. Moreover, at low frequencies $\lim_{\omega \to 0} \tilde{\rho}_{\mathrm{limp}} = \rho_t$ contrary to the rigid frame model (see chapter 5 for the low- and high-frequency limits of the effective density; in particular Equation (5.37), giving the limit at low frequency: $\lim_{\omega \to 0} \tilde{\rho}_{eq} = \sigma/(j\omega)$). In consequence, the difference between the two models is mainly important at low frequencies. This is illustrated in Figure 11.2 for the material of Table 11.2.

Note in passing, that the rigid frame model does not authorize rigid body motion of the material. This is important, for instance in applications when the material is unconstrained (free to move). This effect is mainly important at low frequencies. In such applications, the limp model is preferred. This is key when impedance tube tests are used to derive the acoustical properties of the material from measurement of either the surface impedance or the prorogation constants (especially using transmission measurements).

The limp model is usable when the elasticity of the frame is neglected, either due to the nature of the material (e.g. light fibreglass) or due to the mounting or excitation of the material (e.g. transmission loss of light foam separated by a thin air gap from a supporting plate). As a rule of thumb, when the material is bonded onto a vibrating structure, the rigid frame model should not be used. On the other hand, the limp model can be used when the bulk modulus of frame in vacuum is much smaller than bulk modulus of the fluid in the pores. Beranek (1947) suggested the use of the limp approximation for porous materials satisfying $\left|K_c/K_f\right| < 0.05$ where K_c and K_f are the bulk modulus of frame in vacuum and the bulk modulus of the fluid in the pores, respectively. Doutres *et al.* (2007) used a numerical study comparing a full porous–elastic model with the limp model to derive a criterion for the use of the latter. Two configurations were used: (i) sound absorption of a porous layer backed by a rigid wall, and (ii) sound radiation from a porous layer backed by

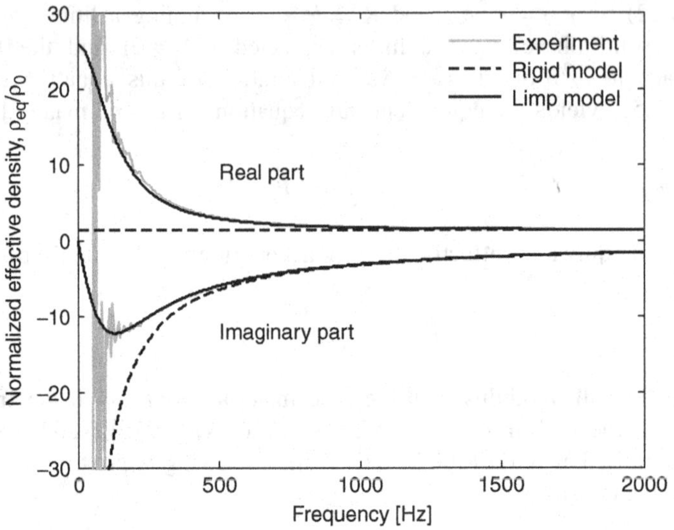

Figure 11.2 Normalized effective density of the soft fibrous material (Table 11.2); comparison between predictions using the limp frame and the rigid frame models. Measurement performed by Panneton (2007) using the method of Utsono *et al.* (1989).

Table 11.2 The parameters used to predict the effective density of the material in Figure 11.2.

Material	Thickness, h (mm)	ϕ	σ (N s/m^4)	α_∞	Λ (μm)	Λ' (μm)	ρ_1 (kg/m^3)
Soft fibrous	50	0.98	25×10^3	1.02	90	180	30

a vibrating wall. In both cases, the materials are assumed of infinite extent. The criterion, named frame structural interaction (FSI), is derived from the displacement ratio of the compressional frame-borne and airborne waves (Equation 6.71): $FSI = (\tilde{\rho}_{\mathrm{limp}}/\tilde{\rho}_c)\tilde{K}_c/\tilde{K}_f$ where $\tilde{\rho}_c = \rho_1 - \tilde{\rho}_{12}/\phi$. Using this frequency-dependent criterion they relaxed the range of validity of Beranek's criterion to $|K_c/K_f| < 0.2$. Approximating K_f by its isothermal value in air, $K_f \approx P_0$ (101.3 kPa), the latter criterion stipulates that the limp model is applicable for materials having a bulk modulus lower than 20 kPa. Recall however that this criterion does not account for the effects of boundary condition and mounting effects. As mentioned above, the limp model can be used for materials having a higher bulk modulus in various specific configurations. This is the case, e.g. for thin light foam decoupled with an air gap from a vibrating structure.

11.3.5 Thin elastic plate

In the case of a thin elastic plate in bending (bending stiffness D; thickness h and mass per unit area m), the harmonic form of the equation of motion is given by:

$$Z_s(\omega)v_3(M') = \sigma_{33}(M') - \sigma_{33}(M) \tag{11.56}$$

where

$$Z_s(\omega) = j\omega m\left(1 - \frac{Dk_t^4}{\omega^2 m}\right)$$

is the mechanical impedance of the panel and $k_t = k\sin\theta$ with k the wave number in the free air. $\sigma_{33}(M)$ and $\sigma_{33}(M')$are respectively, the normal stresses just in front and just behind the plate and v_3 is the normal velocity of the thin plate $v_3 = v_3(M) = v_3(M')$. Using the vector $\mathbf{V}(M) = [\sigma_{33}(M)v_3(M)]^{\mathrm{T}}$ to express the mechanical field in a point M of the plate, the transfer matrix [T^{i}] relating $\mathbf{V}(M)$ and $\mathbf{V}(M')$ is deduced directly from Equation (11.56)

$$[T] = \begin{bmatrix} 1 & -Z_s(\omega) \\ 0 & 1 \end{bmatrix} \tag{11.57}$$

Note that the mechanical impedance of the panel can be written in the equivalent form:

$$Z_s(\omega) = j\omega m\left[1 - \left(\frac{\omega}{\omega_c}\right)^2 \sin^4\theta\right] \tag{11.58}$$

where

$$\omega_c = c^2\sqrt{\frac{m}{D}}$$

is the panel's critical frequency. Damping in the panel can be accounted for by using a complex value of the Young's modulus.

11.3.6 Impervious screens

Impervious screens are usually used to cover or protect acoustic materials. Their modelling using the transfer matrix method depends on their mounting. When they are free to move, that is when there is air on both sides of the screen, they can be simply modelled as a thin plate with negligible stiffness. The mechanical impedance in Equation (11.57) reduces to $Z(\omega) = j\omega m$. When the screen is bonded onto a porous material, the modelling needs to account for the interface forces.

Let M and M' two points close, respectively, to the forward and the backward face of a screen in mechanical contact. Assuming the screen to be flexible with a non-negligible bending stiffness, Newton's law applied to the screen, leads to

$$j\,\omega m v_3(M') = \sigma_{33}^s(M') - \sigma_{33}^s(M) - D\frac{\partial^4 v_3(M')}{j\omega\partial x_1^4} \tag{11.59}$$

$$j\,\omega m v_1(M') = \sigma_{13}^s(M') - \sigma_{13}^s(M) + S\frac{\partial^2 v_1(M')}{j\omega\partial x_1^2} \tag{11.60}$$

In these equations, v_1^s and v_3^s are the x_1 and x_3 components of the velocity at point M, respectively. $\sigma_{33}^{\mathbf{s}}$ and $\sigma_{13}^{\mathbf{s}}$ are the normal and tangential stresses at point M. The quantities

m, D and S are the mass per unit area, the bending stiffness, and the membrane stiffness of the screen, respectively.

The operators $\partial^2/\partial x_1^2$ and $\partial^4/\partial x_1^4$ can be replaced by $-k_t^2$ and k_t^4, respectively. In consequence, the screen governing equations read:

$$j\omega m\left(1-\frac{Dk_t^4}{m\omega^2}\right)v_3^s(M')=\sigma_{33}^s(M')-\sigma_{33}^s(M) \tag{11.61}$$

$$j\omega m\left(1-\frac{Sk_t^2}{m\omega^2}\right)v_1^s(M')=\sigma_{13}^s(M')-\sigma_{13}^s(M) \tag{11.62}$$

The velocity of the screen $\boldsymbol{v}(M)$ is the same on both faces. Using the vector $\boldsymbol{V}^s(M)=[v_1^s(M)\quad v_3^s(\mathrm{M})\quad \sigma_{33}^s(M)\quad \sigma_{13}^s(\mathrm{M})]^{\mathrm{T}}$ to describe the screen variables, Equations (11.61) and (11.62) with equality of the velocity vector at M and M', lead to the following 4×4 transfer matrix for the screen:

$$[T]=\begin{bmatrix}1 & 0 & 0 & 0\\ 0 & 1 & 0 & 0\\ 0 & -Z_s(\omega) & 1 & 0\\ -Z_s'(\omega) & 0 & 0 & 1\end{bmatrix} \tag{11.63}$$

where $Z_s(\omega)=j\omega m(1-Dk_t^4/m\omega^2)$ and $Z_s'(\omega)=j\omega m(1-Sk_t^2/m\omega^2)$.

The interface conditions described in Section 11.4.2 for a solid layer can be used when the screen is in contact with a porous or fluid medium.

In the case of a negligible stiffness impervious screen, the latter can be modelled as a flexible membrane. In this case Equation (11.59) is replaced by

$$j\omega m v_3(M')=\sigma_{33}^s(M')-\sigma_{33}^s(M)+I\frac{\partial^2 v_3(M')}{j\omega\partial x_1^2} \tag{11.64}$$

and in consequence, impedance Z in Equation (11.62) is replaced by $Z_s''(\omega)=j\omega m(1-Ik_t^2/m\omega^2)$ where I is the tension of the screen.

11.3.7 Porous screens and perforated plates

For microporous screens a more refined model must be used to account for the added resistance (proportional to the flow resistance of the screen). For a perforated plate, there is also an important reactance term (Chapter 9). In both cases, when the screen is not in mechanical contact, an equivalent fluid model (rigid or limp) may be used. When, it is in contact, either a poroelastic model or a refined screen model (using four variables) should be used (Ingard 1994; Atalla and Sgard 2007).

11.3.8 Other media

The transfer matrix method can be generalized to account for other domains such as thick plates, orthotropic plates, composite and sandwich panels, transversally isotropic porous materials, etc. In these cases the transfer matrices are formulated in terms of the wave heading (direction of propagation) in the plane (x_1x_3). The example of a thick orthotropic panel is discussed in Chapter 12.

11.4 Coupling transfer matrices

The transfer matrices of various types of layer were evaluated in the previous section. This section is devoted to the continuity conditions between two adjacent layers of different nature. Figure 11.3 illustrates a stratified medium, where two points M_{2k} and M_{2k+1} $(k = 1, n-1)$ are close to each other at each side of a boundary between layers (k) and $(k+1)$. An interface matrix, which depends on the nature of the two layers, must be used to relate the acoustic field vectors $\mathbf{V}^{(k)}(\mathrm{M}_{2k})$ and $\mathbf{V}^{(k+1)}(\mathrm{M}_{2k+1})$. For simplification, the interface matrices are derived for the two first layers of Figure 11.3.

11.4.1 Two layers of the same nature

If the two adjacent layers have the same nature, the continuity conditions are exploited to build a global transfer matrix which describes the acoustic propagation between M_1 and M_4. If the two layers are not porous, the global transfer matrix is simply equal to product of the transfer matrices of the two layers. But, if the two layers are porous, the continuity conditions are affected by the porosities of the layers:

$$\begin{aligned}
&v_1^s(M_2) = v_1^s(M_3)\\
&v_3^s(M_2) = v_3^s(M_3)\\
&\phi_1(v_3^f(M_2) - v_3^s(M_2)) = \phi_2(v_3^f(M_3) - v_3^s(M_3))\\
&\sigma_{33}^s(M_2) + \sigma_{33}^f(M_2) = \sigma_{33}^s(M_3) + \sigma_{33}^f(M_3)\\
&\sigma_{13}^s(M_2) = \sigma_{13}^s(M_3)\\
&\frac{\sigma_{33}^f(M_2)}{\phi_1} = \frac{\sigma_{33}^f(M_3)}{\phi_2}
\end{aligned} \tag{11.65}$$

In this case, the global transfer matrix $[T^p]$ is written as:

$$[T^p] = [T_1^p][I_{pp}][T_2^p] \tag{11.66}$$

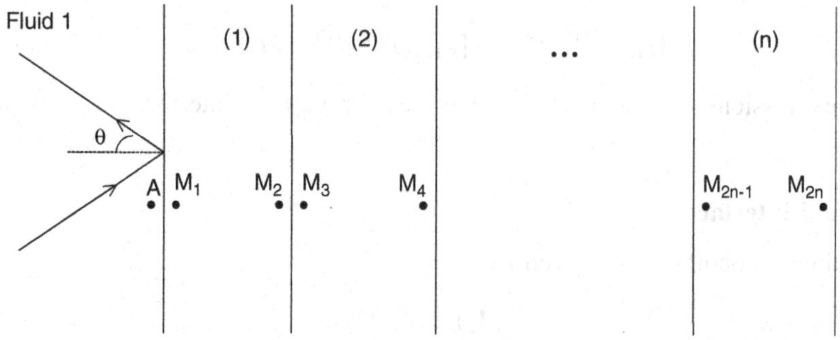

Figure 11.3 Plane wave impinging on a multilayer domain.

where $[T_1^p]$ et $[T_2^p]$ are the transfer matrices of the two porous layers, and $[I_{pp}]$ is a 6×6 interface matrix built from Equations (11.65)

$$[I_{pp}] = \begin{bmatrix} 1 & 0 & 0 & 0 & 0 & 0 \\ 0 & 1 & 0 & 0 & 0 & 0 \\ 0 & \left(1 - \dfrac{\phi_2}{\phi_1}\right) & \dfrac{\phi_2}{\phi_1} & 0 & 0 & 0 \\ 0 & 0 & 0 & 1 & 0 & \left(1 - \dfrac{\phi_1}{\phi_2}\right) \\ 0 & 0 & 0 & 0 & 1 & 0 \\ 0 & 0 & 0 & 0 & 0 & \dfrac{\phi_1}{\phi_2} \end{bmatrix} \tag{11.67}$$

Note that this interface matrix is equal to the 6×6 unit matrix if the two layers have the same porosity.

The transfer matrix of a porous layer using the second Biot representation (Appendix 6.A) is given in Section 10.9 of Chapter 10 in the case of a transversally isotropic medium. The matrix for an isotropic medium is a special case. In this representation the components of $\boldsymbol{V}(z) = [u_x^s, u_z^s, w_z, \sigma_{zz}^t, \sigma_{zx}^t, p]^{\mathrm{T}}$ are equal at each side of the boundary between two porous layers with frames bonded together. The interface matrix is a unit matrix in this representation and the transfer matrix of a layered porous medium is simply the product of the transfer matrices of each layer.

11.4.2 Interface between layers of different nature

When the adjacent layers have different natures, the continuity equations may be used to relate the two interface matrices $[I_{12}]$ and $[J_{12}]$ to the field variable vectors at M_1 and M_2

$$[I_{12}]\mathbf{V}^{(1)}(M_2) + [J_{12}]\mathbf{V}^{(2)}(M_3) = 0 \tag{11.68}$$

Matrices $[I_{12}]$ and $[J_{12}]$ depend on the nature of the two interfacing layers. The number of rows of the two matrices is equal to the number of continuity equations at the interface. These interface matrices relate the field vectors at points M_2 and M_3, by Equation (11.68). Since $\boldsymbol{V}^{(2)}(M_3) = [T^{(2)}]\boldsymbol{V}^{(2)}(M_4)$ where $[T^{(2)}]$ is the transfer matrix of the second layer, the acoustic propagation between the points M_2 and M_4 is expressed by

$$[I_{12}]\boldsymbol{V}^{(1)}(M_2) + [J_{12}][T^{(2)}]\boldsymbol{V}^{(2)}(M_4) = 0 \tag{11.69}$$

The expressions for the interface matrices for various interfaces are given in the following.

Solid–fluid interface

The continuity conditions are given by

$$\begin{aligned} v_3^{\mathrm{s}}(\mathrm{M}_2) &= v_3^{\mathrm{f}}(\mathrm{M}_3) \\ \sigma_{33}^{\mathrm{s}}(\mathrm{M}_2) &= -\mathrm{p}(\mathrm{M}_3) \\ \sigma_{13}^{\mathrm{s}}(\mathrm{M}_2) &= 0 \end{aligned} \tag{11.70}$$

These equations can be rewritten as $[I_{sf}]V^s(M_2) + [J_{sf}]V^f(M_3) = 0$, with

$$[I_{sf}] = \begin{bmatrix} 0 & 1 & 0 & 0 \\ 0 & 0 & 1 & 0 \\ 0 & 0 & 0 & 1 \end{bmatrix}, \qquad [J_{sf}] = \begin{bmatrix} 0 & -1 \\ 1 & 0 \\ 0 & 0 \end{bmatrix} \tag{11.71}$$

Matrices $[I]$ and $[J]$ must be interchanged for a fluid–solid interface. The same matrices can be used for a solid–thin plate interface.

Porous–fluid interface

The continuity conditions are given by

$$\begin{aligned} &(1-\phi)v_3^s(\mathrm{M}_2) + \phi v_3^f(\mathrm{M}_2) = v_3^f(\mathrm{M}_3) \\ &\sigma_{33}^s(\mathrm{M}_2) = -(1-\phi)\mathrm{p}(\mathrm{M}_3) \\ &\sigma_{13}^s(\mathrm{M}_2) = 0 \\ &\sigma_{33}^f(\mathrm{M}_2) = -\phi\mathrm{p}(\mathrm{M}_3) \end{aligned} \tag{11.72}$$

where ϕ is the porosity of the porous layer. These equations can be rewritten in the form $[I_{pf}]V^p(M_2) + [J_{pf}]V^f(M_3) = 0$, with

$$[I_{pf}] = \begin{bmatrix} 0 & (1-\phi) & \phi & 0 & 0 & 0 \\ 0 & 0 & 0 & 1 & 0 & 0 \\ 0 & 0 & 0 & 0 & 1 & 0 \\ 0 & 0 & 0 & 0 & 0 & 1 \end{bmatrix}, \qquad [J_{pf}] = \begin{bmatrix} 0 & -1 \\ (1-\phi) & 0 \\ 0 & 0 \\ \phi & 0 \end{bmatrix} \tag{11.73}$$

Matrices $[I]$ and $[J]$ must be interchanged for a fluid–porous interface.

Solid–porous interface

The continuity conditions are given by

$$\begin{aligned} v_1^s(\mathrm{M}_2) &= v_1^s(\mathrm{M}_3) \\ v_3^s(\mathrm{M}_2) &= v_3^s(\mathrm{M}_3) \\ v_3^s(\mathrm{M}_2) &= v_3^f(\mathrm{M}_3) \\ \sigma_{33}^s(\mathrm{M}_2) &= \sigma_{33}^s(\mathrm{M}_3) + \sigma_{33}^f(\mathrm{M}_3) \\ \sigma_{13}^s(\mathrm{M}_2) &= \sigma_{13}^s(\mathrm{M}_3) \end{aligned} \tag{11.74}$$

which can be rewritten as $[I_{sp}]V^s(M_2) + [J_{sp}]V^p(M_3) = 0$, with

$$[I_{sp}] = \begin{bmatrix} 1 & 0 & 0 & 0 \\ 0 & 1 & 0 & 0 \\ 0 & 1 & 0 & 0 \\ 0 & 0 & 1 & 0 \\ 0 & 0 & 0 & 1 \end{bmatrix}, \qquad [J_{sp}] = -\begin{bmatrix} 1 & 0 & 0 & 0 & 0 & 0 \\ 0 & 1 & 0 & 0 & 0 & 0 \\ 0 & 0 & 1 & 0 & 0 & 0 \\ 0 & 0 & 0 & 1 & 0 & 1 \\ 0 & 0 & 0 & 0 & 1 & 0 \end{bmatrix} \tag{11.75}$$

Matrices $[I]$ and $[J]$ must be interchanged for a porous–solid interface. Note that these matrices can be used to interface an impervious screen in mechanical contact.

Thin plate–porous interface

For a thin plate in contact with a porous layer, the continuity conditions are

$$\begin{aligned} & v_3^s(M_2) = v_3(M_3) \\ & v_3^f(M_2) = v_3(M_3) \\ & \sigma_{33}^s(M_2) + \sigma_{33}^f(M_2) = -p(M_3) \\ & \sigma_{13}^s(M_2) = 0 \end{aligned} \tag{11.76}$$

with $p = -\sigma_{33}$, the pressure at the plate's interface with the porous media. These conditions can be rewritten as $[I_{pi}]V^p(M_2) + [J_{pi}]V^i(M_3) = 0$, with

$$[I_{pi}] = -\begin{bmatrix} 0 & 1 & 0 & 0 & 0 & 0 \\ 0 & 0 & 1 & 0 & 0 & 0 \\ 0 & 0 & 0 & 1 & 0 & 1 \\ 0 & 0 & 0 & 0 & 1 & 0 \end{bmatrix}, \quad [J_{pi}] = \begin{bmatrix} 0 & -1 \\ 0 & -1 \\ 1 & 0 \\ 0 & 0 \end{bmatrix} \tag{11.77}$$

11.5 Assembling the global transfer matrix

A simple product of transfer and interface matrices cannot generally be used to calculate the global transfer matrix of a stratified medium, since most of the interface matrices are not square. On the other hand, Equation (11.69) relates the acoustic field vectors at the right-hand-side boundaries of adjacent layers in the stratified medium. For the medium illustrated by Figure 11.3, this equation leads to the following relations

$$\begin{aligned} & [I_{f1}]V^f(A) + [J_{f1}][T^{(1)}]V^{(1)}(M_2) = 0 \\ & [I_{(k)(k+1)}]V^{(k)}(M_{2k}) + [J_{(k)(k+1)}][T^{(k+1)}]V^{(k)}(M_{2(k+1)}) = 0, k = 1, \ldots, n-1 \end{aligned} \tag{11.78}$$

This set of equations can be rewritten in the form $[D_0]V_0 = 0$, where

$$[D_0] = \begin{bmatrix} [I_{f1}] & [J_{f1}][T^{(1)}] & [0] & \cdots & [0] & [0] \\ [0] & [I_{12}] & [J_{12}][T^{(2)}] & \cdots & [0] & [0] \\ \vdots & \vdots & \vdots & & \vdots & \vdots \\ [0] & [0] & [0] & \cdots & [J_{(n-2)(n-1)}][T^{(n-1)}] & [0] \\ [0] & [0] & [0] & \cdots & [I_{(n-1)(n)}] & [J_{(n-1)(n)}][T^{(n)}] \end{bmatrix} \tag{11.79}$$

and

$$v_0 = [\ v^f(A) \quad v^{(1)}(M_2) \quad v^{(2)}(M_4) \quad \cdots \quad v^{(n-1)}(M_{2n-2}) \quad v^{(n)}(M_{2n})\ ]^T \tag{11.80}$$

Matrix $[D_0]$ is rectangular. However, the global transfer matrix must be square since the physical problem is well posed. At the excitation side, one impedance equation linking the pressure to the normal velocity is missing. At the termination side, impedance conditions relating the field variables are missing. These impedance conditions are three for a porous layer, two for a solid layer and one for a fluid, thin plate or impervious layer. Thus, if N is the dimension of V_0; $[D_0]$ has $(N-4)$ rows if the last layer is porous,

$(N-3)$ rows if this layer is elastic solid, and $(N-2)$ rows if it is fluid (or equivalent fluid), thin plate or an impervious screen. The impedance conditions at the termination side depend closely on the nature of the termination: hard wall or semi-infinite fluid domain.

11.5.1 Hard wall termination condition

If the multilayer is backed by a hard wall (Figure 11.4) and if a component of the field vector $\mathbf{V}^{(n)}(M_{2n})$ is a velocity, then this component is equal to zero (infinite impedance). These conditions may be written in the form $[Y^{(n)}]V^{(n)}(M_{2n}) = 0$, where $[Y^{(n)}]$ and $V^{(n)}$are defined according to the nature of the layer (n).

$$[Y^p] = \begin{bmatrix} 1 & 0 & 0 & 0 & 0 & 0 \\ 0 & 1 & 0 & 0 & 0 & 0 \\ 0 & 0 & 1 & 0 & 0 & 0 \end{bmatrix}, \qquad [Y^s] = \begin{bmatrix} 1 & 0 & 0 & 0 \\ 0 & 1 & 0 & 0 \end{bmatrix}, \qquad [Y^f] = \begin{bmatrix} 0 & 1 \end{bmatrix} \tag{11.81}$$

where the superscript refers to the nature of the layer in contact with the wall: p for porous, s for elastic solid, and f for fluid, equivalent fluid, thin plate or impervious screen. Otherwise, vector $\mathbf{V}^{(n)}$is the field variable vector of the layer in contact with the wall.

Adding the new equations to the previous system, $[D_0]V_0 = 0$, a new system is obtained, whose matrix $[D]$ has $(N-1)$ rows and N columns:

$$[D]V = 0 : [D] = \begin{bmatrix} & [D_0] & & \\ \hdashline [0] & \cdots & [0] & [Y^{(n)}] \end{bmatrix}, \qquad V = V_0 \tag{11.82}$$

11.5.2 Semi-infinite fluid termination condition

If the multilayer is terminated with a semi-infinite fluid layer (Figure 11.5), continuity conditions may be written to relate the vectors $\mathbf{V}^{(n)}(M_{2n})$ and the semi-infinite fluid vector $\mathbf{V}^f(B)$, where B is a point in the semi-infinite medium, close to the boundary. These conditions are expressed as

$$[I_{(n)f}]V^{(n)}(M_{2n}) + [J_{(n)f}]V^f(B) = 0 \tag{11.83}$$

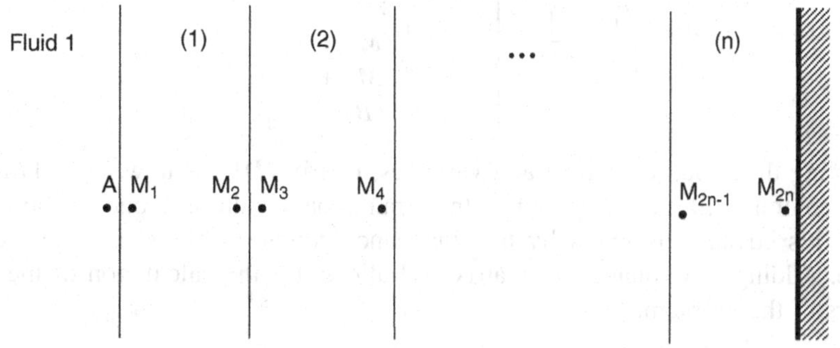

Figure 11.4 A multilayer domain backed by a hard wall.

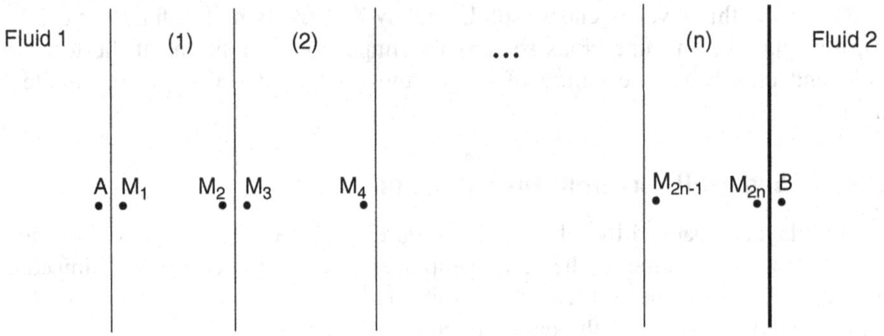

Figure 11.5 A multilayer domain backed by a semi-infinite fluid.

where $V^f(B) = [p(B) v_3^f(B)]^{\mathrm{T}}$ and $[I_{(n)f}]$ and $[J_{(n)f}]$ are interface matrices which depend on the nature of the last layer (n). Furthermore, if Z_B is the characteristic impedance of the semi-infinite fluid, the impedance at point B is given by $Z_B/\cos\theta = p(B)/v_3^f(B)$, or

$$[-1 \quad Z_B/\cos\theta\ ]V^f(B) = 0 \tag{11.84}$$

Equations (11.83) and (11.84) lead to a new matrix system

$$[D]V = 0 : [D] = \left[\begin{array}{cccc|cc} & & & & \multicolumn{2}{c}{[0]} \\ & [D_0] & & & \multicolumn{2}{c}{\vdots} \\ & & & & \multicolumn{2}{c}{[0]} \\ \hline [0] & \cdots & [0] & [I_{(n)f}] & \multicolumn{2}{c}{[J_{(n)f}]} \\ 0 & & \cdots & 0 & -1 & Z_B/\cos\theta \end{array}\right],$$

$$V = \left[\begin{array}{c} V_0 \\ \hline V^f(B) \end{array}\right] = \left[\begin{array}{c} V^f(A) \\ V^{(1)}(M_2) \\ V^{(2)}(M_4) \\ \vdots \\ V^{(n-1)}(M_{2n-2}) \\ V^{(n)}(M_{2n}) \\ V^f(B) \end{array}\right] \tag{11.85}$$

Counting the added equations and variables, matrix $[D]$ has now $(N+1)$ rows and $(N+2)$ columns. In summary, for both termination conditions, one equation is still needed. This equation is given by the impedance condition of the fluid at the excitation side. Adding this equation to matrix $[D]$ allows for the calculation of the acoustic indicators of the problem.

11.6 Calculation of the acoustic indicators

11.6.1 Surface impedance, reflection and absorption coefficients

If a plane acoustic wave impinges upon a stratified medium illustrated by either Figure 11.4 (absorption problem) or Figure 11.5 (transmission problem), at an incidence angle θ, the surface impedance Z_s of the medium is written as $Z_s = p(A)/v_3^f(A)$, or

$$[-1 \quad Z_s]V^f(A) = 0 \tag{11.86}$$

Adding this new equation to the system of Equations (11.82) or (11.85), a new system is formed with a square matrix:

$$\begin{bmatrix} -1 & Z_s & 0 & \cdots & 0 \\ \hline & & [D] & & \end{bmatrix} V = 0 \tag{11.87}$$

The determinant of this matrix is equal to zero, so Z_s is calculated by

$$Z_s = -\frac{\det[D_1]}{\det[D_2]} \tag{11.88}$$

where $\det[D_1]$ (resp. $\det[D_2]$) is the determinant of the matrix obtained when the first column (resp. the second column) has been removed from $[D]$. The reflection coefficient R and the absorption coefficient α are then given by the classical formulas:

$$R = \frac{Z_s \cos\theta - Z_0}{Z_s \cos\theta + Z_0} \tag{11.89}$$

and

$$\alpha(\theta) = 1 - |R|^2 \tag{11.90}$$

In case of a diffuse field excitation, the absorption coefficient is defined as follows:

$$\alpha_d = \frac{\displaystyle\int_{\theta_{min}}^{\theta_{max}} \alpha(\theta)\cos\theta\sin\theta d\theta}{\displaystyle\int_{\theta_{min}}^{\theta_{max}} \cos\theta\sin\theta d\theta} \tag{11.91}$$

where $\alpha(\theta)$ is the absorption coefficient at a given angle of incidence θ, as defined previously, θ_{min} and θ_{max} are the selected diffuse field integration limits, usually 0° and 90°.

11.6.2 Transmission coefficient and transmission loss

When the multilayer is extended by a semi-infinite fluid medium, the transmission coefficient T and the reflection coefficient R are related by

$$\frac{p(A)}{1+R} - \frac{p(B)}{T} = 0 \tag{11.92}$$

Adding this new equation to the system of Equations (11.85), a new system is formed with a $(N{+}2) \times (N{+}2)$ square matrix

$$\left[\begin{array}{ccccc} T & 0 & \cdots & -(1+R) & 0 \\ \hdashline & & [D] & & \end{array}\right] V = 0 \tag{11.93}$$

The determinant of this matrix is equal to zero, so T is calculated by

$$\mathrm{T} = -(1+\mathrm{R})\frac{\det[\mathrm{D_{N+1}}]}{\det[\mathrm{D_1}]} \tag{11.94}$$

where $\det[D_{N+1}]$ is the determinant of the matrix obtained when the $(N{+}1)$th column has been removed from matrix $[D]$.

For a plane wave of incidence θ, the transmission loss is defined by

$$TL = -10\log \tau(\theta) \tag{11.95}$$

where $\tau(\theta) = |T^2(\theta)|$ is the transmission coefficient for the angle of incidence θ. In case of a diffuse field excitation, the transmission loss is defined as:

$$TL_d = -10\log\left[\frac{\displaystyle\int_{\theta_{\min}}^{\theta_{\max}} |\tau(\theta)|^2 \cos\theta \sin\theta \, d\theta}{\displaystyle\int_{\theta_{\min}}^{\theta_{\max}} \cos\theta \sin\theta \, d\theta}\right] \tag{11.96}$$

where $\tau(\theta)$ is the transmission coefficient for a given angle of incidence θ, varying from $\theta_{\min}$ to $\theta_{\max}$.

In the numerical implementation, the diffuse field integration is done numerically. For example, 21-point Gauss–Kronrod rules can be used. Moreover, in the case where the transfer matrix of a layer depends on the wave heading in the (x_1, x_3) plane (e.g. the case of an orthotropic plate), integration over heading needs to be accounted for. An example is given in Chapter 12 where the size effects are investigated.

The solution procedure described in the previous section relies on the numerical evaluation of three determinants at each frequency and/or incidence angle. For a small number of layers, the computation time is not an issue unless a diffuse field indicator is needed where a numerical integration is performed. This may be expensive, especially at high frequencies. Moreover, the evaluation of a determinant may be sensitive to ill-conditioning of the matrix. A variant solution procedure is based on fixing arbitrarily the amplitude of the incident pressure to 1. Consider the system given by either Equation (11.87) or (11.93). In both cases, let the $(N{+}1) \times (N{+}1)$ square matrix obtained by eliminating the first column of matrix $[D]$ be denoted by $[D_1]$. Let Frepresent the vector obtained by multiplying the eliminated first column by -1. Finally, let V_1 represent the variables vector less its first entry. The solution of the $(N{+}1) \times (N{+}1)$ linear system, $[D_1]V_1 = F$ allows for the calculation of the acoustic indicators (depending on the problem at hand):

$$Z_s = \frac{1}{V_1(1)} \tag{11.97}$$

$$T = (1+R)V_1(N) \tag{11.98}$$

This solution procedure is straightforward and necessitates only the solution of a linear system at each frequency and/or incidence angle.

11.6.3 Piston excitation

From a transfer matrix viewpoint, a piston is equivalent to a plate with a fixed normal velocity. The variable vector is given by $V^p(A) = [p(A), v_3^f(A)]^T$ where the velocity is fixed. The continuity conditions between the excitation (the piston) and the first layer are discussed in Section 11.4. Using these interface conditions, the solution approach remains unchanged. The example of a semi-infinite fluid termination is presented here for the sake of illustration. Since the piston's velocity is given, the second column of matrix $[D]$, given in Equation (11.85), is eliminated, resulting in a square matrix denoted by $[D_2]$. Let F represent the vector obtained by multiplying the eliminated second column by -1. Finally, let V_2 represent the variable vector less its second entry. The solution procedure is as follows.

First, the $(N+1) \times (N+1)$ linear system $[D_2]V_2 = F$ is solved for vector V_2. Next, the acoustic variables are calculated:

$$\begin{aligned} p(A) &= V_2(1) \\ p(B) &= V_2(N-1) \\ v(B) &= V_2(N) \end{aligned} \tag{11.99}$$

Using these variables, various vibro-acoustic indicators can be calculated. For example, the mechanical to acoustical conversion factor, defined as the ratio of the power radiated in the receiver domain to the input power is given by

$$\mathrm{TL} = -10\log_{10}\left[\frac{\prod_{\text{radiated,free face}}}{\prod_{\text{input}}}\right] \tag{11.100}$$

with $\prod_{\text{radiated,free face}} = 1/2\,\mathrm{Re}(p(\mathrm{B})v^*(\mathrm{B}))$ and $\prod_{\text{input}} = 1/2\,\mathrm{Re}\,(p(A)v^*(A))$.

The fraction of dissipated power, defined by the difference between input mechanical power and power transmitted to the receiver domain, is given by

$$\mathrm{DP} = -10\log_{10}\left[\frac{\prod_{\text{dissipated}}}{\prod_{\text{input}}}\right] \tag{11.101}$$

with $\prod_{\text{dissipated}} = \prod_{\text{input}} - \prod_{\text{radiated}}$.

Finally, the vibration transmissibility, defined by the ratio of the free face vibration level to the imposed velocity, is given by

$$\mathrm{VT} = -10\log_{10}\left[\frac{\langle \mathrm{v}^2\rangle_{\text{piston}}}{\langle \mathrm{v}^2\rangle_{\text{free-face}}}\right] \tag{11.102}$$

Note that the use of the transfer matrix method in the case of a point load excitation is discussed in Chapter 12. The case of a point source excitation is also briefly discussed in Chapter 12.

11.7 Applications

This section presents application examples of the transfer matrices to the prediction of the absorption and transmission loss of sound packages. Further examples, taking into account size effects, mechanical excitation and/or more complicated structures will be discussed in Chapter 12. The predictions in the following examples are calculated using a computer program implementing the method of the transfer matrices as detailed above. All the examples of Chapters 5 and 6 can be easily reproduced using the transfer matrix method and thus are not reproduced here.

11.7.1 Materials with porous screens

A porous screen is a thin layer of porous material, with a thickness of about 1 mm or less. Measurements and predictions using the TMM of the surface impedance of two materials with porous screens are compared in this section. Thin porous screens are frequently used at the surface or inside sound absorbents. They can improve the absorption with a small increase in the thickness of the sound absorbing material. If the screen is an isotropic porous material, it can be modelled as a layer of isotropic porous material and represented by a transfer matrix. If the screen is anisotropic, approximate descriptions of sound propagation can be used.

In spite of possible simplifications, these descriptions remain complicated, with many parameters that are not easily measurable. As for the case of the glass wool studied in Section 6.5.4, it is generally possible, at normal incidence, to model the anisotropic porous screen as an equivalent isotropic material. The examples herein are restricted to normal incidence. Examples of prediction at oblique incidence are given in Rebillard *et al.* (1992) and Lauriks (1989).

The first material, represented in Figure 11.6, is a plastic foam covered by a sheet of glass wool. The parameters for the foam and the sheet are given in Table 11.3.

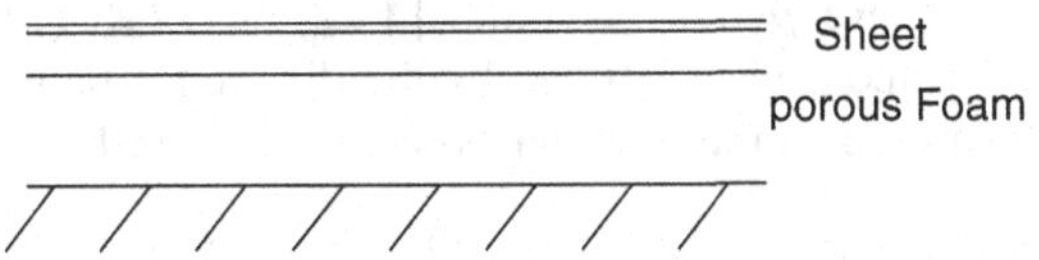

Figure 11.6 A porous material made up of a sheet of glass wool bonded onto a layer of plastic foam. The material is bonded onto a rigid impervious wall.

Table 11.3 The parameters used to predict the surface impedance of the material in Figure 11.6.

Material	Thickness, h (mm)	ϕ	σ (N s/m^4)	α_∞	Λ (μm)	Λ' (μm)	ρ_1 (kg/m^3)	E (Pa)	ν	η_s
Foam	38	0.98	5×10^3	1.1	150	216	33	130×10^3	0.3	0.1
Glass wool	0.45	0.7	$1{\cdot}1 \times 10^6$	1	10	20	660	2.6×10^6	0.3	0.1

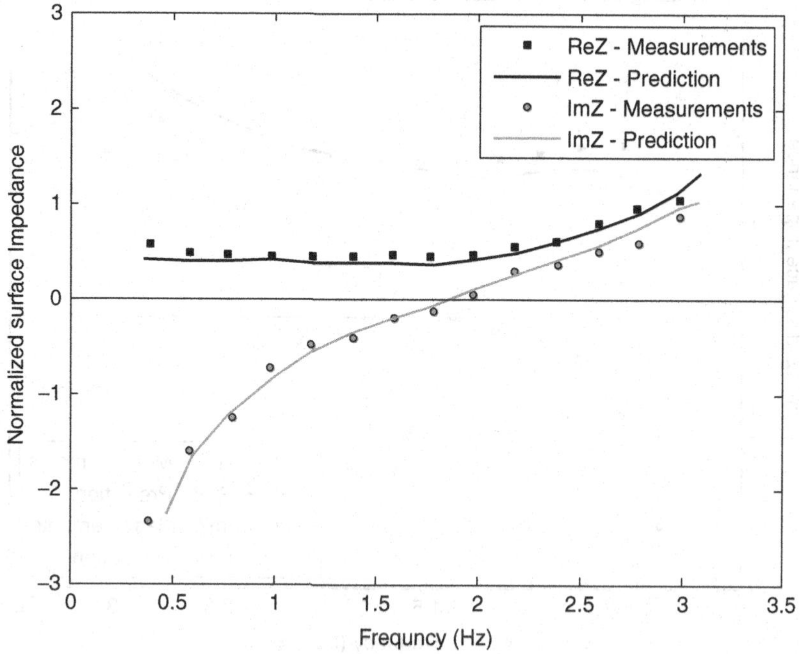

Figure 11.7 Surface impedance Z at normal incidence of the foam described in Table 11.3. Measurements reproduced from Rebillard *et al.* (1992).

The Johnson–Champoux–Allard model, summarized in Section 5.8, is used for the two materials. The characteristic lengths for the glass wool verify $\Lambda' = 2\Lambda$ as predicted by the model of Section 5.3.4. The static bulk modulus of the frame of the sheet is large, about 1 MPa and at normal incidence the sheet can be considered as a rigid framed material, due to its small thickness. The dynamic Young's modulus E has not been measured, but has been chosen sufficiently large for the sheet to behave as a rigid framed material at normal incidence.

The surface impedance is presented in Figure 11.7 for the foam and in Figure 11.8 for the sheet bonded on to the foam. The measured surface impedances have been presented in Rebillard *et al.* (1992). The absorption coefficient of the foam and of the foam plus the facing, are presented in Figure 11.9. The increase in absorption due to the facing is very large at low frequencies for the material considered. The main effect of the facing at medium frequencies is an increase in the real part of the impedance of the foam that is close to the flow resistance σh of the facing. This effect is easily explained. The density of the sheet is very large, and the inertial force almost immobilizes the screen. The velocity of the air does not vary noticeably from one face to the other in the sheet, because of its small thickness; consequently, the pressure difference Δp between both sides of the sheet is equal to $\Delta p = \sigma h v$, v being the velocity of the air above the sheet. The increase in the impedance is $\Delta p/v$ and is equal to the flow resistance σh of the sheet. However, in the case of porous screens with lower density, the screen is not motionless, and a porous elastic or limp model must be used for the screen to predict the surface impedance.

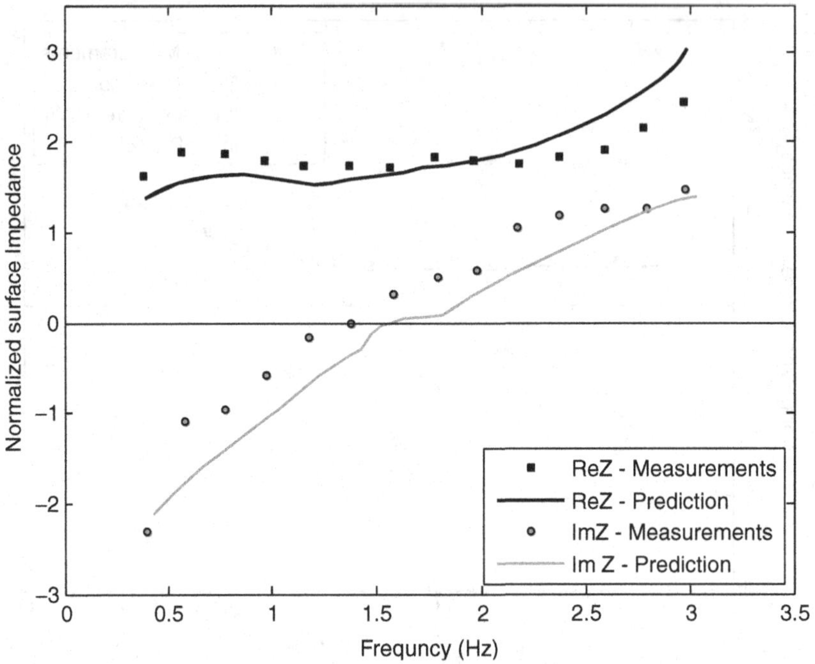

Figure 11.8 Surface impedance Z at normal incidence of the layered material represented in Figure 11.6. Measurements reproduced from Rebillard *et al.* (1992).

The second example of a material with a porous screen is represented in Figure 11.10. Between foam 3 and the blanket, there is a thin layer of high flow resistance created by the gluing of the blanket on to the foam 3. The parameters for the different porous layers are indicated in Table 11.4. As for the first material, the bulk modulus of the screen is large and, due to its small thickness, the screen can be considered to be a rigid framed material. The Young's modulus of the screen has not been measured, but has been set sufficiently large so that the screen behaves as a rigid framed material at normal incidence. The three other layers are modeled as porous-elastic layers. The Johnson–Champoux–Allard model is used for all the layers of the material. The surface impedance at normal incidence is presented in Figure 11.11. The measurement is reproduced from Allard *et al.* (1987). The predicted surface impedance of the material without the screen is also represented in Figure 11.11. There is no measurement related to this prediction, because the screen is inside the material, and the material is damaged when the screen is removed. For this material, the density of the screen is smaller and its flow resistivity larger compared with the screen of the first material, and the screen is not immobilized by the inertial forces and the stiffness of the frame over the range of frequencies for which the measurements were performed. Moreover, the screen is set between two porous layers and its effect on surface impedance is not as simple as for the previous material. The difference between the two predicted impedances cannot be calculated by the simple model which was valid for the case of the first stratified material. The absorption coefficients of the second material

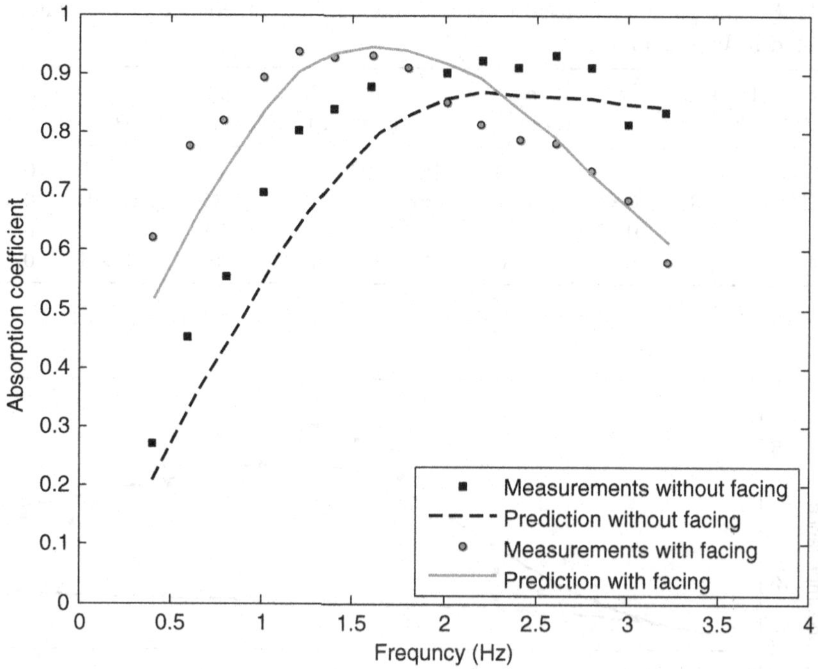

Figure 11.9 The absorption coefficient of the foam with and without facing. The parameters for the foam and the facing are give in Table 11.3. Measurements reproduced from Rebillard *et al.* (1992).

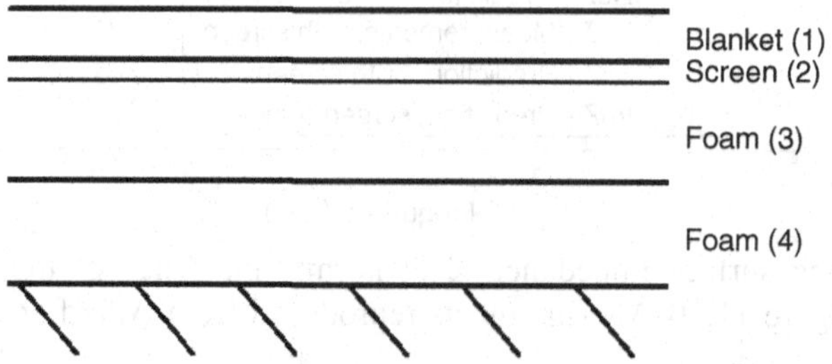

Figure 11.10 A layered material comprising a blanket (1), a screen (2) and two foams (3) and (4). The material is bonded onto a rigid impervious wall.

with and without the screen are presented in Figure 11.12. The predicted increase in the absorption due to the screen is important at low frequencies as in the case of the first material.

The last example consists of a moderately thin resistive screen inserted between a foam and a felt. The properties of the materials are in Table 11.5. The normal incidence absorption of the system is compared with simulations in Figure 11.13 in the configuration where the foam is bonded onto a rigid wall. Both the foams and the felt were

Table 11.4 The parameters used to predict the surface impedance of the material represented in Figure 11.10.

Material	Thickness, h (mm)	ϕ	σ (N s/m^4)	α_∞	Λ (μm)	Λ' (μm)	ρ_1 (kg/m^3)	E (Pa)	ν	η_s
Blanket (1)	4	0.98	34×10^3	1.18	60	86	41	286×10^3	0.3	0.015
Screen (2)	0.8	0.8	$3{\cdot}2 \times 10^6$	2.56	6	24	125	2.6×10^6	0.3	0.1
Foam (3)	5	0.97	87×10^3	2.52	36	118	31	143×10^6	0.3	0.055
Foam (4)	16	0.99	65×10^3	1.98	37	120	16	46.8×10^6	0.3	0.1

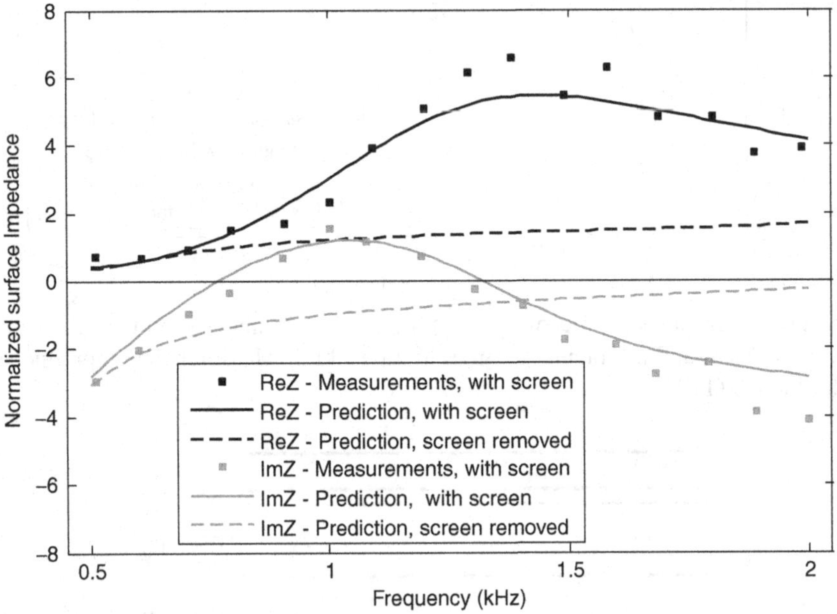

Figure 11.11 The surface impedance Z at normal incidence of the layered material represented in Figure 11.10. Measurements reproduced from Allard *et al.* (1987).

modelled as porous limp materials. The screen was modelled using a rigid porous material following the model of Johnson, Champoux and Allard. As explained by Atalla and Sgard (2007), the characteristic lengths of the screen were derived from the measured flow resistance following $\Lambda = \Lambda' = r = \sqrt{8\eta/\phi\sigma}$. The air flow resistivity and the porosity (or percentage open area) of the screen were measured. The tortuosity was taken equal to $\alpha_\infty(\omega) = 1 + (\varepsilon_e/d)(\tilde{\alpha}_{Felt} + \tilde{\alpha}_{Foam})$ where $\tilde{\alpha}_{Felt}$ and $\tilde{\alpha}_{Foam}$ are the dynamic tortuosity of the foam and the felt, respectively, d, the thickness of the screen and ε_e is taken equal to $0.48\sqrt{\pi r^2}(1 - 1.47\sqrt{\phi} + 0.47\sqrt{\phi^3})$. It is shown in Atalla and Sgard (2007) that this model captures well the behaviour of perforated screens and plates in contact with porous materials. Because of its small thickness, the stiffness and mass of the screen were ignored. The results of Figure 11.13 confirm the validity of this assumption and the used model for the characteristic lengths and tortuosity. Note in passing that the approach

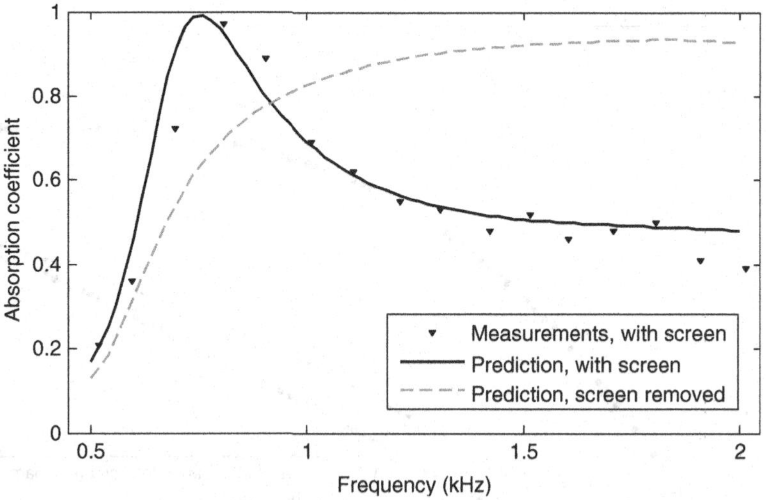

Figure 11.12 The absorption coefficient of the material represented in Figure 11.10. The parameters for the different layers are given in Table 11.4. Measurements reproduced from Allard *et al.* (1987).

Table 11.5 The parameters used to predict the sound absorption of a felt–screen–foam system.

Material	Thickness, h (mm)	ϕ	σ (N s/m^4)	α_∞	Λ (μm)	Λ' (μm)	ρ_1 (kg/m^3)
Felt (1)	19	0.99	23×10^3	1.4	64	131	66
Screen (2)	0.08	0.08	137×10^3	model	model	model	
Foam (3)	27	0.99	10.9×10^3	1.02	100	130	8.8

of Chapter 9 can also be used within the TMM framework to solve problems involving perforated facings.

11.7.2 Materials with impervious screens

The surface impedance of porous layers faced by impervious screens has been studied by Zwikker and Kosten (1942) at normal incidence. The Biot theory has been used by Bolton (1987) to predict the surface impedance of these materials at oblique incidence. Similar work has been performed by Lauriks *et al.* (1990) with the transfer matrices. The use of these matrices gives more flexibility to the method because a stratified porous material can be represented in the same way as only one porous layer by a transfer matrix.

A plastic foam surfaced with an impervious screen is represented in Figure 11.14. The screen is modelled as a perfectly flexible membrane and is represented by a 4×4 transfer matrix to account for shearing by the foam. The surface impedance of the system is presented in Figure 11.15. The parameters that characterize the foam are presented in Table 11.6. The thickness of the screen is equal to 25 μm, and the mass per unit area is equal to 0·02 kg m^{-2}. The flexural stiffness and the stiffness related to an increase of

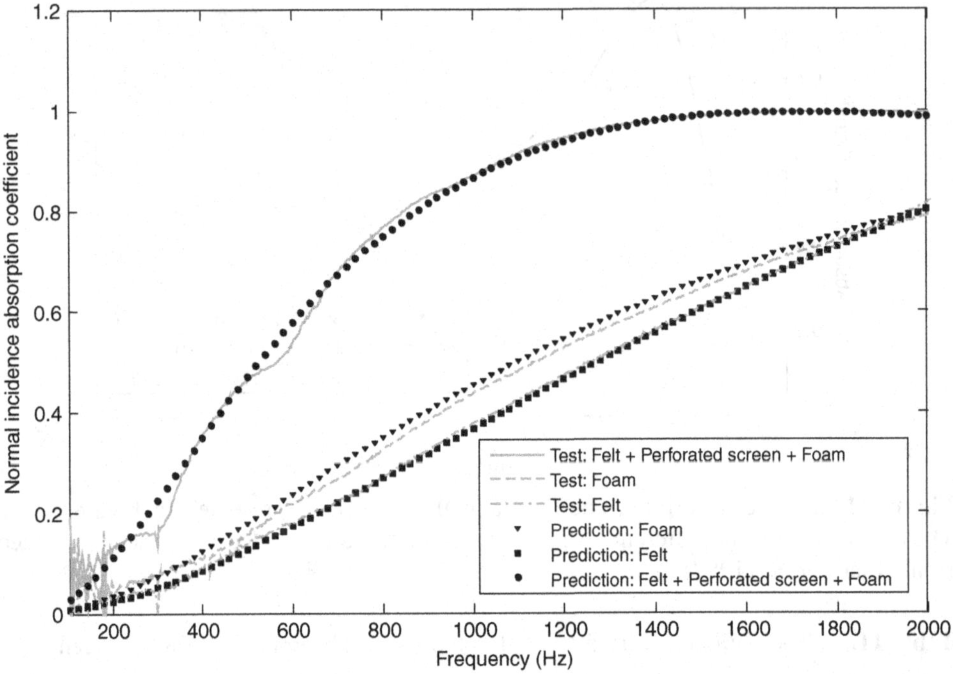

Figure 11.13 The absorption coefficient of a felt–screen–foam system; comparison between measurements and predictions. Measurements reproduced from Atalla and Sgard (2007). Reprinted from Atalla, N. & Sgard, F. Modelling of perforated plates and screens using rigid frame porous models. *J. Sound Vib.* **303** (1–2), 195–208. (2007) with permission from Elsevier.

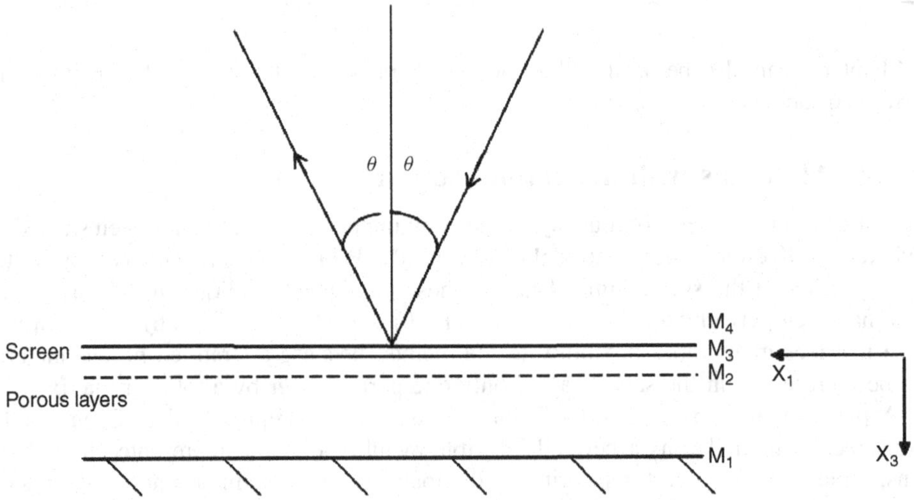

Figure 11.14 A porous material surfaced with an impervious screen and bonded on to a rigid impervious wall.

Table 11.6 The parameters used to predict the surface impedance of the material represented in Figure 11.14.

Material	Thickness, h (mm)	ϕ	σ (N s/m^4)	α_∞	Λ (μm)	Λ' (μm)	ρ_1 (kg/m^3)	E (Pa)	ν	η_s
Foam	20	0.98	22×10^3	1.9	87	146	30	294×10^3	0.2	0.18

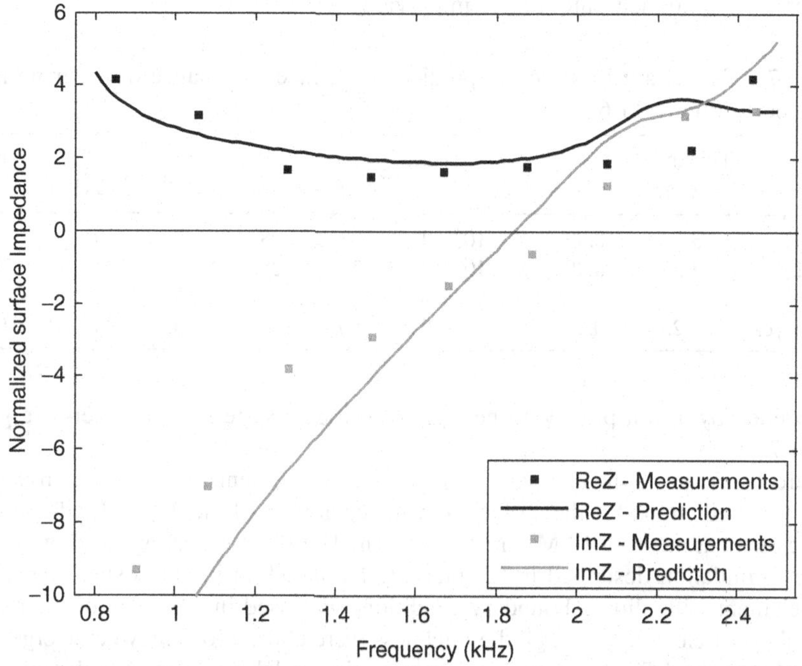

Figure 11.15 Surface impedance Z at an angle of incidence $\theta = 30°$ of foam with an impervious facing, bonded on to a rigid impervious wall.

area have been neglected, while the frequency dependence of the viscous force and of the bulk modulus of the air in the foam have been calculated using the model of Section 5.9.2. The agreement between prediction and measurement is not very good. Nevertheless, the predicted and measured impedances present similar behaviour. A resonance appears close to 2000 Hz and the imaginary part of the impedance increases quickly with frequency. It may be noticed that the measurement is difficult to perform far from the resonance, the reflection coefficient being very close to 1.

Next, the example of a porous material made up of several porous layers with an embedded impervious screen wall is presented. The structure is represented in Figure 11.16. It is made up of a carpet (1, 2), an impervious screen (3) and a fibrous layer (4). The carpet is modelled as a porous material made up of two layers, because the fibres are grouped in small bundles fixed to an impervious surface at the lower face, and are more regularly distributed at the upper face. The screen is 3 mm thick and has a surface density of 6 kg/m^2. Because of the mass density and thickness of the screen

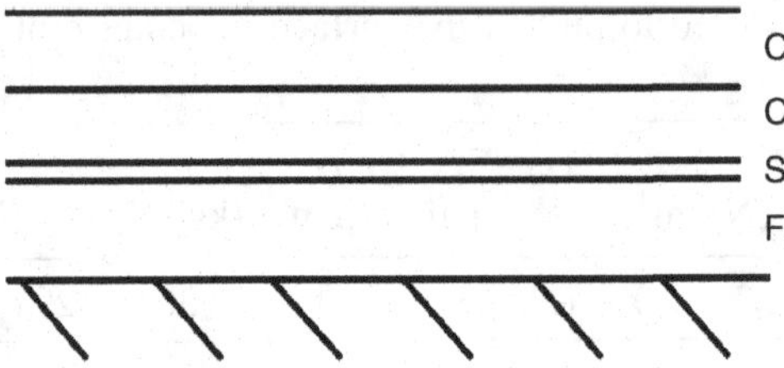

Figure 11.16 A porous material made up of several porous layers with an inner impervious screen and bonded onto a rigid impervious wall.

Table 11.7 The parameters used to predict the surface impedance of the material represented in Figure 11.6.

Material	Thickness, h (mm)	ϕ	σ (N s/m^4)	α_∞	Λ (μm)	Λ' (μm)	ρ_1 (kg/m^3)	E E (Pa)	ν	η_s
Carpet (1)	3.5	0.99	5×10^3	1	23	28	60	20×10^3	0	0.5
Carpet (2)	3.5	0.99	5×10^3	1	23	28	60	20×10^3	0	0.5
Screen	3						2000			
Fibrous layer	1.25	0.98	33×10^3	1.1	50	110	60	100×10^3	0	0.88

it is modelled by a thin plate with negligible stiffness. Material parameters are given in Table 11.7.

Figure 11.17 shows the comparison between measurements, taken from Brouard *et al.* (1994), and predictions for the surface impedance at normal incidence. Predictions using the finite element method (FEM) are also shown. The FEM model is based on the mixed pressure formulation described in Chapter 13. To model an infinite extent material, one quad 4 element with sliding boundary conditions was used in the plane of the materials. The numbers of elements through the thickness were chosen to achieve convergence. For the TMM and the FE predictions, the poroelastic model was used for the carpets and the fibrous layer while a septum model (mass layer) was used for the impervious screen. Good agreement is observed between the two models and the measurements.

11.7.3 Normal incidence sound transmission through a plate–porous system

A layer of the glass wool studied in Section 6.5.4, of thickness 5 cm, is bonded on to a plate of aluminium, of thickness 1 mm. As indicated in Section 6.5.4, the glass wool is anisotropic, and the transmission is calculated only at normal incidence using the properties of Table 11.8. At normal incidence, there is no flexural deformation of the plate, and one only needs the values of the thickness and the density to characterize the plate (the transmission is governed by mass law). The transmission coefficient τ at normal incidence is represented in Figure 11.18, and compared with the transmission through the plate, with the glass wool removed. It appears that the transmission is larger in two intervals of frequencies when the glass wool is present. The frame of the glass

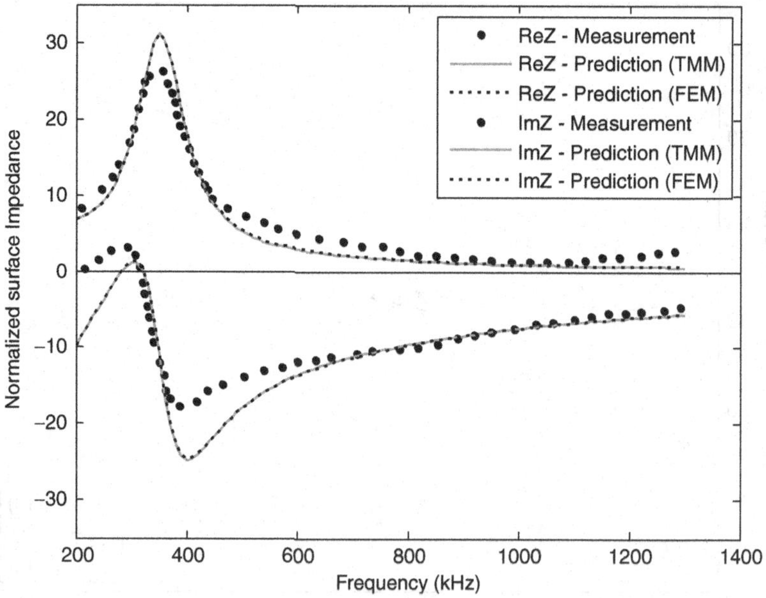

Figure 11.17 Surface impedance Z at an angle of incidence $\theta = 0°$ of a multilayer sound package bonded on to a rigid impervious wall. Measurement vs predictions. Measurements reproduced from Brouard *et al.* (1994).

Table 11.8 The parameters used to predict the results of Figure 11.18.

Material	Thickness h (mm)	ϕ	σ (N s/m^4)	α_∞	Λ (μm)	Λ' (μm)	ρ_1 (kg/m^3)	E (Pa)	ν	η_s
Glass wool	3.8	0.94	40×10^3	1.06	56	110	130	4.4×10^6	0	0.1
Plate	1						2800	$7. \times 10^{10}$	0.3	0.007

wool and the plate can be decoupled by inserting an air gap between the glass wool and the plate. The air gap is simulated by a fluid layer. The new predicted transmission loss is represented in Figure 11.18. The reinforcement of the transmission has decreased noticeably.

It may be noticed that the effect of the decoupling was previously pointed by Shiau *et al.* (1988), for different configurations, and observed by Roland and Guilbert (1990) for the case of the same glass wool bonded onto a plate made of concrete.

11.7.4 Diffuse field transmission of a plate–foam system

The material is a foam of thickness $h = 2 \cdot 54$ cm bonded onto a 0.6 mm aluminium plate. The parameters for the foam and the plate are given in Table 11.9. The transmission loss is predicted using the transfer matrix method. The integration is usually performed up to an angle smaller than 90° (Mulholland *et al.* 1967), and the upper limit for the integration has

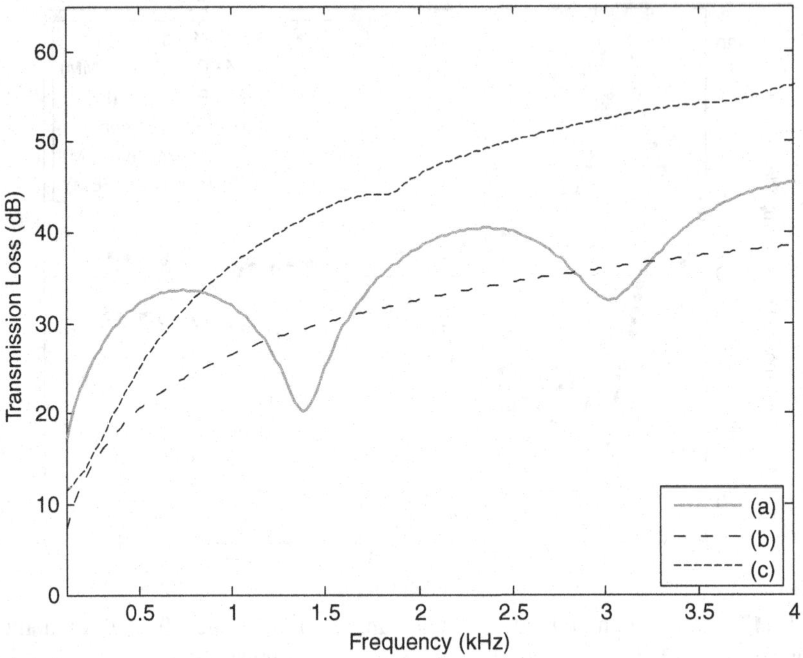

Figure 11.18 (a) The predicted transmission loss coefficient $TL = -10\log_{10}\tau$ at normal incidence of a glass wool bonded on to a plate. (b) Transmission loss of the plate. (c) Transmission loss when the plate and the material are decoupled by an air gap.

Table 11.9 The parameters used to predict the results of Figure 11.19.

Material	Thickness, h (mm)	ϕ	σ (N s/m^4)	α_∞	Λ (μm)	Λ' (μm)	ρ_1 (kg/m^3)	E (Pa)	ν	η_s
Foam	25.4	0.98	6.6×10^3	1.03	200	380	11.2	2.93×10^5	0.2	0.06
Plate	1.6						2800	7.2×10^{10}	0.3	0.007

been chosen as 78°. Prediction is compared in Figure 11.19 with measurement. Despite the finite dimension of the material ($0.8 \times 0.8\text{m}^2$), the agreement between prediction and measurement is good. Chapter 12 presents a correction for the size effects and its experimental validation.

The transmission loss in the opposite direction (TL') has been verified to match the transmission loss (*TL*) in the initial direction. The difference between the measured values of *TL* and the transmission loss coefficient TL' in the opposite direction is smaller than $2 \cdot 5$ db in the range (50–4000 Hz). The identity of *TL* and TL', can be proved by using methods described in Allard (1993).

Several other transmission loss examples in both single wall and double wall configurations, together with diffuse field absorption examples accounting for size effects, will be discussed in Chapter 12.

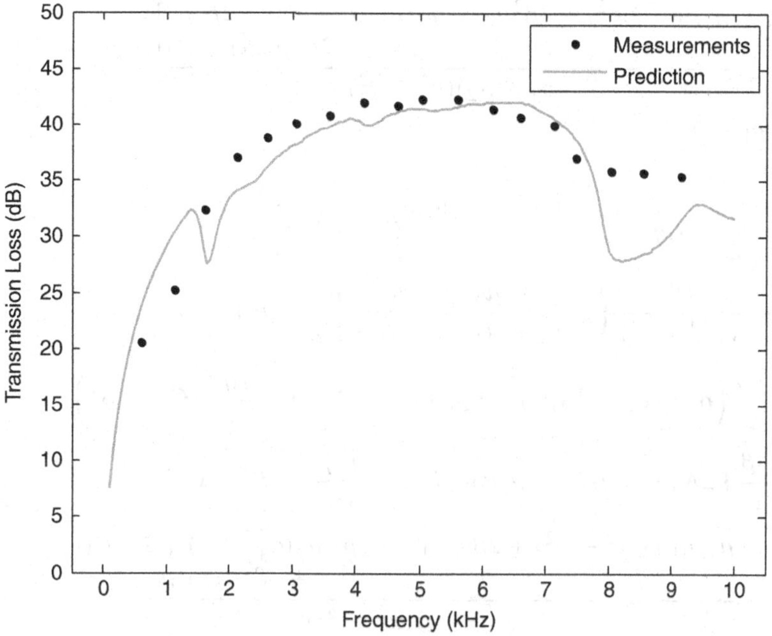

Figure 11.19 The transmission loss for a foam bonded on to a plate, in diffuse field. Prediction vs measurements.

Appendix 11.A The elements T_{ij} of the Transfer Matrix $[T]$

$$T_{11} = \frac{2N\beta^2(p_2D_1 - p_1D_2) - (p_3(C_1D_2 - C_2D_1))}{\Delta} \tag{11.A.1}$$

$$T_{12} = j\beta\frac{\alpha_2q_1[\mu_2(\alpha_3^2 - \beta^2) + 2\beta^2\mu_3] - \alpha_1q_2[\mu_1(\alpha_3^2 - \beta^2) + 2\beta^2\mu_3] + 2\alpha_3q_3\alpha_1\alpha_2(\mu_1 - \mu_2)}{\alpha_1\alpha_2(\mu_1 - \mu_2)(\beta^2 + \alpha_3^2)} \tag{11.A.2}$$

$$T_{13} = j\beta\frac{(\alpha_1q_2 - \alpha_2q_1)}{\alpha_1\alpha_2(\mu_1 - \mu_2)} \tag{11.A.3}$$

$$T_{14} = \beta\omega\frac{[p_1D_2 - p_2D_1 - p_3(D_2 - D_1)]}{\Delta} \tag{11.A.4}$$

$$T_{15} = \frac{j\omega}{\mathrm{N}(\beta^2 + \alpha_3^2)}\left(\frac{\beta^2q_1(\mu_2 - \mu_3)}{\alpha_1(\mu_2 - \mu_1)} + \frac{\beta^2q_2(\mu_1 - \mu_3)}{\alpha_2(\mu_1 - \mu_2)} + \alpha_3q_3\right) \tag{11.A.5}$$

$$T_{16} = \beta\omega\left(\frac{p_2C_1 - p_1C_2 - p_3(C_1 - C_2) + 2N\beta^2(p_2 - p_1)}{\Delta}\right) \tag{11.A.6}$$

$$T_{21} = \frac{j\beta}{\Delta}\left(2N(\alpha_1q_1D_2 - \alpha_2q_2D_1) - \frac{q_3}{\alpha_3}(C_1D_2 - C_2D_1)\right) \tag{11.A.7}$$

$$T_{22} = \frac{p_2[\mu_1(\alpha_3^2 - \beta^2) + 2\beta^2\mu_3] - p_1[\mu_2(\alpha_3^2 - \beta^2) + 2\beta^2\mu_3] + 2\beta^2 p_3(\mu_1 - \mu_2)}{(\mu_1 - \mu_2)(\beta^2 + \alpha_3^2)} \tag{11.A.8}$$

$$T_{23} = \frac{p_1 - p_3}{\mu_1 - \mu_2} \tag{11.A.9}$$

$$T_{24} = -\frac{j\omega}{\Delta}\left(\alpha_1 q_1 D_2 - \alpha_2 q_2 D_1 + \frac{\beta^2 q_3}{\alpha_3}(D_2 - D_1)\right) \tag{11.A.10}$$

$$T_{25} = -\frac{\omega\beta}{N(\beta^2 + \alpha_3^2)}\left(p_1\frac{\mu_2 - \mu_3}{\mu_2 - \mu_1} + p_2\frac{\mu_1 - \mu_3}{\mu_1 - \mu_2} - p_3\right) \tag{11.A.11}$$

$$T_{26} = \frac{j\omega}{\Delta}\left(\alpha_1 q_1(C_2 + 2N\beta^2) - \alpha_2 q_2(C_1 + 2N\beta^2) - \frac{q_3\beta^2}{\alpha_3}(C_1 - C_2)\right) \tag{11.A.12}$$

$$T_{31} = \frac{j\beta}{\Delta}\left(2N(\alpha_1\mu_1 q_1 D_2 - \alpha_2\mu_2 q_2 D_1) - \frac{\mu_3 q_3}{\alpha_3}(C_1 D_2 - C_2 D_1)\right) \tag{11.A.13}$$

$$T_{32} = \frac{-\mu_1 p_1[\mu_2(\alpha_3^2 - \beta^2) + 2\beta^2\mu_3] + \mu_2 p_2[\mu_1(\alpha_3^2 - \beta^2) + 2\beta^2\mu_3] + 2\beta^2\mu_3 p_3(\mu_1 - \mu_2)}{(\mu_1 - \mu_2)(\alpha_3^2 + \beta^2)} \tag{11.A.14}$$

$$T_{33} = \frac{\mu_1 p_1 - \mu_2 p_2}{\mu_1 - \mu_2} \tag{11.A.15}$$

$$T_{34} = \frac{j\omega}{\Delta}\left(-\alpha_1\mu_1 q_1 D_2 + \alpha_2\mu_2 q_2 D_1 + \frac{\beta^2\mu_3 q_3}{\alpha_3}(D_1 - D_2)\right) \tag{11.A.16}$$

$$T_{35} = \frac{-\beta\omega}{N(\alpha_3^2 + \beta^2)}\left(p_1\mu_1\frac{\mu_2 - \mu_3}{\mu_2 - \mu_1} + p_2\mu_2\frac{\mu_1 - \mu_3}{\mu_1 - \mu_2} - p_3\mu_3\right) \tag{11.A.17}$$

$$T_{36} = \frac{j\omega}{\Delta}\left(\mu_1\alpha_1 q_1(C_2 + 2N\beta^2) - \mu_2\alpha_2 q_2(C_1 + 2N\beta^2) - \frac{\beta^2}{\alpha_3}\mu_3 q_3(C_1 - C_2)\right) \tag{11.A.18}$$

$$T_{41} = \frac{2N\beta}{\omega\Delta}[C_1 p_1 D_2 - C_2 p_2 D_1 - p_3(C_1 D_2 - C_2 D_1)] \tag{11.A.19}$$

$$T_{42} = -j\frac{C_1 q_1\alpha_2[\mu_2(\alpha_3^2 - \beta^2) + 2\beta^2\mu_3] - C_2 q_2\alpha_1[\mu_1(\alpha_3^2 - \beta^2) + 2\beta^2\mu_3] - 4N\alpha_3\beta^2\alpha_1\alpha_2(\mu_1 - \mu_2)q_3}{\alpha_1\alpha_2\,\omega(\beta^2 + \alpha_3^2)(\mu_1 - \mu_2)} \tag{11.A.20}$$

$$T_{43} = j\frac{\alpha_2 C_1 q_1 - \alpha_1 C_2 q_2}{\omega\alpha_1\alpha_2(\mu_1 - \mu_2)} \tag{11.A.21}$$

$$T_{44} = \frac{-p_1 C_1 D_2 + p_2 C_2 D_1 - 2N\beta^2 p_3(D_2 - D_1)}{\Delta} \tag{11.A.22}$$

$$T_{45} = \frac{-j\beta}{\beta^2 + \alpha_3^2}\left(\frac{C_1 q_1}{N\alpha_1}\frac{\mu_2 - \mu_3}{\mu_2 - \mu_1} + \frac{C_2 q_2}{N\alpha_2}\frac{\mu_1 - \mu_3}{\mu_1 - \mu_2} - 2q_3\alpha_3\right) \tag{11.A.23}$$

$$T_{46} = \frac{p_1 C_1(C_2 + 2N\beta^2) - p_2 C_2(C_1 + 2N\beta^2) - 2N\beta^2 p_3(C_1 - C_2)}{\Delta} \tag{11.A.24}$$

$$T_{51} = \frac{jN\beta^2}{\Delta\omega}\left(4N\alpha_1 q_1 D_2 - 4N\alpha_2 q_2 D_1 - q_3\frac{\alpha_3^2 - \beta^2}{\beta^2\alpha_3}(C_1 D_2 - C_2 D_1)\right) \tag{11.A.25}$$

$$T_{52} = \frac{\begin{array}{r}2N\beta p_1[\mu_2(\alpha_3^2 - \beta^2) + 2\beta^2\mu_3] - 2N\beta p_2[\mu_1(\alpha_3^2 - \beta^2) + 2\beta^2\mu_3] \\ + 2N\beta p_3(\alpha_3^2 - \beta^2)(\mu_1 - \mu_2)\end{array}}{\omega(\mu_1 - \mu_2)(\beta^2 + \alpha_3^2)} \tag{11.A.26}$$

$$T_{53} = \frac{-2N\beta}{\omega(\mu_1 - \mu_2)}(p_1 - p_2) \tag{11.A.27}$$

$$T_{54} = \frac{2jN\beta}{\Delta}\left(\alpha_1 q_1 D_2 - \alpha_2 q_2 D_1 - \frac{q_3}{2}\frac{\alpha_3^2 - \beta^2}{\alpha_3}(D_2 - D_1)\right) \tag{11.A.28}$$

$$T_{55} = \frac{2\beta^2}{\beta^2 + \alpha_3^2}\left(p_1\frac{\mu_2 - \mu_3}{\mu_2 - \mu_1} + p_2\frac{\mu_1 - \mu_3}{\mu_1 - \mu_2} + p_3\frac{\alpha_3^2 - \beta^2}{2\beta^2}\right) \tag{11.A.29}$$

$$T_{56} = \frac{-2jN\beta}{\Delta}\left[\alpha_1 q_1(C_2 + 2N\beta^2) - \alpha_2 q_2(C_1 + 2N\beta^2) + \frac{q_3}{2}\frac{\alpha_3^2 - \beta^2}{\alpha_3}(C_1 - C_2)\right] \tag{11.A.30}$$

$$T_{61} = \frac{2N\beta D_1 D_2}{\omega\Delta}(p_1 - p_2) \tag{11.A.31}$$

$$T_{62} = -\frac{j}{\omega}\frac{\alpha_2 q_1 D_1[\mu_2(\alpha_3^2 - \beta^2) + 2\beta^2\mu_3] - \alpha_1 q_2 D_2[\mu_1(\alpha_3^2 - \beta^2) + 2\beta^2\mu_3]}{\alpha_1\alpha_2(\mu_1 - \mu_2)(\beta^2 + \alpha_3^2)} \tag{11.A.32}$$

$$T_{63} = \frac{j}{\omega(\mu_1 - \mu_2)}\left(\frac{q_1 D_1}{\alpha_1} - \frac{q_2 D_2}{\alpha_2}\right) \tag{11.A.33}$$

$$T_{64} = \frac{-D_1 D_2}{\Delta}(p_1 - p_2) \tag{11.A.34}$$

$$T_{65} = -\frac{j\beta}{N(\beta^2 + \alpha_3^2)}\left(\frac{q_1 D_1}{\alpha_1}\frac{\mu_2 - \mu_3}{\mu_2 - \mu_1} + \frac{q_2 D_2}{\alpha_2}\frac{\mu_1 - \mu_3}{\mu_1 - \mu_2}\right) \tag{11.A.35}$$

$$T_{66} = \frac{p_1 D_1(C_2 + 2N\beta^2) - p_2 D_2(C_1 + 2N\beta^2)}{\Delta} \tag{11.A.36}$$

In these expressions, the quantities α_i, β, C_i, D_i, p_i, q_i and Δ are equal to, respectively,

$$\alpha_i = k_{i3} \qquad i = 1, 2, 3 \tag{11.A.37}$$

$$\beta = k_t \tag{11.A.38}$$

$$C_i = (P + Q\mu_i)(\beta^2 + \alpha_i^2) - 2N\beta^2 \qquad i = 1, 2 \tag{11.A.39}$$

$$D_i = (R\mu_i + Q)(\beta^2 + \alpha_i^2) \qquad i = 1, 2 \tag{11.A.40}$$

$$p_i = \cos k_{i3}h \qquad i = 1, 2, 3 \tag{11.A.41}$$

$$q_i = \sin k_{i3}h \qquad i = 1, 2, 3 \tag{11.A.42}$$

$$\Delta = D_1(2N\beta^2 + C_2) - D_2(2N\beta^2 + C_1) \tag{11.A.43}$$

Note:

The expressions of the elements T_{ij} are not simple, and it can be easier to evaluate $[T]$ by the use of Equation (11.33) from $[\Gamma(0)]$ and $[\Gamma(h)]^{-1}$ than to write explicit expressions (11.A.1)–(11.A.36) in a program. In order to avoid a matrix inversion, the origin of the x_3 axis can be changed, and the following equation can be used:

$$[T] = [\Gamma(-h)][\Gamma(0)]^{-1} \tag{11.A.44}$$

The matrix elements of $[\Gamma(0)]^{-1}$ are given in Lauriks *et al.* (1990).

References

Allard, J.F., Champoux, Y. and Depollier, C. (1987) Modelization of layered sound absorbing materials with transfer matrices. *J. Acoust. Soc. Amer.* **82**, 1792–6.

Allard, J.F. (1993) *Propagation of sound in Porous media. Modeling sound absorbing materials*. Chapter 11, Elsevier Applied Science, London.

Atalla, N. and Sgard, F. (2007) Modeling of perforated plates and screens using rigid frame porous models. *J. Sound Vib.* **303**(1-2), 195–208.

Beranek, L.L. (1947) Acoustical properties of homogeneous isotropic rigid tiles and flexible blankets. *J. Acoust. Soc. Amer.* **19**, 556–68.

Bolton, J.S. (1987) Optimal use of noise control foams. *J. Acoust. Soc. Amer.* **82**, suppl. 1, 10.

Brekhovskikh, L.M. (1960) *Waves in Layered Media*. Academic Press, New York.

Brouard B., Lafarge, D. and Allard, J.F. (1994) Measurement and prediction of the surface impedance of a resonant sound absorbing structure. *Acta Acustica* **2**, 301–306.

Brouard, B., Lafarge, D. and Allard, J.F. (1995) A general method of modeling sound propagation in layered media. *J. Sound Vib.* **183** (1), 129–142

Depollier, C. (1989) *Théorie de Biot et prédiction des propriétés acoustiques des matériaux poreux. Propagation dans les milieux acoustiques désordonnés*. Thesis, Université du Maine, France.

Doutres O., Dauchez, N., Geneveaux, J.M. and Dazel, 0. (2007) Validity of the limp model for porous materials: A creterion based on Biot theory. *J. Acoust. Soc. Amer.* **122**(4), 2038–2048.

Folds, D. and Loggins, C.D. (1977) Transmission and reflection of ultrasonic waves in layered media. *J. Acoust. Soc. Amer.* **62**, 1102–9.

Ingard K.U. (1994) *Notes on Sound Absorption Technology*. Noise Control Foundation. Arlington Branch Poughkeepsie, NY

Katragadda, S., Lai, H.Y. and Bolton, J.S. (1995) A model for sound absorption by and sound transmission through limp fibrous layers. *J. Acoust. Soc. Amer.* **98**, 2977.

Lauriks, W. (1989) *Onderzoek van de akoestische eigenschappen van gelaagde poreuze materialen*. Thesis, Katholieke Universiteit Leuven, Belgium.

Lauriks, W., Cops, A., Allard, J.F., Depollier, C. and Rebillard, P. (1990) Modelization at oblique incidence of layered porous materials with impervious screens. *J. Acoust. Soc. Amer.* **87**, 1200–6.

Mulholland, K.A., Parbroo, H.D., and Cummings, A. (1967) The transmission loss of double panels, *J. Sound Vib.*, **6**, 324–34.

Panneton, R. (2007) Comments on the limp frame equivalent fluid model for porous media, *J. Acoust. Soc. Amer*. **122**(6), EL 217–222.

Roland, J. and Guilbert, G. (1990) L'amélioration acoustique d'un produit du bâtiment. *CSTB Magazine* **33**, 11–12.

Rebillard, P. *et al.*, (1992) The effect of a porous facing on the impedance and the absorption coefficient of a layer of porous material. *J. Sound Vib*. **156**, 541–55.

Shiau, J.M., Bolton, J.S. and Ufford, D.A. (1988) Random incidence sound transmission through foam-lined panels. *J. Acoust. Soc. Amer* **84**, suppl. 1, 96

Scharnhorst, K.P. (1983) Properties of acoustic and electromagnetic transmission coefficients and transfer matrices of multilayered plates. *J. Acoust. Soc. Amer*., **74**, 1883–6.

Utsuno H., Tanaka T. and Fujikawa T. (1989) Transfer function method for measuring characteristic impedance and propagation constant of porous materials. *J. Acoust. Soc. Amer*, **86**, 637–643.

Zwikker, C. and Kosten, C.W. (1949) *Sound Absorbing Materials*. Elsevier, New York.

12

Extensions to the transfer matrix method

12.1 Introduction

This chapter discusses various extensions and applications of the transfer matrix method (TMM). First, corrections to account for size effects in absorption and transmission loss problems are presented with validation examples. Next, the application of the method to point load excited panels with attached sound packages is discussed.

12.2 Finite size correction for the transmission problem

The classical transfer matrix method assumes a structure of infinite extent. For transmission loss application, the method correlates well with experiments at mid to high frequencies for a large range of flat panels. Discrepancies are however observed at low frequencies, especially for panels of small size. This section presents a simple geometrical correction to account for this 'geometrical' finite size effect (Ghinet and Atalla, 2002). The rationale behind the approach presented is to replace the radiation efficiency in the receiving domain by the radiation efficiency of an equivalent baffled window. This approach is thus strictly valid for planar structures.

12.2.1 Transmitted power

Consider a baffled panel of area S excited by a plane incident wave with heading angles(θ, ϕ)

$$\hat{p}_i(M) = \exp\left[-jk_0(\cos\phi\sin\theta x + \sin\phi\sin\theta y + \cos\theta z)\right] \tag{12.1}$$

Propagation of Sound in Porous Media: Modelling Sound Absorbing Materials, Second Edition J. F. Allard and N. Atalla

In the receiving domain, the acoustic pressure reads:

$$\hat{p}(M) = \hat{p}_{ray}(M) \tag{12.2}$$

Since the panel is baffled, the radiated pressure is given by

$$\hat{p}_{ray}(M) = -\int_S \frac{\partial \hat{p}(M_0)}{\partial n} G(M, M_0)\,\mathrm{d}S(M_0) \tag{12.3}$$

where $G(M, M_0) = \exp[-jk_0 R]/(2\pi R)$, $R = \sqrt{(x - x_0)^2 + (y - y_0)^2}$ and n denotes the outward normal to the radiating surface S pointing into the receiving domain.

Let v_n denote the normal surface velocity, the radiated (transmitted) power is given by

$$\begin{aligned} \Pi_t &= \frac{1}{2}\mathrm{Re}\left(\int_S \hat{p}_{ray}(M) v_n^*(M)\right) \mathrm{d}S(M) \\ &= \frac{1}{2}\mathrm{Re} \iint_S \frac{j}{\omega\rho_0}\frac{\partial \hat{p}}{\partial n}(M_0) G(M, M_0) \frac{\partial \hat{p}^*}{\partial n}(M)\,\mathrm{d}S(M_0) dS(M) \end{aligned} \tag{12.4}$$

As in the classical implementation of the TMM, we assume that the surface impedance, Z_m, seen from the radiating surface is governed by the plane wave assumption and thus is given by $\rho_0 c/\cos\theta$. Using the relationship $\partial\hat{p}/\partial n + jk\beta\hat{p} = 0$ with $\beta = \rho_0 c/Z_m = \cos\theta$ the transmitted power reads

$$\Pi_t = \frac{1}{2}\frac{S\cos^2\theta}{\rho_0 c}\mathrm{Re}\left(\frac{jk_0}{S}\iint_S \hat{p}(M_0) G(M, M_0) \hat{p}^*(M)\,\mathrm{d}S(M_0)\,\mathrm{d}S(M)\right) \tag{12.5}$$

The parietal pressure, on the receiver side, is related to the incident pressure by the pressure transmission coefficient: $\hat{p} = T_\infty \hat{p}_i$. Here T_∞ is calculated using the TMM and in consequence is constant over the surface. The subscript ∞ is used to denote that the panel is assumed of infinite extent and its radiation impedance given by the plane wave approximation. Defining the energy transmission coefficient $\tau_\infty = |T_\infty|^2$, the transmitted power becomes

$$\Pi_t = \left(\frac{1}{2}\frac{S\cos\theta\tau_\infty}{\rho_0 c}\right)\cos\theta\ \mathrm{Re}\left(\frac{jk_0}{S}\iint_S p_i(M_0) G(M, M_0) p_i^*(M)\,\mathrm{d}S(M_0)\,\mathrm{d}S(M)\right) \tag{12.6}$$

The first term of Equation (12.6) represents the classical expression of the transmitted power, assuming an infinite structure. The second term represents a geometrical correction accounting for the finite size effect. It is given by the ratio of the radiation efficiency σ_R of the incident plane wave forced finite structure radiator to the radiation efficiency σ_∞ of the infinite structure:

$$\frac{\sigma_R}{\sigma_\infty} = \sigma_R\cos\theta = \cos\theta\ \mathrm{Re}\left(\frac{jk_0}{S}\iint_S p_i(M_0) G(M, M_0) p_i^*(M)\,\mathrm{d}S(M_0)\,\mathrm{d}S(M)\right) \tag{12.7}$$

Indeed, consider a flat baffled panel forced to vibrate by the incident plane wave

$$v(x, y) = |v| \exp[-jk_0(\cos\phi\sin\theta x + \sin\phi\sin\theta y)] \tag{12.8}$$

Assuming $|v|$ constant over the surface, its radiated power is given by

$$\Pi_{rad} = \frac{1}{2}\mathrm{Re}\left[j\rho_0\omega \int_S \int_S v(M_0)G(M, M_0)v^*(M)\,\mathrm{d}S(M_0)\,\mathrm{d}S(M)\right]$$
$$= \frac{|v|^2}{2}\mathrm{Re}\left[j\rho_0\omega \int_S \int_S \exp[-jk\sin\theta(\cos\phi x_0 + \sin\phi y_0)]G(M, M_0)\right.$$
$$\left.\exp\left[jk_0\sin\theta(\cos\phi x + \sin\phi y)\right]\mathrm{d}x\,\mathrm{d}y\,\mathrm{d}x_0\,\mathrm{d}y_0\right] \quad (12.9)$$

and thus its radiation efficiency is

$$\sigma_R = \frac{\Pi_{rad}}{\rho_0 c S\langle V^2\rangle} = \frac{\mathrm{Re}(Z_R)}{\rho_0 c} \quad (12.10)$$

with

$$Z_R = \frac{j\rho_0\omega}{S}\int_S\int_S \hat{p}_i(M_0)G(M, M_0)\hat{p}_i^*(M)\,\mathrm{d}S(M_0)\,\mathrm{d}S(M)$$
$$= \frac{j\rho_0\omega}{S}\int_S\int_S \exp[-jk_t(\cos\phi x_0 + \sin\phi y_0)]G(M, M_0) \quad (12.11)$$
$$\times \exp[jk_t(\cos\phi x + \sin\phi y)]\,\mathrm{d}x\,\mathrm{d}y\,\mathrm{d}x_0\,\mathrm{d}y_0$$

where $k_t = k_0 \sin\theta$. Note that radiation impedance Z_R only depends on the geometry of the panel and the angles of incidence θ and ϕ.

Using this correction, the transmitted power becomes

$$\Pi_t = \frac{1}{2}\frac{S\cos\theta\tau_\infty}{\rho_0 c}\sigma_R(k_t, \varphi)\cos\theta \quad (12.12)$$

Since the transmitted power for the infinite size panel is given by

$$\Pi_{t,\infty} = \frac{1}{2}\frac{S\cos\theta}{\rho_0 c_0}\tau_\infty \quad (12.13)$$

one obtains

$$\Pi_t = \Pi_{t,\infty}\sigma_R(k_t, \varphi)\cos\theta \quad (12.14)$$

Figure 12.1 plots the expression of the heading averaged geometrical radiation efficiency

$$\bar{\sigma}_R(\theta) = \frac{1}{2\pi}\int_0^{2\pi}\sigma_R\left(k_t\left(\theta\right), \varphi\right)\mathrm{d}\varphi = \frac{1}{2\pi}\int_0^{2\pi}\frac{\mathrm{Re}(Z_R)}{\rho_0 c_0 S}\,\mathrm{d}\varphi \quad (12.15)$$

as a function of incidence angle θ for different frequencies. The corresponding radiation efficiency for an infinite panel is also represented. It is clearly seen that the correction is primarily important for low frequencies and near grazing angles. In consequence the computation algorithm can be modified to apply this correction selectively. Appendix 12.A presents a numerical algorithm to estimate Z_R.

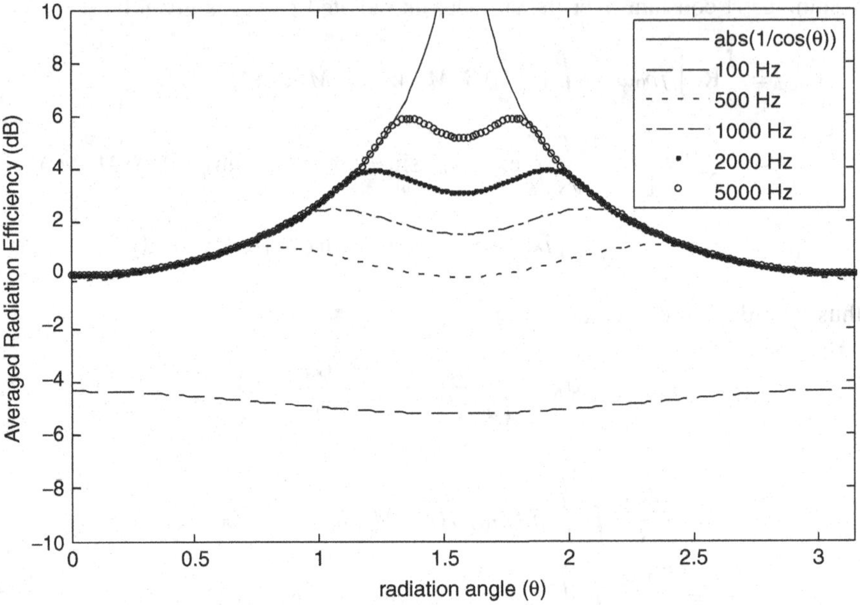

Figure 12.1 Variation of the heading averaged geometrical radiation efficiency with incidence angle at selected frequencies for a 1 × 0.8 m rectangular window.

An alternative finite size correction is given by Villot *et al.* (2001). It is based on the same assumptions and is derived in wave number domain. It is referred to in the literature as a 'windowing' correction. It leads to the following expression for the correction factor:

$$\sigma_R(k_t, \varphi) = \frac{S}{\pi^2} \int_0^{k_0} \int_0^{2\pi} \frac{1 - \cos((k_r \cos\psi - k_t \cos\varphi)L_x)}{[(k_r \cos\psi - k_t \cos\varphi)L_x]^2} \times \frac{1 - \cos((k_r \sin\psi - k_t \sin\varphi)L_y)}{[(k_r \sin\psi - k_t \sin\varphi)L_y]^2} \times \frac{k_0 k_r}{\sqrt{k_0^2 - k_r^2}}\, d\psi\, dk_r \tag{12.16}$$

Comparison of the numerical evaluation of Equation (12.15) using the two approaches leads to similar results. However, the approach presented here is preferred due to its computational efficiency.

In the special case where the trace wave number $k_t = k_0 \sin\theta$ is replaced by the bending wave number of a given structure, Equations (12.10) and (12.11) lead to the radiation efficiency of the structure for this particular wave. For instance, Figure 12.2 gives an example of a 5-mm-thick, simply supported rectangular aluminum plate measuring 1. × 0.8 m and having a critical frequency of 2350 Hz. In this case, $k_t = (\sqrt{\omega}\sqrt[4]{m/D})$ with D the bending stiffness and m the surface mass density. The radiation efficiency obtained is compared with estimation using Leppington's asymptotic formulas (Leppington *et al.* 1982). Good correlation is observed between the two methods.

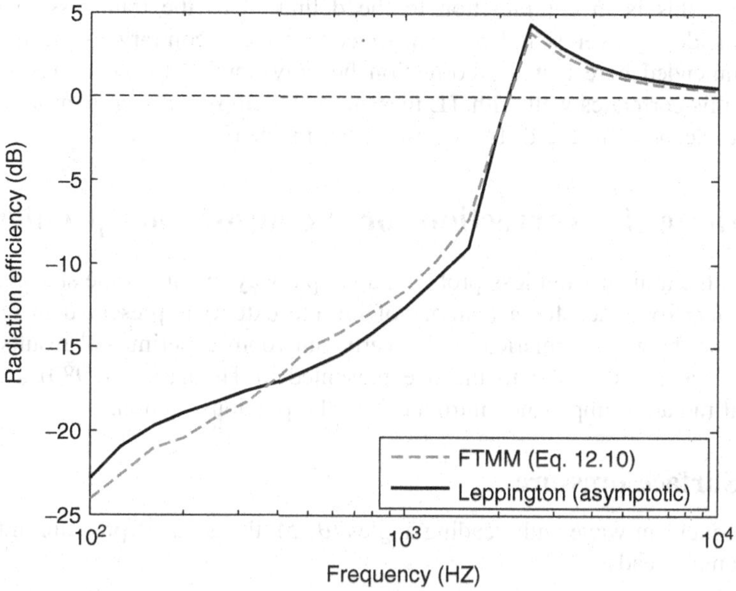

Figure 12.2 Radiation efficiency of a plate in bending. Comparison between Leppington asymptotic formulas and Equation (12.10) with k_t in Equation (12.11) replaced by the plate bending wave number.

12.2.2 Transmission coefficient

By analogy with the infinite case, the transmitted power, Equation (12.12), is written

$$\Pi_t = \frac{1}{2}\frac{S\cos\theta}{\rho_0 c}\tau_f$$

with $\tau_f = |T_f|^2$ the energy transmission coefficient accounting for the finite size effect. It is related to the classical coefficient τ_∞ by

$$\tau_f = \tau_\infty(\sigma_R\cos\theta) \tag{12.17}$$

The associated diffuse field transmission coefficient is given by

$$\tau_{f\_diff} = \frac{\displaystyle\int_0^{\pi/2}\int_0^{2\pi} \tau_f(\theta,\phi)\sin\theta\cos\theta\; \mathrm{d}\phi \mathrm{d}\theta}{\displaystyle\int_0^{\pi/2}\int_0^{2\pi} \sin\theta\cos\theta \mathrm{d}\phi \mathrm{d}\theta} \tag{12.18}$$

Villot *et al.* (2001) suggest the use of the same size correction for the incident power, leading to

$$\tau_f = \tau_\infty(\sigma_R\cos\theta)^2 \tag{12.19}$$

However, this is in contradiction to the definition of the transmission coefficient since the incident power is independent of edge effects, contrary to the input power. It is recommended here that the correction be only applied to the transmitted power. Moreover, this correlates well with TL measurement. However, a correction may still be necessary to account for the diffusiveness of the incident field.

12.3 Finite size correction for the absorption problem

Similarly to the transmission loss problem, a simple way to correct the statistical absorption coefficient (obtained for a material of infinite extent) is presented in order to get a better match between numerical and reverberant room experimental results. The presented formulation is similar to the one presented in Thomasson (1980) and uses the geometrical radiation impedance introduced in the previous section.

12.3.1 Surface pressure

For a plane incident wave with heading angles (θ, ϕ), the acoustic pressure in the emitter (source) domain reads

$$\hat{p}(M) = \hat{p}_b(M) + \hat{p}_{ray}(M) \tag{12.20}$$

where $\hat{p}_b(M)$ is the blocked pressure given by

$$\begin{aligned}\hat{p}_b(M) = &\exp[-jk_0(\cos\phi\sin\theta x + \sin\phi\sin\theta y + \cos\theta z)]\\ &+ \exp[-jk_0(\cos\phi\sin\theta x + \sin\phi\sin\theta y - \cos\theta z)]\end{aligned} \tag{12.21}$$

On the surface S $(z = 0)$, $\partial\hat{p}_b(x, y, 0)/\partial z = 0$ and $\hat{p}_b(x, y, 0) = 2\hat{p}_i(x, y, 0)$.

Assuming that the material has a space-independent normalized admittance $\beta = \rho_0 c/Z_m$ where Z_m is the surface impedance of the material, the impedance condition at the surface reads $\partial\hat{p}/\partial n_{in} = -jk\beta\hat{p}$ where n_{in} is the surface normal vector pointing into the material. Rewriting this relation with the outward normal pointing into the emitter medium, the radiated pressure given by Equation (12.3) becomes

$$\hat{p}_{ray}(M) = -jk\beta\int_S \hat{p}(M_0)G(M, M_0)\,\mathrm{d}S(M_0) \tag{12.22}$$

which reads, on the surface of the material

$$\hat{p}(M) = 2\hat{p}_i(M) - jk\beta\int_S \hat{p}(M_0)G(M, M_0)\,\mathrm{d}S(M_0) \tag{12.23}$$

The pressure at the surface can be written under the form: $\hat{p} = \hat{B}\hat{p}_i = (1 + V_f)\hat{p}_i$ where $\hat{B}$ is assumed constant over the surface. V_f is the equivalent reflection coefficient accounting for the finite size of the sample. It is referred to here by the subscript f in contrast with the classical 'infinite size' reflection coefficient V_∞.

Consider $\delta\hat{p} = \hat{p}_i\delta\hat{B}$ an admissible variation of the parietal pressure field. Multiplying Equation (12.23) by the complex conjugate of this variation and integrating over the material area leads to

$$\int_S \hat{p}(M)\delta\hat{p}^*(M)\,\mathrm{d}S(M) = \int_S 2\hat{p}_i(M)\delta\hat{p}^*(M)\,\mathrm{d}S(M) - jk\beta\int_S\int_S \hat{p}(M_0)G(M,M_0)\delta\hat{p}^*(M)\,\mathrm{d}S(M_0)\,\mathrm{d}S(M) \tag{12.24}$$

that is

$$\int_S \hat{B}\hat{p}_i(M)\hat{p}_i^*(M)\delta\hat{B}^*\,\mathrm{d}S(M) = \int_S 2\hat{p}_i(M)\hat{p}_i^*(M)\delta\hat{B}^*\,\mathrm{d}S(M) - jk\beta\int_S\int_S \hat{B}\hat{p}_i(M_0)G(M,M_0)\hat{p}_i^*(M)\delta\hat{B}^*\,\mathrm{d}S(M_0)\,\mathrm{d}S(M) \tag{12.25}$$

$\delta\hat{B}^*$ being arbitrary, the above equation leads to

$$\hat{B} = \frac{\int_S 2\hat{p}_i(M)\hat{p}_i^*(M)\,\mathrm{d}S(M)}{\int_S \hat{p}_i(M)\hat{p}_i^*(M)\,\mathrm{d}S(M) + jk\beta\int_S\int_S \hat{p}_i(M_0)G(M,M_0)\hat{p}_i^*(M)\,\mathrm{d}S(M_0)\,\mathrm{d}S(M)} \tag{12.26}$$

Using the fact that $|\hat{p}_i| = 1$ (arbitrarily normalized)

$$\hat{B} = \frac{2S}{S + Z_R\beta S} \tag{12.27}$$

where S denotes the area of the material and Z_R is the normalized radiation impedance obtained from Equation (12.11) wherein symbol Z_R denotes a non-normalized impedance.

Finally, using $\hat{p} = \hat{B}\hat{p}_i = (1 + V_f)\hat{p}_i$, the surface pressure is related to the incident pressure by

$$\hat{p}(x,y,0) = \frac{2\hat{p}_i(x,y,0)}{1 + Z_R\beta} = \frac{2\hat{p}_i(x,y,0)Z_A}{Z_A + Z_R} \tag{12.28}$$

with $Z_A = 1/\beta$ the material normalized surface impedance of the material.

In the case where the material is of infinite extent, the normalized radiation impedance $Z_R = 1/\cos\theta$ and the classical formula for the parietal pressure is recovered

$$\hat{p}(x,y,0) = \frac{2\hat{p}_i(x,y,0)Z_A}{Z_A + \dfrac{1}{\cos\theta}} \tag{12.29}$$

12.3.2 Absorption coefficient

The power absorbed at an incidence (θ,ϕ) reads

$$\Pi_{abs,f}(\theta,\phi) = \frac{1}{2}\mathrm{Re}\left[\int_S \hat{p}\hat{v}_n^* dS\right] = \frac{1}{2\rho_0 c}\mathrm{Re}\left[\int_S \hat{p}\frac{\hat{p}^*}{Z_A^*}\,\mathrm{d}S\right] \tag{12.30}$$

Using Equation (12.28)

$$\Pi_{abs,f}(\theta, \phi) = \frac{1}{2}\frac{|\hat{p}_i|^2 S}{\rho_0 c}\frac{4\,\mathrm{Re}\,Z_A}{|Z_A + Z_R|^2} \tag{12.31}$$

On the other hand, the incident power is given by

$$\Pi_{inc}(\theta, \phi) = \frac{1}{2}\frac{|\hat{p}_i|^2 S}{\rho_0 c}\cos\theta \tag{12.32}$$

In consequence, the absorption coefficient for a given incidence (θ, φ) is:

$$\alpha_f(\theta, \phi) = \frac{\Pi_{abs,f}}{\Pi_{inc}} = \frac{1}{\cos\theta}\frac{4\,\mathrm{Re}\,Z_A}{|Z_A + Z_R|^2} \tag{12.33}$$

For an infinite extent material $Z_R(\theta, \phi) = \frac{1}{\cos\theta}$ and the classical incidence absorption formula is recovered

$$\alpha_\infty(\theta) = \frac{4\mathrm{Re}Z_A\cos\theta}{|Z_A\cos\theta + 1|^2} \tag{12.34}$$

The diffuse field incident and absorbed powers are given by

$$\Pi^d_{inc} = \int_0^{2\pi}\int_0^{\pi/2} \Pi_{inc}(\theta)\sin\theta\,\mathrm{d}\theta\,\mathrm{d}\phi \tag{12.35}$$

$$\Pi^d_{abs,f} = \int_0^{2\pi}\int_0^{\pi/2} \Pi_{abs,f}(\theta, \varphi)\sin\theta\,\mathrm{d}\theta\,\mathrm{d}\phi \tag{12.36}$$

The corresponding energy absorption coefficient is

$$\alpha_{\mathrm{f,st}} = \frac{\Pi^d_{abs,f}}{\Pi^d_{inc}} = \frac{\displaystyle\int_0^{2\pi}\int_0^{\pi/2}\frac{4\,\mathrm{Re}\,Z_A}{|Z_A + Z_R|^2}\sin\theta\,\mathrm{d}\theta\,\mathrm{d}\phi}{\displaystyle\int_0^{2\pi}\int_0^{\pi/2}\cos\theta\sin\theta\,\mathrm{d}\theta\,\mathrm{d}\phi} \tag{12.37}$$

A practical approximation to ease the cost of the numerical evaluation of this equation is to replace the radiation impedance $Z_R(\theta, \phi)$ for a given θ by its average over heading angle ϕ

$$Z_{R,avg}(\theta) = \frac{1}{2\pi}\int_0^{2\pi} Z_R(\theta, \phi)\,\mathrm{d}\phi \tag{12.38}$$

The corresponding absorption coefficient becomes

$$\alpha_{f,st,avg} = \frac{\displaystyle\int_0^{\pi/2}\frac{4\,\mathrm{Re}\,Z_A}{|Z_A + Z_{R,avg}|^2}\sin\theta\,\mathrm{d}\theta}{\displaystyle\int_0^{\pi/2}\cos\theta\sin\theta\,\mathrm{d}\theta} \tag{12.39}$$

Note that for large samples and/or at high frequencies (see Figure 12.1), $Z_{R,avg}(\theta) = 1/\cos\theta$, and the classical field incidence absorption formula is again recovered

$$\alpha_{inf,st} = \frac{\int_0^{\pi/2} \frac{4\mathrm{Re}Z_A\cos\theta}{|Z_A\cos\theta + 1|^2}\cos\theta\sin\theta\,d\theta}{\int_0^{\pi/2}\cos\theta\sin\theta\,d\theta} \tag{12.40}$$

Note: A 'true' absorption coefficient can also be obtained by accounting for the size effect in the incident power. 'True' is used here in the sense that its value is always less than one. The idea is to replace the incident power by the input power accounting for the geometric impedance seen by the incident wave

$$\alpha_{f,true} = \frac{\Pi^d_{abs,f}}{\Pi^d_{inc,f}} = \frac{\int_0^{2\pi}\int_0^{\pi/2} \frac{4\mathrm{Re}Z_A}{|Z_A + Z_R|^2}\sin\theta\,d\theta\,d\phi}{\int_0^{2\pi}\int_0^{\pi/2} \frac{4\cos\theta}{|1 + Z_R\cos\theta|^2}\sin\theta\,d\theta\,d\phi} \tag{12.41}$$

Compared with the statistical coefficient, we obtain the following correction factor

$$\frac{\alpha_{f,true}}{\alpha_{f,st}} = \frac{\int_0^{2\pi}\int_0^{\pi/2}\cos\theta\sin\theta\,d\theta\,d\phi}{\int_0^{2\pi}\int_0^{\pi/2} \frac{4\cos\theta}{|1 + Z_R\cos\theta|^2}\sin\theta\,d\theta\,d\phi} = \frac{\pi}{\int_0^{2\pi}\int_0^{\pi/2} \frac{4\cos\theta}{|1 + Z_R\cos\theta|^2}\sin\theta\,d\theta\,d\phi} \tag{12.42}$$

In terms of the heading averaged radiation impedance, it reads

$$\frac{\alpha_{f,true}}{\alpha_{f,st}} = \frac{1}{\int_0^{\frac{\pi}{2}} \frac{8\cos\theta}{|1 + Z_{R,avg}\cos\theta|^2}\sin\theta\,d\theta} \tag{12.43}$$

12.3.3 Examples

Transmission loss of a plate–foam system

The first example considers an aluminium panel with attached foam excited by a diffuse field in a reverberant room. The panel is 1.37×1.64 m and 1 mm thick. The 5.08-cm-thick foam was bonded along its edges to the panel using a double-sided tape. The properties of the materials used are shown in Table 12.1. First, the TL of the bare panel is compared in Figure 12.3 with simulations using the classical TMM and the finite size correction (FTMM). It is clearly seen that the FTMM significantly improves the predictions at low frequencies. Next, Figure 12.4 shows the comparison for the case with the added foam. Again, the comparison is good at low frequencies where size governs the TL. Note that, for much smaller panels, the stiffness of the panel will control its TL at low frequencies (up to the first resonance of the panel) and this is not accounted for in the presented size correction.

Table 12.1 Parameters of the plate–foam system.

Material	Thickness, h (mm)	ϕ	σ (N s/m^4)	α_∞	Λ (μm)	Λ' (μm)	ρ_1 (kg/m^3)	E (Pa)	ν	η_s
Panel	1						2742	6.9×10^{10}	0.3	0.007
Foam 1	variable	0.99	10.9×10^3	1.02	100	130	8.8	80×10^3	0.35	0.14

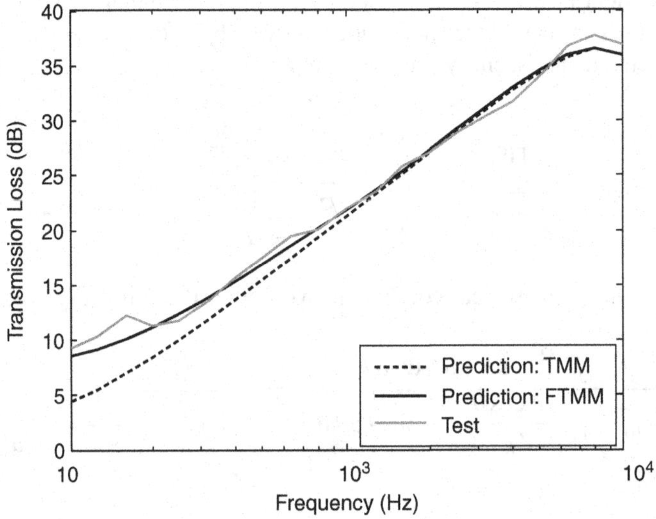

Figure 12.3 Transmission loss of a plate system; effect of size correction.

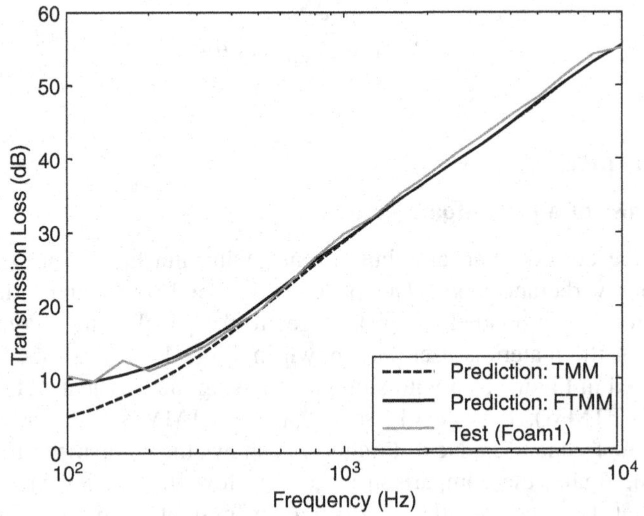

Figure 12.4 Transmission loss of a plate–foam system.

Transmission loss of a plate–wool–plate system

The second example is taken from Villot *et al.* (2001). It consists of the transmission loss of a double wall system made up of a steel plate, a mineral wool and a laminated plate. The system measures 1.3 × 1.3 m and is excited by a diffuse acoustic field. The mineral wool is modelled as a poroelastic layer and the panels as thin plates in bending. The parameters of the three layers are given in Table 12.2. The comparison between tests and predictions is shown in Figure 12.5. Good agreement is observed, especially at low frequencies where size effects are important.

Transmission loss of an orthotropic plate

The next transmission loss example illustrates the use of the TMM for an orthotropic panel. In this case, the properties of the material and thus the structural wave number and the transmission coefficient are heading dependent. The example, taken from Leppington *et al.* (2002), considers a 1.4 × 0.9 m orthotropic panel excited by a diffuse field. The properties of the panel are given in Table 12.3. Note that these physical properties are derived from the following parameters: $D_{11} = 21.34$ Nm; $D_{12}+2D66 = 27.78$ Nm;

Table 12.2 Parameters of the steel wool–laminated plate system.

Material	Thickness, h (mm)	ϕ	σ (N s/m^4)	α_∞	Λ (μm)	Λ' (μm)	ρ_1 (kg/m^3)	E (Pa)	ν	η_s
Steel plate	0.75						7850	2.1×10^{11}	0.3	0.03
Mineral wool	30	0.95	34×10^3	1	40	80	90	40×10^3	0	0.18
Laminated plate	3						1360	6×10^8	0.15	0.15

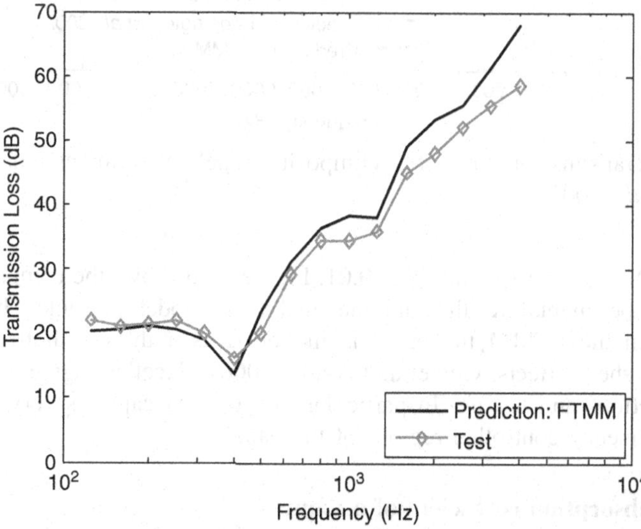

Figure 12.5 Transmission loss of a double wall system system; experimental data taken from Villot *et al.* (2001).

Table 12.3 The parameters of the orthotropic panel.

Parameters	Material from Leppington parameters
Young's modulus along direction 1 E_1 (Pa)	2.0237×10^{12}
Young's modulus along direction 2 E_2 (Pa)	3.1375×10^{13}
Shear modulus in directions 1,2 G_{12} (Pa)	8.8879×10^{11}
Shear modulus in direction 2,3 G_{23} (Pa)	8.8879×10^{11}
Shear modulus in direction 1,3 G_{13} (Pa)	8.8879×10^{11}
Poisson's ratio in direction 1,2 ν_{12}	0.028
Poisson's ratio in direction 2,3 ν_{23}	0.434
Poisson's ratio in direction 1,3 ν_{13}	0.0
Mass density of the material ρ (Kg/m^3)	9740
Structural loss factor of the material η	0.01

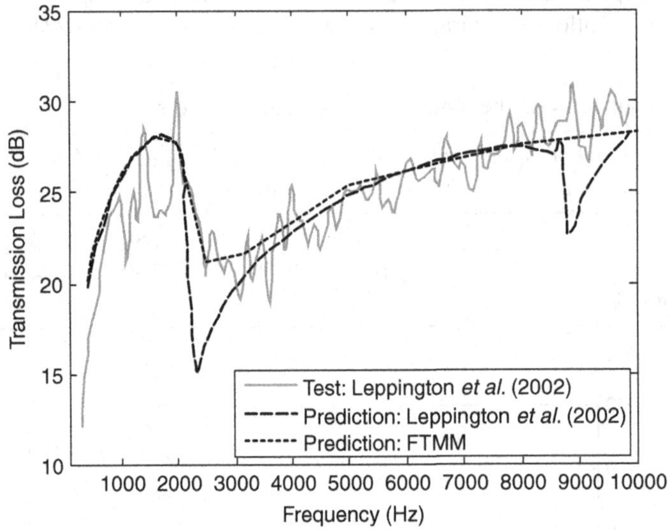

Figure 12.6 Transmission loss of a composite panel; experimental data taken from Leppington *et al.* (2002).

$D_{22} = 330.84$ N; $m = 4.87$ kg/m^2; $\eta = 0.01$. Figure 12.6 shows the comparison between the FTMM, experimental results and the analytical model presented by Leppington *et al.* (2002). In the FTMM the panel is modelled as a thick orthotropic panel (the model includes shear effects; Ghinet and Atalla, 2006). Excellent agreement is observed between predictions and testing. In particular, the FTMM captures very well the mass and critical frequency controlled regions of the panel.

Reverberant absorption coefficient of a foam

This example considers the prediction of the reverberant (sabine) absorption coefficient of the foam presented in Table 12.1. The tests were conducted at the National Research

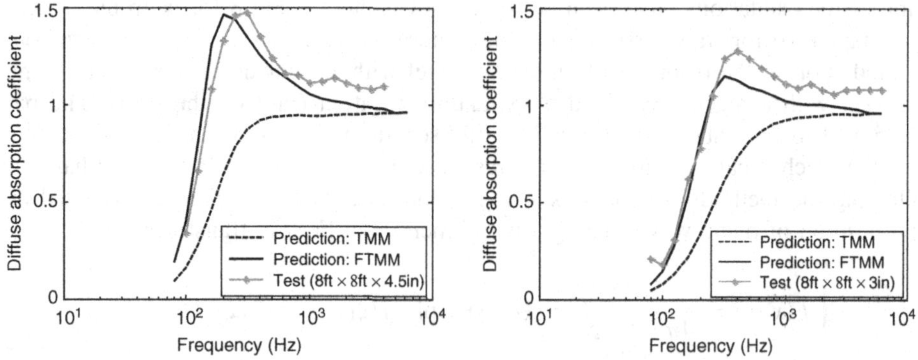

Figure 12.7 Reverberant absorption coefficient of 8 × 8 ft foam samples; tests versus predictions.

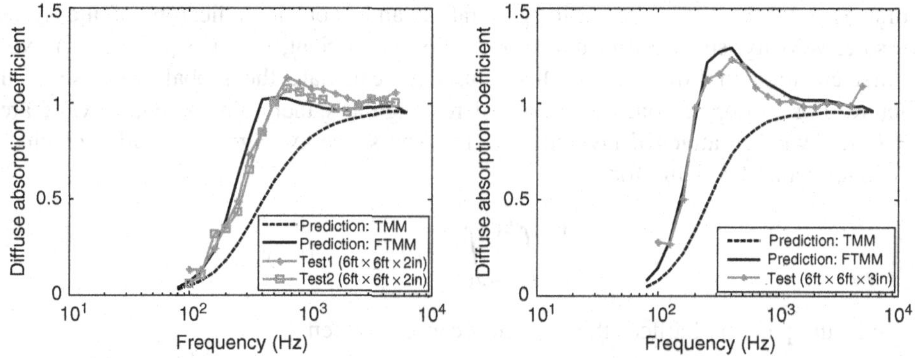

Figure 12.8 Reverberant absorption coefficient of 6 × 6 ft foam samples; tests versus predictions.

Canada (NRC) Laboratory, in Ottawa. The laboratory volume is 250 m^3. The foams were mounted following Mounting A of ASTM C423. Two thicknesses were tested: 3 in (7.64 cm) and 4 in (11.34 cm). The material is modeled as limp. Figure 12.7 shows the comparison between tests and the finite size correction (FTMM). The ideal random incidence absorption (where finite size effect is omitted) is also shown. Overall good agreement is observed for both thicknesses. The same foam was tested in a smaller reverberation room at the Université de Sherbrooke (143 m^3 in volume). The foam test area was 6 × 6 ft (1.829 × 1.829 m). Again, two thicknesses were tested: 2 in (5.04 cm) and 3 in (7.62 cm). The results are shown in Figure 12.8. Again, it is seen that the FTMM is able to capture the overall tendencies of the measured absorption curve.

12.4 Point load excitation

12.4.1 Formulation

This section discusses the use and application of the TMM for a structure-borne excitation. Methods to calculate the response of a porous elastic material to a mechanical or point source excitations have been presented in Chapters 7 and 8 for isotropic materials and Chapter 10 for transversally isotropic materials. Here, we limit the presentation to the prediction of the response of an elastic panel with an attached layered noise control treatment with a mechanical load with random location (rain on the roof). The panel is assumed to be embedded in a rigid baffle separating two semi-infinite fluids. Moreover, the mechanical excitation and the response of the structure are assumed harmonic. Following the methods of Chapters 7 and 8, the load $f(x, y)$ can be represented by a superposition of plane waves, using a two-dimensional Fourier transform

$$\begin{cases} f(x,y) = \dfrac{1}{4\pi^2} \displaystyle\int_{-\infty}^{+\infty} \int_{-\infty}^{+\infty} F(\xi_1,\xi_2) \exp[-j(\xi_1 x + \xi_2 y]\, d\xi_1\, d\xi_2 \\ F(\xi_1,\xi_2) = \displaystyle\int_{-\infty}^{+\infty} \int_{-\infty}^{+\infty} f(x,y) \exp[j(\xi_1 x + \xi_2 y)]\, dx\, dy \end{cases} \tag{12.44}$$

For each wave number component (ξ_1, ξ_2), the transfer matrix method (Chapter 11) is first used to solve for the various vibration and acoustic indicators of the system (pressure, velocity, surface impedance, reflection coefficient, radiated power, ...). Next, the first equation in Equation (12.44) is used to calculate the global response of the structure. For example, consider a baffled rectangular panel (dimension, $L_x \mathrm{x} L_y$; area, $S = L_x L_y$) with an attached layered material. The space averaged quadratic velocity of the excited panel takes the form

$$\langle v^2 \rangle = \frac{1}{2S} \int_0^{L_x} \int_0^{L_y} |v(x,y)|^2\, dx\, dy \tag{12.45}$$

Since the panel is baffled, this equation can be written:

$$\langle v^2 \rangle = \frac{1}{2S} \int_{-\infty}^{\infty} \int_{-\infty}^{\infty} |v(x,y)|^2\, dx\, dy \tag{12.46}$$

with $S = L_x L_y$ the panel's surface. Using the first equation in Equation (12.44), one obtains

$$\langle v^2 \rangle = \frac{1}{8\pi^2 S} \int_{-\infty}^{\infty} \int_{-\infty}^{\infty} v(x,y) \left(\int_{-\infty}^{\infty} \int_{-\infty}^{\infty} v^*(\xi_1,\xi_2) \exp[j(\xi_1 x + \xi_2 y]\, d\xi_1\, d\xi_2 \right) dx\, dy \tag{12.47}$$

Permutation of the order of integration and use of the second equation in Equation (12.44), leads to

$$\langle v^2 \rangle = \frac{\langle F^2 \rangle}{8\pi^2 S} \int_{-\infty}^{+\infty} \int_{-\infty}^{+\infty} |v(\xi_1,\xi_2)|^2\, d\xi_1\, d\xi_2 \tag{12.48}$$

with $\langle F^2 \rangle$ the excitation power spectrum and $v(\xi_1, \xi_2)$ is approximated by the plane wave normal (unit) velocity of the equivalent infinite extent panel with the layered material.

Using the same approach, the acoustic power radiated by the panel into the source domain is given by

$$\Pi_{rad} = \frac{\langle F^2 \rangle}{8\pi^2} \int_0^{2\pi} \int_0^{k_0} |v(\xi, \psi)|^2 \mathrm{Re} Z_{B,\infty}(\xi, \psi) \xi \, d\xi \, d\psi \tag{12.49}$$

Here, integration in polar coordinates defined by $\xi_1 = \xi \cos\psi$ and $\xi_2 = \xi \sin\psi$ is used and

$$Z_{B,\infty}(\xi_1, \xi_2) = \frac{k_0 Z_0}{\sqrt{k_0^2 - (\xi_1^2 + \xi_2^2)}} \tag{12.50}$$

is the radiation impedance of the panel in air (the source domain), k_0 being the acoustic wave number. Other indicators such as the power radiated into the receiver domain can be calculated using the same methodology.

Below the critical frequency of the main structure, Equation (12.49) will lead to poor results. However, as discussed in the previous sections, the transfer matrix method can be extended easily to take into account the panel's size and thus to correct for the radiation efficiency at low frequencies. The application of the FTMM to the calculation of the radiated power transforms Equation (12.49) into:

$$\Pi_{rad} = \frac{\langle F^2 \rangle}{8\pi^2} \int_0^{2\pi} \int_0^{\infty} \frac{Z_0 \sigma_R(k_r, \varphi)}{|Z_{B,\infty}(\xi, \psi) + Z_{S,TMM}(\xi, \psi)|^2} \xi \, d\xi \, d\psi \tag{12.51}$$

The 'finite size' radiation efficiency, $\sigma_R(k_r, \phi)$ is defined by Equations (12.10) and (12.11). $Z_{S,TMM}$ is the impedance of the infinite extent panel with layered material.

12.4.2 The TMM, SEA and modal methods

The TMM can also be used in combination with other methods to account for the effects of a sound package. For instance, in combination with a statistical energy analysis (SEA) model of the point load excited base structure, a light coupling can be assumed between the structure and the sound package. In consequence, the effect of the sound package will be simply represented by an equivalent damping η_{NCT} and an added mass correction. At a given frequency, the dispersion equation of the bare panel (plate, solid, composite, etc.) is solved for the propagating wave number and the latter is imposed on the excited face of the sound package with a pressure-release condition on the rear face. The associated TMM system is then solved to calculate the input power and dissipated powers and in turn the equivalent added damping η_{NCT}. The response of the treated panel is finally recovered from the SEA response of the bare panel using the total damping of the system $\eta_{Tot} = \eta_s + \eta_{NCT} + \eta_{rad}$ with η_{rad} the radiation damping of the panel and η_s its structural damping.

The same methodology can be used in combination of a modal representation of the bare panel's response. The response of the main structure can be written in terms of its modes and the effect of the sound package on each mode (m,n) is replaced by a modal impedance $Z_{mn,NCT}$, calculated using the TMM with a trace wave number $k_{t_{mn}}$, given by the modal wave number of the panel. Once again, it is assumed in this calculation that the receiver face of the multilayer is a pressure release surface. The total modal impedance is written

$$Z_{mn,T} = Z_{mn(Bare\ panel)} + Z_{mn,NCT} \tag{12.52}$$

Here, $Z_{mn(Bare\ panel)}$ is the modal impedance of the bare panel, and $Z_{mn,NCT}$ is calculated by solving for the interface pressure, $P_{\text{interface}}$, at the panel–layered material interface

$$Z_{mn(Bare\ panel)} v_{mn} = F_{mn} - \int_S P_{\text{interface}}(x, y)\varphi_{mn}(x, y)\,\mathrm{d}x\,\mathrm{d}y \tag{12.53}$$

where $\varphi_{mn}(x, y)$, v_{mn}, F_{mn} are the mode shapes, modal velocity and the associated modal force of the bare panel, respectively. The modal impedance $Z_{mn,NCT}$ is related to $P_{\text{interface}}$ via

$$Z_{mn,NCT} V_{mn} = \int_S P_{\text{interface}}\phi_{mn}(x, y)\,\mathrm{d}x\,\mathrm{d}y \tag{12.54}$$

In the modal basis, $P_{\text{interface}}$ can be written in the form:

$$P_{\text{interface}} = \sum_{m,n} P_{mn}\phi_{mn}(x, y) \tag{12.55}$$

Since the modes are orthogonal, this simplifies to

$$Z_{mn,NCT} = \frac{P_{mn}}{v_{mn}} N_{mn} \tag{12.56}$$

where N_{mn} is the norm of the panel's modes

$$N_{mn} = \int_S \phi_{mn}^2(x, y)\,\mathrm{d}x\,\mathrm{d}y \tag{12.57}$$

The ratio P_{mn}/v_{mn} is obtained from the solution of the transfer matrix of the layered material following the methods of Chapter 11.

12.4.3 Examples

In the following, examples of flat plates with attached sound packages are used to demonstrate the performance of the methods described in the previous section. The sound package is made up of a layer of fibre or foam, with or without a cover panel. The performance is measured by comparison with the finite element method. In the FE predictions, the plates are modelled using Cquad4 elements while the fibre and foams are modelled using brick8 porous elements (see Chapter 13). The plates are assumed to be baffled for acoustic radiation. To compare various estimation of the radiated power, the calculations are done using: (i) the geometrical correction (FTMM) in the wave approach, (ii) the asymptotic expressions of the radiation efficiency developed by Leppington *et al.* (1982) in the SEA and modal approaches and (iii) the Rayleigh integral in the FEM. Due to the cost of the FE calculation, the results are averaged using five randomly selected point force positions to approach a rain-on-the-roof type of excitation. In addition, due to the nature of the methods used, the results are frequency band averaged using one-third octave bands. In both the modal approach and FEM, the plate is assumed simply supported.

Table 12.4 The parameters of the plate–fibre system.

Material	Thickness, h (mm)	ϕ	σ (N s/m^4)	α_∞	Λ (μm)	Λ' (μm)	ρ_1 (kg/m^3)	E (Pa)	ν	η_s
Panel	3						2742	6.9×10^{10}	0.3	0.01
Limp foam	30	0.99	10.9×10^3	1.02	100	130	8.8			

The first example considers a 1 m × 1 m 3 mm aluminium plate. The properties of the plates are given in Table 12.4. A comparison of the space averaged quadratic velocity and radiated power of the bare panel computed using the three presented approaches and the FEM are given in Figures 12.9 and 12.10. An excellent agreement is observed for the quadratic velocity using the three methods and the FEM. Both the FEM and the modal approach depict modal fluctuations at lower frequencies (recall that data are presented in one-third octave bands). The results of the classical TMM (infinite panel radiation efficiency) and the FTMM (infinite panel expression with size correction of the radiation efficiency) are shown in Figure 12.10 for the radiated power. As expected the TMM mainly captures correctly the radiated power above the critical frequency. However, using the FTMM, good agreement is observed over the whole frequency range. In this simple case (bare panel), the good agreement observed between the three approaches and the FEM is expected. Still the results corroborate the validity of the FTMM in estimating the radiation efficiency of a flat plate compared with the classical Leppington asymptotic method and Rayleigh integral methods.

Next, the same plate with an attached limp porous layer is investigated. The properties of the layer are given in Table 12.4. In the FE predictions, the limp layer is modelled using brick8 equivalent fluid elements (see Chapter 13 for the finite element modelling of porous media). A comparison of the quadratic velocity and the power radiated into the

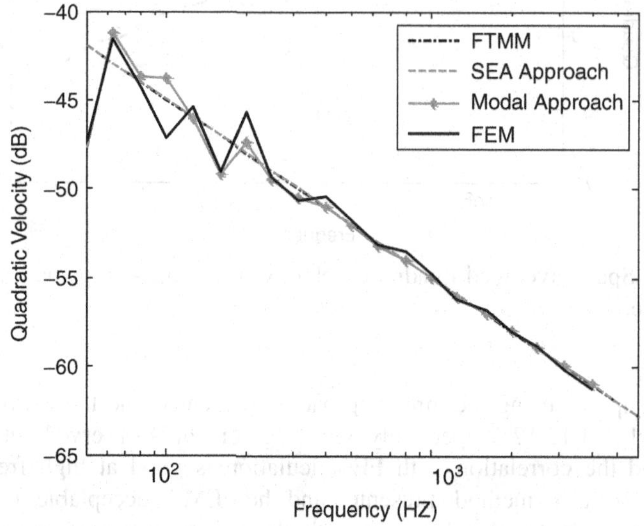

Figure 12.9 Quadratic velocity of a panel excited by a point force.

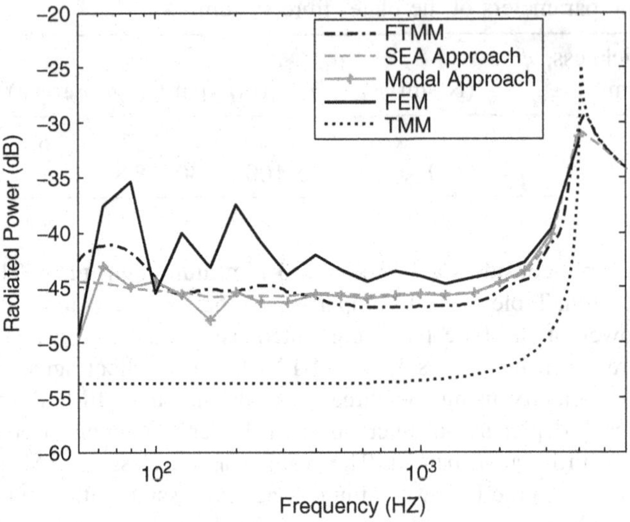

Figure 12.10 Radiated power of a plate excited by a point force.

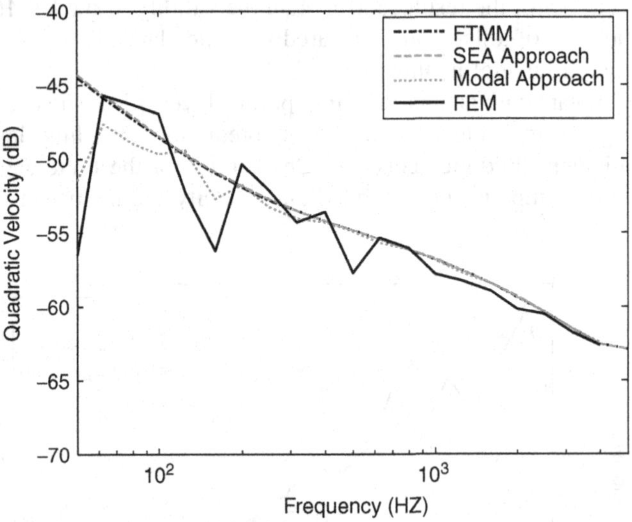

Figure 12.11 Space averaged quadratic velocity of a plate–limp foam system excited by a point force.

source side, computed using the three approaches presented and the FEM are illustrated in Figures 12.11 and 12.12, respectively. Good agreement is observed between the three approaches, and the correlation with FE calculation is good at high frequencies. The comparison between the methods presented and the FEM is acceptable, keeping in mind the assumptions made for the different methods and the different approaches used to estimate the radiation efficiency.

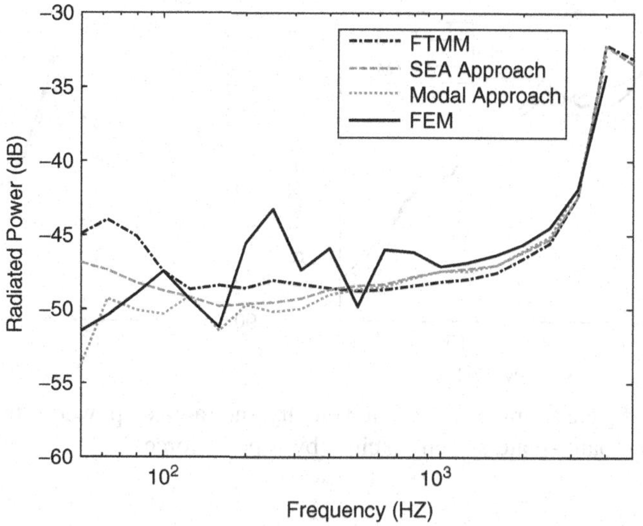

Figure 12.12 Radiated power of a plate–limp foam system excited by a point force.

The final example considers a 0.9 × 0.6 m double plate system made up of foam sandwiched between a steel panel and a limp heavy layer. Two cases are considered: limp foam and elastic foam. The first configuration depicts light coupling between the two panels while the second involves a stronger coupling. The properties and thicknesses of the three layers are given in Table 12.5. The three methods are compared with the finite element method. In the latter, the foam is modelled using the formulations described in Chapter 13. Figures 12.13 and 12.14 show the comparison between FEM and the FTMM for the quadratic velocity and power radiated by the two plates, for the limp foam and the elastic foam, respectively. Note that the SEA and the modal approaches, as presented, cannot be used to estimate the velocity and radiated power at the receiver side. Their results are not shown, but they have been found to correctly capture the vibration response and radiated power of the excited panel for both the limp and elastic foams. The modal method leads to the best correlation with FEM (both assume simply supported panels). On the receiver side, the comparison is found acceptable between the FTMM and the FEM in both cases, especially at higher frequencies. More discrepancies are observed for the foam case; still the main tendencies are well captured.

Table 12.5 The parameters of the plate–fibre system.

Material	Thickness, h (mm)	ϕ	σ (N s/m^4)	α_∞	Λ (μm)	Λ' (μm)	ρ_1 (kg/m^3)	E (Pa)	ν	η_s
Panel 1	1						7800	2.1×10^{10}	0.3	0.007
Limp foam	25.4	0.9	20000	1.6	12	24	30	1300	0	0.4
Foam	25.4	0.99	10.9×10^3	1.02	100	130	8.8	80×10^3	0.35	0.14
Heavy layer (PVC)	1						1000			

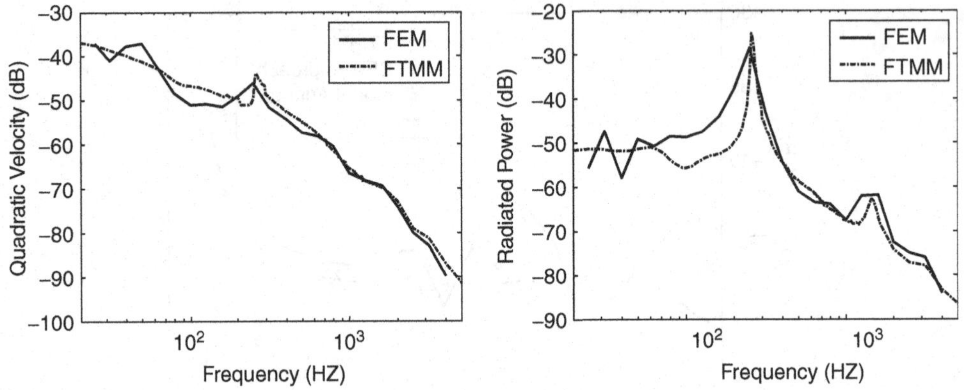

Figure 12.13 Space averaged quadratic velocity and radiated power of the receiver plate of a plate–limp foam–plate system excited by a point force.

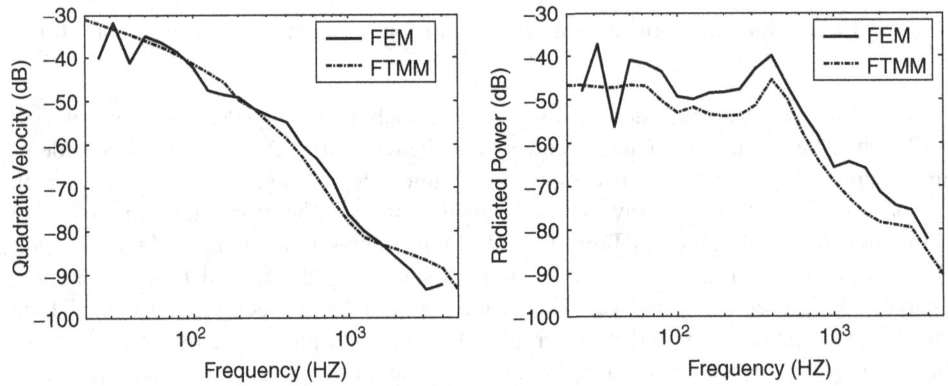

Figure 12.14 Space averaged quadratic velocity and radiated power of the receiver plate of a plate–foam–plate system excited by a point force.

This correlation between FEM and FTMM is qualitatively acceptable and can be used to estimate the insertion loss of a sound package. Figure. 12.15 shows the comparison for the structure-borne insertion loss (SBIL) of the plate–foam–plate system using the FTMM for two damping values of the bare panel: $\eta_1 = 0.007$ and $\eta_1 = 0.3$. The highly damped panel may represent for instance a metal polymer sandwich panel (laminated steel). In addition, the airborne insertion loss (ABIL), calculated using the same method, is shown. The SBIL is obtained by taking the ratio of the acoustical–mechanical conversion efficiencies obtained with and without the material (or, in db, taking the difference)

$$SBIL = 10\log_{10}\left(\frac{\Pi_{\text{input}}}{\Pi_{\text{radiated}}}\right)_{\text{with NCT}} - 10\log_{10}\left(\frac{\Pi_{\text{input}}}{\Pi_{\text{radiated}}}\right)_{\text{bare}} \tag{12.58}$$

The acoustical–mechanical conversion efficiency is defined by taking the ratio of the acoustic radiated power to the mechanical input power: $\Pi_{\text{radiated}} / \Pi_{\text{input}}$. The latter is

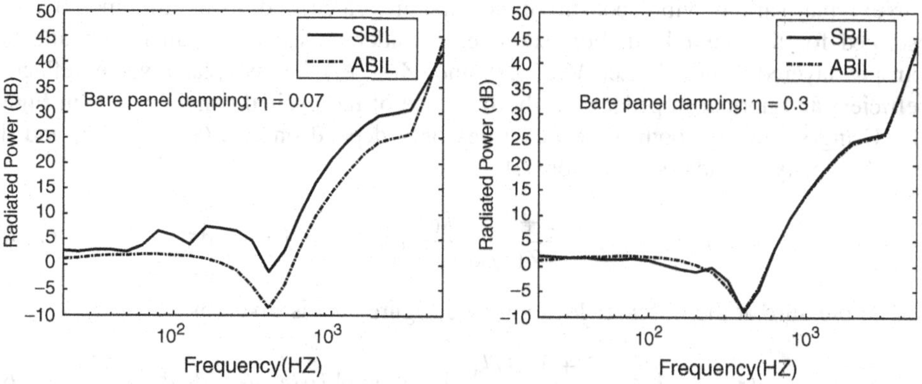

Figure 12.15 Airborne (ABIL)versus, structure-borne (SBIL) of a plate–foam–plate system for two damping values of the bare panel.

calculated from the force and velocity at the excitation point. It is important to use the real part of the complex product which stands for the input energy (effective power). This definition of the SBIL is similar to the classical ABIL obtained by subtracting the transmission loss (TL) of the bare panel from the TL of the same panel covered with the trim material

$$ABIL = TL_{\text{with NCT}} - TL_{\text{bare}} \tag{12.59}$$

While airborne insertion loss is based solely on acoustic excitations (diffuse acoustic field), the structure-borne insertion loss is based on point loads excitation randomly positioned on the panel.

Figure 12.15 shows that the SBIL depicts the same behaviour as the ABIL. It is higher than the ABIL for the lightly damped panel and similar for the highly damped panel. The difference between the two is in particular important at low frequencies and near the double wall resonance of the system. This difference diminishes with the damping of the main structure (including damping added by the sound package). This is in line with the experimentally observed similarity between ABIL and SBIL for damped systems (Nelisse *et al.* 2003). This also justifies to some extent the current SEA practice which uses ABIL to correct both the resonant and nonresonant transmission paths in a panel with an attached sound package under various excitations.

12.5 Point source excitation

Using the methods of Chapters 7 and 8, the application of the TMM in the case of a point source excitation is uncomplicated. First, the Sommerfeld representation is used to decompose the incident pressure into plane waves (Equation 7.2 and Figure 7.1)

$$\begin{aligned} p(R) &= \frac{-j}{2\pi}\int_{-\infty}^{\infty}\int_{-\infty}^{\infty}\frac{\exp[-j(\xi_1 x + \xi_2 y + \mu|z_2 - z_1|)]}{\mu}\,\mathrm{d}\xi_1\,\mathrm{d}\xi_2 \\ \mu &= \sqrt{k_0^2 - \xi_1^2 - \xi_2^2},\ \mathrm{Im}\ \mu \le 0,\ \mathrm{Re}\ \mu \ge 0 \end{aligned} \tag{12.60}$$

Next the various vibroacoustic indicators are calculated using the methodology described for the point load. For example, consider an isotropic panel with attached isotropic layered material. Let $V(\xi_1, \xi_2)$ and $Z(\xi_1, \xi_2)$ be the plane wave reflection coefficient and normal impedance at the free face of the excited panel ($z_2 = 0$ in Figure 7.1). Using symmetry, both these quantities only depend on $\xi = (\xi_1^2 + \xi_2^2)^{1/2}$, and the normal velocity v reads (see Section 7.1)

$$v(r) = -j \int_0^\infty \frac{1 + V(\xi/k_0)}{Z(\xi/k_0)\mu} \mathrm{J}_0(r\xi) \exp[j\mu z_1]\xi \, d\xi \tag{12.61}$$

The pressure at the observer location (see Figure 7.1) is given by

$$p(r) = -j \int_0^\infty \frac{1 + V(\xi/k_0)}{\mu} \mathrm{J}_0(r\xi) \exp[j\mu(z_1 + z_2)]\xi \, d\xi \tag{12.62}$$

Methods to evaluate Equations (12.61) and (12.62) are discussed in Chapter 7 in the context of a porous layer excited by a point source.

12.6 Other applications

As demonstrated in the previous sections, the transfer matrix method can be used to compute vibroacoustic indicators of practical interest, such as air-borne insertion loss, structure-borne insertion loss and damping added by a sound package. In turn these indicators, with the diffuse field absorption and the transmission loss, can be used within SEA framework to account for the effect of sound package in full system configurations, such as cars or aircraft (Pope *et al.* 1983, Atalla *et al.* 2004). Assuming that the trim (sound package) covers the area A_t of the panel and the remaining area $A_b = A - A_t$ is bare, the noise reduction of the panel is given by:

$$NR = 10 \log 10 \left(\frac{\bar{\alpha} A_t + \left[\tau_f + \tau_r \frac{\bar{\eta}}{\eta_{rad}} \right] A_b}{\tau_t \tau_b A_t + \tau_b A_b} \right) \tag{12.63}$$

where $\bar{\eta}$ is the sum of the space and band averaged radiation loss factor, η_{rad}, of the panel and its averaged structural loss factor, η_s; τ_f is the field incidence non-resonant transmission coefficient (mass controlled), τ_r is the diffuse field resonant transmission coefficient, $\tau_b = \tau_f + \tau_r$, τ_t is the transmission coefficient of the trim, and finally $\bar{\alpha}$ is the random incidence trim absorption coefficient. Both $\bar{\alpha}$, τ_t and the trim's contribution to $\bar{\eta}_2$ are estimated numerically using the TMM. For curved panels, the sound package is simply 'unwrapped' and the TMM is used to calculate the sound package absorption, transmission loss, insertion loss, absorption and added damping. This is a current limitation of the method. Its generalization to curved sound packages is still an open issue. Recent finite element and experimental studies show that curvature should be accounted for in the calculation of the insertion loss of a sound package (Duval *et al.* 2008). One can argue, however, that curvature effect will be mainly important before the ring frequency of the panel. This explains why the use of the flat panel assumption is sufficient for aircraft applications. For highly curved panels, such as wheel houses in an automotive, an accurate approach should be used. Similarly to the TMM-modal discussion of the previous section, one possible outcome is to combine FEM and SEA to handle, approximately the energy exchange (coupled loss factor) between a trimmed structure (modelled using finite element) and a fluid sub-system (modelled using SEA).

Appendix 12.A: An algorithm to evaluate the geometrical radiation impedance

Start from the expression of the geometrical radiation impedance:

$$Z = \frac{j\rho_0\omega}{S}\int_0^{L_x}\int_0^{L_y}\int_0^{L_x}\int_0^{L_y}\begin{pmatrix}\exp[-jk_t(x\cos\varphi + y\sin\varphi)]G(x,y,x',y')\\ \exp[jk_t(x'\cos\varphi + y'\sin\varphi)]\end{pmatrix}\mathrm{d}x\mathrm{d}y\mathrm{d}x'\mathrm{d}y' \tag{12.A.1}$$

where $k_t = k_0 \sin\theta$, k_0 is the acoustic wave number, ρ_0 is the density of the fluid, L_x and L_y are respectively the length and the width of the structure.

Using the change of variables $\alpha = \frac{2x}{L_x}$, $\beta = \frac{2y}{L_y}$

$$R = \frac{L_x}{2}\left[(\alpha - \alpha')^2 + \frac{1}{r^2}(\beta - \beta')^2\right]^{\frac{1}{2}} \tag{12.A.2}$$

where r is defined by : $r = L_x / L_y$, and :

$$Z = j\rho_0\omega\frac{L_y}{16\pi}\int_0^2\int_0^2\int_0^2\int_0^2 F_n(\alpha,\beta,\alpha',\beta')\mathrm{K}(\alpha,\beta,\alpha',\beta')\,\mathrm{d}\alpha\,\mathrm{d}\beta\,\mathrm{d}\alpha'\,\mathrm{d}\beta' \tag{12.A.3}$$

with

$$\mathrm{K}(\alpha,\beta,\alpha',\beta') = \frac{\exp[-jk_0R]}{\left[(\alpha-\alpha')^2 + \frac{(\beta-\beta')^2}{r^2}\right]^{1/2}} \tag{12.A.4}$$

and

$$F_n(\alpha,\beta) = \exp\left[-j\frac{k_tL_x}{2}\left[(\alpha-\alpha')\cos\varphi + \frac{1}{r}(\beta-\beta')\sin\varphi\right]\right] \tag{12.A.5}$$

To reduce the order of integration, the following change of variables is used

$$\begin{cases} u = \alpha - \alpha' \\ v = \alpha' \end{cases} \text{and} \begin{cases} u' = \beta - \beta' \\ v' = \beta' \end{cases} \tag{12.A.6}$$

Considering the variable α (the same formula can be written for β), a symbolic form of this change of variable is

$$\int_0^2\int_0^2 \mathrm{d}\alpha\,\mathrm{d}\alpha' \to \int_0^2 \mathrm{d}u\int_0^{2-u}\mathrm{d}v + \int_{-2}^0\mathrm{d}u\int_{-u}^0\mathrm{d}v \tag{12.A.7}$$

If K and F_n are rewritten in terms of u and u'

$$\mathrm{K}(u,u') = \frac{\exp\left[j\frac{k_0 L_x}{2}\left[u^2+\frac{u'^2}{r^2}\right]^{1/2}\right]}{\left[u^2+\frac{u'^2}{r^2}\right]^{1/2}};\ F_n(u,u') = \exp\left[-j\frac{k_t L_x}{2}\left[u\cos\varphi + \frac{u'}{r}\sin\varphi\right]\right] \tag{12.A.8}$$

Equation (12.A.6) leads to

$$\int_0^2\left(\int_0^{2-u} \mathrm{d}v\right)\mathrm{K}(u,u')\,\mathrm{d}u + \int_{-2}^0\left(\int_{-u}^2 \mathrm{d}v\right)\mathrm{K}(u,u')\,\mathrm{d}u = 2\int_0^2 (2-u)\mathrm{K}(u,u')\,\mathrm{d}u \tag{12.A.9}$$

Using these transformations the radiation impedance becomes:

$$Z = j\rho_0\omega\frac{L_y}{4\pi}\int_0^2\int_0^2 (2-u)(2-u')\mathrm{K}(u,u')F_n(u,u')\,\mathrm{d}u\,\mathrm{d}u' \tag{12.A.10}$$

Finally, the change of variables $u \to u+1$ and $u' \to u'+1$ is used to transform the integration domain into [−1,1] :

$$Z = j\rho_0\omega\frac{L_y}{4\pi}\int_{-1}^1\int_{-1}^1 (1-u)(1-u')\mathrm{K}(u+1,u'+1)F_n(u+1,u'+1)\,\mathrm{d}u\,\mathrm{d}u' \tag{12.A.11}$$

This last integral can be evaluated numerically.

References

Atalla, N., Ghinet, S. and Haisam, O. (2004) Transmission loss of curved sandwich composite panels. *18th ICA*, Kyoto.

Duval, A., Dejaeger, L., Baratier, J. and Rondeau, J-F (2008) Structureborne and airborne Insertion Loss simulation of trimmed curved and flat panels using Rayon-VTM-TL: implications for the 3D design of insulators. *Automobile and Railroad Comfort* – November 19–20, Le Mans, France.

Ghinet, S. and Atalla, N. (2002) Vibro-acoustic behaviour of multi-layer orthotropic panels. *Canadian Acoustics* **30**(3) 72–73.

Ghinet, S. and Atalla, N. (2006) Vibro-acoustic behaviours of flat sandwich composite panels. *Transactions of Canadian Soc. Mech. Eng J*. **30**(4).

Leppington, F.G., Broadbent, E.G., & Heron, K.H. (1982) The acoustic radiation efficiency from rectangular panels. Proc. R. Soc. London. **382**, 245.

Leppington, F. G, Heron, K. H. & Broadbent, E. G. (2002) Resonant and non-resonant noise through complex plates. *Proc. R. Soc. Lond*. **458**, 683–704.

Nelisse, H., Onsay, T. and Atalla, N. (2003) Structure borne insertion loss of sound package components. *Society of Automotive Engineers*, SAE paper 03NVC-185, Traverse City, Michigan.

Pope, L.D., Wilby, E.G., Willis, C.M. and Mayes, W.H. (1983) Aircraft Interior Noise Models: Sidewall trim, stiffened structures, and Cabin Acoustics with floor partition. *J. Sound Vib*. **89**(3), 371–417.

Thomasson, S. I. (1980) On the absorption coefficient. *Acustica* **44**, 265–273.

Villot, M., Guigou, C. and Gagliardini, L. (2001) Predicting the acoustical radiation of finite size multi-layered structures by applying spatial windowing on infinite structures. *J. Sound Vib*., **245** (3), 433–455.

13

Finite element modelling of poroelastic materials

13.1 Introduction

The vibroacoustic performance of finite multilayer systems containing poroelastic materials is of the utmost importance for noise control in automobiles, aircraft and several other engineering applications. In the absence of absorbing materials, the vibroacoustic response of complex multilayer structures are classically modelled using the finite element and the boundary element methods. To account for absorbing media, finite element formulations for sound absorbing materials have been developed. They range from simple approaches using equivalent fluid models (Craggs 1978, Beranek and Ver 1992, Panneton *et al.* 1995) to sophisticated approaches based on the Biot theory (Kang and Bolton 1995, Johansen *et al.* 1995, Coyette and Wynendaele 1995, Panneton and Atalla 1996, 1997, Atalla *et al.* 1998, 2001a). The latter approaches are mainly based either on the classical displacement $(\boldsymbol{u}^s, \boldsymbol{u}^f)$ formulation of Biot's poroelasticity equations (Chapter 6) or the mixed displacement–pressure $(\boldsymbol{u}^s, p)$ formulation (Appendix 6.A). However, it has been shown that, while accurate, these formulations have the disadvantage of requiring cumbersome calculations for large finite element models and spectral analyses. To alleviate these difficulties, alternative numerical implementations have been investigated. This chapter reviews various finite element formulations for the modelling of poroelastic materials with an emphasis on the mixed pressure–displacement formulation. The presentation is not comprehensive, but aims at giving the reader a general view of the subject. The generic problem of interest deals with the prediction of the vibroacoustic response (dynamic and acoustic response) of multilayered structures made up of elastic, poroelastic and acoustic media. The poroelastic material may be bonded or unbonded to the structure. The classical assumptions concerning linear acoustic, elastic, and

Propagation of Sound in Porous Media: Modelling Sound Absorbing Materials, Second Edition J. F. Allard and N. Atalla

poroelastic wave propagation are assumed. Also, the air contained in the porous medium is at rest. The presentation will be limited to time harmonic behaviour ($\exp(j\omega t)$).

13.2 Displacement based formulations

The coupled poroelasticity equations (6.54)–(6.56) can be rewritten in the compact form

$$\begin{cases} div\ \sigma^s(\boldsymbol{u}^s, \boldsymbol{u}^f) + \omega^2\tilde{\rho}_{11}\boldsymbol{u}^s + \omega^2\tilde{\rho}_{12}\boldsymbol{u}^f = 0 \\ div\ \sigma^f(\boldsymbol{u}^s, \boldsymbol{u}^f) + \omega^2\tilde{\rho}_{22}\boldsymbol{u}^f + \omega^2\tilde{\rho}_{12}\boldsymbol{u}^s = 0 \end{cases} \tag{13.1}$$

Using the solid and fluid displacement vectors $(\boldsymbol{u}^s, \boldsymbol{u}^f)$ as primary variables, the weak integral form of the poroelasticity equations, reads

$$\begin{aligned} &\int_\Omega (\sigma^s_{ij}(\boldsymbol{u}^s, \boldsymbol{u}^f)e^s_{ij}(\delta\boldsymbol{u}^s) - \tilde{\rho}_{11}\omega^2 u^s_i \delta u^s_i - \tilde{\rho}_{12}\omega^2 u^f_i \delta u^s_i)\,\mathrm{d}\Omega - \\ &\qquad\qquad \int_\Gamma \sigma^s_{ij}(\boldsymbol{u}^s, \boldsymbol{u}^f)n_j \delta u^s_i\,\mathrm{d}S = 0 \\ &\int_\Omega (\sigma^f_{ij}(\boldsymbol{u}^s, \boldsymbol{u}^f)e^f_{ij}(\delta\boldsymbol{u}^f) - \tilde{\rho}_{22}\omega^2 u^f_i \delta u^f_i - \tilde{\rho}_{12}\omega^2 u^s_i \delta u^f_i)\,\mathrm{d}\Omega - \\ &\qquad\qquad \int_\Gamma \sigma^f_{ij}(\boldsymbol{u}^s, \boldsymbol{u}^f)n_j \delta u^f_i\,\mathrm{d}S = 0 \\ &\qquad\qquad\qquad\qquad \forall(\delta\boldsymbol{u}^s, \delta\boldsymbol{u}^f) \end{aligned} \tag{13.2}$$

where $\delta\boldsymbol{u}^s$ and $\delta\boldsymbol{u}^f$ denote admissible variations of $\boldsymbol{u}^s$ and $\boldsymbol{u}^f$, respectively, and where Ω and Γ denote the medium domain and its boundary (Figure. 13.1). For the finite element implementation of Equation (13.2), an analogy with three-dimensional elastic solid elements is used; however, this time, six degrees of freedom per node are necessary: three displacement components of the solid phase and three displacement components of the fluid phase. Moreover, because of the viscous and thermal dissipation mechanisms, the system matrices are frequency dependent. In consequence, for large 3-D multilayer

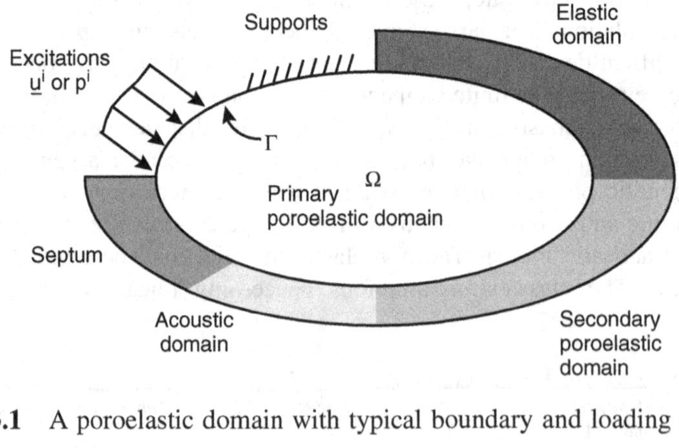

Figure 13.1 A poroelastic domain with typical boundary and loading conditions.

structures, this formulation has the disadvantage of requiring cumbersome calculations for large finite element models and spectral analyses. To alleviate these difficulties, a mixed displacement–pressure formulation has been developed (Appendix 6.A). Note in passing that similar variants of Equation (13.1), using the solid phase displacement $\boldsymbol{u}^s$ and the displacement flux $\boldsymbol{w} = \phi(\boldsymbol{u}^f - \boldsymbol{u}^s)$, have been developed and implemented in the finite element context (Coyette and Pelerin 1994, Coyette and Wynendaele 1995, Johansen *et al.* 1995, Dazel 2005). The use of the displacement flux simplifies certain coupling conditions compared with the classical $(\boldsymbol{u}^s, \boldsymbol{u}^f)$ formulation, but their numerical performances are similar.

13.3 The mixed displacement–pressure formulation

Recall Equation (6.A.22):

$$\begin{cases} div\ \hat{\sigma}^s(\boldsymbol{u}^s) + \omega^2 \tilde{\rho} \boldsymbol{u}^s + \tilde{\gamma}\ \boldsymbol{grad}\ p = 0 \\ \Delta p + \omega^2 \dfrac{\tilde{\rho}_{22}}{\tilde{R}} p - \omega^2 \dfrac{\tilde{\rho}_{22}}{\phi^2} \tilde{\gamma}\ div\ \boldsymbol{u}^s = 0 \end{cases} \tag{13.3}$$

The weak integral equations associated with Equation (13.3) are given by Atalla *et al.* (1998)

$$\int_\Omega \hat{\sigma}^s_{ij}(\boldsymbol{u}^s) e^s_{ij}(\delta \boldsymbol{u}^s)\, d\Omega - \omega^2 \int_\Omega \tilde{\rho} u^s_i \delta u^s_i\, d\Omega - \int_\Omega \tilde{\gamma} \frac{\partial p}{\partial x_i} \delta u^s_i\, d\Omega - \int_\Gamma \hat{\sigma}^s_{ij}(\boldsymbol{u}^s) n_j \delta u^s_i\, dS = 0$$

$$\int_\Omega \left[\frac{\phi^2}{\omega^2 \tilde{\rho}_{22}} \frac{\partial p}{\partial x_i} \frac{\partial(\delta p)}{\partial x_i} - \frac{\phi^2}{\tilde{R}} p\, \delta p \right] d\Omega - \int_\Omega \tilde{\gamma} \frac{\partial(\delta p)}{\partial x_i} u^s_i\, d\Omega + \int_\Gamma \left[\tilde{\gamma}\, u_n - \frac{\phi^2}{\tilde{\rho}_{22}\omega^2} \frac{\partial p}{\partial n} \right] \delta p\, dS = 0 \qquad \forall(\delta \boldsymbol{u}^s, \delta p) \tag{13.4}$$

Here, Ω and Γ refer to the poroelastic domain and its bounding surface (Figure 13.1). $\boldsymbol{\delta u}^s$ and δp are admissible variations of the solid phase displacement vector and the interstitial fluid pressure of the poroelastic medium, respectively. $\boldsymbol{n}$ is the unit outward normal vector around the bounding surface Γ, and subscript $\boldsymbol{n}$ denotes the normal component of a vector. To simplify the formulation and its coupling with elastic, porous and acoustic media, Atalla *et al.* (1998) used the fact that, for the majority of porous materials used in acoustics, the bulk modulus of the porous material is negligible compared with the bulk modulus of the material from which the skeleton is made: $K_b/K_S \ll 1$, thus: $\phi(1 + \tilde{Q}/\tilde{R}) = 1 - K_b/K_s \cong 1$. Using this assumption, the coupling of the $(\boldsymbol{u}^s, p)$ formulation with elastic, acoustic and porous elastic media becomes simple. A detailed derivation and discussion of these coupling conditions are given by Debergue *et al.* (1999). It is shown that in this formulation, a poroelastic medium couples naturally with acoustic and poroelastic media, and couples through a classical fluid–structure coupling matrix with elastic media (solids, septum, etc.). A variant of this formulation that eliminates recourse to the approximation $K_b/K_S \ll 1$ and couples naturally with elastic and poroelastic media, thus making the numerical implementation accurate and easier, was proposed by Atalla *et al.* (2001a). Using Equations (6.A.14), (6.A.19) and (6.A.20), Equation. (13.4) can be rewritten

$$\int_\Omega \hat{\sigma}^s_{ij}(\boldsymbol{u}^s)e^s_{ij}(\delta\boldsymbol{u}^s)\,\mathrm{d}\Omega - \omega^2\int_\Omega \tilde{\rho}u^s_i\delta u^s_i\,\mathrm{d}\Omega - \int_\Omega \tilde{\gamma}\frac{\partial p}{\partial x_i}\delta u^s_i\,\mathrm{d}\Omega - \int_\Gamma \sigma^t_{ij}(\boldsymbol{u}^s,\mathrm{p})\mathrm{n}_j\delta u^s_i\,\mathrm{d}S$$
$$-\int_\Gamma \phi\left(1+\frac{\tilde{Q}}{\tilde{R}}\right)\mathrm{p}\delta u^s_n\,\mathrm{d}S = 0$$
$$\int_\Omega\left[\frac{\phi^2}{\omega^2\tilde{\rho}_{22}}\frac{\partial p}{\partial x_i}\frac{\partial(\delta p)}{\partial x_i} - \frac{\phi^2}{\tilde{R}}p\delta p\right]\mathrm{d}\Omega - \int_\Omega \tilde{\gamma}\frac{\partial(\delta p)}{\partial x_i}u^s_i\,\mathrm{d}\Omega - \int_\Gamma \phi(u^f_n - u^s_n)\delta p\,\mathrm{d}S$$
$$-\int_\Gamma \phi\left(1+\frac{\tilde{Q}}{\tilde{R}}\right)u^s_n\delta p\,\mathrm{d}S = 0$$
$$\forall(\delta\boldsymbol{u}^s,\delta p) \qquad (13.5)$$

where $\boldsymbol{\sigma}^t_{ij} = \boldsymbol{\sigma}^s_{ij} - \phi p\delta_{ij}$ are the components of the total stress defined in Appendix 6.A. Assuming a homogeneous medium, application of the divergence theorem to the last surface integrals of Equation (13.5) with the use of the vector identity, $\nabla\cdot(\beta b) = \nabla\cdot b + \nabla\beta\cdot b)$, lead to the following set of equations

$$\int_\Omega \hat{\sigma}^s_{ij}(\boldsymbol{u}^s)e^s_{ij}(\delta\boldsymbol{u}^s)\,\mathrm{d}\Omega - \omega^2\int_\Omega \tilde{\rho}u^s_i\delta u^s_i\,\mathrm{d}\Omega - \int_\Omega\left(\tilde{\gamma}+\phi\left(1+\frac{\tilde{Q}}{\tilde{R}}\right)\right)\frac{\partial p}{\partial x_i}\delta u^s_i\,\mathrm{d}\Omega$$
$$-\int_\Omega \phi\left(1+\frac{\tilde{Q}}{\tilde{R}}\right)\mathrm{p}\delta u^s_{i,i}\,\mathrm{d}S - \int_\Gamma \sigma^t_{ij}(\boldsymbol{u}^s,\mathrm{p})\mathrm{n}_j\delta u^s_i\,\mathrm{d}S = 0$$
$$\int_\Omega\left[\frac{\phi^2}{\omega^2\tilde{\rho}_{22}}\frac{\partial p}{\partial x_i}\frac{\partial(\delta p)}{\partial x_i} - \frac{\phi^2}{\tilde{R}}p\delta p\right]\mathrm{d}\Omega - \int_\Omega\left(\tilde{\gamma}+\phi\left(1+\frac{\tilde{Q}}{\tilde{R}}\right)\right)\frac{\partial(\delta p)}{\partial x_i}u^s_i\,\mathrm{d}\Omega$$
$$-\int_\Omega \phi\left(1+\frac{\tilde{Q}}{\tilde{R}}\right)\delta\mathrm{p}u^s_{i,i}\,\mathrm{d}S - \int_\Gamma \phi(u^f_n - u^s_n)\delta p\,\mathrm{d}S = 0$$
$$\forall(\delta\boldsymbol{u}^s,\delta p) \qquad (13.6)$$

Noting that $\tilde{\gamma}+\phi(1+\tilde{Q}/\tilde{R}) = \phi/\tilde{\alpha}$ and $\tilde{\rho}_{22} = \tilde{\alpha}\rho_0$ where $\tilde{\alpha}$ is the dynamic tortuosity, and summing the two previous equations, the new expression of the $(\boldsymbol{u}^s, p)$ weak formulation reads (the displacement and pressure dependence of the stress components are taken out to alleviate the presentation)

$$\int_\Omega [\hat{\sigma}^s_{ij}\delta e^s_{ij} - \omega^2\tilde{\rho}u^s_i\delta u^s_i]\,\mathrm{d}\Omega + \int_\Omega\left[\frac{\phi^2}{\tilde{\alpha}\rho_0\omega^2}\frac{\partial p}{\partial x_i}\frac{\partial(\delta p)}{\partial x_i} - \frac{\phi^2}{\tilde{R}}p\delta p\right]\mathrm{d}\Omega$$
$$-\int_\Omega \frac{\phi}{\tilde{\alpha}}\delta\left(\frac{\partial p}{\partial x_i}u^s_i\right)\mathrm{d}\Omega - \int_\Omega \phi\left(1+\frac{\tilde{Q}}{\tilde{R}}\right)\delta(\mathrm{p}u^s_{i,i})\,\mathrm{d}S$$
$$-\int_\Gamma \sigma^t_{ij}\mathrm{n}_j\delta u^s_i\,\mathrm{d}S - \int_\Gamma \phi(u^f_n - u^s_n)\delta p\,\mathrm{d}S = 0 \quad \forall(\delta u^s_i,\delta p) \qquad (13.7)$$

This form shows that the coupling between the two phases is volumetric and of two natures: (i) kinetic (inertial), $\int_\Omega (\phi/\tilde{\alpha})\delta(u_i^s \partial p/\partial x_i)\,\mathrm{d}\Omega$, and (ii) potential (elastic), $\int_\Omega \phi(1+\tilde{Q}/\tilde{R})\delta(pu_{i,i}^s)\,\mathrm{d}\Omega$. It is shown in the next section that this new formulation couples naturally to elastic and porous media.

13.4 Coupling conditions

Considering the generic problem depicted in Figure 13.1, the coupling conditions applied on the bounding surface Γ are of four types : (i) poroelastic–elastic, (ii) poroelastic–acoustic, (iii) poroelastic–poroelastic, and (iv) poroelastic–septum. In the weak formulation, Equation (13.7), the porous medium couples to other media through the following boundary terms:

$$I^p = -\int_\Gamma \sigma_{ij}^t n_j \delta u_i^s \,\mathrm{d}S - \int_\Gamma \phi(u_n^f - u_n^s)\delta p \,\mathrm{d}S \qquad (13.8)$$

This section presents the expression of the coupling conditions for various media interfaces together with two loading conditions, imposed surface pressure and imposed surface displacement, using the $(\underline{u}, p)$ formulation. Finally, the case where the poroelastic media are inserted in a waveguide is considered.

13.4.1 Poroelastic–elastic coupling condition

The elastic medium is described in terms of its displacement vector $\boldsymbol{u}^e$. Let Ω^e and Γ^e denote its volume and its boundary. Under linear elastodynamics and harmonic oscillations assumptions and using displacement as the structure variable, the weak integral form of the structure governing equation is given by

$$\int_{\Omega^e} [\sigma_{ij}^e \delta e_{ij}^e - \omega^2 \tilde{\rho} u_i^e \delta u_i^e]\,\mathrm{d}\Omega - \int_{\Gamma^e} \sigma_{ij}^e \mathrm{n}_j \delta u_i^e \,\mathrm{d}S = 0 \quad \forall \delta u_i^e \qquad (13.9)$$

where σ_{ij}^e and e_{ij}^e are the components of the structure stress and strain tensors, ρ_e is the structure density, n_i are the components of the outward normal vector to the surface, and δu_i^e is an arbitrary admissible variation of u_i^e. The surface integral

$$I^e = -\int_{\Gamma^e} \sigma_{ij}^e \mathrm{n}_j \delta u_i^e \,\mathrm{d}S \qquad (13.10)$$

represents the virtual work done by external forces applied on the surface of the structural domain.

When the weak formulation of the poroelastic medium is combined with that of the elastic medium, the boundary integrals of the assembly can be rewritten

$$I^p + I^e = -\int_\Gamma \sigma_{ij}^t n_j \delta u_i^s \,\mathrm{d}S - \int_\Gamma \phi(u_n^f - u_n^s)\delta p \,\mathrm{d}S + \int_\Gamma \sigma_{ij}^e n_j \delta u_i^e \,\mathrm{d}S \qquad (13.11)$$

The positive sign of the third term is due to the direction of the normal vector $\boldsymbol{n}$ which is inward to the elastic medium. The coupling conditions at the interface Γ are given by Equation (11.74)

$$\begin{cases} \sigma^t_{ij}n_j = \sigma^e_{ij}n_j \\ u^f_n - u^s_n = 0 \\ u^s_i = u^e_i \end{cases} \tag{13.12}$$

The first equation ensures the continuity of the total normal stresses at the interface. The second equation expresses the fact that there is no relative mass flux across the impervious interface. The third equation ensures the continuity of the solid displacement vectors. Substituting Equation (13.12) into Equation (13.11), one obtains : $I^p + I^e = 0$; this equation shows that the coupling between the poroelastic and the elastic media is natural. Only the kinematic boundary condition $\boldsymbol{u}^s = \boldsymbol{u}^e$ will have to be explicitly imposed on Γ. In a finite element implementation, this may be done automatically through assembling between the solid phase of the porous media and the elastic media.

13.4.2 Poroelastic–acoustic coupling condition

The acoustic medium is described in terms of its pressure field p^a. Let Ω^a and Γ^a denote its volume and its boundary. Using pressure as the fluid variable and assuming harmonic oscillations, the weak integral form of the fluid system is given by

$$\int_{\Omega^a} \left(\frac{1}{\rho_0 \omega^2} \frac{\partial p^a}{\partial x_i} \frac{\partial (\delta p^a)}{\partial x_i} - \frac{1}{\rho_0 c_0^2} p^a \delta p^a \right) \mathrm{d}\Omega - \frac{1}{\rho_0 \omega^2} \int_{\Gamma^a} \frac{\partial p^a}{\partial n} \delta p^a \, \mathrm{d}S = 0 \quad \forall \delta p^a \tag{13.13}$$

where δp^a an arbitrary admissible variation of p^a, ρ_0 is the density in the acoustic domain, c_0 is the speed of sound in the acoustic medium, and $\partial/\partial n$ is the normal derivative. The surface integral

$$I^a = -\frac{1}{\rho_0 \omega^2} \int_{\Gamma^a} \frac{\partial p^a}{\partial n} \delta p^a \, \mathrm{d}S \tag{13.14}$$

represents the virtual work done by the internal pressure at the boundary of the acoustic domain due to an imposed motion on its surface.

Combining the weak formulations of the poroelastic and acoustic media, the boundary integrals can be rewritten

$$I^p + I^a = -\int_\Gamma \sigma^t_{ij} n_j \delta u^s_i \, \mathrm{d}S - \int_\Gamma \phi(u^f_n - u^s_n)\delta p \, \mathrm{d}S + \int_\Gamma \frac{1}{\rho_0 \omega^2} \frac{\partial p^a}{\partial n} \cdot \delta p^a \, \mathrm{d}S \tag{13.15}$$

The positive sign of the last term of $I^p + I^a$ is due to the direction of the normal vector $\boldsymbol{n}$, inwards to the acoustic medium. The coupling conditions at the interface Γ are given by Equation (11.72):

$$\begin{cases} \sigma^t_{ij} n_j = -p^a n_i \\ \dfrac{1}{\rho_0 \omega^2} \dfrac{\partial p^a}{\partial n} = (1-\phi)u^s_n + \phi u^f_n = u^s_n + \phi(u^f_n - u^s_n) \\ p = p^a \end{cases} \tag{13.16}$$

The first equation ensures the continuity of the normal stresses on Γ. The second equation ensures the continuity between the acoustic displacement and the total poroelastic displacement on Γ. The third equation refers to the continuity of the pressure across the boundary. Substitution of Equation (13.16) into Equation. (13.15) leads to

$$I^p + I^a = \int_\Gamma \delta(p^a u_s^n)\,\mathrm{d}S \tag{13.17}$$

This equation shows that the poroelastic medium will be coupled to the acoustic medium through the classical structure–cavity coupling term. In addition, the kinematic boundary condition $p = p^a$ needs to be explicitly imposed on Γ. Once again, in a finite element implementation, this latter condition is performed automatically through assembling.

13.4.3 Poroelastic–poroelastic coupling condition

Let subscripts 1 and 2 denote the primary and secondary poroelastic media, respectively. Both media are described in terms of their solid phase displacement vector $\boldsymbol{u}$ and pore fluid pressure p. Combining the weak integral formulations of both poroelastic media, the boundary integrals can be rewritten

$$\begin{aligned} I_1^p + I_2^p = &-\int_\Gamma \sigma_{1ij}^t n_j \delta u_{1i}^s\,\mathrm{d}S - \int_\Gamma \phi_1(u_{1n}^f - u_{1n}^s)\delta p_1\,\mathrm{d}S \\ &+ \int_\Gamma \sigma_{2ij}^t n_j \delta u_{2i}^s\,\mathrm{d}S - \int_\Gamma \phi_2(u_{2n}^f - u_{2n}^s)\delta p_2\,\mathrm{d}S \end{aligned} \tag{13.18}$$

The opposite signs between the two first terms and the two last terms are due to the direction of the normal vector $\boldsymbol{n}$, chosen outwards to the primary poroelastic medium. The coupling equations at the interface Γ are given by Equation (11.65)

$$\begin{cases} \sigma_{1ij}^t n_j = \sigma_{2ij}^t n_j \\ \phi_1(u_{1n}^f - u_{1n}^s) = \phi_2(u_{2n}^f - u_{2n}^s) \\ u_{1i}^s = u_{2i}^s \\ p_1 = p_2 \end{cases} \tag{13.19}$$

The first condition ensures the continuity of the total normal stresses. The second equation ensures the continuity of the relative mass flux across the boundary. The two last equations ensure the continuity of the solid phase displacement and pore fluid pressure fields across the boundary, respectively. Using these boundary conditions, the boundary integral reduces to $I^p + I^e = 0$; this equation shows that the coupling between the two poroelastic media is natural. Only the kinematic relations $\boldsymbol{u}_1^s = \boldsymbol{u}_2^s$ and $p_1 = p_2$ will have to be explicitly imposed on Γ. In a finite element implementation, this may be done automatically through assembling.

13.4.4 Poroelastic–impervious screen coupling condition

When the stiffness of the screen is important, the latter can be modelled as a thin plate. Here, the impervious screen is assumed thin and limp with a surface density m. It is represented by the surface domain Γ^m and described in terms of its displacement vector $\boldsymbol{u}^m$. One side of the screen is in contact with the poroelastic domain (i.e. surface $\Gamma^{m-} = \Gamma$), and the other side supports a given load (i.e. surface Γ^{m^+}). The hypothesis of a thin layer supposes that Γ^{m} and Γ are virtually the same. The principle of virtual work applied to the screen leads to

$$I^m = -\int_{\Gamma^m} \omega^2 m u_i^m \cdot \delta u_i^m \, \mathrm{d}S - \int_{\Gamma^{m-}} t_i^- \delta u_i^m \, \mathrm{d}S - \int_{\Gamma^{m+}} t_i^+ \delta u_i^m \, \mathrm{d}S = 0 \tag{13.20}$$

The first term corresponds to the virtual work of the inertial forces while the last two terms correspond to the virtual work of the exterior traction forces ($\boldsymbol{t}^+$ and $\boldsymbol{t}^-$). Combining the weak integral formulations of the poroelastic media and the limp screen, the boundary integrals can be rewritten

$$\begin{aligned} I^p + I^m = &\int_\Gamma \sigma_{ij}^t n_j \delta u_i^s \, \mathrm{d}S - \int_\Gamma \phi(u_n^f - u_n^s)\delta p \, \mathrm{d}S \\ &- \int_{\Gamma^m} \omega^2 m (u_i^m \delta u_i^m) \, \mathrm{d}S - \int_{\Gamma^{m-}} t_i^- \delta u_i^m \, \mathrm{d}S - \int_{\Gamma^{m+}} t_i^+ \delta u_i^m \, \mathrm{d}S \end{aligned} \tag{13.21}$$

The exterior forces acting on the limp screen are written as follows

$$\begin{cases} t_i^- = \sigma_{ij}^t n_j & \text{on } \Gamma^{m-} \\ t_i^+ = F_i & \text{on } \Gamma^{m+} \end{cases} \tag{13.22}$$

where $\boldsymbol{F}$ is the traction force vector applied on the septum. Furthermore, since the screen is assumed thin and impervious, the following conditions have to be verified at the screen–poroelastic interface Γ^{m-}

$$\begin{cases} u_n^f - u_n^s = 0 \\ u_i^s = u_i^m \end{cases} \tag{13.23}$$

The first relation indicates that there is no mass flux across the boundary Γ, and the second relation expresses the continuity of the displacement vectors. By considering the hypothesis of a thin layer, $\Gamma = \Gamma^{m-} = \Gamma^{m+}$ and using these boundary conditions, the coupling term can be rewritten

$$I^p + I^S = -\int_\Gamma \omega^2 m (u_i^s \delta u_i^s) \, \mathrm{d}S - \int_\Gamma F_i \delta u_i^s \, \mathrm{d}S \tag{13.24}$$

Therefore, the effect of the impervious screen on the poroelastic medium consists in an added mass and excitation terms. Note that when the screen separates two poroelastic media, the discontinuity of the pressure field at the screen interface should be accounted for explicitly in the numerical implementation.

13.4.5 Case of an imposed pressure field

In the case of an imposed pressure field p^0 applied on Γ, used for instance to simulate an acoustic plane wave (in this case the blocked pressure is used), the boundary conditions are

$$\begin{cases} \sigma_{ij}^t n_j = -p^i n_i \\ p = p^i \end{cases} \tag{13.25}$$

which express the continuity of the total normal stress and the continuity of the pressure through the interface Γ. Since the pressure is imposed, the admissible variation δp will fall to zero. Consequently, the surface integrals of Equation (13.8) simplify to

$$I^p = \int_\Gamma p^0 \delta u_n^s \, dS \tag{13.26}$$

This equation indicates that, in addition to the kinematic condition $p = p^0$ on Γ, a pressure excitation on the solid phase, given by Equation (13.8), needs to be applied.

13.4.6 Case of an imposed displacement field

In the case of an imposed displacement field $\boldsymbol{u}^0$ applied on Γ, used for instance to simulate a piston motion, the boundary conditions are

$$\begin{cases} u_n^i = (1-\phi)u_n^s + \phi u_n^f \\ u_i^0 = u_i^s \end{cases} \Rightarrow \begin{cases} u_n^f - u_n^s = 0 \\ u_i^s = u_i^0 \end{cases} \tag{13.27}$$

The first condition expresses the continuity of the normal displacements between the solid phase and the fluid phase. The second equation expresses the continuity between the imposed displacement vector and the solid phase displacement vector. Since the displacement is imposed, the admissible variation $\delta \boldsymbol{u}^s = \boldsymbol{0}$. Consequently, the surface integrals of Equation (13.8) simplify to

$$I^p = 0 \tag{13.28}$$

This equation indicates that for an imposed displacement, the boundary conditions reduce to the kinematic condition $\boldsymbol{u}^s = \boldsymbol{u}^0$ on Γ.

13.4.7 Coupling with a semi-infinite waveguide

We consider here the coupling of a poroelastic medium with an infinite waveguide. An example of application is the prediction of the absorption and transmission through a nonhomogenous material made from patches of porous, solid and air media (Sgard 2002, Atalla *et al.* 2003). The example of double porosity material is detailed in Section 13.9.5.

Let p^a, ρ_0 and c_0 denote the acoustic pressure, the density and speed of sound in the waveguide, respectively. For a porous material placed in the wave guide, as represented in Figure 13.2, the continuity conditions at the front Γ_1^p and rear Γ_2^p porous interfaces are given by Equation (13.8)

$$I_i^p = -\int_{\Gamma_i^p} \sigma_{ij}^t n_j \delta u_i^s \, \mathrm{d}S - \int_{\Gamma_i^p} \phi(u_n^f - u_n^s)\delta p \, \mathrm{d}S \qquad i = 1,\ 2 \tag{13.29}$$

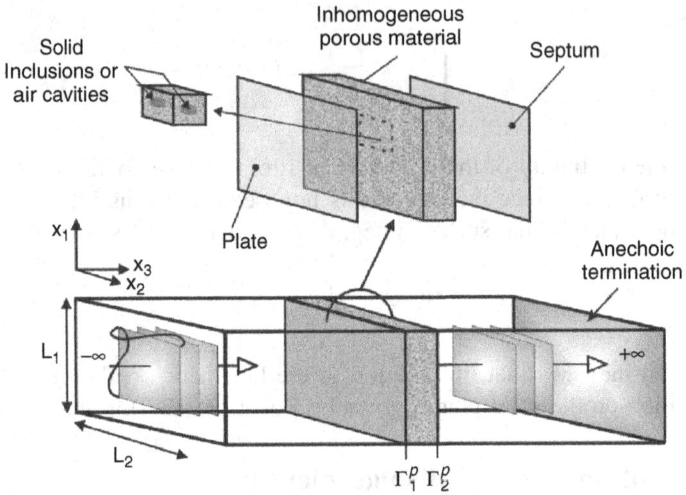

Figure 13.2 A complex porous component in a waveguide (Sgard 2002).

Using interface conditions, Equation (13.16), leads to

$$I_i^p = \int_{\Gamma_i^p} \delta(pu_n)\,\mathrm{d}S - \int_{\Gamma_i^p} \frac{1}{\rho_0\omega^2}\frac{\partial p^a}{\partial n}\delta p\,\mathrm{d}S \tag{13.30}$$

The first term amounts to the calculation of a classical coupling matrix. The second term will be rewritten using the orthogonal modes of the waveguide. The pressure p^a on the emitter side is written as the sum of the blocked pressure p_b satisfying

$$\left.\frac{\partial p_b}{\partial n}\right|_{\Gamma_1^p} = 0$$

and the pressure p_{rad} radiated from the surface of the poroelastic medium

$$p^a = p_b + p_{rad} \tag{13.31}$$

For a normal mode excitation of amplitude p_0 the complex amplitude of p_b reduces to $2p_0$. The radiated pressure may then be expressed in terms of the orthogonal normal modes φ_{mn} in the waveguide

$$p_{rad}(\mathbf{x}) = \sum_{m,n} B_{mn}\varphi_{mn}(x_1, x_2)e^{jk_{mn}x_3} \tag{13.32}$$

where $\mathbf{x} = (x_1, x_2, x_3)$. For a waveguide with a rectangular cross section $L_1 \times L_2$

$$\varphi_{mn}(x_1, x_2) = \cos\left(\frac{m\pi x_1}{L_1}\right)\cos\left(\frac{n\pi x_2}{L_2}\right) \tag{13.33}$$

and

$$k_{mn}^2 = k^2 - \left(\frac{m\pi}{L_1}\right)^2 - \left(\frac{n\pi}{L_2}\right)^2 \tag{13.34}$$

B_{mn} are modal amplitudes obtained from the modes orthogonality properties

$$\int_{\Gamma_1^p} p_{rad}\varphi_{mn}(x_1, x_2)\,\mathrm{d}S_x = \int_{\Gamma_1^p} (p^a - p_b)\varphi_{mn}(x_1, x_2)\,\mathrm{d}S_x \tag{13.35}$$

which leads to

$$B_{mn} = \frac{1}{N_{mn}} \int_{\Gamma_1^p} (p^a - p_b)\varphi_{mn}(x_1, x_2)\,\mathrm{d}S_x \tag{13.36}$$

where

$$N_{mn} = \int_{\Gamma_1^p} |\varphi_{mn}(x_1, x_2)|^2\,\mathrm{d}S_x \tag{13.37}$$

is the norm of mode (m, n).

The calculation for the second term of I_2^p is carried out in the same way, taking into account the fact that Equation (13.31) reduces to $p^a = p_{rad}$. Finally,

$$\begin{aligned} I_i^p = &-\int_{\Gamma_i^p} \delta(pu_n)\,\mathrm{d}S - \frac{1}{j\omega}\int_{\Gamma_i^p}\int_{\Gamma_i^p} A(\boldsymbol{x},\boldsymbol{y})p(\boldsymbol{y})\delta p(\boldsymbol{x})\,\mathrm{d}S_x\,\mathrm{d}S_y \\ &+ \frac{\epsilon}{j\omega}\int_{\Gamma_i^p}\int_{\Gamma_i^p} A(\boldsymbol{x},\boldsymbol{y})p_b(\boldsymbol{y})\delta p(\boldsymbol{x})\,\mathrm{d}S_x\,\mathrm{d}S_y \quad i = 1, 2 \end{aligned} \tag{13.38}$$

where $\varepsilon = 1$ if $i = 1$ and 0 if $i = 2$. In Eq (13.38), $\boldsymbol{x} = (x_1, x_2)$ and $\boldsymbol{y} = (y_1, y_2)$ are two points on the interface. $A(\boldsymbol{x}, \boldsymbol{y})$ is an admittance operator given by

$$A(\mathbf{x}, \mathbf{y}) = \sum_{mn} \frac{k_{mn}}{\rho_0\omega N_{mn}}\varphi_{mn}(x_1, x_2)\varphi_{mn}(y_1, y_2) \tag{13.39}$$

Equation (13.38) has the advantage of depicting the coupling with the waveguide in terms of radiation admittance in the emitter and receiver media, together with a blocked-pressure loading. To see the radiation effect, note for example that the power radiated (transmitted) into the receiving part of the waveguide is given by

$$\Pi_2 = \frac{1}{2}\mathrm{Re}\left[-j\omega\int_{\Gamma_2^p} p^a u_n^{a^*}\,\mathrm{d}S\right] \tag{13.40}$$

This can be written

$$\Pi_2 = \frac{1}{2}\mathrm{Re}\left[\int_{\Gamma_2^p}\int_{\Gamma_2^p} p^a(\mathbf{x})A^*(\mathbf{x}, \mathbf{y})p^{a^*}(\mathbf{y})\,\mathrm{d}S_x\,\mathrm{d}S_y\right] \tag{13.41}$$

Note that at low frequencies (below the cutoff frequency of the waveguide), higher modes lead to a purely imaginary admittance of an inertance type. These modes are evanescent and do not radiate in the tube.

For completion note that, if the porous medium is replaced by an elastic medium (e.g. the case of a porous medium sandwiched between two plates; a medium made up from patches of solid and porous elements, ...), the coupling boundary term of the elastic medium, Equation (13.10), reads

$$\begin{aligned} I_i^e = \int_{\Gamma_i^e} p^a \delta u_n^e \,\mathrm{d}S = j\omega \int_{\Gamma_i^e} \int_{\Gamma_i^e} u_n^e(y) Z(\boldsymbol{x}, \boldsymbol{y}) \delta u_n^e(\boldsymbol{x}) \,\mathrm{d}S_y \,\mathrm{d}S_x \\ + \varepsilon \int_{\Gamma^p} \int_{\Gamma^p} p_b(y) \delta u_n^e(\boldsymbol{x}) \,\mathrm{d}S_y \,\mathrm{d}S_x = 0 \end{aligned} \tag{13.42}$$

$Z(\boldsymbol{x}, \boldsymbol{y})$ is an impedance operator given by

$$Z(\boldsymbol{x}, \boldsymbol{y}) = \sum_{mn} \frac{\rho_0 \omega}{k_{mn} N_{mn}} \varphi_{mn}(x_1, x_2) \varphi_{mn}(y_1, y_2) \tag{13.43}$$

Equation (13.38) depicts the coupling with the waveguide in terms of radiation impedance in the emitter and receiver media, together with a blocked-pressure loading. Note that this equation holds if the elastic structure is replaced by a septum.

13.5 Other formulations in terms of mixed variables

To eliminate the drawbacks of the $(\boldsymbol{u}^s, \boldsymbol{u}^f)$ formulation, Göransson (1998) presented a symmetric $(\boldsymbol{u}^s, p, \varphi)$ formulation wherein the fluid is described by both the fluid pressure and the fluid displacement scalar potential. In particular, this formulation assumes the fluid displacement to be rotational free which is not true due to the strong inertial and viscous coupling between the two phases. Using a simple plate–foam–plate example, Hörlin (2004) showed, as a consequence of this rotational free assumption, that this formulation overestimates the dissipation associated with the relative motion between the fluid and the frame and in consequence the viscous damping. To eliminate this limitation, Hörlin (2004) proposed an extension based on four variables: frame displacement, acoustic pore pressure, scalar potential and the vector potential of the fluid displacement. The associated coupling conditions and convergence studies in the context of both h and p elements implementation are also given. Numerical examples show that this formulation leads to the same solutions as the previously described $(\boldsymbol{u}^s, \boldsymbol{u}^f)$ and $(\boldsymbol{u}^s, p)$ formulations. In consequence, the latter is preferred for its computational efficiency. Note that variants on the $(\boldsymbol{u}^s, p)$ formulation can be found in Hamdi *et al.* (2000) and Dazel (2005).

13.6 Numerical implementation

The numerical implementation of the presented formulations presented, using the finite element method, is classical and is discussed in several references (e.g. Atalla *et al.* 1998). For example, using classical finite element notations, the discretized form of Equation (13.7) leads to the following linear system

$$\begin{bmatrix} [Z] & -[\tilde{C}] \\ -[\tilde{C}]^T & [A] \end{bmatrix} \begin{Bmatrix} u^s \\ p \end{Bmatrix} = \begin{Bmatrix} F^s \\ F^f \end{Bmatrix} \tag{13.44}$$

where $\{u\}$ and $\{p\}$ represent the solid phase and the fluid phase global nodal variables, respectively. $[Z] = -\omega^2[\tilde{M}] + [\hat{K}]$ is the mechanical impedance matrix of the skeleton with $[\tilde{M}]$ and $[\hat{K}]$ equivalent mass and stiffness matrices:

$$\int_{\Omega} \tilde{\rho} u_i^s \delta u_i^s \, d\Omega = \langle \delta u^s \rangle [\tilde{M}] \{u^s\} \tag{13.45}$$

$$\int_{\Omega} \hat{\sigma}_{ij}^s \delta e_{ij}^s \, d\Omega = \langle \delta u^s \rangle [\hat{K}] \{u^s\} \tag{13.46}$$

$[A] = [\tilde{H}]/\omega^2 - [\tilde{Q}]$ is the acoustic admittance matrix of the interstitial fluid with $[\tilde{H}]$ and $[\tilde{Q}]$ equivalent kinetic and compression energy matrices for the fluid phase

$$\int_{d\Omega} \frac{\phi^2}{\tilde{\rho}_{22}} \frac{\partial p}{\partial x_i} \frac{\partial (\delta p)}{\partial x_i} \, d\Omega = \langle \delta p \rangle [\tilde{H}] \{p\} \tag{13.47}$$

$$\int_{\Omega} \frac{\phi^2}{\tilde{R}} p \delta p \, d\Omega = \langle \delta p \rangle [\tilde{Q}] \{p\} \tag{13.48}$$

$[\tilde{C}] = [\tilde{C}_1] + [\tilde{C}_2]$ is a volume coupling matrix between the skeleton and the interstitial fluid variables

$$\int_{\Omega} \frac{\phi}{\tilde{\alpha}} \delta \left(\frac{\partial p}{\partial x_i} u_i^s \right) d\Omega = \langle \delta u^s \rangle [\tilde{C}_1] \{p\} + \langle \delta p \rangle [\tilde{C}_1]^T \{u^s\} \tag{13.49}$$

$$\int_{\Omega} \phi \left(1 + \frac{\tilde{Q}}{\tilde{R}} \right) \delta (p u_{i,i}^s) \, dS = \langle \delta u^s \rangle [\tilde{C}_2] \{p\} + \langle \delta p \rangle [\tilde{C}_2]^T \{u^s\} \tag{13.50}$$

$\{F^s\}$ is the surface loading vector for the skeleton

$$\int_{\Gamma} T_i \delta u_i^s \, dS = \langle \delta u^s \rangle \{F^s\} \tag{13.51}$$

where $T_i = \sigma_{ij}^t n_j$ represent the components on the coordinate axis x_i of the total stress vector T. Finally $\{F^f\}$ is the surface kinematic coupling vector for the interstitial fluid

$$\int_{\Gamma} \phi (u_n^f - u_n^s) \delta p \, dS = \langle \delta p \rangle \{F^f\} \tag{13.52}$$

In the above equations, { } denotes a vector and ⟨ ⟩ its transpose.

The system of equations (13.44) is first solved in terms of the porous solid phase nodal displacements and interstitial nodal pressures. Next, the vibroacoustic indicators of interest are calculated.

The main limitation of the numerical implementations of these formulations resides in their computational cost. Typical implementations use linear and quadratic elements. Due to the highly dissipative nature of the domains and the existence of different wavelength scales, classical mesh criteria used for elastic domains and modal techniques are not strictly applicable. A discussion of meshing criteria and convergence can be found in (Panneton 1996, Dauchez *et al.* 2001, Rigobert *et al.* 2003, Hörlin, 2004). As an example, it is found that, for linear elements, as many as 12 elements for the smallest 'Biot' wavelength may be required, in some applications, for convergence! Several authors studied alternatives to alleviate the computational cost. Examples of techniques include the use of selective modal analysis (Sgard *et al.* 1997), use of complex modes (Dazel *et al.* 2002; Dazel 2005), plane wave decomposition (Sgard 2002), adaptive meshing (Castel 2005), axisymmetric implementations (Kang *et al.* 1999, Pilon 2003) and finally use of hierarchical elements (Hörlin *et al.* 2001, Hörlin 2004, Rigobert 2001, Rigobert *et al.* 2003, Langlois 2003). The latter authors show that *p*-implementations allow for quick convergence and accurate results. Still, the classical linear and quadratic elements are the widely used due to the simplicity of their implementations and their generality (for instance handling complex geometries with discontinuities; examples include foams with embedded masses or cavities and domains made up of a patchwork of several materials; see Sgard *et al.* 2007).

An interesting implementation based on the use of the mixed formulation is given by Hamdi *et al.* (2000). Consider the practical configuration wherein a typical porous component is attached to a master structure along a part Γ^e of its boundary and is coupled on the remaining part Γ^a to an acoustic cavity, as shown in Figure 13.3. Using the notations of Section 13.4, the weak integral form of the porous component accounting for the coupling with the master structure and the cavity is given by

$$\begin{aligned}&\int_{\Omega^p} [\hat{\sigma}^s_{ij}\delta e^s_{ij} - \omega^2\tilde{\rho} u^s_i \delta u^s_i]\,\mathrm{d}\Omega + \int_{\Omega^p}\left[\frac{\phi}{\omega^2\tilde{\alpha}\rho_0}\frac{\partial p}{\partial x_i}\frac{\partial(\delta p)}{\partial x_i} - \frac{\phi^2}{\tilde{R}}p\delta p\right]\mathrm{d}\Omega \\ &- \int_{\Omega^p}\frac{\phi}{\tilde{\alpha}}\delta\left(\frac{\partial p}{\partial x_i}u^s_i\right)\mathrm{d}\Omega - \int_{\Omega^p}\phi\left(1+\frac{\tilde{Q}}{\tilde{R}}\right)\delta(pu^s_{i,i})\,\mathrm{d}\Omega \\ &+ \int_{\Gamma^a}\delta(u_n p(\mathbf{x}))\,\mathrm{d}S_x - \int_{\Gamma^a} W_n \delta p\,\mathrm{d}S - \int_{\Gamma^e} T_i \delta u^s_i\,\mathrm{d}S = 0 \qquad \forall(\delta u^s_i, \delta p)\end{aligned} \tag{13.53}$$

In this equation, $T_i = \sigma^t_{ij} n_j$ are the components of the total stress vector at the structure–porous interface Γ^e and W_n is the normal acoustic displacement at the interface Γ^a with the cavity. Note that the last two terms of Equation (13.53) represent the energy exchanged between the porous component and its surrounding environment. The first term represents the acoustic energy absorbed by the porous component and the second term the mechanical energy absorbed by the porous component from the master structure. The FEM discretization of Equation (13.53) allows for the computation of the mixed impedance matrix of the porous component by eliminating all its internal degrees

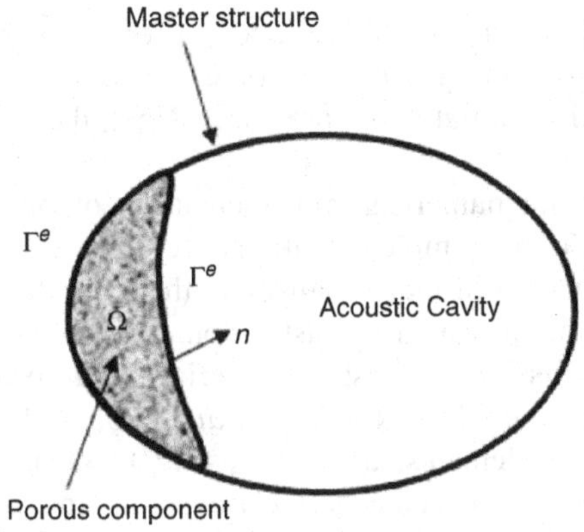

Figure 13.3 Porous component attached to a master structure–cavity system.

of freedom (dof) and keeping only the dof attached to the master structure and the acoustic cavity. In addition, to allow for incompatible meshes, Equation (13.53) is augmented using Lagrange multipliers to enforce the continuity of the pressure and the displacement at the interface Γ^a and Γ^e, respectively. The resulting condensed impedance matrix is added to enrich and couple the modal impedance matrices of the master structure and the cavity. This formulation has the advantage of considerably simplifying the modelling effort by: (i) not increasing the size of the global master structure-cavity model and (ii) allowing for the direct and independent modelling of the porous component, the master structure and the acoustic cavity. Using this methodology, the accurate solution of a fully trimmed vehicle is possible (Anciant *et al.* 2006). An example is discussed in Section 13.9.7.

13.7 Dissipated power within a porous medium

The classical vibroacoustic indicators (kinetic energy, quadratic velocity, quadratic pressure, absorption coefficient, transmission loss, etc.) can be easily calculated from the solution of the formulations presented. One important feature of the FE implementation is the ability to break down the dissipated power within a poroelastic material in terms of the relative contributions of the viscous, thermal and structural effects. The expressions of these contributions are well known for the $(\boldsymbol{u}^s, \boldsymbol{u}^f)$ formulations (Dauchez *et al.*, 2002). For the $(\boldsymbol{u}^s, p)$ formulations, an initial derivation of the expression of these powers was given by Sgard *et al.* (2000). Recently, a formal derivation was given by Dazel *et al.* (2008). Dazel started from the $(\boldsymbol{u}^s, \boldsymbol{u}^f)$-based expressions to derive both the dissipated powers and stored energy expressions for the $(\boldsymbol{u}^s, p)$ formalism. He highlighted in particular the proper interpretation of Sgard *et al.* (2000) initial expressions. In this section, and for the sake of conciseness, the simple derivation of Sgard *et al.* (2000) is recalled.

With the following particular choice for the admissible functions, $\delta \boldsymbol{u}^s = -j\omega \boldsymbol{u}^{s*}$ for the solid-phase displacement vector and $\delta p = -j\omega p^*$ for the fluid-phase interstitial pressure, where f^* denotes the complex conjugate of f, the weak integral form is rewritten

$$\underbrace{-j\omega \int_\Omega \hat{\sigma}^s_{ij}(\boldsymbol{u}^s) : \underline{\underline{\epsilon}}^s(\boldsymbol{u}^{s*})\,\mathrm{d}\Omega}_{\Pi^s_{elas}} + \underbrace{j\omega^3 \int_\Omega \tilde{\rho} u^s_i u^{s*}_i\,\mathrm{d}\Omega}_{\Pi^s_{iner}}$$

$$+ \underbrace{j\omega \int_\Omega \frac{\phi^2}{\tilde{R}} p\, p^*\,\mathrm{d}\Omega}_{\Pi^f_{elas}} \underbrace{- j\omega \int_\Omega \frac{\phi^2}{\tilde{\alpha}\rho_0\omega^2} \frac{\partial p}{\partial x_i}\frac{\partial p^*}{\partial x_i}\,\mathrm{d}\Omega}_{\Pi^f_{iner}}$$

$$\underbrace{+ j\omega \int_\Omega \frac{\phi}{\tilde{\alpha}} \left(\frac{\partial p}{\partial x_i} u^{s*}_i + \frac{\partial p^*}{\partial x_i} u^s_i \right) \mathrm{d}\Omega + j\omega \int_\Omega \phi \left(1 + \frac{\tilde{Q}}{\tilde{R}} \right) (p u^{s*}_{i,i} + p^* u^s_{i,i})\,\mathrm{d}\Omega}_{\Pi^{sf}_{coup}}$$

$$\underbrace{+ j\omega \int_\Gamma \sigma^t_{ij} \mathrm{n}_j u^{s*}_i\,\mathrm{d}S\,\mathrm{d}S + j\omega \int_\Gamma \phi(u^f_n - u^s_n) p^*\,\mathrm{d}S}_{\Pi^s_{exc}} = 0 \qquad (13.54)$$

This provides the following power balance equation:

$$\Pi^s_{elas} + \Pi^s_{iner} + \Pi^f_{elas} + \Pi^f_{iner} + \Pi^{fs}_{coup} + \Pi^f_{exc} = 0 \tag{13.55}$$

where Π^s_{elas}, Π^s_{iner} represent the power developed by the internal and inertia forces in the solid phase in vacuum, respectively; Π^f_{elas}, Π^f_{iner} represent the power developed by the internal and inertia forces in the interstitial fluid, respectively; Π^{fs}_{coup} represents the power exchanged between the two phases; finally Π^f_{exc} represents the power developed by external loading.

The time-averaged power dissipated within the porous medium can be subdivided into contributions from powers dissipated through structural damping in the skeleton, viscous and thermal effects: $\Pi_{diss} = \Pi^s_{diss} + \Pi^v_{diss} + \Pi^t_{diss}$. The time-averaged power dissipated through structural damping is obtained from the imaginary part of Π^s_{elas}

$$\Pi^s_{diss} = \frac{1}{2}\mathrm{Im}\left[\omega \int_\Omega \hat{\sigma}^s_{ij}(\boldsymbol{u}^s) : \underline{\underline{\epsilon}}^s(\boldsymbol{u}^{s^*})\, \mathrm{d}\Omega\right] \tag{13.56}$$

The power dissipated through viscous effects is obtained from $\Pi^s_{iner} + \Pi^f_{iner} + \Pi^{fs}_{coup}$:

$$\Pi^v_{diss} = -\frac{1}{2}\left[\omega^3 \int_\Omega \mathrm{Im}(\tilde{\rho})u^s_i u^{s^*}_i\, \mathrm{d}\Omega - \int_\Omega \mathrm{Im}\left(\frac{\phi^2}{\tilde{\alpha}\rho_0\omega^2}\right)\frac{\partial p}{\partial x_i}\frac{\partial p^*}{\partial x_i}\, \mathrm{d}\Omega\right.$$
$$\left. + 2\int_\Omega \mathrm{Im}\left(\frac{\phi}{\tilde{\alpha}}\right)\mathrm{Re}\left(\frac{\partial p}{\partial x_i}u^{s^*}_i\right)\mathrm{d}\Omega\right] \tag{13.57}$$

Note that in deriving the above equation, the fact that

$$\mathrm{Im}\left(\frac{\tilde{Q}}{\tilde{R}}\right) = 0$$

for all materials has been used. Finally, the power dissipated through thermal effects is obtained from Π^f_{elas}

$$\Pi^t_{diss} = -\frac{1}{2}\omega \int_\Omega \mathrm{Im}\left(\frac{\phi^2}{\tilde{R}}\right) p\ p^*\, \mathrm{d}\Omega \tag{13.58}$$

Section 13.9.3 presents an example illustrating the use of these expressions.

13.8 Radiation conditions

Numerical approaches based on the formulations described provide efficient tools to solve problems where the porous material is coupled to elastic structures and finite extent acoustic cavities. When the porous domain is in contact with free air, the surface impedance should be calculated accurately to account for the coupling between the porous medium and air medium. When the air medium is bounded this is done easily through the coupling conditions. The difficulty arises when the porous domain is in contact with an infinite domain. The acoustic radiation of a porous medium into an unbounded fluid medium has usually been neglected. The classical approximation for modelling free field radiation of porous materials assumes the total stress tensor and the interstitial pressure at the radiation surface to be zero (Debergue *et al.* 1999). In this case, it is assumed that the surface impedance of the solid phase is much higher than the impedance of the surrounding acoustic medium, and in consequence the porous system is assumed to be vibrating in vacuum. Other methods use a plane wave approximation in the same manner as is classically done in the transfer matrix method (Chapter 11). This is mainly acceptable at high frequencies, but is clearly erroneous at low frequencies. These approximations can be alleviated for specific problems. In the case of radiation of a porous medium in a waveguide, the radiation impedance can be calculated accurately by expressing the radiated pressure in terms of the modal behaviour of the waveguide (Section 13.4.7). In the case of a thin porous plate under flexural vibration, Horoshenkov and Sakagami (2000) considered the absorption and transmission problems and presented a parametric study on the influence of the porous plate parameters on its absorption. Takahashi and Tanaka (2002) considered the same problem and presented an analytical model to calculate the radiation impedance of a thin porous plate in flexure; in particular, they discussed the effects of plate permeability on its radiation damping based on numerical examples. Atalla *et al.* (2006) presented a formulation, based on the mixed displacement–pressure formulation, for evaluating the sound radiation of baffled poroelastic media in the special case where the material surface is baffled. It expresses the free field condition using Rayleigh's integral, in terms of an added admittance matrix and a solid phase–interstitial pressure coupling term. The approach is general and easy to implement. It can handle situations such as planar multilayer systems with various excitations. Numerical results were presented to illustrate the accuracy of the method. A recall of this formulation is presented here.

Consider the coupling of a planar baffled porous domain with a semi-infinite fluid (Figure 13.4). In the $(\boldsymbol{u}^s, p)$ formulation, the porous medium couples to the semi-infinite fluid medium through the boundary term given by Equation (13.8). Since at the free surface: $\sigma_{ij}^t n_j = -p n_i$, Equation (13.8) becomes:

$$I^p = \int_\Gamma \delta\left(p u_n^s\right) \mathrm{d}S - \int_\Gamma \left[\phi\left(u_n^f - u_n^s\right) + u_n^s\right] \delta p \, \mathrm{d}S \tag{13.59}$$

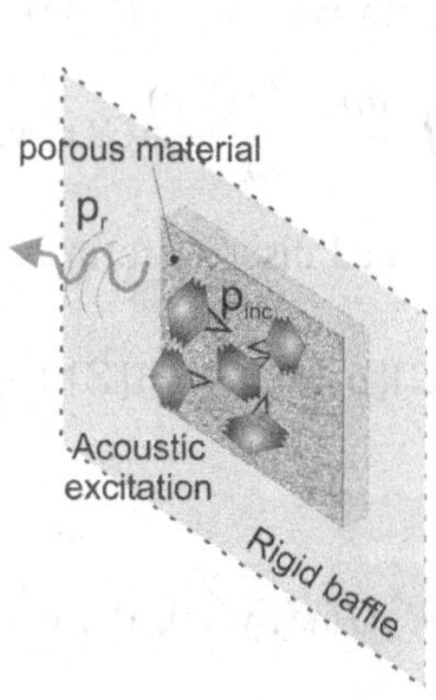

Figure 13.4 Radiation from a flat baffled poroelastic material.

The continuity of the normal displacement at the radiating surface,

$$\phi(u_n^f - u_n^s) + u_n^s = \frac{1}{\rho_0\omega^2}\frac{\partial p^a}{\partial n}$$

leads to

$$I^p = \int_\Gamma \delta\left(pu_n^s\right)\,\mathrm{d}S - \frac{1}{\rho_0\omega^2}\int_\Gamma \frac{\partial p^a}{\partial n}\delta p\,\mathrm{d}S \tag{13.60}$$

In the semi-infinite domain, the acoustic pressure p^a is the sum of the blocked pressure p_b and the radiated pressure p_r. So at the surface $\partial p^a/\partial n = \partial p_r/\partial n$, and in the numerical implementation context, the discrete form associated to the second term of Equation (13.60) reads

$$\frac{-1}{\rho_0\omega^2}\int_\Gamma \frac{\partial p^a}{\partial n}\delta p\,\mathrm{d}S = \frac{-1}{\rho_0\omega^2}\langle\delta p\rangle[C]\left\{\frac{\partial p_r}{\partial n}\right\} \tag{13.61}$$

where [C] is the classical coupling matrix given by the assembling of the elemental matrices ($\Gamma = \cup\Gamma^e$):

$$[C] = \sum_e \int_{\Gamma^e} \langle N^e\rangle\{N^e\}\,\mathrm{d}S^e \tag{13.62}$$

with $\{N^e\}$ denoting the vector of the used surface element shape functions, $\langle N^e\rangle$ its transpose and $\sum_e$ the assembling process.

The planar porous material being inserted into a rigid baffle, the radiated acoustic pressure is related to the normal velocity via Rayleigh's integral

$$p_r(x,y,z) = -\int_\Gamma \frac{\partial p_r(x',y',0)}{\partial n} G(x,y,z,x',y',0)\,\mathrm{d}S' \tag{13.63}$$

where $G(x,y,z,x',y',0) = e^{-jkR}/2\pi R$ is the baffled Green's function, $k = \omega/c_0$, is the acoustic wave number in the fluid, c_0, the associated speed of sound and R is the distance between point (x,y,z) and $(x',y',0)$: $R = \sqrt{(x-x')^2 + (y-y')^2 + z^2}$.

The integral form associated with Equation (13.63) is given by

$$\int_\Gamma p_r(x,y,0)\delta p\mathrm{d}S = -\int_\Gamma\int_\Gamma \frac{\partial p_r(x',y',0)}{\partial n} G(x,y,0,x',y',0)\delta p\mathrm{d}S\mathrm{d}S' \tag{13.64}$$

Using Equation (13.62), the associated discrete form is

$$\langle\delta p\rangle[C]\{p_r\} = -\langle\delta p\rangle[Z]\left\{\frac{\partial p_r}{\partial n}\right\} \tag{13.65}$$

with,

$$[Z] = \sum_e\sum_{e'}\int_{\Gamma^e}\int_{\Gamma^{e'}} \langle N^e\rangle G(x^e, y^e, 0, x^{e'}, y^{e'}, 0)\{N^{e'}\}\,\mathrm{d}S^e\,\mathrm{d}S^{e'} \tag{13.66}$$

Since $\langle \delta p \rangle$ is arbitrary, one gets

$$\left\{ \frac{\partial p_r}{\partial n} \right\} = -[Z]^{-1}[C]\{p_r\} \tag{13.67}$$

Substituting Eq. (13.67) into Eq. (13.61) and recalling that on the interface $p = p^a = p_b + p_r$, the discrete form of Eq. (13.60) reads finally

$$I^p = \langle \delta u_n^s \rangle [C]\{p\} + \langle \delta p \rangle [C]^T \{u_n^s\} - \frac{1}{j\omega}\langle \delta p \rangle [A]\{p - p_b\} \tag{13.68}$$

where

$$[A] = \frac{1}{j\omega\rho_0}[C][Z]^{-1}[C] \tag{13.69}$$

is an admittance matrix. In consequence, the radiation of the porous medium into the semi infinite fluid amounts to an admittance term added to the interface interstitial pressure degrees of freedom and to additional interface coupling terms between the solid phase and the interstitial pressure (first two terms in Equation 13.68). Note that the last term involving p_b is the excitation term and disappears in the case of free radiation. In the numerical implementation, the discretized form Equation (13.44) with $\{F^s\} = 0$ combined with Equation (13.68) leads to the following linear system

$$\begin{bmatrix} -\omega^2[\tilde{M}] + [\hat{K}] & -[\tilde{C}] + [C] \\ -[\tilde{C}]^T + [C]^T & \dfrac{[\tilde{H}]}{\omega^2} - \dfrac{[A]}{j\omega} - [\tilde{Q}] \end{bmatrix} \begin{Bmatrix} u^s \\ p \end{Bmatrix} = \begin{Bmatrix} 0 \\ F^f \end{Bmatrix} \tag{13.70}$$

with

$$\{F^f\} = \frac{1}{j\omega}[A]\{p_b\} \tag{13.71}$$

The system of equations (13.70) is first solved in terms of the porous solid phase nodal displacements and interstitial nodal pressures. Next, the vibroacoustic indicators of interest can be calculated.

13.9 Examples

13.9.1 Normal incidence absorption and transmission loss of a foam: finite size effects

The first example considers an absorption problem and consists of a 0.5 m × 0.5 m × 5.08 cm rectangular sample of foam backed by a rigid wall and excited by a plane wave. Two load cases are considered: normal incidence $(0°, 0°)$ and oblique plane wave $(45°, 0°)$. The properties of the foam are given in Table 13.1. A mesh of 22 × 22 × 15 linear brick poroelastic elements is used for the foam for the normal incidence while a larger mesh of 32 × 32 × 15 elements is used for the oblique plane wave. These meshes have been selected to ensure convergence of calculations. The meshing criteria are classical for the plate and are functions of the Biot's wavelengths for the foam (Chapter 6).

Table 13.1 The parameters of the foam.

Porosity, ϕ	Flow resistivity, σ ($\mathrm{N\,m^{-4}\,s}$)	Tortuosity, α_∞	Viscous charact. dim. (m) Λ	Thermal charact. dim. (m) Λ'	Density of frame, ρ_1 ($\mathrm{kg\,m^{-3}}$)	Young's Modulus, (Pa)	Poisson coefficient, ν	Structural damping, η
0.99	12569	1.02	0.000078	0.000192	8.85	93348	0.44	0.064

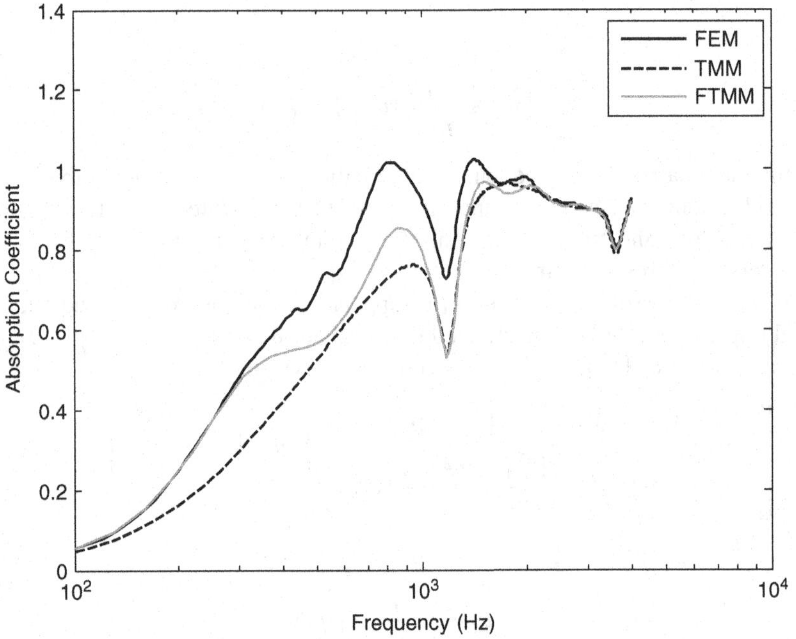

Figure 13.5 Absorption coefficient of 2-in-thick foam slab of dimensions 0.5 × 0.5 m excited by a normal incidence plane wave (Atalla *et al.* 2006.)

Figures 13.5 and 13.6 show the corresponding absorption coefficients. The latter are calculated using a power balance method:

$$\alpha(\theta, \varphi, \omega) = \frac{\Pi_{diss}}{\Pi_{inc}} \tag{13.72}$$

In this expression Π_{diss} denotes the power dissipated in the foam and Π_{inc} denotes the incident power. The results obtained using the transfer matrix method (TMM) are also shown for comparison. As expected, it is clearly seen that, at higher frequencies, the finite element (FEM) calculation asymptotes to the infinite extent configuration since the normalized radiation impedance reduces to 1. The finite size correction of chapter 12 is also used here to compute the curve referred to as FTMM in Figures 13.5 and 13.6. As expected it captures the initial slope of the absorption curve which is governed by the size of the tested sample. Note that as the sample becomes larger, the FEM and FTMM

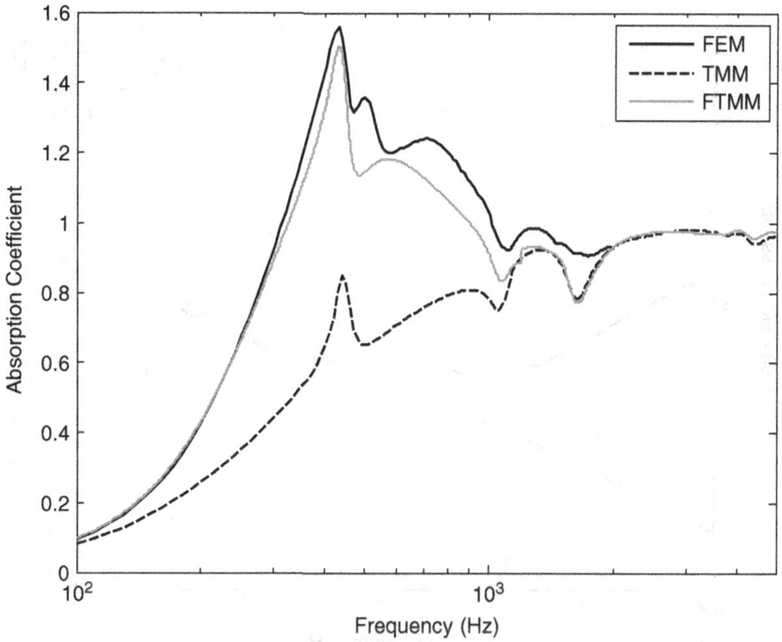

Figure 13.6 Absorption coefficient of 2-in-thick foam slab of dimensions 0.5 × 0.5 m excited by an oblique plane wave (Atalla *et al.* 2006).

curves asymptote towards the TMM result. The same results are obtained for an oblique incidence plane wave, Figure 13.6.

Next, we consider the transmission loss of the same foam sample, which is now between two semi-infinite media. The results are shown in Figure 13.7 for an oblique incidence plane wave excitation (45°, 0°) and compared to the TMM results. Once again, and as expected, the two methods lead to similar results at high frequencies. At low frequencies, the size effects lead to an increase in the transmission loss. Again, the finite size correction of Chapter 12 is shown to lead to acceptable corrections at low frequencies (within 1 dB). Again, for large samples, the results will asymptote towards the TMM result.

13.9.2 Radiation effects of a plate–foam system

Usually, the effect of radiation damping of foam is neglected when the latter radiates in free field. To check this assumption, the following example considers a plate–foam system excited by a point force. The material properties and dimensions of the plate and the foam are listed in Table 13.1. The frequency range of interest is the range below 800 Hz. The plate is meshed using 24 × 15 thin shell elements and the porous material is meshed using 24 × 15 × 9 linear brick poroelastic elements. The plate is assumed simply supported. The foam is bonded onto the plate, clamped (bonded to a hard baffle) along its edges and has a free face which can radiate into a semi infinite space. The space averaged quadratic velocity $\langle v^2 \rangle$ of the plate is shown in Figure 13.8 for three configurations of the free-face boundary condition: (i) using the radiation impedance

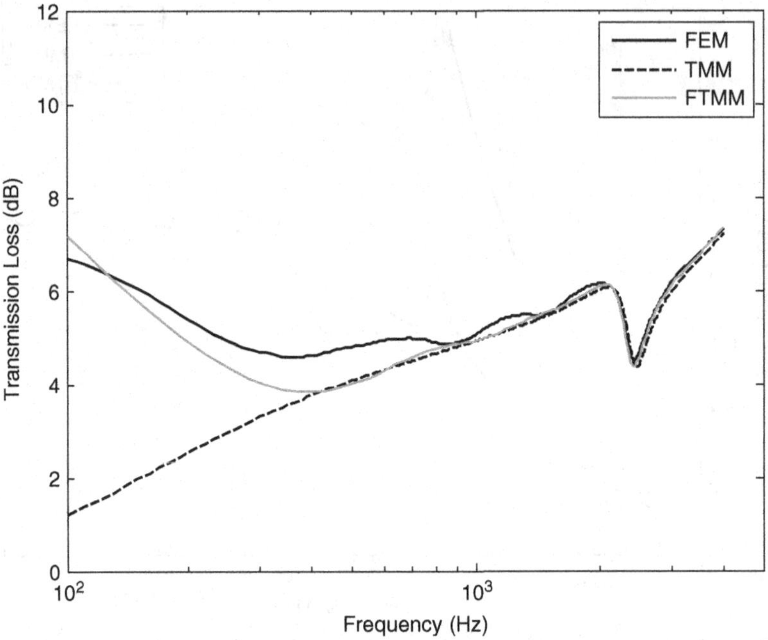

Figure 13.7 Transmission loss of a 2-in-thick foam slab of dimensions 0.5 × 0.5 m excited by an oblique incidence plane wave (Atalla *et al.* 2006).

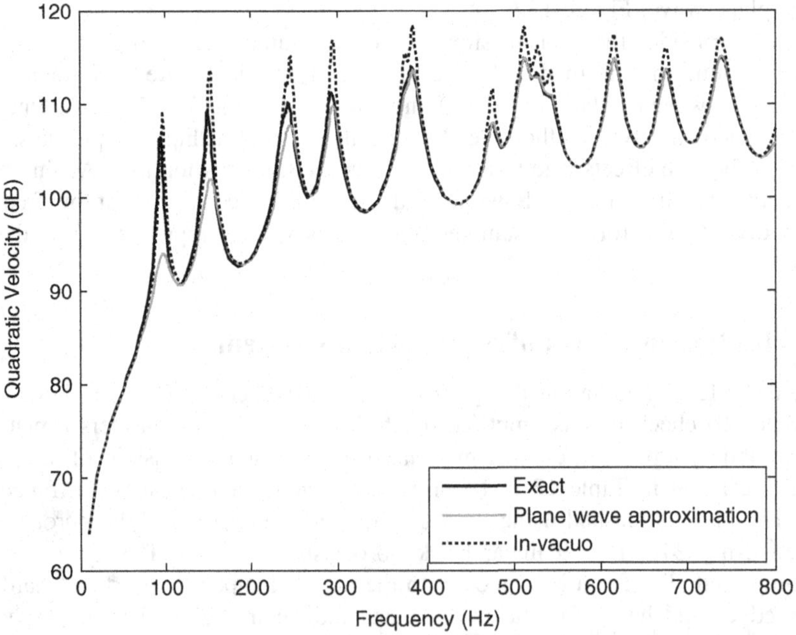

Figure 13.8 Mean-square velocity of a plate–foam system radiating in air.

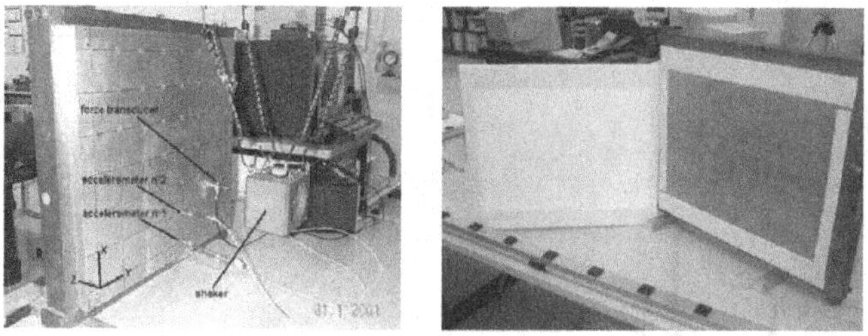

(a) (b)

Figure 13.9 Experimental set-up used to illustrate the damping added by the foam. (a) Excitation system; (b) plate with foam attached using a double-sided tape (shown before assembly).

condition of Section 13.8 (exact approach), (ii) pressure release condition (i.e. $p = 0$ on the free face) and (iii) assuming a plane wave radiation condition ($Z_R = \rho_0 c$). It is clearly seen that the three results are similar except at the resonances where the effect of radiation damping is noticeable, especially for low-frequency modes. Moreover, it is seen that the exact approach and the plane wave approximation lead to similar results at higher frequencies. However, the plane wave approximation clearly overestimates the radiation damping at low frequencies. This result highlights the small effects of the radiation efficiency of the foam. Calculations using other foam materials with different properties (especially several ranges of flow resistivity) lead to similar conclusions.

13.9.3 Damping effects of a plate–foam system

The vibration of a plate–foam system is studied both numerically and experimentally. Figure 13.9 shows the experimental set-up. It consists of a 0.480 m × 0.420 m × 3.175 mm aluminium panel glued to a thin frame to simulate a simple support. The mechanical properties of the panel are: Young's modulus 7.0×10^{10} N/m^2; density = 2742 kg/m^3; Poisson's ratio 0.33; loss factor 0.01. The panel is excited mechanically using a shaker. The normal velocity of the panel is measured using accelerometers positioned on a grid made up of 56 points uniformly distributed over the surface of the panel. A 7.62-cm (3-inch) foam is added to one side of the panel. The foam is attached to the panel along its edges using a double-sided tape, (see Figure 13.9b). The foam is thus not perfectly bonded onto the panel since the double tape creates a thin air space between the panel and the foam. Figure 13.10 shows the measured space averaged quadratic velocity of the panel with and without the attached foam. Significant reduction of the vibration level is observed, in particular at high frequencies. This behaviour is typical of panels with attached noise control treatments.

To investigate the nature of this damping, numerical simulations are used. To achieve confidence in the code, predictions are compared to simulation in Figure 13.11. The acoustical and mechanical properties of the melamine foam used in the model are given in Table 13.1. Good agreement is observed, keeping in mind that: (i) no attempt has been made to select the modal damping of the first plate's modes in the predictions, (ii) in the simulations the foam is assumed perfectly bonded onto the panel and (iii) the mechanical properties of the foam have been measured at low frequencies using a quasi-static method (Langlois *et al.* 2001) and are assumed constant over the whole frequency range.

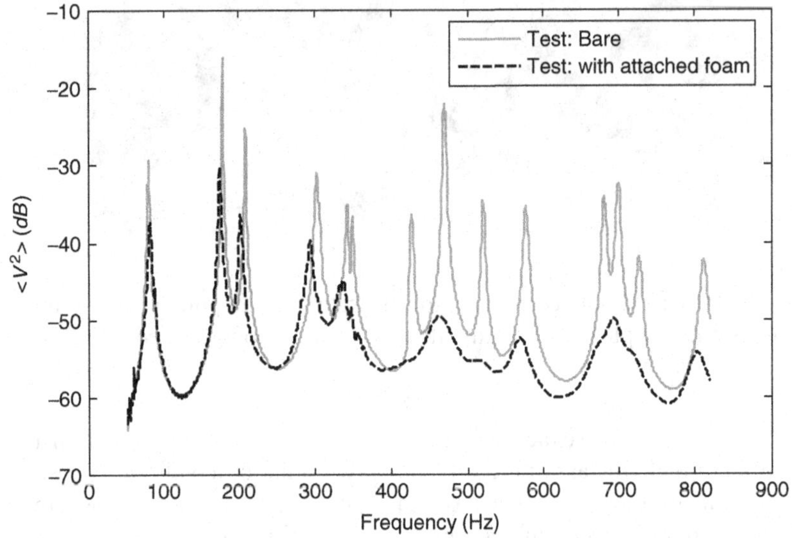

Figure 13.10 Effect of the foam layer on the space averaged quadratic velocity of the plate.

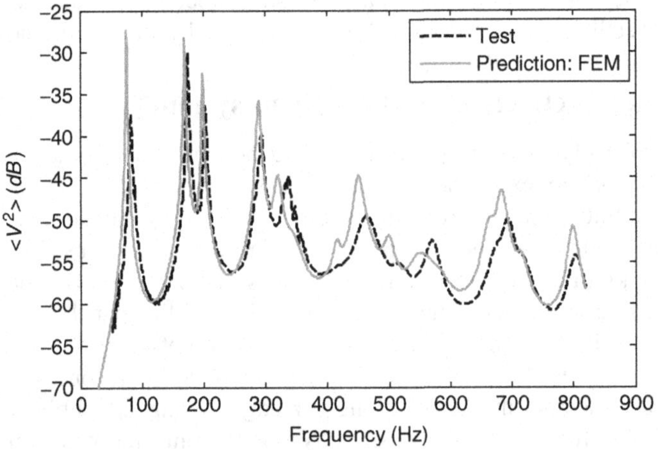

Figure 13.11 Comparison between measured and predicted space averaged quadratic velocity of the plate with attached foam layer.

Figure 13.12 presents the contribution to the dissipated power within the foam panel system by the different dissipation mechanisms

$$\begin{cases} \Pi_{tot} = \Pi_{diss\_plate} + \Pi_{diss\_foam} \\ \Pi_{diss\_foam} = \Pi_{struct} + \Pi_{viscous} + \Pi_{thermal} \end{cases} \tag{13.73}$$

The different dissipated powers are computed numerically using the expressions presented in Section 13.7. Radiation damping is neglected as explained in the previous

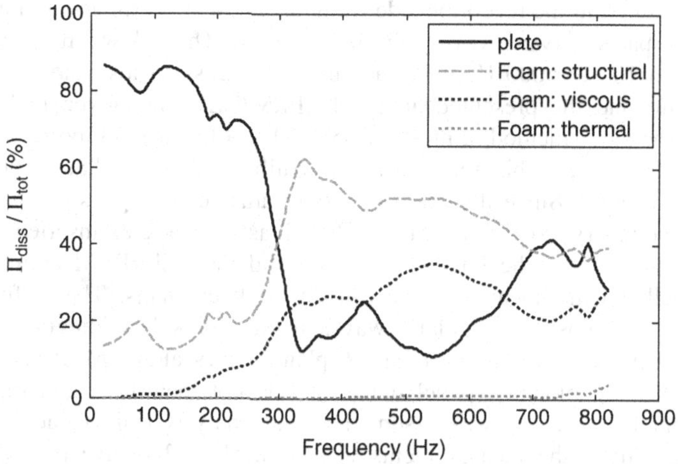

Figure 13.12 Relative contribution of the various dissipation mechanisms in the panel–foam system.

example. It is observed that at low frequencies, the damping is dominated by losses within the panel. At higher frequencies, the structural damping of the frame of the foam dominates the losses followed by the viscous losses in the foam. Note that dissipation by thermal effects is negligible. This is a consequence of the fact the foam is assumed free at its edges and one of its faces. The results of Figure 13.12 confirm the large dissipation seen at high frequencies and thus confirm the damping added by the foam attached to the panel. If the foam is modelled as a solid layer, at low frequencies both the full poroelastic computation and the solid models will lead to same results. However, at higher frequencies the solid model will overestimate the space averaged quadratic velocity of the response. This stresses the importance of viscous damping at higher frequencies. This importance depends strongly on the nature of the foam or fibre system and its attachment to the panel. Further discussion on this subject can be found in Dauchez *et al.* (2002) and Jaouen *et al.* (2005).

13.9.4 Diffuse transmission loss of a plate–foam system

A transmission loss test was performed on a flat aluminium panel with attached foam. The panel dimensions were 1.64 m × 1.19 m × 1.016 mm and the foam dimensions were 1.64 m × 1.19 m × 7.62 cm. The foam is attached along its edges to the panel using double-sided tape. The tests were performed at the Groupe Acoustique de l'Université de Sherbrooke (GAUS) transmission loss facility. The facility utilizes a semi-anechoic–reverberant transmission loss suite. The reverberation room dimensions are 7.5 m × 6.2 m × 3 m with a Schroëder frequency of 200 Hz and a reverberation time of 5.3 s at 1000 Hz. The free volume of the semi- anechoic chamber is 6 m × 7 m × 3 m with an operational frequency range from 200 Hz to 80 kHz. The plate is secured in a mounting window between the two chambers. The intensity technique is used to determine the transmission loss. The technique follows closely standard ISO 15186-1: 2000. The reverberation chamber is excited using six loudspeakers and sound power is captured using a microphone on a rotating boom. On the anechoic side, the sound intensity is measured in the reception side using an automated arm and an intensity probe with a 12 mm spacer between two $^1/_2$-in microphones. This allows measurements to be carried out from 100 to 5000 Hz. The measured transmission loss and predictions with both the FTMM and the presented approach (FEM/BEM) are given in Figures 13.13 and 13.14. In the latter method, a mesh of $45 \times 34 \times 9$ linear brick poroelastic elements was used for the foam. This mesh was compatible with the plate's mesh (45×34 Quad4 shell elements). Since the foam was only attached along its edges to the plate, the coupling boundary condition was modelled as an air gap inserted between the two components for both the FTMM approach and the FEM/BEM approach. The air gap was modelled with linear acoustic 8-node brick elements. The diffuse field was modelled as a superposition of plane waves using a GAUSS integration scheme of 6×6 points (6 plane waves along θ and 6 plane waves along ϕ). It should be noted at this stage that the mounted panel damping was not measured, and that a nominal modal damping ratio of 3% was assumed in the analysis (this value is justified by edge damping). Since the measurements were done in 1/3 octave bands, the FTMM results were calculated at the band centre frequencies while the FEM/BEM results were calculated at 1000 frequency points using a logarithmic frequency step (Figure 13.13) and converted to 1/3 octave bands (Figure 13.14). It is seen that the FTMM leads to a very good agreement throughout the frequency range of the test (up to 5000 Hz), apart from a slight overestimation at low frequencies (below 500 Hz). This corroborates the effectiveness of this method for predicting transmission loss of multilayer systems over the whole frequency range (see Chapter 12). Equally, the FEM method shows excellent comparison, except at higher frequencies; where it diverges as a consequence of the mesh used. However, at these high frequencies there is no need for this deterministic

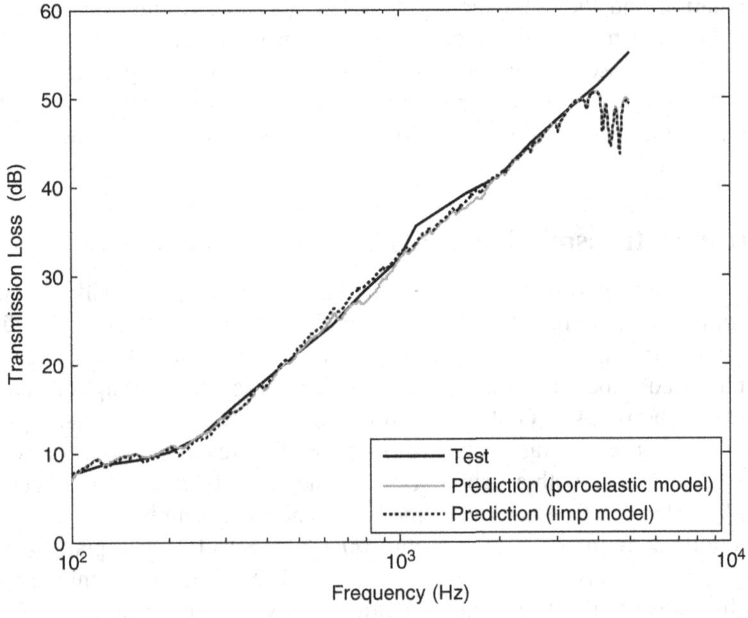

Figure 13.13 Transmission loss of a plate-foam system. Tests versus FEM predictions: narrow band comparison.

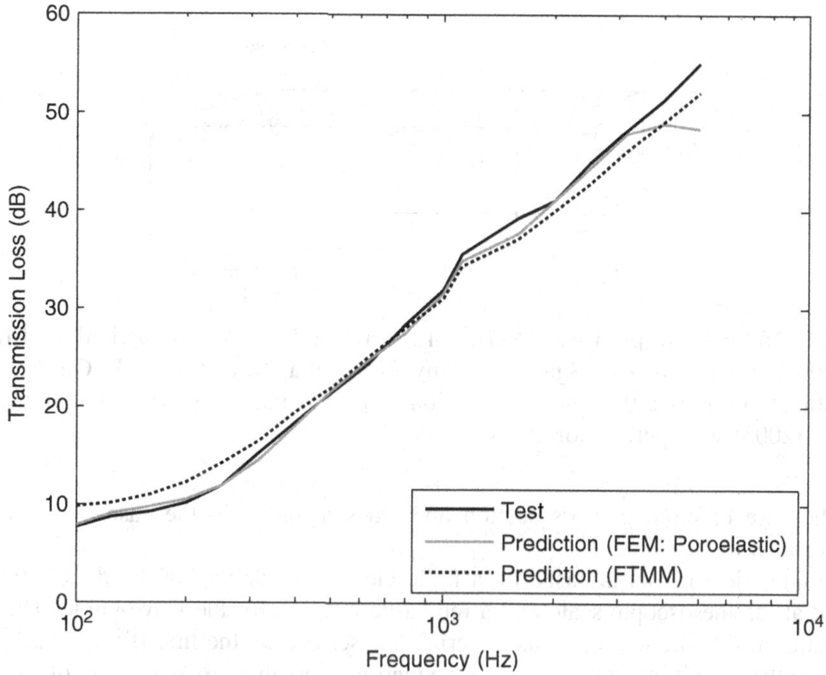

Figure 13.14 Transmission loss of a plate–foam system. Tests versus FTMM and FEM predictions; 1/3 octave band comparison.

and expensive method: the FTMM and even the TMM does an excellent job. Finally, it is worth mentioning that for this particular problem (unbonded foam) similar results can be obtained using a limp model for the foam thus diminishing considerably the computational effort (Figure 13.13).

13.9.5 Application to the modelling of double porosity materials

As discussed in Chapter 5, the concept of meso-perforations in appropriately chosen porous media can help enhance their sound absorption performance. The meso-perforated materials are also referred to as 'double porosity materials' since they are made up of two interconnected networks of pores of different characteristic size. Several theoretical, numerical and experimental studies have been done on this subject. Sgard *et al.* (2005) give an excellent review of these studies and establish practical design rules to develop optimized noise control solutions based on this concept. This section is limited to an example explaining the use of finite element methods to simulate these materials.

The general configuration of the problem is depicted in Figure 13.15. It consists of a meso-perforated porous material placed in a waveguide with rigid walls, acoustically excited by a plane wave. The rear of the material is terminated by a rigid wall. Let the porous material be characterized by its porosity ϕ_m, its static flow resistivity σ_m, its geometric tortuosity $\alpha_{\infty m}$, its viscous and thermal characteristic lengths Λ_m and Λ'_m, and its thermal permeability $\Theta_m(0)$. The perforations are assumed rectangular or cylindrical and are characterized by a perforation rate (also named meso-porosity) ϕ_p, a radius

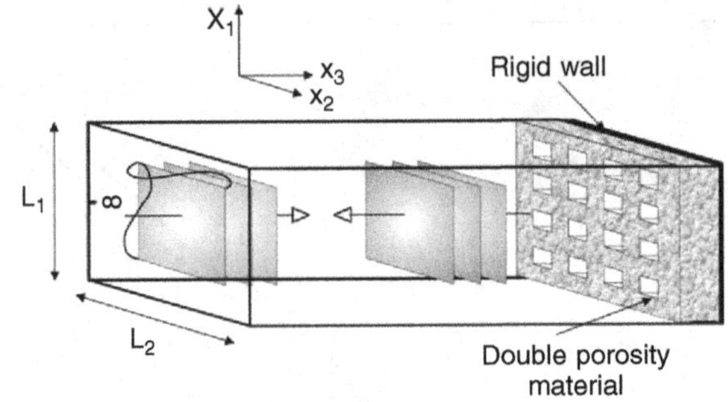

Figure 13.15 Double porosity material in a waveguide backed by a rigid wall (Sgard *et al.* 2005). Reprinted from Sgard, F., Olny, X., Atalla, N. & Castel, F. On the use of perforations to improve the sound absorption of porous materials. *Applied acoustics* **66**, 625-651 (2005) with permission from Elsevier.

R in the case of circular cross section and sides a and b in the case of rectangular cross-sections.

The numerical model is based on a finite element modelling of the double porosity material at the mesoscopic scale and a modal description for the wave-guide. The wave propagation inside the microporous material is described by the first Biot–Allard theory and inside the meso-pores by Helmholtz's equation. Note that, given the size of the holes, the associated viscous characteristic frequency is very low and the dissipation can thus be considered as negligible, justifying the use of Helmholtz's equation (established for a perfect gas) in the meso-pores. The weak integral forms associated with each domain are then discretized using finite elements. The coupling between the double porosity material and the waveguide is accounted for explicitly using the modal behaviour of the waveguide. Atalla *et al.* (2001b) provide the details in the simplified case where the skeleton of the porous medium is assumed motionless. The following presentation accounts for the elasticity of the material.

Combining Equations (13.8), (13.17) and (13.38), the weak integral form governing the acoustic behaviour of a porous layer coupled with the modal behaviour of the waveguide of cross-section $L_1 \times L_2$, can be written

$$\begin{aligned}
&\int_{\Omega^p} [\hat{\sigma}^s_{ij}\delta e^s_{ij} - \omega^2\tilde{\rho}u^s_i\delta u^s_i]\,\mathrm{d}\Omega + \int_{\Omega^p}\left[\frac{\phi}{\omega^2\tilde{\alpha}\rho_0}\frac{\partial p}{\partial x_i}\frac{\partial(\delta p)}{\partial x_i} - \frac{\phi^2}{\tilde{R}}p\delta p\right]\mathrm{d}\Omega \\
&\quad - \int_{\Omega^p}\frac{\phi}{\tilde{\alpha}}\delta\left(\frac{\partial p}{\partial x_i}u^s_i\right)\mathrm{d}\Omega - \int_{\Omega^p}\phi\left(1+\frac{\tilde{Q}}{\tilde{R}}\right)\delta(pu^s_{i,i})\,\mathrm{d}\Omega \\
&\quad + \int_{\Gamma^p}\delta(u_n p(\mathbf{x}))\,\mathrm{d}S_x - \frac{1}{j\omega}\int_{\Gamma^p}\int_{\Gamma^p}A(x,y)p(y)\delta p(\mathbf{x})\,\mathrm{d}S_y\,\mathrm{d}S_x \\
&\quad + \frac{\varepsilon}{j\omega}\int_{\Gamma^p}\int_{\Gamma^p}A(x,y)p_b(y)\delta p(\mathbf{x})\,\mathrm{d}S_y\,\mathrm{d}S_x = 0 \qquad \forall(\delta u^s_i,\delta p)
\end{aligned} \tag{13.74}$$

For a double porosity medium, the waveguide is in contact with both the poroelastic material and the perforations. Fluid cavities (holes) in contact with the waveguide are modelled in the same way as poroelastic patches. The associated weak integral form

(Equation 13.13) can be written

$$\begin{aligned}\int_{\Omega^a} \frac{1}{\rho_0\omega^2}\frac{\partial p^a}{\partial x_i}\cdot\frac{\partial(\delta p^a)}{\partial x_i} - \frac{1}{\rho_0 c_0^2} p^a\delta p^a)\,\mathrm{d}\Omega \\ - \frac{1}{j\omega}\int_{\Gamma^a}\int_{\Gamma^a} A(\mathbf{x},\mathbf{y})p(\mathbf{y})\delta p^a(\mathbf{x})\,\mathrm{d}S_y\,\mathrm{d}S_x \\ + \frac{1}{j\omega}\int_{\Gamma^a}\int_{\Gamma^a} A(\mathbf{x},\mathbf{y})p_b(\mathbf{y})\delta p^a(\mathbf{x})\,\mathrm{d}S_y\,\mathrm{d}S_x = 0 \quad \forall\,\delta p^a,\end{aligned} \tag{13.75}$$

where Γ^a denotes the front interface between the fluid cavity part and the waveguide.

For a given number of kept normal modes φ_{mn} in the wave-guide, Equations (13.74) and (13.75) are discretized using the finite element method and solved for the solid phase displacement, interstitial pressure variables in the porous medium and acoustic pressure in the holes. From nodal solutions, the normal incidence absorption coefficient can be computed using a power balance (Equation 13.72) or the sum of the powers dissipated inside the heterogeneous porous medium. The expressions of the powers dissipated in the porous materials by structural damping, viscous and thermal effects are explained in Section 13.7.

To illustrate the validity of the described model, a case taken from Atalla *et al.* (2001b) is considered. For the normal incidence sound absorption prediction (Figure 13.15), it is sufficient to consider one representative cell instead of the whole perforated material, provided that the perforation network is periodic. A rockwool sample made up of a generic square cell of side $L_c = 0.085m$ and a central perforation is considered. The acoustic and mechanical properties of the material are given in Table 13.2. Sound absorption measurements have been performed using a 1.2-m-long standing wave tube of 0.085 × 0.085 m square cross section (Olny and Boutin 2003). The samples were slightly constrained around their edges in order to avoid leaks, but no sealant was used. In the experiments, a hole of radius 0.016 m has been cut out from the material. On the other hand, the numerical model assumes an equivalent area square hole with side $a = 0.0283$ m. Linear Hexa 8 (brick) elements were used for both the porous and mesoporous (hole) domains.

Figure 13.16 shows the comparison between simulation (analytical and numerical model) and measurements for a thickness of 5.75 cm. Analytical simulations using the model of Olny and Boutin (2003) are also presented. Based on the chosen perforation size and sample side length, a double porosity of $\phi_p = 0.11$ is used in the analytical model. Recall from Chapter 5 that in the analytical model, the porous material is assumed rigid. Moreover, the diffusion function was chosen to fit the experiments. Contrary to the analytical model, calculations in the numerical model are based on Biot's theory with no assumption or parameter adjustment. That is, no assumption is made on the motion of the skeleton and in consequence the influence of the solid phase on the acoustic absorption is assessed. Excellent agreement is found for the numerical model. It is observed in

Table 13.2 The parameters used to predict the absorption coefficients in Figure 13.16.

Material	ϕ	σ (N.s/m^4)	α_∞	Λ (μm)	Λ' (μm)	$\Theta_m(0)$ (m^2)	ρ_1 (kg/m^3)	E (Pa)	ν	η_s
Rockwool	0.94	135 000	2.1	49	166	3.3×10^{-9}	130	4400	0	0.1

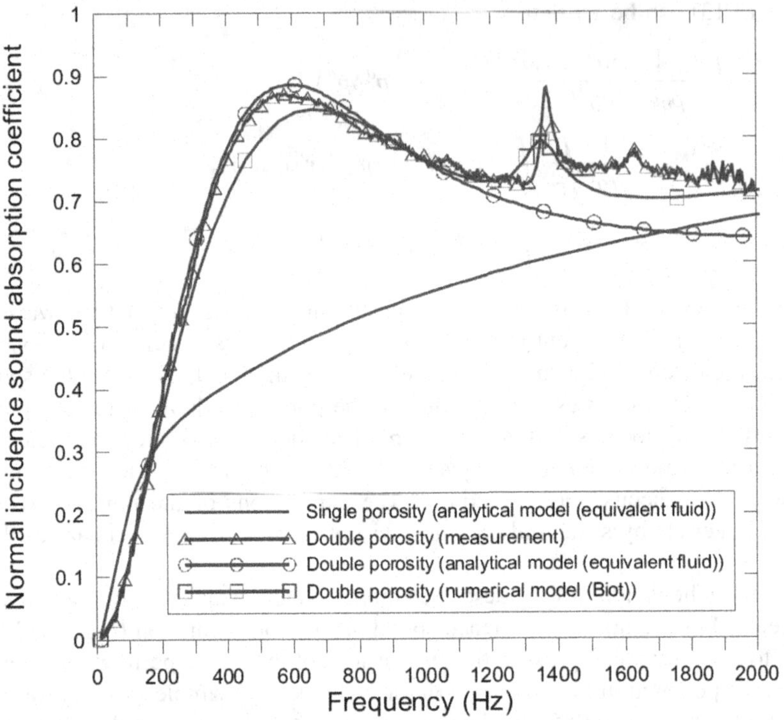

Figure 13.16 Normal incidence absorption coefficient. Comparison between analytical fluid equivalent model, numerical Biot model and measurement for 5.75-cm-thick rock wool (Sgard *et al.* 2005). Reprinted from Sgard, F., Olny, X., Atalla, N. & Castel, F. On the use of perforations to improve the sound absorption of porous materials. *Applied acoustics* **66**, 625-651 (2005) with permission from Elsevier.

particular that the numerical model is able to capture the skeleton resonance occurring around 1350 Hz. Also, this figure demonstrates the important increase in the absorption coefficient of the perforated material compared with the corresponding single porosity value. Another advantage of the numerical model is the illustration of the physics derived from the homogenization theory, namely the pressure diffusion effect. Figure 13.17 shows an example of the variation of the pressure at the mesoscale for the studied example. It is seen that the wavelength in the microporous domain is of the same order as the mesoscopic size.

The finite element approach avoids most of the hypothesis of the analytical model and can be used to deal with more complex configurations. In particular, the modeling at the mesoscopic scale does not require that the size of the hole be small with respect to the dimensions of the microporous material. This assumption is however necessary for the analytical model to be able to homogeneize the heterogeneous porous material. The numerical model can thus account for the diffraction by the hole. Also the handling of complex shaped holes can be implemented straightforwardly using appropriate meshing tools contrary to the analytical model which is limited to simple shapes. Finally, the numerical model allows for a mixture of arbitrary located microporous materials and gas patches together with a coupling with other subdomains (vibrating structures, air gaps,

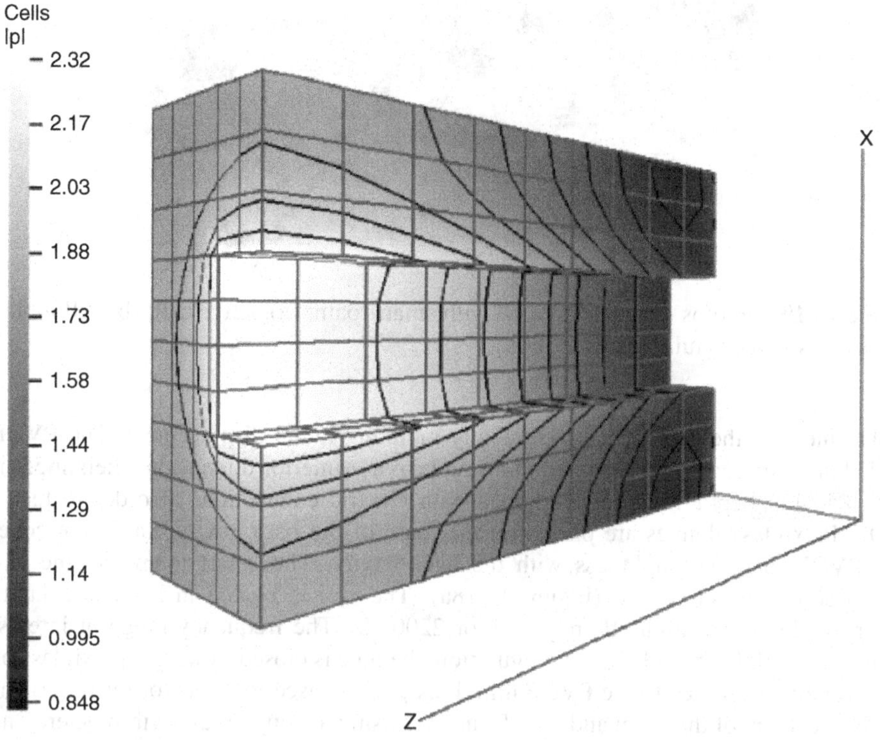

Figure 13.17 Illustration of the diffusion effect in the substrate material (Sgard *et al.* 2005). Reprinted from Sgard, F., Olny, X., Atalla, N. & Castel, F. On the use of perforations to improve the sound absorption of porous materials. *Applied acoustics* **66**, 625-651 (2005) with permission from Elsevier.

etc...). Note that the numerical model is not restricted to the calculation of absorption of single heterogeneous porous materials placed into a rectangular standing wave tube. It can also deal with the computation of transmission performance of complex multilayered systems involving plates, septum, air gaps, conventional and heterogeneous porous materials as illustrated in Figure 13.2. Several examples can be found in Sgard *et al.* (2005 and 2007).

13.9.6 Modelling of smart foams

This example, taken from Leroy (2008), illustrates the use of the $(\boldsymbol{u}^s, p)$ formulation for modeling smart foams. A smart foam combines the passive dissipation capability of a foam in the high frequency range and the active absorption ability of an actuator (generally piezoelectric PVDF) in the low-frequency range. This results in a passive/active absorption control device that can efficiently operate over a broad range of frequencies. The 3D finite element model of a smart foam prototype is presented here with its experimental validation.

The configuration is made up of a half-cylinder of melamine foam covered with a PVDF film (Figure 13.18a). The curved shape of the PVDF ensures coupling of in-plane

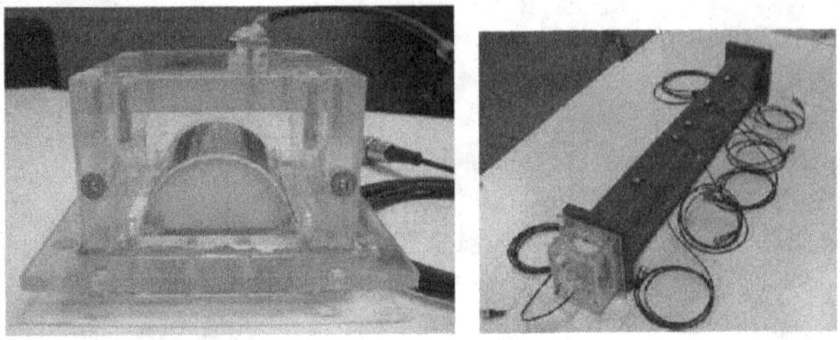

(a) (b)

Figure 13.18 Photos of the active cell with smart foam; (a) active cell; (b) cell mounted in the measurement tube (LeRoy 2008).

displacement of the film with its efficiently radiating radial displacement. The PVDF is bonded onto the foam and mounted in a small cavity (interior dimensions: height 55 mm; width 85 mm; depth 110 mm) equipped with electric connections in order to feed the PVDF. Plexiglass flanges are placed on the foam of the back cavity that is not covered with PVDF to ensure tightness with the back cavity. The smart foam and the cavity form a so-called active cell (Figure 13.18a). The cell is placed in a rectangular tube (Figure 13.18b) with a cutoff frequency of 2200 Hz. The frequency range of interest is [100–1500 Hz]. In the radiation configuration, the tube is closed by a rigid plexiglass plate and a voltage is applied to the PVDF film. Plexiglass is used to allow for measurement of the displacement of the foam and PVDF surfaces using a Doppler laser vibrometer. On the other hand, the acoustic pressures at both ends of the tube are measured. For absorption measurements, a loudspeaker is installed in place of the plexiglass plate. The absorption measurement is done using the Chung and Blaser (1980) method with several microphone pairs in the tube. This measurement gives information on the passive behaviour of the cell.

Finite element model

The model uses quadratic poroelastic elements with $(\boldsymbol{u}^s, p)$ formulation, as well as elastic, fluid and piezoelectric elements. The weak integral formulation of the porous media is given in Equation (13.7). However, since the measurement of the foam properties revealed anisotropy, a simplified orthotropic model was assumed for the mechanical properties of the foams. Because of the difficulty of measuring the anisotropy of the acoustic parameters (flow resistivity, porosity, tortuosity, viscous and thermal characteristic lengths) they are assumed the same for all three directions. The Young's moduli for the three directions have been obtained by measuring the absorption coefficients of a cubic foam sample oriented along the orthotropic directions. The three Young's moduli have then been adjusted in order to fit the resonance peak of the absorption coefficient with the measured resonance peak of the absorption. Special care was given to the mounting of the foam in the tube since it is known to affect the absorption resonance (a controlled air gap was left around the foam lateral surface).

The used PVDF film (made by Measurement Specialties Inc.) has 28-μm thickness and Cu–Ni electrodes. Its properties were taken from published data. Since the melamine is highly porous, its surface was conditioned with a heat-reactivatable membrane before the PVDF was bonded onto it using double-sided tape. This method allows for a durable and controlled bonding. This bonding layer adds mass and stiffness to the PVDF film, and thus was accounted for in the finite element model as an isotropic thin elastic layer. The properties of this layer (Young's modulus, Poisson ratio, structural loss factor) were identified using a parametric study comparing numerical results and experimental measurements of the radiated pressure from a smart foam placed in a small rectangular cavity (interior dimensions: height 20 mm; width 64 mm; depth 78 mm). These parameters have been found to have a negligible influence on the passive absorption of the smart foam.

The modelling of the PVDF film is plate-like. The associated weak formulation, in discretized form, is given by (Leroy 2008):

$$\begin{aligned}
&\int_{\Gamma^{pi}} \Bigg[\underbrace{\{\delta\varepsilon_m\}^T[H_m]\{\varepsilon_m\}}_{\text{work of the in-plane elastic forces}} + \underbrace{\{\delta\chi\}^T[H_f]\{\chi\}}_{\text{work of the bending elastic forces}} \\
&\quad + \underbrace{\{\delta\gamma\}^T[H_c]\{\gamma\}}_{\text{work of the shearing elastic forces}} \Bigg] \mathrm{d}S - \underbrace{\int_{\Gamma^{pi}} \omega^2\{\delta u\}^T[\rho I]\{u\}\,\mathrm{d}S}_{\text{work of the inertial forces}} \\
&\quad - \underbrace{\int_{\Gamma^{pi}} [\{\delta\varepsilon_m\}^T\{e_c\}E_z + \delta E_z\{e_c\}^T\{\varepsilon_m\} + \delta E_z\varepsilon_{33}E_z]h\,\mathrm{d}S}_{\text{work of the piezoelectric and dielectric forces}} \\
&\quad - \underbrace{\int_{\Gamma^{pi}} \{\delta u\}^T\{\sigma\}_n\,\mathrm{d}S}_{\text{work of the elastic external forces}} + \underbrace{\int_{\Gamma^{\Phi}} \delta\Phi D_z\,\mathrm{d}S}_{\text{work of the electrical external forces}} = 0 \quad \forall(\delta u, \delta\Phi)
\end{aligned} \tag{13.76}$$

where Γ^{pi} denotes the surface of the piezoelectric domain and Γ^{Φ} the surface over which the electric charge is applied. The first line of this equation depicts the elastic, plate-like, behaviour of the film. $\langle\varepsilon_m\rangle$ is the in-plane strain field of in-plane type, $\langle\chi\rangle$ is the curvature vector and $\langle\gamma\rangle$ is the shear strain vector. In order to use this plate element in 3D problems, a drilling degree of freedom of rotation is added, resulting in six degrees of freedom for the elastic variables. Matrices $[H_m]$, $[H_f]$ and $[H_c]$ represent the in-plane stiffness matrix, the bending stiffness matrix and the shearing stiffness matrix, respectively. h, ρ_{pi}, E and ν represent the thickness, density, Young's modulus and Poisson ratio of the PVDF film, respectively. The third line of Equation (13.76) describes the piezoelectric behaviour of the film. It assumes that the electric field is applied across the film's thickness, $E_z = \mathbf{\Phi}/h$ where $\mathbf{\Phi}$ is the electrical potential. This potential is the selected electrical variable in this 2D formulation. The dielectric permittivity matrix $[\varepsilon_d]$ reduces to the component of the permittivity along the z axis ε_{33}. The matrix of the piezoelectric coupling coefficients $[e]$ becomes a column vector

$$\{e_c\} = \begin{Bmatrix} e_{31} \\ e_{32} \\ 0 \end{Bmatrix} \tag{13.77}$$

Finally, the last line of Equation (13.76) describes the external loads. Again, because of 2D modeling, the electrical displacement vector reduces to a scalar, noted D_z. It is observed that coupling of the piezoelectric with a porous or elastic domain is natural (see Section 13.4).

Numerical versus experimental results

Two configurations are studied: (i) active radiation and (ii) passive absorption. In the radiation configuration, the tube is closed by a rigid plate and a 100 V is applied to the PVDF film. Figure 13.19 presents the comparison between the measured and simulated

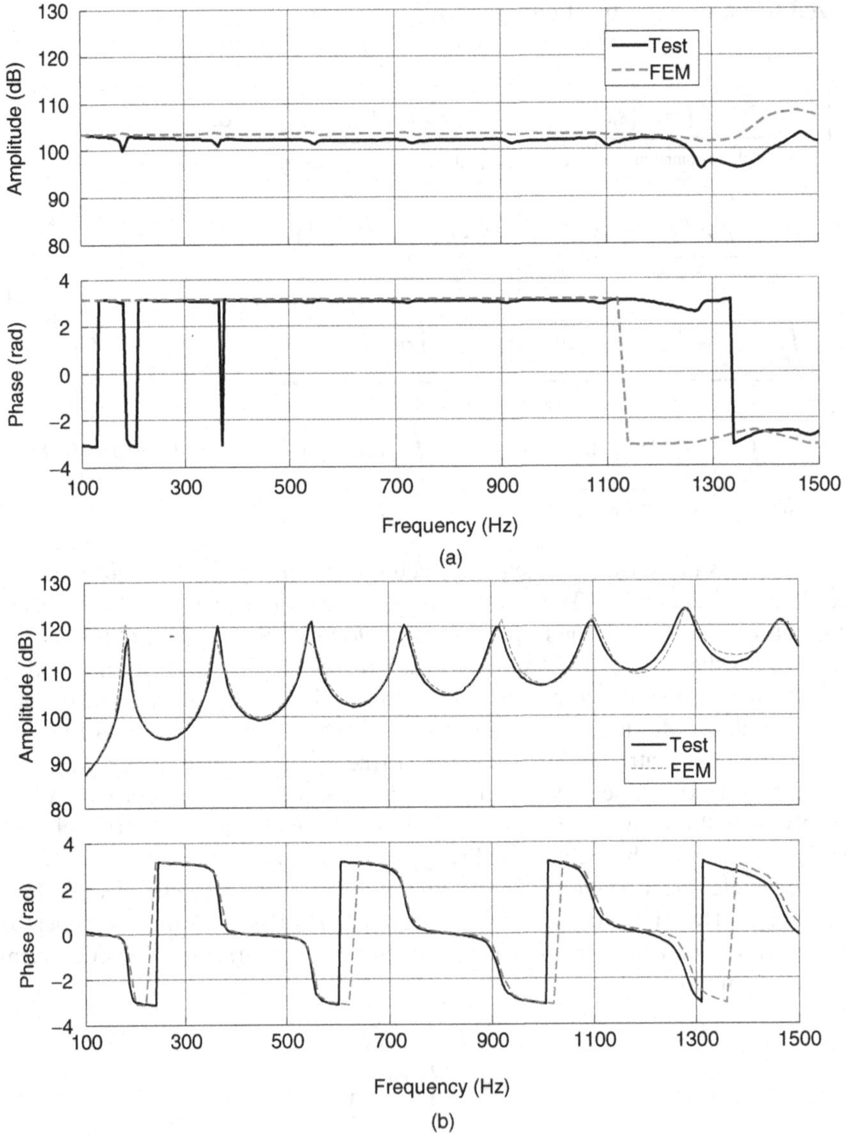

Figure 13.19 Acoustic pressure: (a) in the back cavity and (b) near the rigid end of the tube for 100 V applied to the PVDF film (LeRoy 2008).

pressure radiated by the smart foam. The comparison is excellent. The pressure in the back cavity is constant up to 1250 Hz (Figure 13.19a) where the second resonance mode of the foam–PVDF system begins to affect the displacement field. At the other end of the tube (Figure 13.19b), the radiated pressure is controlled by the modal behaviour of the tube. All the peaks correspond to tube resonances. Note that the pressure level is around 100 dB which is quite important. In the back cavity, it is around 103 dB.

Figure 13.20a, b presents the comparison for the displacement at the centre of the free surface of the foam and the centre of the PVDF film, respectively. It is observed that the model follows closely the trends in the measured displacement data (Figure 13.20a). However, there are few discrepancies. For the foam displacement, the model seems to slightly overestimate the amplitude (Figure 13.20a). The peak of the experimental data is well represented by the numerical model, but it is located at a slightly higher frequency. For the PVDF displacement, the FE model predicts the amplitude very well, but does not clearly capture the shape of the peak (Figure 13.20b); this may be due to the uncertainty on either the measurement position or the properties of the film.

In the passive absorption configuration, the measured absorption coefficient of the smart foam is compared with predictions in Figure 13.21. Overall, the comparison is very good. The main differences are observed at frequencies lower than 200 Hz and higher than 1000 Hz. At low frequencies, possible leaks in the tube may have disturbed the measurements. Moreover, at these low frequencies, the relative influence of the other dissipation mechanisms in the tube (speaker suspension, join, wall...) is important. At frequencies higher than 1000 Hz, the uncertainties on the microphone position become important. Moreover, the modal behaviour of the foam–PVDF system is significant and affects the absorption. And since the structural behaviour above 1000 Hz is not perfectly predicted by the model, the predicted absorption coefficient is slightly different from measurement.

Despite various uncertainties on: (i) the properties of the foam, the PVDF film and the bonding layer (glue) between the two; and (ii) the fabrication of the smart foam prototype, the comparison between the model and the measurements is very good. The results clearly demonstrate the validity of the model. It constitutes a useful platform to simulate and optimize various configurations of smart foams. A thorough discussion is given by Leroy (2008).

13.9.7 An industrial application

Finally, an example of an industrial application, courtesy of Faurecia (Duval *et al.* 2008), is shown. The vibroacoustic response of a car floor module, excited by a shaker was measured at Faurecia labs. The module was fixed in the separation wall of a reverberant room coupled to a movable absorbing concrete-walled receiver cavity. The floor was excited by a shaker positioned at a front reinforcement beam location and its radiation measured using microphones in the receiver movable concrete cavity. To achieve a clamped boundary condition, 3-mm-thick plates were used to connect through continuous welding, the floor perimeter to a 100×100 mm^2 metal frame filled with concrete and fixed to the horizontal separation wall of the coupled reverberant rooms. These additional plates have been 3D modelled and integrated in the FEM model.

The simulations, conducted at Faurecia, use a sub-structuring based implementation of the $(\boldsymbol{u}^s, p)$ formulation. This implementation, described in Hamdi *et al.* (2000)

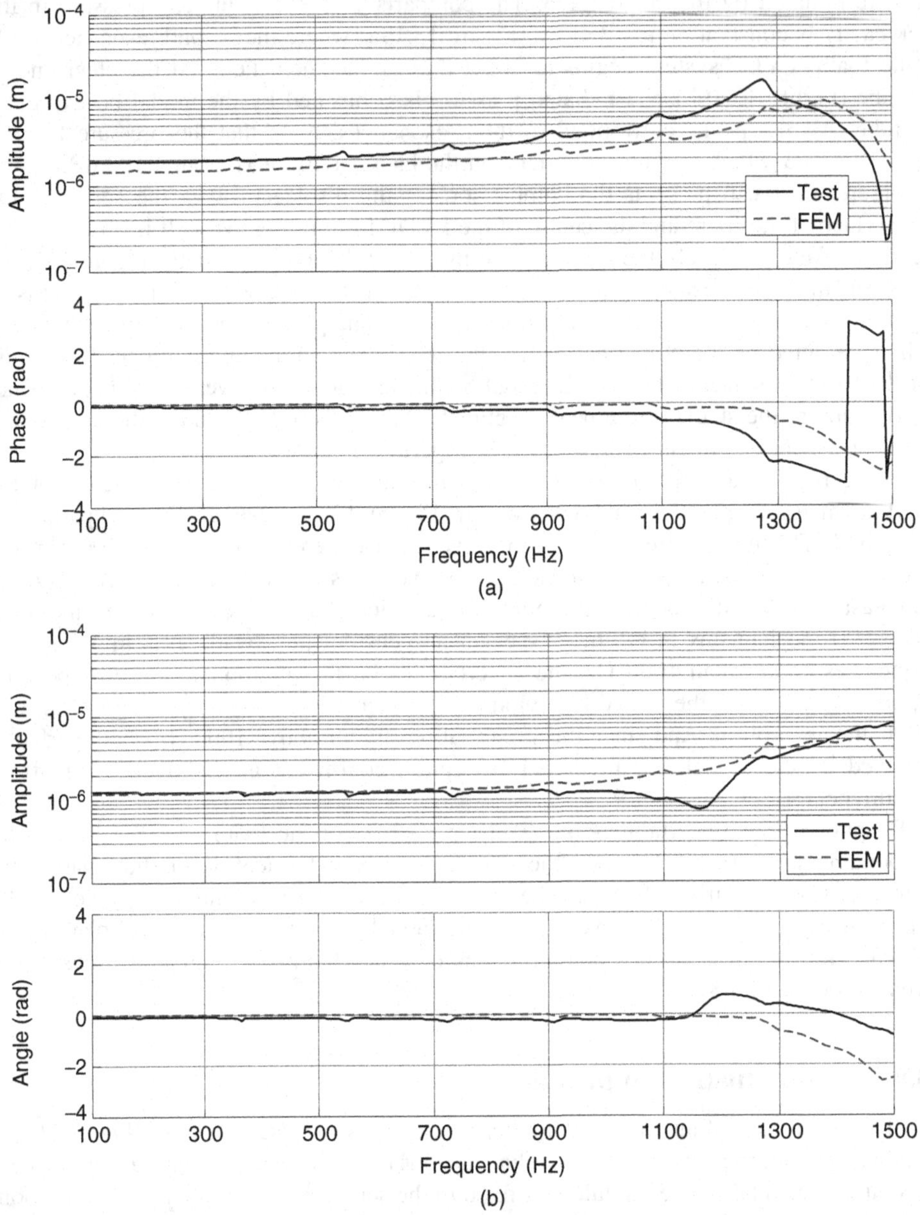

Figure 13.20 Normal displacement at the centre of: (a) the foam free surface and (b) the PVDF surface for 100 V applied to the PVDF film (LeRoy 2008).

and Omrani *et al.* (2006) has been described in Section 12.6 for the radiation problem (structure–cavity configuration of Figure 13.3). For the transmission loss problem, it uses first the finite element method to calculate the mechanical impedance of the trim components to be superimposed to the modal mechanical impedance of the panel structure. Next, a boundary element method (BEM) is used to calculate the radiation impedance

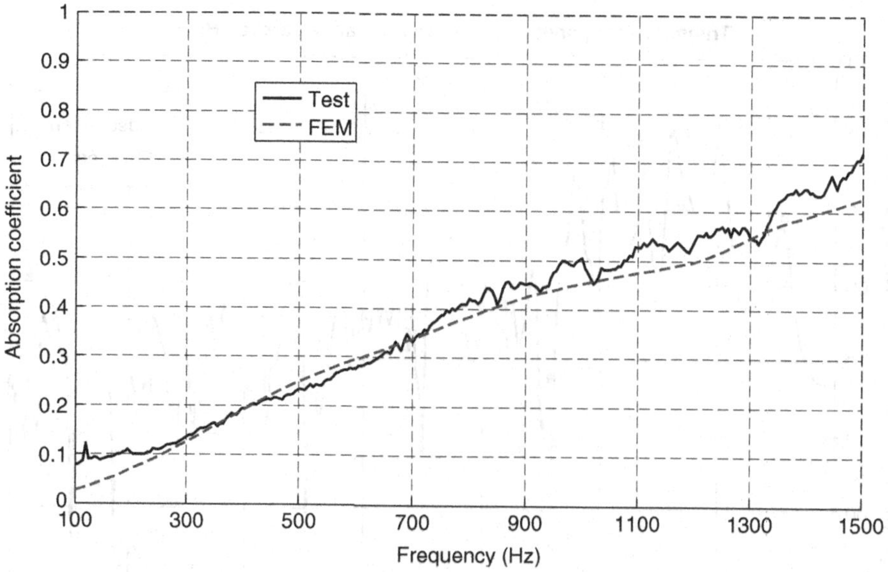

Figure 13.21 Passive absorption of the foam (LeRoy 2008).

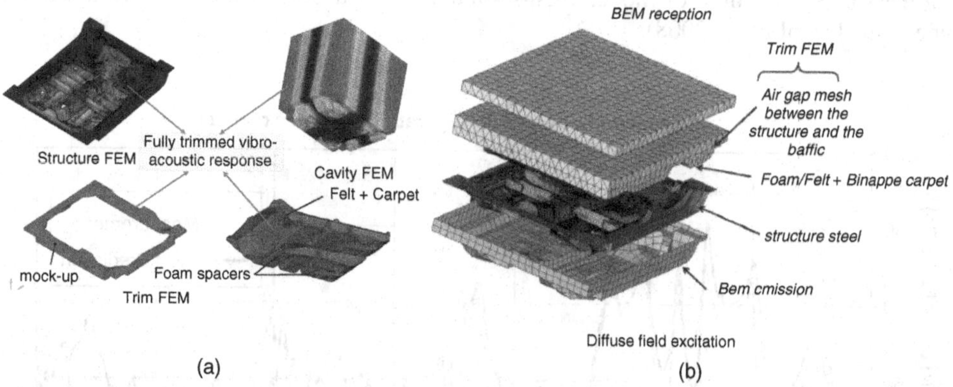

Figure 13.22 Example of the modelling of a fully trimmed car floor panel: (a) radiation problem, the floor panel is mechanically excited and radiates in a cavity; (b) transmission loss problem (Duval *et al.* 2008).

of the panel structure and the acoustic power radiated by the structure in the receiving media. This approach allows for incompatible meshes and authorizes the solution of large systems such as a fully trimmed vehicle (Anciant *et al.* 2006).

The examples of two types of floor insulator are shown here: (i) an absorption type, called bi-permeable concept, with an airflow resistive carpet superposed on a two-layer system made up from a soft spring felt on top of foam spacers; and (ii) an insulation type, with a 2 kg/m^2 heavy layer inserted between the airflow resistive carpet and the soft felt spring–foam spacer system. A graphic description of the meshed components is given in Figure 13.22. Figure 13.22a describes the radiation configuration while Figure 13.22b

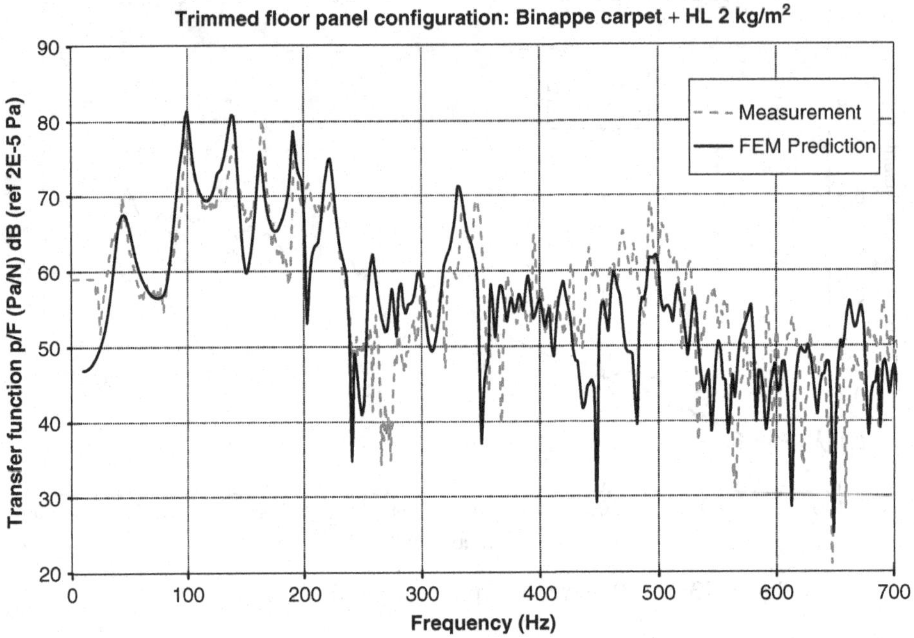

Figure 13.23 Example of model versus simulation of a floor panel with an insulation type trim (Duval *et al.* 2008).

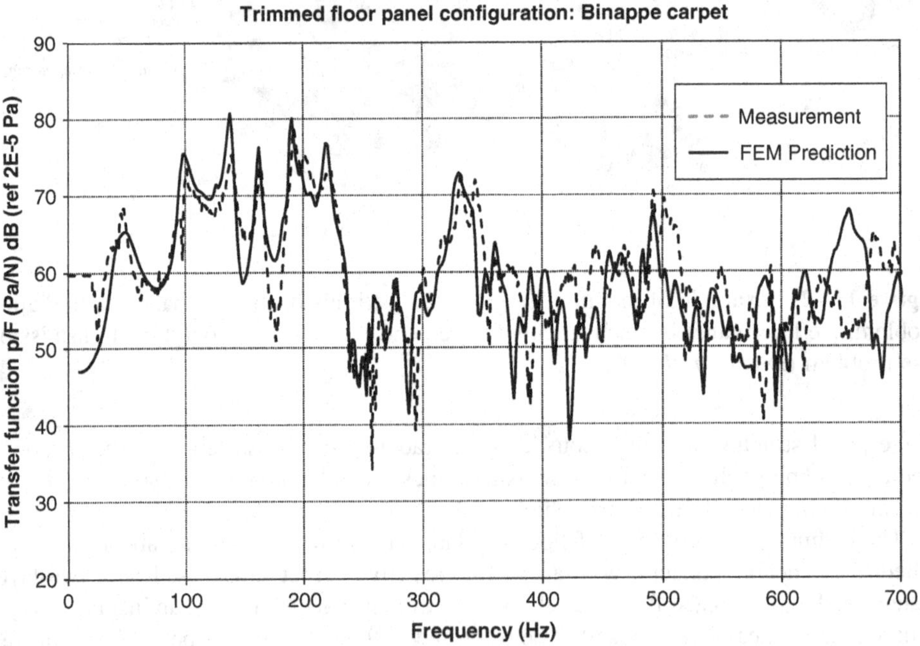

Figure 13.24 Example of model versus simulation of a floor panel with an absorption type trim (Duval *et al.* 2008).

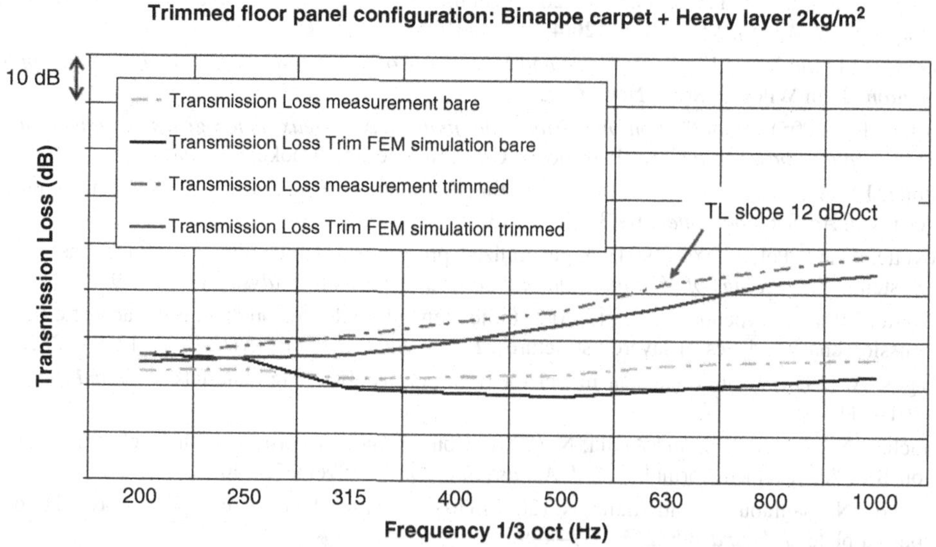

Figure 13.25 Example of model versus simulation of the transmission loss of a floor panel with an absorption type trim (Duval *et al.* 2008).

describes the transmission loss case. An example of comparison between measurements and simulations for the point-to-point pressure over force FRF (radiation case) are presented in Figure 13.23 and Figure 13.24. The simulations are very good, up to 700 Hz, for both the absorption and insulation configurations. The correlation could have been improved if the very thin decoupling air gaps between the felt and the heavy layer were explicitly modelled (this is corroborated by systematic investigations on flat samples). Figure 13.25 shows an example of comparison between measurements and simulations of the transmission loss of the insulation configuration. Again, acceptable correlation is observed. These examples demonstrate the good prediction of acoustic radiation and transmission through the trims computed by FEM simulation, as well as the good prediction of the damping induced by the trims to the structure. They also prove the importance of good coverage and contact between the trims and the structure.

References

Anciant, M., Mebarek, L., Zhang, C. and Monet-Descombey, J. (2006) Full trimmed vehicle simulation by using Rayon-VTM. JSAE.

Atalla N., Panneton, R. and Debergue, P. (1998) A mixed displacement pressure formulation for poroelastic materials. *J. Acoust. Soc. Amer*. **104**, 1444–1452.

Atalla N., Hamdi, M.A. and Panneton, R. (2001a) Enhanced weak integral formulation for the mixed (u,p) poroelastic equations. *J. Acoust. Soc. Am*. **109**(6), 3065–3068.

Atalla N., Sgard, F., Olny, X. and Panneton, R. (2001b) Acoustic absorption of macro-perforated porous materials. *J. Sound Vib*. **243**(4), 659–678.

Atalla, N., Amedin, C.K., Atalla, Y., Sgard, F. and Osman, H. (2003) Numerical modeling and experimental investigation of the absorption and transmission loss of heterogeneous porous materials. *Proceedings of 10th International congress on Sound and Vibration*, ICSV12, pp 4673–4680.

Atalla, N., Sgard, F. and Amedin, C.K. (2006) On the modeling of sound radiation from poroelastic materials. *J. Acoust. Soc. Amer.* **120**(4), 1990–1995.

Beranek, I.I. and Vér, I.L. (1992) *Noise and Vibration Control Engineering. Principles and Application*. John Wiley & Sons, New York.

Castel, F. (2005) *Modélisation numérique de matériaux poreux hétérogènes. Application à l'absorption' basse fréquence*. Phd thesis, Université de Sherbrooke, Canada.

Chung, J.Y and Blaser, D.A. (1980) Transfer function method of measuring in-duct acoustic properties. *J. Acoust. Soc. Amer*. **68**(3), 907–913.

Coyette, J. and Pelerin, Y. (1994) A generalized procedure for modeling multi-layer insulation Systems. *Proceedings of 19th International Seminar on Modal Analysis*, 1189–1199.

Coyette, J.P. and Wynendaele, H. (1995) A finite element model for predicting the acoustic transmission characteristics of layered structures. *Proceedings of INTER-NOISE 95*, 1279–1282.

Craggs, A. (1978) A finite element model for rigid porous absorbing materials. *J. Sound Vib*. **61**, 101–111.

Dauchez, N., Sahraoui, S. and Atalla, N. (2001) Convergence of poroelastic finite elements based on Biot displacement formulation. *J. Acoust. Soc. Amer*. **109**(1), 33–40.

Dauchez, N., Sahraoui, S. and Atalla, N. (2002) Dissipation mechanisms in a porous layer bonded onto a plate. *J. Sound Vib*. **265**, 437–449.

Dazel, O., Sgard, F., Lamarque, C.-H. and Atalla, N. (2002) An extension of complex modes for the resolution of finite-element poroelastic problems. *J. Sound and Vib*. **253**(2), 421–445.

Dazel O. (2005) *Synthèse modale pour les matériaux poreux*. PhD Thesis, INSA de Lyon.

Dazel, O., Sgard, F., Beckot, F-X. and Atalla, N. (2008) Expressions of dissipated powers and stored energies in poroelastic media modeled by {u,U} and {u,P} formulations, *J. Acoust. Soc. Amer*. **123**(4), 2054–2063.

Debergue P, Panneton, R. and Atalla, N. (1999) Boundary conditions for the weak formulation of of the mixed (u,p) poroelasticity problem. *J. Acoust. Soc. Amer*. **106**(5), 2383–2390.

Duval, A., Baratier, J., Morgenstern, C., Dejaeger, L., Kobayashi, N. and Yamaoka, H. (2008) Trim FEM simulation of a dash and floor insulator cut out modules with structureborne and airborne excitations. Acoustics 08, Joint ASA, SFA and Euronoise Meeting, Paris.

Göransson, P. (1998) A 3-D symmetric finite element formulation of the Biot equations with application to acoustic wave propagation through an elastic porous medium. *Int. J. Num. Meth*. **41**, 67–192.

Hamdi, M.A, Atalla, N., Mebarek, L. and Omrani, A. (2000) Novel mixed finite element formulation for the analysis of sound absorption by porous materials. *Proceedings of 29th International Congress and Exhibition on Noise Control Engineering*, InterNoise, Nice, France.

Hörlin, N.E, Nordström, M and Göransson, P. (2001) A 3-D hierarchical FE formulation of Biot's equations for elastoacoutsic modeling of porous media. *J. Sound Vib*. **254**(4), 633–652.

Hörlin, N.E. (2004) *Hierarchical finite element modeling of Biot's equations for vibro-acoustic modeling of layered poroelastic media*. PhD Thesis, KTH Royal Institute of Technology, Stockholm, Sweden.

Horoshenkov, K.V. and Sakagami, K. (2000) A method to calculate the acoustic response of a thin, baffled, simply supported poroelastic plate. *J Acoust. Soc. Amer*. **110**(2), 904–917.

Jaouen, L., Brouard, B., Atalla, N. and Langlois, C. (2005) A simplified numerical model for a plate backed by a thin foam layer in the low frequency range. *J. Sound Vib*. **280**(3–5), 681–698.

Johansen, T.F., Allard, J.F. and Brouard, B. (1995) Finite element method for predicting the acoustical properties of porous samples. *Acta Acustica* **3**, 487–491.

Kang, Y.J. and Bolton, J.S. (1995) Finite element modeling of isotropic elastic porous materials coupled with acoustical finite elements. *J. Acoust. Soc. Amer*. **98**, 635–643.

Kang, Y., Gardner, B. and Bolton, J. (1999) An axisymmetric poroelastic finite element formulation. *J. Acoust. Soc. Amer.*, **106**(2), 565–574.

Langlois C., Panneton R. and Atalla N. Polynomial relations for quasi-static mechanical characterization of isotropic poroelastic materials. *J. Acoust. Soc. Amer.* **110**(6), 3032–3040.

Langlois, C. (2003) Modélisation des problèmes vibroacoustiques de basses fréquences par elements finis, MSc Thesis, Université de Sherbrooke, Canada.

Leroy, P. (2008) *Les mousses adaptatives pour l'amélioration de l'absorption acoustique: modélisation, mise en œuvre, mécanismes de contrôle*. PhD Thesis, Universite de Sherbrooke.

Omrani, A., Mebarek, L. and Hamdi, M.A. (2006) Transmission loss modeling of trimmed vehicle components, *Proceedings of ISMA 2006*, Leuven, Belgium.

Olny, X. and Boutin, C. (2003) Acoustic wave propagation in double porosity media, *J. Acoust. Soc. Amer.* **114**, 73–89.

Panneton, R., Atalla, N. and Charron, F. (1995) A finite element formulation for the vibro-acoustic behaviour of double plate structures with cavity absorption. *Can. Aero. Space J.* **41**, 5–12.

Panneton, R. (1996) *Modélisation numérique par éléments finis des structures complexes absorbantes*. PhD Thesis, Université de Sherbrooke, Québec, Canada.

Panneton, R. and Atalla, N. (1996) Numerical prediction of sound transmission through multilayer systems with isotropic poroelastic materials. *J. Acoust. Soc.Amer.* **100**(1), 346–354.

Panneton, R. and Atalla, N. (1997) An efficient finite element scheme for solving the three-dimensional poroelasticity problem in acoustics. *J. Acoust. Soc. Amer.* **101**(6), 3287–3298.

Pilon, D., (2003) *Modélisation axisymétrique de la formulation* $\{u, P\}$ *pour l'étude de l'influence des conditions aux limites d'échantillons poreux sur leur caractérisation: application au tube de Kundt et au rigidimètre*. MSc Thesis, Université de Sherbrooke, Canada.

Rigobert, S. (2001) *Modélisation par éléments finis des systèmes élasto-poro-acoustiques couplées: éléments hiérarchiques, maillages incompatibles, modèles simplifiés*. PhD Thesis, Université de Sherbrooke, Canada.

Rigobert, S., Atalla, N. and Sgard, F. (2003) Investigation of the convergence of the mixed displacement pressure formulation for three-dimensional poroelastic materials using hierarchical elements. *J. Acoust. Soc. Amer.* **114**(5), 2607–2617.

Sgard F., Atalla, N. and Panneton, R. (1997) A modal reduction technique for the finite-element formulation of Biot's poroelasticity equations in acoustics. *134th ASA Meeting, San Diego*.

Sgard F., Atalla, N. and Nicolas, J. (2000) A numerical model for the low-frequency diffuse field sound transmission loss of double-wall sound barriers with elastic porous lining. *J. Acoust. Soc. Amer.* **108**(6), 2865–2872.

Sgard, F. (2002) *Modélisation par éléments finis des structures multi-couches complexes dans le domaine des basses Fréquences*. HDR Thesis, Université Claude Bernard Lyon I – INSA de Lyon, France.

Sgard, F., Olny, X., Atalla, N. and Castel, F. (2005) On the use of perforations to improve the sound absorption of porous materials. *Applied acoustics* **66**, 625–651.

Sgard, F., Atalla, N. and Amedin, C.K., (2007) Vibro-acoustic behavior of a cavity backed by a plate coated with a meso-heterogeneous porous material. *Acta Acustica* **93**(1), 106–114.

Takahashi, D. & and M. Tanaka, M. (2002) Flexural vibration of perforated plates and porous elastic materials under acoustic loading. *J. Acoust. Soc. Amer.* **112**, 1456–1464.

Index

Absorption coefficient, 20, 36, 263, 289–291, 328
Admittance, 288, 319
Added length/mass, perforated facings, 188–191, 208–209
Angle of specular reflection, 138, 140
Anisotropic highly porous media with slanted pores, 39, 43
Anisotropic materials, 67–69, 213–241
Attenuation, 16–17
Audible frequency range, 167

Bessel function, 47, 48, 52, 56, 139, 172
Biot theory
  independent displacement fields, 169–170, 223–224
  inertial forces, 117–118
  limp material, 251–253
  other representations
    Dazel representation, 131–132
    mixed representation, 132–134
    second representation, 131, 215–216
  Biot waves
    compressional wave, 120–122, 168, 180
    shear wave, 122–123, 168, 173, 177, 176
    transversally isotropic poroelastic media (in), 217–222
  porous material having an elastic frame, 111–136
  rigid framed material, 61–65
  stress-strain relations, 112–116, 215–216
  surface impedance, normal incidence, 126–131
  wave equations, 119, 120, 216–217
Boundary conditions, 127–128, 169–170, 223–224, 227, 232, 313–320
Brewster angle, 149, 154
Bulk modulus
  air
    cylindrical pores, 50–65
    poroelastic isotropic media, 114, 124
    rigid framed materials, 83–89
  elastic solid the frame is made of, 114
  fluid, 10
  frame, 113, 116

Champoux Allard model, 84–87
Characteristic lengths
  thermal, 80, 104–106
    measurement, 81–82
  viscous, 79–80
    calculation for cylinders, 106–107
    measurement, 81, 82
Characteristic sizes
  macroscopic, 91, 95–96

Propagation of Sound in Porous Media: Modelling Sound Absorbing Materials, Second Edition J. F. Allard and N. Atalla

Characteristic sizes (*continued*)
mesoscopic, 96
micropore structure, 96
microscopic, 91
Characteristic values, 92
Circular cross-section, cylindrical tube, 45–53
Circular holes, impedance, 194–198
Circularly perforated facings, 205–211
Complex exponential representation, 26
Complex s plane, 139–140
Compressional waves
Biot wave, 120–122
elastic solid, 11–13
fluid, 10–11
Coupling
conditions, 313–320
inertial, 68
piezoelectric, 341
potential, 112
transfer matrices, 257
Cut, 139, 141, 151
Cutoff frequency, 178
Cylindrical pore (porous material with), 45–72
cross-section varieties, 54
arbitrary, 60–61
circular, 45–48, 52–53
hexagonal, 55, 57
rectangular, 55–57
slits, 48–50, 52–53
triangular, 55, 57

Damping
added, 304, 330
effects, 331–333
modal
radiation, 297, 325, 329
structural, 255, 324, 333
thermal, 333
equivalent, 297
viscous, 85, 123,333
Dazel representation, 131–132
Decoupling
frequency, 126, 251
partial, 123, 126, 173, 177, 219, 220, 226
Deformable media, 2–5
strain in, 2–4
stress in, 4–5
Delany-Bazley laws, 20, 22–23, 29, 42
Density, *see* effective density
Diffuse field
absorption coefficient, 263, 304
incident power, 290
transmission coefficient, 287
transmission loss, 264, 275, 304, 334
Dimensionless macroscopic space variable, 91
Dimensionless microscopic space variable, 91
Dimensionless numbers, 92–93
Dimensionless quantities, 92–93
asymptotic expansion of, 94
Double porosity media
asymptotic development method for, 97–98
finite element modelling, 335–339
definition, 95
high permeability contrast, 99–102
low permeability contrast, 98–99
orders of magnitude for realistic, 96–97

Effective density
arbitrary cross-section cylindrical pores, 60–61, 64–65
circular cross-section cylindrical pores, 47–48, 57–59
oblique pores, 69–70
rigid-framed media, 83–89
slits, 48–50, 59–60
transversally isotropic media, 217, 227
Elastic isotropic frame, Biot theory, 111–135
Elastic isotropic frame excitation
circular and line stress field, 172–173
normal unit stress source, 168, 170, 171
plane wave excitation, 169–172
point source in air, 179–182

Rayleigh wave and mode excitation, 173–182
Elastic isotropic media, 5–8
Elastic transversally isotropic frame, Biot theory, 213–241
Elastic transversally isotropic frame excitation
mechanical excitation, 227–228, 230–232
Rayleigh wave excitation, 232, 236
sound source in air, 223–225
Elastic solids, wave equations, 11–13
compressional waves, 12–13, 178
transverse waves, 13, 178
Elastic transversally isotropic solids/elastic frames in vacuum, 214–215
stress-strain relations, 214
wave equations, 214–215
Electric dipole field, 152
Energy
kinetic energy and tortuosity, 78–79
potential energy in Biot theory, 112–113
Equations relating the dimensionless quantities, 93
Error function, 152
asymptotic development, 152
Evanescence, 146, 147

Facings, *see* Screens
Fibrous material surface impedance, 129–131
Finite Element Method (modelling of poroelastic materials)
coupling conditions, 313–320
damping, 324, 331
matrix
admittance matrix, 321, 327
coupling matrix, 311, 318, 321, 326, 341
impedance matrix, 322
mass matrix, 321
mechanical impedance matrix, 320
stiffness matrix, 321, 341
dielectric permittivity matrix, 341
modelling of double porosity materials, 335–339
modelling of smart foams, 339
power dissipation within a porous material, 323–324
radiation condition, 324–327
weak formulations
displacement-displacement, 310
mixed displacement-pressure, 311
other, 320
Floquet theorem, 205
Flow resistivity, 21–22, 57–60, 191–192
anisotropic materials, 68–69, 227
Delany-Bazley laws, 22–25
perforated facings, 182, 194
Fluids
acoustic impedance, layer backed by an impervious wall, 18–19, 23–25
layer equivallent to a porous layer, 89
multilayered, 35–36
oblique incidence acoustic impedance, 15–43
transversally isotropic, 39–41
unbounded, 15, 17
attenuation, 16–17
travelling waves in, 15–16
wave equations, 10–11
Fourier transform, 167, 173, 183–184
Frequency, *see* also resonance
critical, 255, 297
cutoff, 194
decoupling, 126, 25
diffusion, 102
high and low frequency approximation, 55–60, 70–72, 75, 83
quarter length resonance, 129
transition, 85, 89, 97
viscous characteristic, 102, 336

Gedanken experiments, 113–116
Glass wool, 115–116

Governing equations
adherance condition, 92
air state, 92
heat conduction, 92
mass balance, 92
Navier-Stokes, 91
thermal boundary conditions, 92
Grazing incidence, 156, 158, 160
Green's function, 326
Grounded dielectric, 148

Hankel functions, 139, 140
Hankel transform, 167, 172, 183–184, 232
Heat exchange in cylindrical tube, 50–53
Helmholtz resonator, 203–205
Hexagonal cross-section, cylindrical tube, 54
Highly porous material, acoustic impedance, 15–27, 29–43
Delany-Bazley laws, 22–23
normal incidence, 15, 27
oblique incidence, 29–43
transversally isotropic, 39–43
Homogenization for periodic structures, 91–95
Hydrostatic pressure, 8
Hydraulic radius, 61, 65

Impedance
characteristic impedance, 16, 17, 19, 22, 26, 66, 139
mechanical impedance, 255, 320
modal impedance, 297–298
radiation impedance, 284–285, 290, 297, 305, 320, 325
surface impedance
fluids, and highly porous materials, 15–43
identical pores, perpendicular to the surface, 65–67
multilayered fluids, 35–36
poroelastic materials
isotropic, 126–131, 170–171
transversally isotropic, 224
porous materials with perforated facings, 194–205, 205–211
rigid framed porous media, 87, 139
slanted pores, 71
Inertia, and perforated facings, 187–194
Inertial forces in Biot theory, 117–118
Inhomogeneous plane waves, 29–31
Insertion loss, 302–304
Isotropic elastic media, 1–13
matrix representation, 245–248
Isotropic fluids, 15–25, 29–39
impedance at normal incidence, 15–25
impedance at oblique incidence, 29–39

Johnson *et al.* model, 84–85

Kinetic energy and tortuosity, 78–79
Kronecker delta, 5

Lafarge simplified model, 83–84, 85–87
Lamb dispersion curves, 178
Lamé coefficients, 5, 6
Laplacian operators, 2, 9
Lateral wave, 142, 144
Limp frame model, 251–253
Locally reacting medium, 35, 67, 145, 151, 155
Longitudinal strain, 7
Loss factor, 304–305

Macroscopic size, 91
Membrane equation, 255–256
Microscopic size, 91
Mixed representation, 132–134
Modes, 167, 177, 182
higher order, 194, 204
norm, 298, 319
panel, 298
waveguide, 318, 319, 321, 337
Modes and resonances 177–179

Monopole field, 137–139; *see* also Point source
reflected, 140, 151, 156, 157
Sommerfeld representation, 137–139
Multilayered fluids, 19

Newton equation, 9
Nonlocally reacting medium, 151, 156
Normal incidence, 19–20, 65–67, 69–71
fluids, 15–27
impedance variation, 17–18
multilayered fluids, 19
oblique pores, 69–71
perforated facings (porous layers with), 187–194
circular holes, 187–192
square holes, 192–194
surface impedance, 15–27, 65–67, 71, 87, 16–131, 194–205
Numerical distance, 151, 163, 165

Oblique incidence
absorption coefficient, 36–37
circular holes (facings with), 205–211
added length, 208–209
facing/material boundary impedance, 205–208
fluids, 29–43
reflection coefficient, 36, 39
reflection-refraction, 31–33
isotropic poroelastic media, 168–172
locally reacting materials, 67
rigid framed materials, 89
square holes (facings with), 210–211
added length, 211
surface impedance, 29–43, 139, 170, 209, 224
transversally isotropic poroelastic media, 223–225
Oblique pores, 69–71
effective density, 69–70
impedance, 71
Operators, vectors, 1–2

Panel
baffled, 283–284, 296
critical frequency, 255
curved, 304
damping, 255,302–303, 331
mechanical impedance, 255
orthotropic, 256, 293
perforated, 187–212
radiation, 284–285
sandwich, 256, 302
Passage path, *see* path of steepest descent
Passage path method, *see* steepest descent method
Path of steepest descent, 141
Perforated faced porous materials, 187–212
Circular aperture
added mass/length, 188–191
design of, 202–203
flow resistance, 191–192
Helmoltz resonator, 203–205
inertial effect, 187–191
square aperture, 192–194, 198, 210–211
stratified materials, 194, 198
Permeability
dynamic thermal, 74–75
dynamic viscous, 74
static thermal, 75–78
static viscous, 74–76
Perpendicular-oriented pores, 57–61, 65–67
circular cross-section, 57–59
impedance, 65–67
slit, 59–60
Phase velocity, 11
Piston excitation, 265–317
Plate, *see* panel
Point load, 295–297
Point source
above rigid framed materials, 137–165
above poroelastic materials, 179–182

Point source (*continued*)
modelling using the Transfer Matrix Method, 303
Poisson ratio, 6, 116, 130, 168, 180, 214
Pole, 139, 145, 146, 150, 151, 152, 153, 156, 158, 161, 167, 174, 176,
contribution, 139, 150–151
localization
radial dependence of the reflected field, 153–154
vertical dependence of the total pressure, 155–156
main pole, 146, 153
plane waves associated with, 146, 150
pole subtraction method, 151–153
trajectory, 145, 146
Porosity, 20, 21
Porous layer
finite thickness, 65–67, 71, 87, 129–131, 151, 160, 177–182, 223
semi-infinite, 140, 141, 142, 144, 145, 148, 151, 152, 154, 158, 173
thin, 145, 146, 147, 151, 154, 157, 159, 162
Potential coupling term, 112–115
Power,
absorbed, 289–290
balance equation, 324, 332
dissipated 265
within a porous medium, 323–324
incident, 290
input, 265
radiated, 265, 285, 297
structural
transmitted, 283, 285–287
Pressure, hydrostatic, 8
Pride *et al.* model, 83, 85, 87

Quasi-static regime, 168

Radiation, 284–285, 288–291, 324–327, 329
Random incidence, *see* Diffuse field
Rayleigh integral, 326
Rayleigh wave
modified (in layers of finite thickness), 176–177
Rayleigh wavelength, 177
empirical formula, 173
semi-infinite layers (in), 173–176
transversally isotropic poroelastic media (in), 232–235
Reference integral method, 151, 161
Reflection coefficient, 19–20, 36–37, 138, 139, 145, 148, 149, 152, 161, 170, 225, 263, 288
Representative elementary volume, 91
Resonance, 76, 129, 130, 177, 179, 204, 205
Riemann sheet
physical, 140, 151, 174, 177, 227, 231
second, 140

Scalar displacement potential, 10, 11, 12
Scalar velocity potential, 120, 168
Screens
impervious, 255, 259, 261, 271–274, 315–316
perforated (*see* also Perforated faced porous materials), 205, 270
porous, 34, 266–271
resistive, 200–201, 269
Second representation (Biot theory), 133–134, 215–216
Separation of scales
macroscopic/mesoscopic, 96
separation parameter, 96
macroscopic/microscopic, 91
separation parameter, 91
mesoscopic/microscopic, 96
separation parameter, 96
Shear waves, 13, 122–123, 180, 246
Smart foams (modelling), 339
Snell-Descarte laws, 32–33
Sommerfeld integral, 139, 232

Sommerfeld representation, monopole
field, 137–139, 167, 223, 225
Stationary point, 141, 142
Statistical Energy Analysis (SEA), 297
Steepest descent method, 140–145, 151, 157
Stress-strain relations
Biot theory, isotropic poroelastic media, 112, 116
Biot theory, transversally isotropic poroelastic media, 215, 216
isotropic elastic media, 5–8
transversally isotropic elastic media, 214
Structure-borne excitation, *see* point load excitation
Superposition, of waves, 17, 334
Surface impedance
normal incidence, 15–27, 34–35, 65–67, 71, 87, 126–131, 194–205
oblique incidence, 29–43, 67, 138, 170–171, 205–217, 224, 226, 237
Surface wave, 148

Tamura method, 148, 150
T. M. electromagnetic surface waves, 148
Tortuosity
classical, 78, 79
in alternating cylindrical pores, 103–104
in transversally isotropic poroelastic materials, 227
in transversally isotropic rigid framed materials, 67–69
dynamic, *see* dynamic tortuosity
quasi-static, 82
Transfer Matrix Method (TMM)
coupling matrices, 257–260
finite size correction (FTMM)
absorption, 288–291
point load excitation,
transmission, 283–288
excitation,
acoustic, 263–265
piston, 265
point load, 295–297
point source, 303–304
termination conditions, 260–261
Transfer matrix representation
fluid, 244–245
impervious screens, 255–256
isotropic poroelastic media, 247–251
isotropic solid media, 245–247
porous screens, 256
thin plate, 254–255
transversally isotropic poroelastic media, 236–238
Transmission coefficient, 263–264, 274, 284, 287, 293, 304
Transmission loss, *see* transmission coefficient
Transversally isotropic poroelastic media
mechanical excitation, 227–228
Rayleigh poles and Rayleigh waves, 232, 236
sound source in air (above), 222–227
stress-strain relations, 215–216
wave equations, 216–217
Trapping constant, 76–78
Two-dimensional Fourier transform, 137, 183

Vector displacement potential, 11–13, 122
Vector velocity potential, 169
Vector operator, 1, 2
Viscosity, 46

Wave equations
elastic isotropic solid, 11–13
elastic transversally isotropic solid, 214–215
isotropic fluid, 10–11
isotropic poroelastic material, 119–120, 131–134
isotropic rigid and limp framed materials, 251–254

Wave equations (*continued*)
transversally isotropic fluid, 39–41
transversally isotropic poroelastic material, 216–217
Waveguide,
coupling with simple porosity coupling, 317–320
coupling with double porosity media, 336–337
Wave number,
acoustic, 11, 15, 30
bending, 286
elastic solid, 12–13, 245–246
poroelastic, 121–122
Rayleigh, 173
Wilson model, 85–88

Young modulus, 6, 214

Zenneck wave, 152